CMOS DATA CONVERTERS
FOR COMMUNICATIONS

THE KLUWER INTERNATIONAL SERIES IN ENGINEERING AND COMPUTER SCIENCE

ANALOG CIRCUITS AND SIGNAL PROCESSING
Consulting Editor: **Mohammed Ismail**. *Ohio State University*

Related Titles:

ANALOG SIGNAL GENERATION FOR BIST OF MIXED-SIGNAL INTEGRATED CIRCUITS
   G.W. Roberts, A.K. Lu
   ISBN: 0-7923-9564-6
INTEGRATED FIBER-OPTIC RECEIVERS
   A. Buchwald, K.W. Martin
   ISBN: 0-7923-9549-2
MODELING WITH AN ANALOG HARDWARE DESCRIPTION LANGUAGE
   H.A. Mantooth, M. Fiegenbaum
   ISBN: 0-7923-9516-6
LOW-VOLTAGE CMOS OPERATIONAL AMPLIFIERS: Theory, Design and Implementation
   S. Sakurai, M. Ismail
   ISBN: 0-7923-9507-7
ANALYSIS AND SYNTHESIS OF MOS TRANSLINEAR CIRCUITS
   R. Wiegerink
   ISBN: 0-7923-9390-2
COMPUTER-AIDED DESIGN OF ANALOG CIRCUITS AND SYSTEMS
   L.R. Carley, R.S. Gryurcsik
   ISBN: 0-7923-9351-1
HIGH-PERFORMANCE CMOS CONTINUOUS-TIME FILTERS
   J. Silva-Martínez, M. Steyaert, W. Sansen
   ISBN: 0-7923-9339-2
SYMBOLIC ANALYSIS OF ANALOG CIRCUITS: Techniques and Applications
   L.P. Huelsman, G.G. Gielen
   ISBN: 0-7923-9324-4
DESIGN OF LOW-VOLTAGE BIPOLAR OPERATIONAL AMPLIFIERS
   M.J. Fonderie, J.H. Huijsing
   ISBN: 0-7923-9317-1
STATISTICAL MODELING FOR COMPUTER-AIDED DESIGN OF MOS VLSI CIRCUITS
   C. Michael, M. Ismail
   ISBN: 0-7923-9299-X
SELECTIVE LINEAR-PHASE SWITCHED-CAPACITOR AND DIGITAL FILTERS
   H. Baher
   ISBN: 0-7923-9298-1
ANALOG CMOS FILTERS FOR VERY HIGH FREQUENCIES
   B. Nauta
   ISBN: 0-7923-9272-8
ANALOG VLSI NETWORKS
   Y. Takefuji
   ISBN: 0-7923-9273-6

CMOS DATA CONVERTERS FOR COMMUNICATIONS

by

Mikael Gustavsson
GlobeSpan, Inc.

J. Jacob Wikner
Ericsson Components AB

and

Nianxiong Nick Tan
GlobeSpan, Inc.

KLUWER ACADEMIC PUBLISHERS
BOSTON / DORDRECHT / LONDON

A C.I.P. Catalogue record for this book is available from the Library of Congress.

ISBN 0-7923-7780-X

Published by Kluwer Academic Publishers,
P.O. Box 17, 3300 AA Dordrecht, The Netherlands.

Sold and distributed in North, Central and South America
by Kluwer Academic Publishers,
101 Philip Drive, Norwell, MA 02061, U.S.A.

In all other countries, sold and distributed
by Kluwer Academic Publishers,
P.O. Box 322, 3300 AH Dordrecht, The Netherlands.

Printed on acid-free paper

All Rights Reserved
© 2000 Kluwer Academic Publishers, Boston
No part of the material protected by this copyright notice may be reproduced or
utilized in any form or by any means, electronic or mechanical,
including photocopying, recording or by any information storage and
retrieval system, without written permission from the copyright owner.

Printed in the Netherlands.

To the late Vanja Gustavsson
Mikael Gustavsson

To Ulrica and my family
J Jacob Wikner

To Regina and Diana
N. Nick Tan

CONTENTS

Preface   xv
Abbreviations and Acronyms   xix

1 CHARACTERIZATION OF DATA CONVERTERS   1

- 1.1 Introduction   1
- 1.2 The Ideal Data Converter   1
- 1.3 Quantization   6
- 1.4 Static Performance   9
  - 1.4.1 Differential Nonlinearity (DNL)
  - 1.4.2 Integral Nonlinearity (INL)
  - 1.4.3 Offset Error
  - 1.4.4 Gain Error
  - 1.4.5 Monotonicity
- 1.5 Dynamic Performance   13
  - 1.5.1 Settling Errors in DACs
  - 1.5.2 Glitches in DACs
  - 1.5.3 Clock Feedthrough (CFT) in DACs
  - 1.5.4 Sampling Time Uncertainty in DACs
  - 1.5.5 Dynamic Errors in ADCs
- 1.6 Frequency Domain Measures   17
  - 1.6.1 Signal-to-Noise Ratio (SNR)
  - 1.6.2 Spurious Free Dynamic Range (SFDR)
  - 1.6.3 Harmonic Distortion (HDk)
  - 1.6.4 Total Harmonic Distortion (THD)
  - 1.6.5 Signal-to-Noise and Distortion Ratio (SNDR)
  - 1.6.6 Effective Number Of Bits (ENOB)
  - 1.6.7 Peak SNDR
  - 1.6.8 Dynamic Range (DR)
  - 1.6.9 Effective Resolution Bandwidth (ERB)
  - 1.6.10 Inter-Modulation Distortion (IMD)
  - 1.6.11 SNR with Multi-Tone Input
  - 1.6.12 Multi-Tone Power Ratio (MTPR)
- 1.7 Summary   25

2 DATA CONVERTER REQUIREMENTS FOR COMMUNICATIONS 27

- 2.1 Introduction 27
- 2.2 Line Coding and Modulation 28
  - 2.2.1 Baseband Codes
  - 2.2.2 Passband Codes
  - 2.2.3 Excess Bandwidth
- 2.3 Channel Capacity and Error Probability 33
  - 2.3.1 Channel Capacity
  - 2.3.2 Error Probability
- 2.4 Duplexing Method 38
- 2.5 Multi-Carrier Systems 40
- 2.6 Minimum Data Converter Requirements 41
  - 2.6.1 DAC for Baseband
  - 2.6.2 DAC for Passband
  - 2.6.3 DAC for DMT
  - 2.6.4 ADC Requirements
  - 2.6.5 Influence of the Duplexing Methods
- 2.7 Optimum Data Converters for ADSL 48
  - 2.7.1 Optimum ADCs
  - 2.7.2 Optimum DACs
- 2.8 ADC Requirements for Wideband Radio 55
  - 2.8.1 IF Sampling
  - 2.8.2 ADC Requirements
- 2.9 Summary 58

3 OVERVIEW OF HIGH-SPEED A/D CONVERTER ARCHITECTURES 61

- 3.1 Introduction 61
- 3.2 Sample-and-Hold (S/H) Circuits 61
  - 3.2.1 Sampling-Time Uncertainty
  - 3.2.2 Thermal Noise
  - 3.2.3 Nonlinearity
- 3.3 Flash Converters 64
  - 3.3.1 Static Errors
  - 3.3.2 Dynamic Errors
  - 3.3.3 Signal Dependent Delay in Comparators
  - 3.3.4 Meta-Stability
- 3.4 Semi-Flash Converters 67
  - 3.4.1 Static Errors
  - 3.4.2 Dynamic Errors
- 3.5 Folding and Interpolating Converters 70
- 3.6 Pipelined Converters 71
  - 3.6.1 Static Errors
  - 3.6.2 Dynamic Errors

Contents   ix

    3.7   Oversampling Sigma-Delta ADCs   72
    3.8   Low Speed Converters   73
    3.9   Time-Interleaved (Parallel) Converters   75
           3.9.1   Offset Errors
           3.9.2   Gain Errors
           3.9.3   Phase Skew Errors
    3.10 Reported Performance   77
    3.11 Summary   81

4   OVERVIEW OF D/A CONVERTER ARCHITECTURES   87

    4.1   Introduction   87
    4.2   Nyquist-Rate D/A Converters   87
           4.2.1   Nyquist-Rate DACs
           4.2.2   Nyquist-Rate DACs with Oversampling
    4.3   Binary Weighted DAC Architecture   91
           4.3.1   Current-Steering DAC
           4.3.2   R-2R Ladder DAC
           4.3.3   Charge Redistribution DAC
    4.4   Thermometer Coded DAC Architecture   95
    4.5   Encoded DAC Architecture   96
    4.6   Hybrid DAC Architecture   98
    4.7   Low-Speed DAC Architecture   100
           4.7.1   Algorithmic DAC Architecture
           4.7.2   Switched-Current Algorithmic DAC
    4.8   Pipelined DAC Architecture   102
    4.9   Oversampling D/A Converter (OSDAC)   104
           4.9.1   Interpolator and Interpolation Filters
           4.9.2   Sigma-Delta Modulator
           4.9.3   M-bit DAC
           4.9.4   Continuous-Time Filter
    4.10 Special Improvement Techniques   114
           4.10.1 Dynamic Randomization and Element Matching
    4.11 DAC Comparison   119
    4.12 Summary   121

5   OVERVIEW OF CIRCUIT TECHNIQUES   125

    5.1   Introduction   125
    5.2   Sample-and-Hold Circuits   125
    5.3   Switched-Current Techniques   126
           5.3.1   Mismatch
           5.3.2   Speed
           5.3.3   Transmission Errors
           5.3.4   Noise

		5.3.5	Clock Feedthrough Errors	
		5.3.6	Aperture Errors	
	5.4	Switched-Capacitor Techniques		132
		5.4.1	Mismatch	
		5.4.2	Speed	
		5.4.3	Transmission Errors	
		5.4.4	Noise	
		5.4.5	Aperture and Clock Feedthrough Errors	
	5.5	Summary		137

6 ANALOG FUNCTIONAL BLOCKS   139

	6.1	Introduction		139
	6.2	The Multiplying DAC		139
	6.3	MDACs in SI Pipelined ADCs		140
	6.4	Performance Limitations in SI MDACs		143
		6.4.1	Speed	
		6.4.2	Residue Amplification	
		6.4.3	Distributing Currents to Several Outputs	
		6.4.4	Time-Interleaved Stage	
		6.4.5	D/A Conversion	
		6.4.6	Noise	
		6.4.7	Dynamic Range	
	6.5	Comparisons of SI MDACS		153
	6.6	SC MDACs		156
		6.6.1	Effect of Finite Opamp Gain and Parasitic Capacitors	
		6.6.2	Speed	
	6.7	Noise In SC MDACs		159
		6.7.1	kT/C Noise	
		6.7.2	Thermal Noise in Opamp	
		6.7.3	Input Referred Thermal Noise of SC Amplifier I	
		6.7.4	Input Referred Thermal Noise of SC Amplifier II	
		6.7.5	Thermal Noise in MDAC	
	6.8	SI Integrator		165
		6.8.1	Speed	
		6.8.2	Noise	
	6.9	SC Integrator		169
		6.9.1	Speed	
		6.9.2	Finite Opamp Gain	
		6.9.3	Effect of Finite Bandwidth and Finite DC Gain	
		6.9.4	Noise	
	6.10	Summary		174

7 BASIC ANALOG CIRCUIT DESIGN 177

- 7.1 Introduction 177
- 7.2 Operational Amplifiers 177
  - 7.2.1 Slew Rate vs. Unity-Gain Bandwidth
  - 7.2.2 Current-Mirror Based Opamp
  - 7.2.3 Folded-Cascode Opamp
  - 7.2.4 Telescopic Opamp
  - 7.2.5 Gain Boosting Technique
  - 7.2.6 Common-Mode Feedback
- 7.3 Voltage Comparators 187
  - 7.3.1 Amplifier-Type Comparator
  - 7.3.2 Latch-Type Comparator
  - 7.3.3 Metal Stability and Error Probability
  - 7.3.4 Offset Cancellation Techniques
  - 7.3.5 Circuit Example
- 7.4 Current Comparators 197
- 7.5 Voltage and Current References 199
  - 7.5.1 Voltage Reference Generation
  - 7.5.2 Current Reference Generation
  - 7.5.3 Reference Buffer
- 7.6 Summary 202

8 LOW-VOLTAGE ANALOG TECHNIQUES 205

- 8.1 Introduction 205
- 8.2 Impact of Scaling 205
- 8.3 Low Voltage Opamps 206
  - 8.3.1 Compensation Techniques
  - 8.3.2 Simulations
  - 8.3.3 Common-Mode Feedback
- 8.4 Clock Voltage Doublers 220
  - 8.4.1 High Voltage Generator
  - 8.4.2 Bootstrap Switching
  - 8.4.3 Level Conversion Circuit
- 8.5 Switched Opamp Technique 224
- 8.6 Summary 226

9 PIPELINED A/D CONVERTERS 229

- 9.1 Introduction 229
- 9.2 The Pipelined Converter 229
  - 9.2.1 Sub ADC
  - 9.2.2 Sub DAC
  - 9.2.3 Residue and S/H Amplifier

		9.2.4 Digital Output	
	9.3	Digital Correction	235
		9.3.1 Output Codes and Decoding	
		9.3.2 Modified Converter	
	9.4	Digital Calibration	242
		9.4.1 Other Calibration Techniques	
	9.5	Errors in Pipelined ADCs	245
		9.5.1 ADC Errors	
		9.5.2 S/H Amplifier Errors	
		9.5.3 DAC Errors	
	9.6	Impact of Non-Ideal SC Circuits	248
		9.6.1 Finite Opamp Gain	
		9.6.2 Noise	
		9.6.3 Limited Bandwidth	
		9.6.4 Limited Slew Rate	
		9.6.5 Matching	
		9.6.6 How to Choose Stage Resolution	
	9.7	Summary	254

10 TIME-INTERLEAVED A/D CONVERTERS 257

	10.1	Introduction	257
	10.2	Principle of Time-Interleaved ADCs	257
	10.3	Errors in Time-Interleaved ADCs	260
		10.3.1 Offset Errors	
		10.3.2 Gain Errors	
		10.3.3 Phase Skew Errors	
	10.4	Error Reduction Techniques	267
		10.4.1 Two-Rank Sample-and-Hold	
		10.4.2 Averaging	
		10.4.3 Randomization	
		10.4.4 Calibration	
		10.4.5 Filter Banks	
	10.5	Passive Sampling Technique	270
		10.5.1 Sampling in SC Time-Interleaved ADCs	
		10.5.2 Improved Passive Sampling Technique	
		10.5.3 Effect of Parasitic Capacitors	
	10.6	Realization of the Improved Sampling Technique	280
		10.6.1 M-Channel S/H Circuit	
		10.6.2 Opamp Sharing Technique	
	10.7	Derivations	282
		10.7.1 Output Spectrum	
		10.7.2 Phase Skew Errors	
		10.7.3 Gain Errors	
	10.8	Summary	288

11 OVERSAMPLING A/D CONVERTERS 291

11.1 Introduction 291
11.2 Basics of Oversampling Sigma-Delta Converters 291
11.3 Oversampled Sigma-Delta Converters for High Signal Bandwidths 293
    11.3.1 2-2 Fourth Order Cascaded Modulator
    11.3.2 Improved 2-2 Cascaded Modulator
    11.3.3 2-1-1 Cascaded Modulator
    11.3.4 Cascaded Modulator with Multi-Bit Quantizer
11.4 SC Implementation 299
    11.4.1 Integrator Modifications
11.5 Non-Ideal Effects 303
    11.5.1 Finite Opamp Gain
    11.5.2 Finite Bandwidth
    11.5.3 Non-linear Settling
    11.5.4 Capacitor Mismatch
    11.5.5 Gain Errors in Cascaded Modulators
    11.5.6 Thermal Noise
    11.5.7 Comparator Errors
    11.5.8 Non-linear Effects
11.6 Design Example 311
11.7 Summary 319

12 MODELING OF NYQUIST D/A CONVERTERS 321

12.1 Introduction 321
12.2 Errors in Current-Steering DACs 322
    12.2.1 Major Error Sources in Current-Steering DACs
12.3 Output Resistance Variations 323
    12.3.1 DNL and INL vs. output resistance
    12.3.2 SNDR vs. Output Resistance
    12.3.3 SFDR vs. Output Resistance
    12.3.4 Influence of Parasitic Resistance
12.4 Current Source Mismatch 335
    12.4.1 SNDR vs. Mismatch
    12.4.2 SFDR vs. Mismatch
    12.4.3 Impact of Correlated and Graded Matching Errors
12.5 Influence of Circuit Noise 342
12.6 Influence of Dynamic Behavior 345
    12.6.1 Settling Error
    12.6.2 Bit Skew and Glitches
12.7 Summary 350

13 IMPLEMENTATION OF CMOS CURRENT-STEERING D/A CONVERTERS 353

13.1 Introduction 353
13.2 Current-Steering DAC Topologies 354
    13.2.1 Array Structure
    13.2.2 Segmented Structures
    13.2.3 Current Cell Matrix Structures
13.3 Practical Design Considerations 358
    13.3.1 Implementation of Current Sources
    13.3.2 Current Switches
    13.3.3 Digital Circuits
    13.3.4 Current Source Calibration
13.4 A CMOS Current-Steering DAC Chipset 366
    13.4.1 Unit Current Sources and Source Array
    13.4.2 Current Switches
    13.4.3 Digital Circuits
    13.4.4 Chip Implementations
13.5 Chipset Measurements 370
    13.5.1 Measurement Results
    13.5.2 Conclusions and Comparison
13.6 Summary 376

PREFACE

"In the telecommunication era, analog design is not just a matter of isolated self-satisfying circuit tweaking, it is above all a system planning."

Digital radio and high-speed internet access have created a great demand on high-performance data converters. The conventional view of data converters as numerical conversion blocks does not suffice for this kind of communication applications. From a communication system's perspective, the bit error rate (BER) requirement determines the signal-to-noise ratio (SNR) requirement for a given modulation and duplexing scheme. Therefore, when we design data converters for a communication system, frequency-domain measures such as the SNR are of great importance. In theory, any known distortion components in a transmit channel of a communication system can be corrected by measuring it through the use of a pilot signal or during the initialization phase. But in practice, it is difficult to predict the distortions due to the time-varying interference of adjacent channels in most communication systems. Therefore, the distortion and inter-modulation of data converters are of great importance as well. The use of discrete multi-tone (DMT) modulation for digital subscriber line (DSL) and the use of orthogonal frequency division multiplexing (OFDM) for wideband radio further complicate the data converter design in that even the single-tone frequency-domain measures do not suffice.

Most universities teach analog courses as circuit design courses with little interaction with communication or signal processing systems. Most analog designers can only design circuits specified by system engineers without much understanding of the system, and most system engineers have not much knowledge about analog imperfections. This combination usually results in non-optimal system solutions. If analog designers with the system knowledge participate in the system specification and system partitioning, we can really design the optimal system with judicious trade-offs between analog and digital, software and hardware. By realizing the industrial and educational need of analog designers with system knowledge, the Microelectronics Research Center of Ericsson started analog activities with close cooperations with Linköping University in 1995. One of the results is this book.

CMOS data converters for communications addresses analog design for communication systems. It distinguishes itself from other data converter books by emphasizing system-related aspects of data converter designs and frequency-domain performance of data converters. However, this is not a book written by system designers rather than by circuit designers. It therefore also covers circuit designs and implementations of

CMOS data converters, based on measured chips.

CMOS data converters for communications bridges the gap between the communication system requirements and the mixed signal design. It derives data converter requirements from the communication system specifications such as line codes, modulation schemes, and duplexing methods, etc. It also gives examples of relating asymmetrical DSL (ADSL) and wideband ratio systems to the data converter requirements.

CMOS data converters for communications can be used as a reference book by analog circuit designers to understand the data converter requirements for communications. It can also be used by communication system designers to understand the difficulties of certain performance requirements on data converters. This book reflects the authors many year philosophy in both industrial practice and academic education. In the telecommunication era, analog design is not just a matter of isolated self-satisfying circuit tweaking, it is above all a system planning. To prepare analog students for the new challenge, this book can be an excellent resort.

This book consists of 13 chapters.

In chapter 1, we discuss the characterization of data converters with emphasis on frequency-domain measures. It also discusses the multi-tone power ratio (MTPR) measure for DMT or OFDM applications. It is shown that multi-carrier modulations usually have a higher dynamic range requirement on data converters due to the instantaneous summation of many carriers resulting in a large peak to average ratio (PAR).

In chapter 2, we first briefly discuss the fundamentals of digital communication systems and introduce the relevant terminologies such as bit error rate (BER), line codes, modulation schemes, and duplexing methods, etc. Then we derive the minimum data converter requirements as a function of the bit error rate, the line code, and the modulation. The influence of duplexing methods is discussed as well. By considering the transmit and receive power and the background noise/interference power, we also derive the optimum data converter requirements for ADSL and wideband radio systems.

Chapters 3 and 4 are devoted to the overview of data converter architectures. These two chapters cover the most data converter architectures and discuss their suitabilities for communications. Chapter 3 presents an overview of high-performance analog-to-digital converters (ADCs) while chapter 4 presents an overview of high-performance digital-to-analog converters (DACs).

Chapters 5~8 are devoted to analog circuits and circuit techniques before we discuss data converter designs and implementations.

Chapter 5 briefly compares the two existing circuit techniques, the switched-capacitor (SC) and switched-current (SI) technique. Although the SI technique offers the simplicity advantages among others, it is advised to use the SC technique for high-performance ADCs due to the higher dynamic range and lower distortion of SC circuits.

For most ADCs, sample-and-hold (S/H) amplifiers have a great impact on the frequency-domain performance such as the SNR and distortion, especially for sub-sampled

ADC systems where the signal frequency is beyond the Nyquist bandwidth determined by the sampling rate. S/H amplifiers are a special case of multiplying DACs used in pipelined ADCs and a special case of integrators used in oversampling ADCs. Chapter 6 discusses the design and characterization of these functional blocks such as multiplying DACs and integrators.

Chapter 7 discusses the circuit building blocks needed for data converters including operational amplifiers (opamps), comparators, and references, etc. It highlights how to design high-gain high-speed opamps and fast low-power comparators. It also covers common-mode feedback (CMFB) techniques and practical issues such as bandgap reference generation and reference buffering.

Chapter 8 is an extension of chapter 7, focusing on low-voltage building blocks in order to operate analog circuits at a supply voltage of 2.5 V or even lower. It includes low voltage (2.5 V) high-speed opamps, voltage doublers, and high-voltage generators, etc.

Chapters 9~11 deal with ADC designs and implementations.

Chapter 9 is devoted to pipelined ADCs, from architecture to design and implementation. It discusses different pipelining techniques, different coding techniques, digital correction techniques, and digital calibration techniques.

Chapter 10 is about how to increase the speed of ADCs by time-interleaving or parallelism. The major limitation of time-interleaved ADCs is the phase skews in the different channels which introduces distortion. A global S/H circuit at the input would eliminate this limitation but the speed and accuracy would be limited by the opamp used in the S/H circuit. Besides discussions on the conventional methods, it also presents a passive global sampling technique to increase the speed beyond the opamp limitation and reduce the phase skew distortion by 8~10 dB.

Chapter 11 covers oversampling ADCs. It discusses different trade-offs between single-stage and multi-stage architectures, between single-bit and multi-bit quantizers. A design example of a fourth-order oversampling ADC is also presented. It highlights the architecture selection as well as the SC design and implementation.

Chapter 12 discusses the modeling of Nyquist DACs for communications. Through this modeling, it is possible to relate the frequency-domain requirements to actual design parameters and therefore provide a guideline for circuit designers.

Chapter 13 presents the actual design and implementation of a current-steering DAC chipset, including high-speed (> 50 MHz) and ultra-low voltage (1.5 V) chips. Based on the discussed architecture and improved current source matching, it is demonstrated possible to deliver a spurious dynamic range (SFDR) over 75 dBc in standard digital CMOS process and achieve an SFDR over 65 dBc with a single 1.5-V supply.

This book is based on the analog research conducted at the Microelectronics Research Center of Ericsson and at the Linköping University. First we would like to thank Dr. Gunnar Björklund, director of the Microelectronics Research Center for initializing and supporting the analog activities at the research center. We would also like to thank

Prof. Lars Wanhammar for his 'digital' support of the analog activities in his group and for educating and helping us in different aspects of digital signal processing. We would like to thank many of our colleagues at Ericsson and at Linköping university for their contributions. Dr. Bengt Jonsson, Helge Stenström, and Dr. Svante Signell of Ericsson Radio Systems contributed significantly by working with us on ADC projects. We appreciate the help from Peter Petersson of Ericsson Radio System in measuring some of the DAC chips. The help from Pierre Dalheim-Lander, Niklas U. Andersson, and K. Ola Andersson in DAC implementations was very valuable. We thank Yonghong Gao of Royal Institute of Technology, Sweden, for the help in the design of oversampling DACs. We would like to acknowledge the invaluable discussions with Marc Delvoux of GlobeSpan, Inc., concerning general digital transmission techniques. We also would like to thank Anders Ihlström of GlobeSpan, Inc., for valuable discussions on oversampling ADC designs.

Last but not the least, we would like to thank our families for their love, patience, and support.

ABBREVIATIONS AND ACRONYMS

2B1Q	Two Binary 1 Quaternary
$\Sigma\Delta$	Sigma-Delta
A/D	Analog to Digital
ADC	Analog-to-Digital Converter
ADSL	Asymmetrical digitial subscriber line
AGC	Automatic Gain Control
ASK	Amplitude-Shift Keying
AWGN	Additive White Gaussian Noise
BER	Bit Error Rate
BP	Band pass
BPSK	Binary Phase-Shift Keying
CAP	Carrierless Amplitude Phase (modulation)
CFT	Clock Feedthrough
CNR	Carrier-to-Noise Ratio
CMFB	Common-Mode Feedback
D/A	Digital to Analog
DAC	Digital-to-Analog Converter
DEM	Dynamic Element Matching
DMT	Discrete Multi-Tone
DNL	Differential Non-Linearity
DR	Dynamic Range
DSL	Digital Subscriber Line
EC	Echo Cancellation
ENOB	Effective Number Of Bits
ERB	Effective Resolution Bandwidth
FDD	Frequency-Division Duplexing
FDM	Frequency-Division Multiplexing
FIR	Finite-length Impulse Response
FM	Frequency Modulation
FSK	Freqeuncy-Shift Keying
GMSK	Gaussian Minimum-Shift Keying

GSM	Global System for Mobile Communication
HDSL	High data-rate Digital Subscriber Line
HDTV	High-Definition Television
HP	High Pass
IF	Intermediate Frequency
IIR	Infinite-length Impulse Response
IMD	Inter-Modulation Distortion
INL	Integral Non-Linearity
ISDN	Integrated Service Digital Network
LP	Low Pass
LSB	Least Significant Bit
MDAC	Multiplying Digital-to-Analog Converter
MSB	Most Significant Bit
MTPR	Multi-Tone Power Ratio
OFDM	Orthogonal Frequency Division Multiplexing
Opamp	Operational Amplifier
OSADC	Oversampling Analog-to-Digital Converter
OSDAC	Oversampling Digital-to-Analog Converter
PAM	Phase Amplitude Modulation
PAR	Peak-to-Average Ratio
PCM	Pulse Code Modulation
PGA	Programmable Gain Amplifier
PSD	Power Spectral Density
PSK	Phase-Shift Keying
QAM	Quadrature Amplitude Modulation
QPSK	Quadrature Phase-Shift Keying
RF	Radio Frequency
RSD	Redundant Signed Digit
S/H	Sample-and-Hold
SC	Switched Capacitor
SDD	Space-Division Duplexing
SFDR	Spurious-Free Dynamic Range
SFG	Signal Flow Graph
SI	Switched Current
SNR	Signal-to-Noise Ratio
SNDR	Signal-to-Noise-and-Distortion Ratio
SR	Slew Rate
TDD	Time-Division Duplexing

TDM	Time-Division Multplexing
THD	Total Harmonic Distortion
VDSL	Very high data-rate Digital Subscriber Line
XDSL	all/any digital subscriber line

1 CHARACTERIZATION OF DATA CONVERTERS

1.1 INTRODUCTION

When a data converter is used in telecommunications applications it is important to know the limitations of the converter and how they affect the performance of the entire system. Therefore measures to characterize the converters are needed. In this chapter we investigate some basic properties of data converters and introduce a number of performance measures that are commonly used. In Sec. 1.2 we review the fundamentals of data converters while quantization noise is considered in Sec. 1.3. Performance measures and error sources are treated in Sec. 1.4 - Sec. 1.6.

1.2 THE IDEAL DATA CONVERTER

The analog-to-digital converter (ADC or A/D converter) viewed as a black box, shown in Fig. 1-1 a), takes an analog input signal usually in the form of a voltage or a

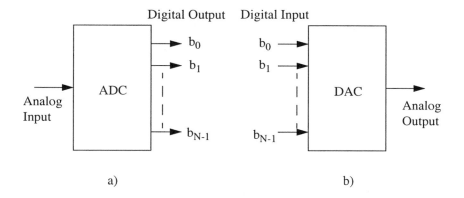

Figure 1-1 The ADC (a) and DAC (b) viewed as a black box.

current and converts it to a digital output signal. The digital-to-analog converter (DAC or D/A converter) has the opposite function and converts a digital input signal to an analog output signal, as shown in Fig. 1-1 b). The digital signal is a binary coded representation of the analog signal using N bits. There are many different ways of coding the output signal [1]. Some of the most common coding schemes for $N = 4$ are shown in Table 1-1. The maximum number of codes when using a N bit code is 2^N.

Table 1-1 Digital Coding Schemes

Decimal	Sign-Magnitude	Two's-Complement	Offset Binary	One's-Complement	Gray
+7	0111	0111	1111	0111	1000
+6	0110	0110	1110	0110	1001
+5	0101	0101	1101	0101	1011
+4	0100	0100	1100	0100	1010
+3	0011	0011	1011	0011	1110
+2	0010	0010	1010	0010	1111
+1	0001	0001	1001	0001	1101
+0	0000	0000	1000	0000	1100
-0	1000	(0000)	(1000)	1111	1100
-1	1001	1111	0111	1110	0100
-2	1010	1110	0110	1101	0101
-3	1011	1101	0101	1100	0111
-4	1100	1100	0100	1011	0110
-5	1101	1011	0011	1010	0010
-6	1110	1010	0010	1001	0011
-7	1111	1001	0001	1000	0001
-8	-	1000	0000	-	0000

The leftmost bit of the digital word is usually called the most significant bit (MSB) and the rightmost bit is called the least significant bit (LSB). The size of the LSB compared to the total code range is sometimes referred to as the *resolution* [1] of the converter. Hence, the resolution of an N bit converter is $1/2^N$. For simplicity it is sometimes more convenient to refer to the resolution as being N bits. In some types of ADCs, e.g., flash and folding ADCs, other codes are used internally which use more than N bits to represent 2^N codes. These codes are usually converted to a more convenient representation at the output of the converter. The offset binary code is commonly used in data converters. We assume in the following that the digital code, X_d, is

$$X_d = (b_{N-1}, b_{N-2}, \ldots, b_2, b_1, b_0) \tag{1-1}$$

where b_{N-1} is the MSB and b_0 is the LSB, and that all the bits are 0 or 1. The analog value of the data converter is denoted X_a. The transfer function of the data converter is a staircase function as illustrated in Fig. 1-2, where the ideal transfer function of a

1.2 The Ideal Data Converter

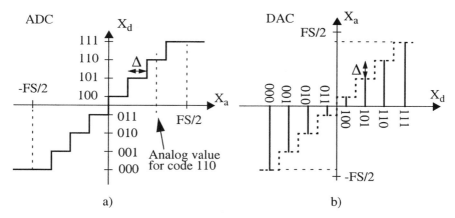

Figure 1-2 The ADC (a) and DAC (b) transfer functions for N = 3.

3-bit ADC is shown in Fig. 1-2a) and the transfer function of an ideal DAC in Fig. 1-2b). In both cases it has been assumed that an offset binary code is used. Notice that the ADC output saturates for large input values. Therefore the input signal should never be in the saturated range since this gives rise to large conversion errors. The step size Δ is equal to the analog value of the LSB and the conversion range without causing saturation is referred to as the Full Scale (FS) range of the converter. Hence the step size is

$$\Delta = \frac{FS}{2^N} \tag{1-2}$$

Here we have assumed that the analog signal range is [-FS/2, +FS/2] but it is not uncommon to assume that it is [0, +FS].

It is convenient to introduce a number of symbols based on the transfer function of the ideal converter. The digital codes are numbered by an index $k = 0, ..., 2^N - 1$, such that $k = 0$ corresponds to the code representing the smallest analog value and $k = 2^N - 1$ corresponds to the largest analog value. For the binary offset code the index k can be calculated as

$$k = \sum_{l=0}^{N-1} b_l \cdot 2^l \tag{1-3}$$

The k-th digital code is denoted $X_{d,k}$, while the corresponding analog value is denoted $X_{a,k}$. This is illustrated in Fig. 1-3. For the ADC it is useful to introduce the analog transition levels as the analog value where the output signal changes between two digital codes. The transition levels are denoted $X_{t,k}$. The number of transition levels is $2^N + 1$. The first and last transition levels are $X_{t,0} = -FS/2$ and $X_{t,2^N} = FS/2$.

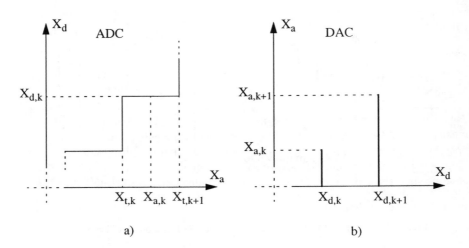

Figure 1-3 Notation for digital codes and analog levels in a) an ADC and b) a DAC.

To calculate the analog value that corresponds to a certain digital code, the reference voltage or current of the data converter must be known. The reference determines the actual analog value and is often equal to FS in volt or ampere. The analog value assuming an **offset binary code** can be calculated as

$$X_{a,k} = \Delta \cdot \sum_{l=0}^{N-1} 2^{l-1} \cdot (-1)^{(1-b_l)} \qquad (1\text{-}4)$$

The analog value calculated above corresponds to the output of an ideal DAC or the midpoint between two transition levels in an ideal ADC (see Fig. 1-3).

To clarify the notation we give a short example. Assume we have a 3-bit converter and that $FS = 2$. According to (1-2) the step size is

$$\Delta = \frac{FS}{2^N} = \frac{2}{2^3} = \frac{1}{4} \qquad (1\text{-}5)$$

The digital code 101 which is the 6-th code corresponds to $k = 5$ (k starts at 0). This means that $X_{d,5} = 101$, $b_2 = 1$, $b_1 = 0$ and $b_0 = 1$. According to (1-4) the analog value can now be calculated as

$$\begin{aligned} X_{a,5} &= \Delta \cdot \sum_{l=0}^{3-1} 2^{l-1} \cdot (-1)^{(1-b_l)} = \\ &= \frac{1}{4} \cdot \left(2 \cdot (-1)^0 + 1 \cdot (-1)^1 + \frac{1}{2} \cdot (-1)^0\right) = \frac{3}{8} \end{aligned} \qquad (1\text{-}6)$$

1.2 The Ideal Data Converter

The **signed-magnitude code** uses the MSB as a sign bit. If $b_{N-1} = 0$ the analog value is positive while if $b_{N-1} = 1$ the analog value is negative. The analog value is calculated as

$$X_{a,k} = \Delta \cdot (-1)^{b_{N-1}} \cdot \sum_{l=0}^{N-2} 2^l \cdot b_l \tag{1-7}$$

There are now two codes representing the zero value, $0\ldots00$ and $10\ldots00$. This means that there are only $2^N - 1$ analog levels. The signed-magnitude code may have operand sign-dependent operations which introduces extra control logic and computation time in the hardware [2]. However, the signed-magnitude code allows the use of some special architectures for e.g. binary weighted current-steering DACs [3].

For the **one's complement code** we have

$$X_{a,k} = \Delta \cdot \left(-b_{N-1} \cdot (2^{N-1} - 1) + \sum_{l=0}^{N-2} 2^l \cdot b_l \right) \tag{1-8}$$

Also in this case we have two codes for the zero value, $0\ldots00$ and $1\ldots11$.

The **two's complement code** also utilizes a sign bit in the MSB position. The analog value is given by

$$X_{a,k} = \Delta \cdot \left(-2^{N-1} \cdot b_{N-1} + \frac{1}{2} + \sum_{l=0}^{N-2} 2^l \cdot b_l \right) \tag{1-9}$$

The two's complement code has good properties for arithmetic operations and is therefore used in several applications.

The **thermometer code** is used in flash converters. With this coding the number of ones in the code determines the corresponding analog value. This is illustrated in Table 1-2 showing the thermometer codes for $N = 4$ and the corresponding index k.

Table 1-2 Thermometer codes for $N = 4$

k	Thermometer code	Walking one
4	1111	1000
3	0111	0100
2	0011	0010
1	0001	0001
0	0000	0000

This table also shows the walking one code which is used in some data converters.

The number of codes that can be represented with the thermometer code is only $N+1$ and it is therefore rarely used in digital signal processing (DSP).

It is a general practice to choose the code representation that is most suitable for a specific data converter. The code can easily be converted to a representation that is more suitable for the algorithms in the DSP.

1.3 QUANTIZATION

The number of codes in a data converter is approximately 2^N, if an effective coding is used. Hence only a limited number of analog amplitude levels can appear in the data converter and there is an inherent conversion error. The quantization error of a 3 bit ADC is shown in Fig. 1-4. We see that the amplitude of the error is $\pm\Delta/2$ and that it

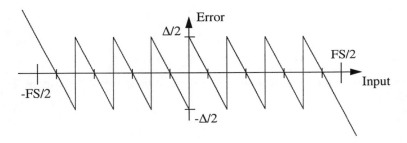

Figure 1-4   Quantization error for a 3 bit ADC.

is divided into 2^N segments. In each segment the error changes linearly as a function of the input signal. If the resolution of the data converter is high it is normally reasonable to assume that the conversion error can be treated as white noise having equal probability of lying anywhere within the range $\pm\Delta/2$. For noise signals we can calculate the power (mean square value) as [4]

$$P_n(t) = x_{rms}^2(t) = \int_{-\infty}^{\infty} e^2 \cdot p(e,t) de \qquad (1\text{-}10)$$

where $x(t)$ is the noise signal, $p(e,t)$ the probability density function and e is used as an integration variable. In this case we assume that the noise has a uniform probability density function and that it is not a function of time, i.e.

$$p(e,t) = \begin{cases} \dfrac{1}{\Delta}, & -\dfrac{\Delta}{2} < e < \dfrac{\Delta}{2} \\ 0, & \text{all other } e \end{cases} \qquad (1\text{-}11)$$

1.3 Quantization

The power, i.e. the mean square value, can now be calculated as [5]

$$P_n = \int_{-\Delta/2}^{\Delta/2} e^2 \cdot \frac{1}{\Delta} de = \frac{\Delta^2}{12} \qquad (1\text{-}12)$$

The same result can be achieved by using a slightly different approach which may be more intuitive for DACs [6]. Assume that we sweep through all the codes sequentially as a function of time, i.e., sweeping a ramp, with period T as shown in Fig. 1-5. Each

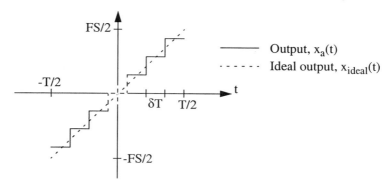

Figure 1-5   Input ramp to determine quantization noise of a DAC.

code is represented during a certain time interval given by $\delta T = T/2^N$. The actual analog output, $x_a(t)$, is a staircase function while the ideal output signal, $x_{ideal}(t)$, is a ramp function. We can now find the error power (mean square value) as

$$P_n = x_{err,rms}^2 = \frac{1}{T}\int_0^T [x_a(t) - x_{ideal}(t)]^2 dt \qquad (1\text{-}13)$$

where $x_a(t) - x_{ideal}(t)$ corresponds to the quantization error. The error has the same sawtooth shape as in Fig. 1-4 but now as a function of time. Since the same pattern is repeated 2^N times it is sufficient to calculate the power within only one of the segments. The quantization error in the time interval $t \in [0, \delta T]$ is

$$x_a(t) - x_{ideal}(t) = -\frac{t \cdot \Delta}{\delta T} + \frac{\Delta}{2} \qquad (1\text{-}14)$$

The error power can now be calculated as

$$P_n = \frac{1}{\delta T}\int_0^{\delta T} \left[-\frac{t \cdot \Delta}{\delta T} + \frac{\Delta}{2}\right]^2 dt = \frac{\Delta^2}{\delta T}\int_0^{\delta T}\left[\left(\frac{t}{\delta T}\right)^2 - \frac{t}{\delta T} + \frac{1}{4}\right]^2 dt = \frac{\Delta^2}{12} \qquad (1\text{-}15)$$

The result is the same as in (1-12).

The above results are valid if all the quantization levels are uniformly spaced. Assuming that the input signal is statistically equally distributed over the whole input amplitude range, this uniform quantization intuitively provides the minimum error. If the signal is not equally distributed over the amplitude range it may be better to use non-uniform quantization. This means that the step size is not equal for all codes in the data converter. Normally a higher resolution is used for small input signals. An example of a transfer function with non-uniform quantization is shown in Fig. 1-6. With this

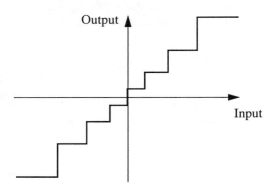

Figure 1-6   Transfer function with non-uniform quantization.

method some amplitude levels are quantized with larger errors, but statistically a minimum quantization noise is achieved since the probability for these amplitude levels is lower. Non-uniform quantization is widely used when transmitting voice signals.

The quantization noise is a fundamental limitation in data converters that represents a lower limit on how small the error power can be. Due to circuit imperfections the total error will always be larger in an actual implementation. To evaluate the performance of a data converter it is convenient to introduce a number of performance measures.

The performance measures can be divided into two groups, static and dynamic measures [3, 7, 8]. A common source of static errors is component mismatch in the implementation. The differential non-linearity (DNL) and integral non-linearity (INL) are often used as static performance measures. The dynamic performance is determined by signal-dependent errors as non-linear slewing, clock feedthrough (CFT), glitches, settling errors, etc. Both static and dynamic properties can be investigated in the frequency domain [7]. The signal-to-noise ratio (SNR), total harmonic distortion (THD), spurious-free dynamic range (SFDR) and signal-to-noise-and-distortion ratio (SNDR) are commonly used as dynamic performance measures. These measures are often determined in single-tone measurements but in many telecommunications applications a multi-tone signalling system is used and static measures or even single-tone measurements may not give all the information needed to characterize the converter.

1.4 STATIC PERFORMANCE

Due to non-ideal circuit elements in the actual implementation of a data converter the code transition points in the transfer function will be moved as illustrated in Fig. 1-7.

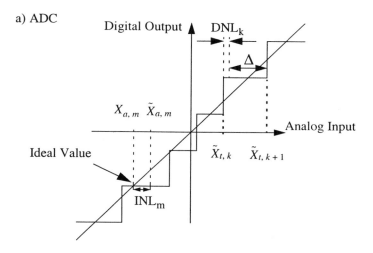

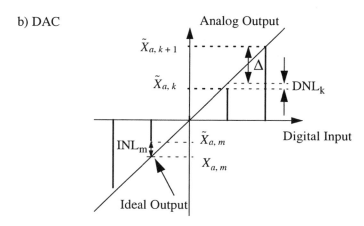

Figure 1-7 Non-ideal transfer function with INL and DNL errors (a) ADC and (b) DAC.

To distinguish between the actual and ideal values in the data converters, all actual values are indicated with a ~. This means that $X_{a,k}$ corresponds to the ideal analog value for digital code $X_{d,k}$ while $\tilde{X}_{a,k}$ corresponds to the actual value.

1.4.1 Differential Nonlinearity (DNL)

The step size in the non-ideal data converter deviates from the ideal size Δ and this error is called the differential nonlinearity (DNL) error. For a DAC the DNL can be defined as the difference between two adjacent analog outputs minus the ideal step size, i.e.

$$\mathrm{DNL}_k = \tilde{X}_{a,k+1} - \tilde{X}_{a,k} - \Delta \tag{1-16}$$

The DNL is often normalized with respect to the step size to get the relative error, i.e.

$$\mathrm{DNL}_k = \frac{\tilde{X}_{a,k+1} - \tilde{X}_{a,k} - \Delta}{\Delta} \tag{1-17}$$

The above definitions are often most practical for DACs since the analog values can be directly measured at the output. For ADCs it may however be more practical to define the DNL based on the difference between transition points, since this corresponds to the result of a histogram test [3]. The normalized DNL of the ADC can be expressed as

$$\mathrm{DNL}_k = \frac{\tilde{X}_{t,k+1} - \tilde{X}_{t,k} - \Delta}{\Delta} \tag{1-18}$$

1.4.2 Integral Nonlinearity (INL)

The total deviation of an analog value from the ideal value is called integral nonlinearity (INL). The normalized INL can be expressed as [9]

$$\mathrm{INL}_k = \frac{\tilde{X}_{a,k} - X_{a,k}}{\Delta} \tag{1-19}$$

for both ADCs and DACs. The relation between INL and DNL is given by

$$\mathrm{INL}_k = \sum_{l=1}^{k} \mathrm{DNL}_l \tag{1-20}$$

The nonlinearity errors are usually measured using a low frequency input signal to exclude dynamic errors appearing at high signal frequencies. The DNL and INL are therefore usually used to characterize the static performance. In some applications offsets and linear gain errors are acceptable and it is then common to specify the INL with respect to a best fit line rather than to the ideal transfer function [1]. Hence, offsets and linear gain errors will not appear in the INL.

1.4.3 Offset Error

The offset, X_{offset}, of the converter can be found by minimizing the expression

1.4 Static Performance

$X_{a,k} - \tilde{X}_{a,k} - X_{offset}$ for all k, with the least square method. To find the minimum we first find where the derivative with respect to X_{offset} is zero, i.e.

$$\frac{\partial}{\partial X_{offset}} \sum_{k=0}^{2^N-1} (\tilde{X}_{a,k} - X_{a,k} - X_{offset})^2 = 0 \qquad (1\text{-}21)$$

which gives

$$X_{offset} = \frac{1}{2^N} \cdot \sum_{k=0}^{2^N-1} (\tilde{X}_{a,k} - X_{a,k}) \qquad (1\text{-}22)$$

This corresponds to a minimum and the offset is thus given by (1-22). We see that the offset corresponds to the average of all the errors in the converter. To eliminate the offset from the INL calculations, the offset should be subtracted from all the analog values, $\tilde{X}_{a,k}$.

1.4.4 Gain Error

The gain can be linear or non-linear as illustrated in Fig. 1-8. Compared to the ideal

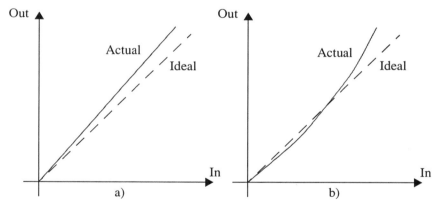

Figure 1-8 Characteristics of a) linear and b) non-linear gain error.

straight line, the actual output has a linear gain error (Fig. 1-8a) and also non-linearity (Fig. 1-8b). Linear gain error does not introduce distortion as long as the output signal does not clip. The actual output with a linear gain and offset error can be written as

$$\tilde{X}_a = A \cdot X_a + X_{offset} \qquad (1\text{-}23)$$

where A is the gain error while the actual output for a non-linear gain can be expressed as

$$\tilde{X}_a = A_1 \cdot X_a + A_2 \cdot X_a^2 + A_3 \cdot X_a^3 + \ldots + X_{offset} \tag{1-24}$$

The non-linear errors can sometimes be reduced by using pre-distortion [10].

The actual gain and the offset, A and X_{offset}, are found by using the least square method. We first find where the derivative with respect to X_{offset} and A are zero, i.e.

$$\frac{\partial}{\partial A} \sum_{k=0}^{2^N-1} [\tilde{X}_{a,k} - (X_{offset} + A \cdot X_{a,k})]^2 = 0 \tag{1-25}$$

and

$$\frac{\partial}{\partial X_{offset}} \sum_{k=0}^{2^N-1} [\tilde{X}_{a,k} - (X_{offset} + A \cdot X_{a,k})]^2 = 0 \tag{1-26}$$

By using (1-25) and (1-26) the gain and offset can be shown to be

$$A = \frac{\langle \tilde{X}_a \cdot X_a \rangle - \langle \tilde{X}_a \rangle \cdot \langle X_a \rangle}{\langle X_a^2 \rangle - \langle X_a \rangle^2} \tag{1-27}$$

and

$$X_{offset} = \langle \tilde{X}_a \rangle - A \cdot \langle X_a \rangle \tag{1-28}$$

where $\langle X \rangle$ indicates a mean value, i.e.

$$\langle \tilde{X}_a \rangle = \frac{1}{2^N} \sum_{k=0}^{2^N-1} \tilde{X}_{a,k}, \; \langle X_a \rangle = \frac{1}{2^N} \sum_{k=0}^{2^N-1} X_{a,k}, \ldots \tag{1-29}$$

Here we assume that the analog mid-point in the ideal transfer function is 0, which implies that the mean value $\langle X_{a,k} \rangle = 0$. The above expressions can then be simplified as

$$A = \frac{\langle \tilde{X}_{a,k} \cdot X_{a,k} \rangle}{\langle X_{a,k}^2 \rangle} \tag{1-30}$$

and

$$X_{offset} = \langle \tilde{X}_{a,k} \rangle \tag{1-31}$$

The line $X_{offset} + A \cdot X_a$ is the best-fit straight line with respect to the actual output values, $\tilde{X}_{a,k}$. The compensated values, can now be used to find the DNL and INL

1.5 Dynamic Performance

errors that are not affected by offset and linear gain errors. The INL can for instance be calculated as

$$INL_k = \frac{\tilde{X}_{a,k} - (A \cdot X_{a,k} + X_{offset})}{A \cdot \Delta} \quad (1\text{-}32)$$

1.4.5 Monotonicity

If the analog amplitude level of the converter increases with increasing digital code, the converter is monotonic. An example of a non-monotonic DAC is shown in Fig. 1-9. Non-monotonicity in an ADC results in missing output codes that never appears for

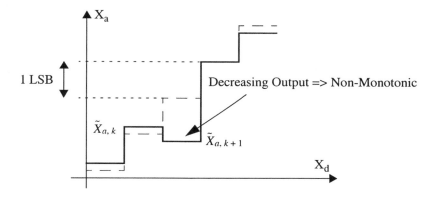

Figure 1-9   A non-monotonic DAC.

any analog input signal. Monotonicity is guaranteed if the deviation from the best-fit straight line is less than half a LSB [3], i.e.

$$|INL_k| \leq \frac{1}{2} \text{LSB for all } k \quad (1\text{-}33)$$

This implies that the DNL errors are less than one LSB [3], i.e.

$$|DNL_k| \leq 1 \text{ LSB for all } k \quad (1\text{-}34)$$

It should be noted that the above relations are sufficient to guarantee monotonicity, but it is possible to have a monotonic converter that does not meet the relations in (1-33) and (1-34). There are some data converter architectures that are monotonic by design, e.g. a thermometer coded DAC.

1.5   DYNAMIC PERFORMANCE

In addition to the static errors that are caused by mismatch in the components in the data converter, several other error sources will appear when the input signal change rapidly. These dynamic errors are often signal and frequency dependent and increases with signal amplitude and frequency. They appear in both ADCs and DACs but are usually more critical in DACs since the shape of the analog wave form determines the performance.

1.5.1 Settling Errors in DACs

In the previous sections the data converters have been regarded as discrete-time circuits that operate on analog values only at discrete time instants. For the ADC this is true, but for the DAC, however, we must also consider the shape of the analog waveform at the output. A number of dynamic effects arise when the output signal is changed between two samples. These dynamic error sources will have a large impact on the DAC performance, especially at high clock and signal frequencies. The ADC is affected by dynamic errors as well, but as long as the final value at the end of the sampling period has a low enough error the performance is not degraded.

When the input of the DAC is changed, the analog output should ideally change from the ideal start value, $X_{a,k}$, to the ideal final value, $X_{a,m}$, see Fig. 1-10. Due to circuit

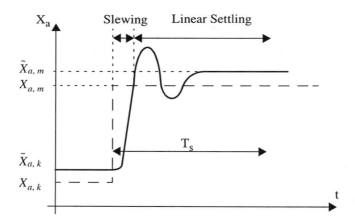

Figure 1-10 Actual output signal and ideal output signal (dashed) of a DAC.

imperfections the actual start and final values are $\tilde{X}_{a,k}$ and $\tilde{X}_{a,m}$ respectively. The output signal of an actual DAC can not change its value instantly. The time it takes for the output to settle within a certain accuracy of the final value, for instance 0.1%, is called the settling time, T_s, and determines the highest possible speed of the circuit. The settling can be divided in two phases, a non-linear slewing phase and a linear settling phase. The slewing phase should be as small as possible since it both increases the settling time and introduces distortion in the analog waveform. The slewing is normally caused by a too small bias current in the circuit driving the output and is therefore increased for large steps when more current is needed.

1.5 Dynamic Performance

There may be additional dynamic error sources in the DAC that can both change the final value and the shape of the waveform, such as glitches and clock feedthrough (CFT).

1.5.2 Glitches in DACs

Glitches occur when the switching time of different bits in a binary weighted DAC is unmatched. For a short period of time a false code could appear at the output. For example if the code transition is

$$0111\ldots 111 \rightarrow 1000\ldots 000$$

and the MSB switches faster than the LSBs, the code $11\ldots 111$ may be present for a short time. This code represents the maximum value and hence a large glitch appears at the output. The glitch adds a signal dependent error to the output signal, that degrades the performance. The effect on the output signal is determined by the energy of the glitch. If the glitch is modeled as a pulse, as shown in Fig. 1-11, with a certain

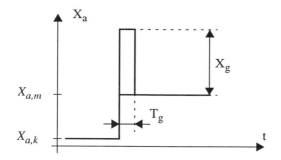

Figure 1-11 Glitch modeled as a pulse with height X_g and duration T_g.

amplitude height, X_g, and with a time duration, T_g, the normalized average power, P_g, of the glitch distributed over the shortest possible code duration, i.e., the clock period, $T_s = 1/f_s$, is

$$P_g = X_g^2 \cdot \frac{T_g}{T_s} \tag{1-35}$$

Assume that the maximum peak glitch amplitude, i.e., the amplitude of the MSB, is

$$X_{g,max} = 2^{N-1} \cdot \Delta \tag{1-36}$$

which gives the maximum glitch power over one clock cycle

$$P_{g,max} = 2^{2N-2} \cdot \Delta^2 \cdot \frac{T_g}{T_s} \tag{1-37}$$

This should be compared with the power of the quantization noise, P_Q, during the same period of time

$$P_Q = \frac{\Delta^2}{12} \tag{1-38}$$

The power of the glitch should be smaller than the quantization in order not to decrease the SNR. Hence

$$P_{g,max} = 2^{2N-2} \cdot \Delta^2 \cdot \frac{T_g}{T_s} < \frac{\Delta^2}{12} = P_Q \tag{1-39}$$

which gives a bound on the time duration of the glitch as

$$T_g < \frac{T_s}{3 \cdot 2^{2N}} \tag{1-40}$$

It should be noted that the glitches are difficult to model accurately and the above result is only a coarse approximation. Sometimes the glitch impulse is also specified by the glitch area with the unit $pV \cdot s$ [3].

1.5.3 Clock Feedthrough (CFT) in DACs

Due to capacitive coupling in switches the clock (or digital switching signals) affects the analog output signal [11]. The clock feedthrough (CFT) can introduce both harmonic distortion and distortion tones at multiples of the clock signal.

For current-steering DACs the CFT error can be modelled in a similar way as glitches, while in e.g. SC DACs the CFT will give an error in the final value. The CFT is reduced when reducing the capacitive coupling and therefore the switch transistor sizes should be small to decrease the size of the parasitic capacitances. However, with a smaller transistor, the on-resistance increases which may degrade the performance due to an increased settling time.

1.5.4 Sampling Time Uncertainty in DACs

Due to noise and other non-ideal effects in the circuit the time between two samples will change. This sampling time variation gives an error in the output that is determined by the size of the output step and the time variation. The average power of this error can be calculated as

$$P_{T_\delta} = X_\delta^2 \cdot \frac{T_\delta}{T_s} \tag{1-41}$$

where X_δ is the step size, T_δ the sampling time error and T_s the sampling period. The step size X_δ is determined by the difference between two consecutive samples, i.e.

$$X_\delta = X((n+1) \cdot T_s) - X(nT_s) \tag{1-42}$$

For a sinusoidal signal the step size is proportional to the signal frequency, and hence

1.6 Frequency Domain Measures

the error in (1-41) will increase at higher signal frequencies since the step size gets larger. The largest possible step size is

$$X_\delta = \Delta \cdot 2^N \tag{1-43}$$

which gives the maximum error power

$$P_{T_\delta, max} = \Delta^2 \cdot 2^{2N} \cdot \frac{T_\delta}{T_s} \tag{1-44}$$

The error power should be smaller than the quantization noise and therefore we have

$$\Delta^2 \cdot 2^{2N} \cdot \frac{T_\delta}{T_s} < \frac{\Delta^2}{12} \tag{1-45}$$

which gives an upper bound on the sampling time error

$$T_\delta < \frac{T_s}{3 \cdot 2^{N+2}} \tag{1-46}$$

1.5.5 Dynamic Errors in ADCs

The dynamic errors in ADCs are mainly caused by the same effects as in DACs, but only their effect on the final value at the end of the sampling period is important. This means that it does not matter if the settling is non-linear or if there are glitches as long as the remaining settling error is small enough. The settling time is, however, in most cases increased when the settling is slew rate limited. The effect of dynamic errors in different ADC architectures is treated in more detail in chapter 3.

1.6 FREQUENCY DOMAIN MEASURES

For data converters used in communications applications, the INL and DNL are not sufficient to characterize the performance. It is more convenient to characterize the performance in the frequency domain using measures as the signal-to-noise ratio (SNR) and spurious-free dynamic range (SFDR). The performance is usually determined by using a single-tone sinusoidal input signal, but sometimes dual-tone [12] or multi-tone measurements are more informative. Since several communication standards use multi-tone modulation the single-tone measures may not be sufficient to characterize converters for multi-tone applications [7]. In Fig. 1-12 we show a typical FFT spectrum of a 14-bit non-ideal ADC when the input signal is a single-tone sinusoidal. The input signal appears as the fundamental in the FFT spectrum and the quantization error generates a white noise floor. The nonlinearities in the ADC cause harmonic tones to appear above the noise floor where some of the harmonics may be folded from higher frequencies due to the sampling process. Based on the FFT in Fig. 1-12 a number of measures to characterize dynamic performance of ADCs can be defined.

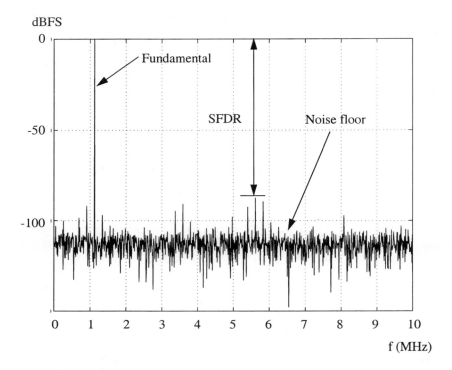

Figure 1-12 FFT spectrum of non-ideal 14 bit ADC.

1.6.1 Signal-to-Noise Ratio (SNR)

A sinusoidal signal is often used to characterize a data converter. It is therefore interesting to calculate the ideal signal-to-noise ratio of a data converter using such an input signal. The maximum amplitude without causing saturation of a sinusoidal input signal is $\Delta \cdot 2^{N-1}$ and the average power of the sine wave is given by

$$P_s = \frac{(\Delta \cdot 2^{N-1})^2}{2} \tag{1-47}$$

The signal-to-noise ratio (SNR) of an ideal ADC with a sinusoidal input signal can now be calculated as [3]

$$SNR = \frac{P_s}{P_n} = \frac{(\Delta \cdot 2^{N-1})^2/2}{\Delta^2/12} = 1.5 \cdot 2^{2N} \tag{1-48}$$

or expressed in decibels

$$SNR_{dB} = 10 \cdot \log\left(\frac{P_s}{P_n}\right) = 6.02 \cdot N + 1.76 \, \text{dB} \tag{1-49}$$

1.6 Frequency Domain Measures

Hence the SNR_{dB} is increased by 6 dB for every additional bit in the converter. The representation using decibels is more convenient. For simplicity the subindex dB is not used in the following. It should be noted that (1-49) only holds if the input is a full scale sine wave. The effect of having a non-sinusoid input signal is discussed in Sec. 1.6.11. The SNR of an actual data converter including non ideal effects can be determined by measuring the output signal. For single tone measurements the SNR is the ratio of the power of the fundamental and the total noise power within a certain frequency band, excluding the harmonic components, i.e.

$$SNR = 10 \cdot \log\left(\frac{\text{Signal Power}}{\text{Total Noisefloor Power}}\right) \tag{1-50}$$

1.6.2 Spurious Free Dynamic Range (SFDR)

The spurious free dynamic range (SFDR) is the ratio of the power of the signal and the power of the largest spurious within a certain frequency band. SFDR is usually expressed in dBc as

$$SFDR(dBc) = 10 \cdot \log\left(\frac{\text{Signal Power}}{\text{Largest Spurious Power}}\right) = 10 \cdot \log\left(\frac{X_1^2}{X_s^2}\right) \tag{1-51}$$

where X_1 is the rms value of the fundamental and X_s the rms value of the largest spurious. In some cases it is more convenient to express the SFDR with the full scale input (dBFS) as reference rather than the input signal, i.e.

$$SFDR(dBFS) = 10 \cdot \log\left(\frac{\left(\frac{FS}{2\sqrt{2}}\right)^2}{X_s^2}\right) \tag{1-52}$$

The relation between SFDR in dBFS and dBc is given by

$$SFDR(dBc) = SFDR(dBFS) - 10 \cdot \log\left(\frac{X_1^2}{\left(\frac{FS}{2\sqrt{2}}\right)^2}\right) \tag{1-53}$$

1.6.3 Harmonic Distortion (HD_k)

The harmonic distortion with respect to the k-th harmonic (HD_k) is the power ratio between the k-th harmonic and the fundamental as

$$HD_k = 10 \cdot \log\left(\frac{k\text{:th Harmonic Power}}{\text{Signal Power}}\right) =$$
$$= 10 \cdot \log\left(\frac{X_k^2}{X_1^2}\right) \qquad (1\text{-}54)$$

where X_1 is the rms value of the fundamental and X_k the rms value of the k-th harmonic component. With the above definition the harmonic distortion is a negative number. Sometimes HD_k is defined as

$$HD_k = 10 \cdot \log\left(\frac{X_1^2}{X_k^2}\right) \qquad (1\text{-}55)$$

which is positive. Thus there is a sign difference between the two definitions.

1.6.4 Total Harmonic Distortion (THD)

The total harmonic distortion (THD) is the ratio of the total harmonic distortion power and the power of the fundamental in a certain frequency band, i.e.

$$THD = 10 \cdot \log\left(\frac{\text{Total Harmonic Distortion Power}}{\text{Signal Power}}\right)$$
$$= 10 \cdot \log\left(\sum_{k=2}^{\infty} X_k^2 / X_1^2\right) \qquad (1\text{-}56)$$

where X_1 is the rms value of fundamental and X_k the rms value of the k-th harmonic component. Since there is an infinite number of harmonics the THD is usually calculated using the first 10-20 harmonics or until the harmonics can not be distinguished from the noise floor. The THD is sometimes defined as

$$THD = 10 \cdot \log\left(\frac{\text{Signal Power}}{\text{Total Harmonic Distortion Power}}\right) \qquad (1\text{-}57)$$

The only difference to the expression in (1-56) is the sign.

1.6.5 Signal-to-Noise and Distortion Ratio (SNDR)

The signal-to-noise and distortion ratio (SNDR) is the ratio of the power of the fundamental and the total noise and distortion power within a certain frequency band, i.e.

$$SNDR = 10 \cdot \log\left(\frac{\text{Signal Power}}{\text{Noise and Distortion Power}}\right) \qquad (1\text{-}58)$$

1.6.6 Effective Number Of Bits (ENOB)

The effective number of bits (ENOB) is a measure based on the SNDR of an ADC with a full scale sinusoidal input signal, according to (1-49). The ENOB is determined by

$$ENOB = \frac{SNDR - 1.76}{6.02} \quad (1\text{-}59)$$

where SNDR correspond to the actual (non-ideal) value.

It should be noted that all the measures above are both frequency and signal amplitude dependent. At low input amplitude levels the ADC performance is usually limited by the quantization noise while the distortion will limit the performance at higher signal levels. For communications applications it is common to specify the SNDR as a function of the input amplitude as shown in Fig. 1-13.

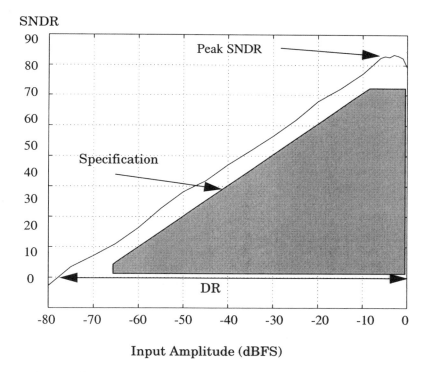

Figure 1-13   SNDR vs. input frequency.

1.6.7 Peak SNDR

The maximal SNDR in Fig. 1-13 is referred to as the peak SNDR and is usually located slightly below the full scale input (0 dBFS in the plot). For oversampling ADCs the peak SNDR may be located significantly below the full scale input. The requirements from the application is specified as an area in the plot that must be met by the ADC (the shaded area in Fig. 1-13).

1.6.8 Dynamic Range (DR)

The range from full scale (FS) to the smallest detectable signal (usually SNDR = 0) is called the dynamic range (DR) of the converter, i.e.

$$DR = 10 \cdot \log\left(\frac{\text{Maximum signal power}}{\text{Smallest signal power}}\right) \qquad (1\text{-}60)$$

In communications it is not uncommon to define the smallest signal as the signal which gives a certain SNR > 0.

1.6.9 Effective Resolution Bandwidth (ERB)

An important parameter for the data converter is the signal bandwidth that can be handled. The bandwidth is limited by the analog bandwidth of the input circuits in the ADC as well as the maximal sampling frequency of the converter. The input signal frequency must be smaller than the Nyquist frequency (half the sampling frequency) to avoid aliasing in conventional applications of ADCs. The signal bandwidth can be larger than the Nyquist frequency for sub-sampling ADCs. To specify the frequency behaviour of the converter it is common to plot the SNDR, SFDR or SNR as function of input frequency as illustrated in Fig. 1-14. The effective resolution bandwidth is the

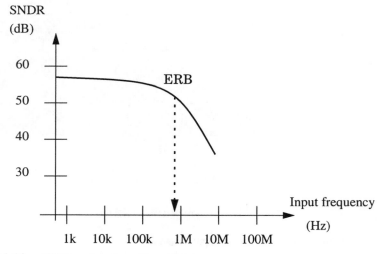

Figure 1-14   SNDR as function of input frequency.

input frequency where the SNDR has dropped 3 dB (or ENOB 1/2 bit). A well designed wideband ADC for sub-sampling applications can have an effective bandwidth well above the Nyquist frequency.

1.6.10 Inter-Modulation Distortion (IMD)

Inter-modulation distortion (IMD) appears when the input is a multi-tone signal. Assume that two tones with the frequencies f_1 and f_2 are applied to the converter with sampling rate f_s. Intermodulation distortion will appear at the frequencies

$$(k \cdot f_1 + m \cdot f_2) \bmod (f_s/2) \tag{1-61}$$

where k and m are integer numbers, and further $k \neq 0$, $m \neq 0$, and $f_1 \neq f_2$. The inter-modulation distortion is calculated as

$$IMD = 10 \cdot \log\left(\frac{\sum X_{k,m}^2}{X_0^2}\right) \tag{1-62}$$

where X_0 is the rms value of the fundamental and $X_{k,m}$ is the rms value of the tones at the frequencies given by (1-61). For some multi-tone applications the tone frequencies are multiples of a specific fundamental frequency and hence the intermodulation terms will interfere with other tones, see Sec. 1.6.12.

1.6.11 SNR with Multi-Tone Input

For multi-tone measurements we also use the peak-to-average ratio (PAR) or crest factor. The PAR gives information on how the signal is distributed over the amplitude range. A low PAR indicates a more uniform distribution, which is advantageous in most cases. The PAR can be calculated as

$$PAR = \frac{\text{peak amplitude}}{\text{rms value}} \tag{1-63}$$

For a sinusoidal signal, $PAR = \sqrt{2}$. A sinusoidal input signal is common when testing converters but in many practical communications applications there are more than one sinusoidal tone in the input signal. In such applications it is convenient to use the peak-to-average ratio (PAR) when calculating the SNR. For an N bit converter with input range FS the average power, P_s, is

$$P_s = \frac{1}{(PAR)^2} \cdot \left(\frac{FS}{2}\right)^2 = \frac{1}{(PAR)^2} \cdot (\Delta \cdot 2^{N-1})^2 \tag{1-64}$$

making the SNR of an ideal ADC

$$SNR = 10 \cdot \log\left(\frac{\frac{1}{(PAR)^2} \cdot (\Delta \cdot 2^{N-1})^2}{\Delta^2/12}\right) =$$
$$= 6.02 \cdot N + 4.77 - 20 \cdot \log(PAR) \tag{1-65}$$

We see from (1-65) that a small PAR is preferred since it maximizes the SNR. The relation between the SNR of a single-tone measurement and a multi-tone measurement is

$$SNR_{multi} = SNR_{single} - 20 \cdot \log\left(\frac{PAR}{\sqrt{2}}\right) \qquad (1\text{-}66)$$

1.6.12 Multi-Tone Power Ratio (MTPR)

It is difficult to find the distortion for multi-tone transmission schemes (DMT, OFDM, etc.), since tones often are multiples of a fundamental frequency. This implies that distortion terms, harmonics, are added to information carriers. A method to find out the distortion is to apply a number of tones (that are multiples of a fundamental, ω_0), all with the same amplitude, A. Some tones are left out, and the distortion terms that occur at these positions determine the quality and the multi-tone power ratio (MTPR). The MTPR is defined as

$$\text{MTPR} = 10 \cdot \log\left(\frac{A^2/2}{\text{Power of tones at left-out frequency positions}}\right) \qquad (1\text{-}67)$$

This is also depicted in Fig. 1-15, where 25 tones have been applied. At two frequen-

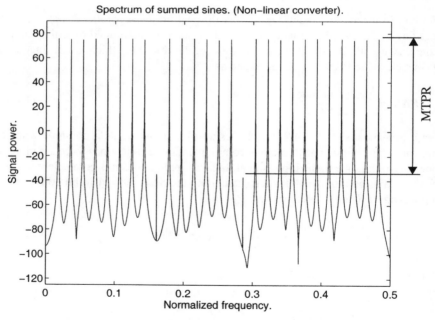

Figure 1-15 Illustration of the impact of MTPR. Two tones are left out and the non-linearity can be found by observing the power of the tones appearing at the left-out positions.

cies tones have been excluded. The non-linearity of the converter introduces harmonics at these positions, and MTPR can be determined by the power of these harmonics. It should be noted that MTPR is a good measure for harmonic distortion. In certain data converters the non-harmonic may exceed the harmonic distortion. This means that $SFDR < MTPR$.

1.7 SUMMARY

In this chapter we have discussed the fundamentals and characterization of data converters. The performance of the data converter is limited by both static and dynamic errors. The static errors are usually caused by mismatch in the circuit elements of the converter and will limit the accuracy of the conversion at low speed. The static performance is often described by the differential-non-linearity (DNL) and the integral-non-linearity (INL). Quite often, offset and linear gain errors are acceptable. Therefore the these errors are sometimes removed before the DNL and INL are calculated. The static linearity measures are rarely sufficient to characterize the performance of data converters in communications. The dynamic errors, as settling errors glitches, etc. will typically increase at higher sampling and signal frequencies. It is usually easier to analyze the effect of these errors in the frequency domain. A number of frequency domain measures as SNDR, SFDR and ENOB were introduced. These measures are normally derived from the output spectrum when a sinusoidal input signal is used. In some applications it may be necessary to use several sinusoidal tones to get all the information. A common measure in such measurements is multi-tone power ratio (MTPR). We showed that the power-to-average ratio (PAR) of the input signal affects the maximum SNR of the converter. A small PAR is preferable but a multi-tone signal normally have a larger PAR than a single-tone signal.

REFERENCES

[1] D. H. Sheingold, *Analog-Digital Conversion Handbook*, Prentice-Hall, 1986.

[2] I. Koren, *Computer Arithmetic Algorithms*, Prentice-Hall, Englewood Cliffs, NJ, USA, 1993, ISBN 0-13-151952-2.

[3] R. J. van de Plassche, *Integrated Analog-to-Digital and Digital-to-Analog Converters*, Kluwer Academic Publishers, Boston, MA, USA, 1994.

[4] P. J. Fish, *Electronic Noise and Low Noise Design*, The Macmillan Press ltd, 1993.

[5] J. C. Candy and G. C. Temes, *Oversampling Delta-Sigma Data Converters*, IEEE Press, New York, NY, USA, 1992, ISBN 0-87942-281-5.

[6] D. A. Johns and K. Martin, *Analog Integrated Circuit Design*, John Wiley & Sons, Inc. 1997, ISBN 0-471-14448-7.

[7] P. Hendriks, "Specifying Communication DACs," *IEEE Spectrum*, vol. 34, no. 7, pp. 58-69, July 1997.

[8] P. Hendriks, "Tips for Using High-Speed DACs in Communications Design", *IEEE Electronic Design*, no. 2, pp. 112-8, Jan. 1998.

[9] J. B. Simoes, J. Landeck and C. M. B. A. Correia, "Nonlinearity of a Data-Acquisition System with Interleaving/Multiplexing", IEEE Trans. Instrumentation and Measurement, vol. 46, no. 6, pp. 1274-79, Dec. 1997.

[10]　A. Netterstrom, "Using Digital Pre-Distortion to Compensate for Analog Signal Processing Errors," in *Proc. of the IEEE 1990 Int'l Radar Conf.*, pp. 243-8, Arlington, VA, USA, May 7-10, 1990.

[11]　C. Toumazou, J. B. Hughes, and N. C. Battersby, *Switched-Currents: an Analogue Technique for Digital Technology*, Peter Peregrinus, Stevenage, UK, 1993.

[12]　M. Benkaïs, S. Le Masson, and P. Marchegay, "A/D Converter Characterization by Spectral Analysis in 'Dual-Tone' Mode," *IEEE Trans. on Instrumentation and Measurement*, vol. 44, no. 5, p. 940-4, Oct. 1995.

2 DATA CONVERTER REQUIREMENTS FOR COMMUNICATIONS

2.1 INTRODUCTION

Data converters are crucial building blocks in modern communications systems where DSPs are extensively used. A D/A converter is usually needed at the transmit side and an A/D converter is usually needed at the receive side. The most critical criteria in choosing such data converters are that the data converters shall not degrade the SNR and that the data converters shall not introduce any spurs or distortions. Depending on the actual communication system, the data converter requirements vary dramatically. There are many system-related issues that can influence the data converter requirements, among which modulation schemes, duplexing methods, and analog pre-processing are most important and less understood by data converter designers.

In this chapter, we will briefly discuss communication system-related issues that have strong impacts on data converter requirements. The data converter requirements can also be derived from the standard for a given system and this is also detailed in this chapter in treating asymmetrical digital subscriber line (ADSL) systems. Following this introduction, we will brief line coding and modulation in Sec. 2.2. Both baseband and passband codes will be discussed. We will also touch upon the excess bandwidth concept since it has significant impact on the peak-to-average ratio (PAR). In Sec. 2.3, we will discuss the relationship between the bit rate, SNR, and error probability. The maximum bit rate for a given channel is referred to channel capacity. In Sec. 2.4, we will discuss the duplexing methods. In Sec. 2.5, we will discuss multi-carrier systems and their impact on the PAR. In Sec. 2.6, we will apply the system-related constrains to deriving the minimum requirement on data converters for digital transmission while Sec. 2.7 is devoted to finding the optimum data converters for ADSL. Sec. 2.8 will address A/D converter requirements for wideband radio. In all discussions, we will concentrate on SNR without explicitly discussing distortions. The SNR should be treated as SNDR and the number of bits should be treated as the effective number of bits for data converters in this chapter. The only difference between the noise and distortion generated in a data converter is that oversampling has no impact on the distortion while it reduces the noise power spectral density.

Notice that we do not intend to treat the communication systems in detail. We only touch upon some of key parameters and concepts so that data converter designers can understand their impacts. For some wired communication systems and most radio sys-

28 Chapter 2. Data Converter Requirements for Communications

tems, coding is used that generally reduces the error probability or reduces the SNR requirement for a given error probability. Also depending on the system trade-off, certain design margin is needed. These two factors can be easily taken into consideration by modifying the required SNR value. Since they are system-specific, we will not include them in the following discussions.

2.2 LINE CODING AND MODULATION

Line coding is the conversion of abstract symbols into real, temporal waveforms to be transmitted in the baseband while modulation is the process where the message information is added to a carrier according to radio terminology [1]. However, for wired digital transmission, e.g, for xDSL (x stands for different types of digital subscriber line), people tend to mix the terminologies of line coding and modulation. For xDSL, we sometimes refer to modulation as the conversion of bit streams into equivalent analog signals that are suitable for the transmission line [2]. The reason is that modulation is usually done digitally for xDSL while modulation is done at radio frequencies (RF) for radio communication.

In this book, we do not confuse readers with different jargons. We specify line codes as baseband or passband codes. The baseband codes are referred to the codes that have energy at DC while the passband codes are referred to the codes that do not have energy at DC, although baseband codes can be modulated to a carrier as well.

2.2.1 Baseband Codes

One of the widely used baseband codes is the phase amplitude modulation (PAM) code. A b-bit PAM has $M = 2^b$ equally spaced levels symmetrically placed about zero (b=1, 2 ...). A b-bit PAM is often referred to PAM-M or M-ary PAM. In Fig. 2-1 we

Figure 2-1 An 8-PAM constellation diagram.

show a 3-bit or 8 PAM. The assignment of the b-bit information to the $M = 2^b$ possible signal amplitudes can be done in different ways. The preferred mapping or assignment is called Gray coding in which the adjacent signal amplitudes differ by only one binary digit [3]. If a PAM modulates a carrier, it is usually called amplitude-shift keying (ASK) in digital communication.

As discussed in Chapter 1, the peak-to-average ratio (PAR) has a strong impact on the data converter requirements. Assume that the distance between two adjacent levels is d, the peak amplitude of an M-ary PAM is then $d \cdot (M-1)/2$. Under the assumption of equal probability of occurrence for all the levels, the *rms* amplitude is given by (if M is even)

2.2 Line Coding and Modulation

$$\sqrt{\left\{\left(\frac{d}{2}\right)^2 + \left(d + \frac{d}{2}\right)^2 + \ldots + \left(\frac{M}{2} \cdot d - \frac{d}{2}\right)^2\right\} \cdot \frac{2}{M}}$$

$$= \frac{d}{2} \cdot \sqrt{\frac{M^2 - 1}{3}}$$

(2-1)

Therefore, the peak-to-average ratio is given by

$$PAR = \left\{\frac{d}{2} \cdot (M-1)\right\} / \left\{\frac{d}{2} \cdot \sqrt{\frac{M^2 - 1}{3}}\right\}$$

$$= \sqrt{\frac{3 \cdot (M-1)}{M+1}} = \sqrt{\frac{3 \cdot (2^b - 1)}{(2^b + 1)}}$$

(2-2)

The simulated PAR for PAM is shown in Fig. 2-2. The PAR for PAM approaches 1.73

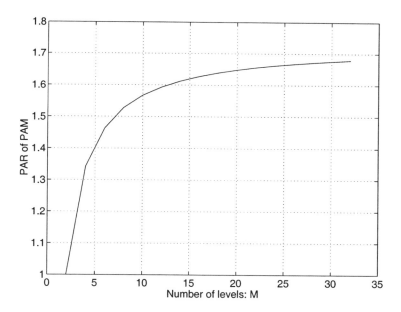

Figure 2-2   Peak-to-average ratio of PAM.

when the number of levels increases. The 4-PAM, or two bits per quaternary (2B1Q) used in integrated service digital network (ISDN) and high-bit-rate digital subscriber line (HDSL) has a PAR of about 1.34.

Binary PAM or 2-PAM is widely used in both radio communication and xDSL due to its simplicity. There are two major classes of binary line codes: level codes and transition codes. Level codes carries information in the voltage level while transition codes carry information in the change in level appearing in the line code waveform. There are many variations based on 2-PAM. Interested readers are referred to [1,2,3].

It is not unusual that PAM codes can have an odd number of levels. It can include a level having a value of 0. For instance, PAM-3 is used in Ethernet 100BASE-TX and PAM-5 is used in Ethernet 100BASE-T2 and 1000BASE-T [4]. In this case, the code mapping is more complex [4]

Pulse code modulation (PCM) codes are widely used in communication for transmitting voice signals. A PCM signal is obtained from the PAM signal by encoding each value into a digital word. A PCM signal can be thought of as a serial representation of a PAM signal. The actual transmission of PCM codes is binary. In order to reduce the data converter requirement, compression such as A law or μ law is used for voice communication. Interested readers are referred to [5].

2.2.2 Passband Codes

Passband codes have no energy at DC. The widely used passband code is the quadrature amplitude modulation (QAM) code. A QAM signal is constructed by the summation of an in-phase signal (I) and a quadrature signal (Q), given by

$$I + Q = \varphi(t) \cdot \cos(\omega_c \cdot t) - \varphi(t) \cdot \sin(\omega_c \cdot t) \tag{2-3}$$

where ω_c is the carrier frequency and $\varphi(t)$ is a real-time pulse like a sinc or a square-root raised cosine pulse that is determined by the digital data stream. The multiplication of the pulse by a cosine and sine moves the energy away from the DC to the carrier frequency and different pulse shapes have different properties such as bandwidth requirement, inter-symbol interference, etc.

The QAM codes are two-dimensional. With a b-bit QAM there are $M = 2^b$ symbols in the constellation. A 6-bit or 64-QAM constellation is shown in Fig. 2-3. Assume

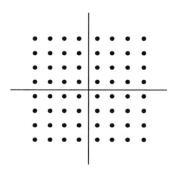

Figure 2-3   A 64-QAM constellation.

that the distance between two neighboring symbols on the x- or the y-axis is d, an $M = 2^b$ QAM constellation have the following magnitudes if b is even

2.2 Line Coding and Modulation

$$\frac{d}{2} \cdot \sqrt{2}, \frac{d}{2} \cdot \sqrt{1+3^2}, \ldots \frac{d}{2} \cdot \sqrt{1+\left(2^{\frac{b}{2}}-1\right)^2}, \frac{d}{2} \cdot \sqrt{3^2+1}, \frac{d}{2} \cdot \sqrt{3^2+3^2}$$

$$\ldots \frac{d}{2} \cdot \sqrt{3^2+\left(2^{\frac{b}{2}}-1\right)^2}, \ldots \frac{d}{2} \cdot \sqrt{\left(2^{\frac{b}{2}}-1\right)^2+1},$$

$$\ldots \frac{d}{2} \cdot \sqrt{\left(2^{\frac{b}{2}}-1\right)^2+\left(2^{\frac{b}{2}}-1\right)^2}$$

Under the assumption of equal probability of occurrence for all the levels, the *rms* magnitude is given by

$$\frac{d}{2} \cdot \sqrt{\sum_{k=1}^{\frac{b}{2}} \sum_{j=1}^{\frac{b}{2}} \{(2^j-1)^2 + (2^k-1)^2\} \cdot \frac{4}{2^b}}$$

$$= \frac{d}{2} \cdot \sqrt{\frac{2(2^b-1)}{3}}$$

(2-4)

The maximum signal magnitude is

$$\frac{d}{2} \cdot \sqrt{\left(2^{\frac{b}{2}}-1\right)^2 + \left(2^{\frac{b}{2}}-1\right)^2} = \frac{d}{2} \cdot \sqrt{2} \cdot \left(2^{\frac{b}{2}}-1\right)$$

(2-5)

Therefore, the peak-to-average ratio is given by

$$PAR = \sqrt{\frac{3 \cdot \left(2^{\frac{b}{2}}-1\right)}{\left(2^{\frac{b}{2}}+1\right)}} \cdot PAR_c = \sqrt{\frac{3 \cdot (\sqrt{M}-1)}{(\sqrt{M}+1)}} \cdot PAR_c$$

(2-6)

where PAR_c is the peak-to-average ratio of the carrier. If the carrier is a sine wave, PAR_c is equal to 1.4.

In Fig. 2-4 we show the peak-to-average ratio as a function of the number of levels M. When the number of levels M increases to infinity, the PAR approaches the maximum value of 2.45.

Carrierless amplitude and phase (CAP) modulation is considered as a special case of QAM. If the carrier frequency is not significantly larger than the bandwidth, the carrier modulation in QAM is superficial because a judicious choice of two DC-free functions can realize the same function. Compared with QAM, CAP simplifies the transmitter implementation [6].

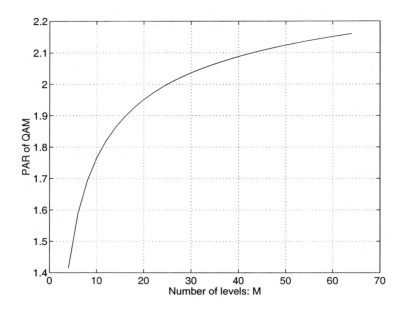

Figure 2-4   Peak-to-average ratio of QAM.

If we keep the signal amplitude constant and only change the phase according to the digital bit stream, we have MPSK, or M-ary phase-shift-keying. Binary PSK (BPSK, M = 2) and quadrature PSK (QPSK, M = 4) are widely used in radio communications. BPSK is equivalent to 2 PAM (if modulated) or 2 QAM, and QSPK is equivalent to 4 QAM.

We can also use the bit information to modulate the carrier frequencies. It is called frequency-shifting keying (FSK). Binary FSK (BFSK) is widely used in radio communications.

2.2.3   Excess Bandwidth

When we transmit data successively, interferences between the successive symbols deteriorate the performance. To minimize the interferences between successive symbols (intersymbol interferences), we can use Nyquist pulses that are orthogonal to one another. Another possibility to minimize the intersymbol interferences is to introduce a controlled amount of interference at the transmitter, which can be removed at the receiver. This technique is called partial-response signalling. Interested readers are referred to [1].

There are many Nyquist pulses, and the best known are the raised-cosine pulses, which decay with $1/t^3$ instead of $1/t$ as in sinc pulses. The fast decay in time is crucial since it reduces the timing-phase errors in the sampling clock of the receiver.

Suppose that the symbol period is T, or the symbol rate is $f_s = 1/T$. The excess bandwidth is defined as the bandwidth relative to the symbol rate, i.e, $(1 + \alpha) \cdot f_s$, where α is the fraction that defines the excess bandwidth (α is usually between 0 and 1). Notice that sometimes the excess bandwidth is defined as $(1 + \alpha) \cdot 0.5 \cdot f_s$ if the spectrum is symmetrical.

With a larger excess bandwidth, the pulses ring less and it is easier to receive the signal, but the channel usage is not effective, and vice versa. The reason that we touch upon the excess bandwidth concept here is due to its influence on the peak-to-average ratio.

It is very tedious and complicated to derive the *PAR* as a function of the excess bandwidth, since it is dependent on the pulse shapes. Due to the excess bandwidth, the PAR usually increases. The smaller the excess bandwidth is, the larger the increase in the PAR is. It is not unusual for the *PAR* to increase by more than 50% due to the effect of the excess bandwidth. Also notice that with a smaller excess bandwidth, the bandwidth requirement on data converters should be reduced. This advantage can seldom be utilized since it requires a relatively complex digital decimation filter with a fractional decimation ratio.

2.3 CHANNEL CAPACITY AND ERROR PROBABILITY

2.3.1 Channel Capacity

In communication, a terminology called channel capacity is used to describe the maximum transmit throughput and to compare the efficiency of a certain coding vs. this theoretical limitation. If the channel has a bandwidth of *BW*, the maximum number of bits per second that the channel can support, or the channel capacity, is given [3] by

$$C = BW \cdot \log 2(1 + SNR) \qquad (2\text{-}7)$$

where *SNR* is the received signal to noise ratio. Normalizing against the bandwidth, we have the channel capacity in bits per second per Hz, given [7] by

$$c = \log 2(1 + SNR) \qquad (2\text{-}8)$$

Obviously, if we have *M* levels in the code, the number of bits per second per Hz is equal to $\log_2(M)$. In Fig. 2-5, we show the channel capacity vs. the SNR. It is seen that for every 1-bit increase in the channel capacity, we need to increase the SNR by 3 dB.

The above conclusion is drawn based on the assumption of optimum codes that necessarily requires infinite complexity and infinite encoding/decoding delay [8]. To achieve the same number of bits per Hertz, the SNR requirement increases above the theoretical SNR depending on many factors such as modulation schemes and coding. The actual data rate is smaller than this theoretical value and it is given by

$$\bar{c} = \log 2\left(1 + \frac{SNR}{\Gamma}\right) \qquad (2\text{-}9)$$

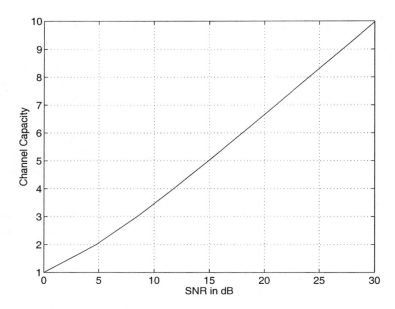

Figure 2-5   Channel capacity vs. SNR.

where Γ is referred to as the gap which is always larger than 1. The gap is a function of the error probability, modulation, and coding. With coding, we can reduce the gap. The difference in the gap Γ between the coded and un-coded transmission is sometimes referred to as coding gain. The readers are referred to [1,2,3,7].

For baseband codes that are one-dimensional, the channel capacity is defined [2] as

$$c = \frac{1}{2} \cdot \log 2(1 + SNR) \qquad (2\text{-}10)$$

For baseband transmission, the channel capacity increases by 1 bit/second/Hz if we increase the SNR by 6 dB.

2.3.2 Error Probability

When we receive the signal, we need to re-construct the symbol constellation. Intuitively, the larger the minimum distance between two symbols, the easier it is to reconstruct the constellation. In the presence of noise, some symbols will not be able to be re-constructed correctly. Therefore, the error probability, or the symbol error rate is determined by the minimum symbol distance and the noise. The larger the minimum distance between two symbols is, the easier it is to recover the information. To achieve the same minimum distance, the MPSK usually needs more signal energy than M-ary QAM for M larger than 4. The larger M is, the more extra energy is needed. With the same energy, QAM usually have better symbol error probability than PSK, depending on M.

2.3 Channel Capacity and Error Probability

In order to evaluate the symbol error rate, we have to introduce the Q function that is widely used in telecommunications. The quantity $Q(x)$ is the probability that a unit variance zero-mean Gaussian random variable exceeds the value in the argument, x, i.e.,

$$Q(x) = \int_x^\infty \frac{1}{\sqrt{2\pi}} \cdot e^{-\frac{u^2}{2}} du \qquad (2\text{-}11)$$

For an M-ary PAM, the error probability is given [2] by

$$P_{PAM} = 2 \cdot \left(1 - \frac{1}{M}\right) \cdot Q\left(\sqrt{SNR} \cdot \sqrt{\frac{3}{M^2 - 1}}\right) \qquad (2\text{-}12)$$

Therefore, we have the SNR in dB for an M-ary PAM given by

$$SNR_{PAM} = 20 \cdot \log\left\{Q^{-1}\left(\frac{1}{2 \cdot \left(1 - \frac{1}{M}\right)} \cdot P_{PAM}\right)\right\} + 10 \cdot \log\left(\frac{M^2 - 1}{3}\right) \qquad (2\text{-}13)$$

We plot the SNR requirement as a function of M for different error probabilities in Fig. 2-6. The gap Γ can be easily found by using equations (2-9) and (2-13). It is given by

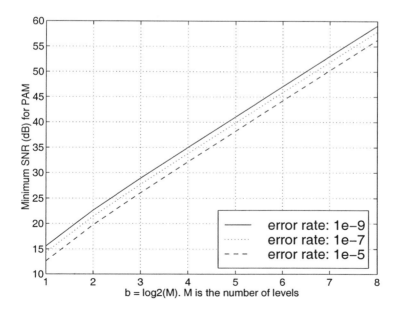

Figure 2-6 Minimum SNR for un-coded M-ary PAM for different error probabilities.

$$\Gamma = 20 \cdot \log \left\{ Q^{-1} \left(\frac{1}{2 \cdot \left(1 - \frac{1}{M}\right)} \cdot P_{PAM} \right) \right\} - 4.77 \qquad (2\text{-}14)$$

In Fig. 2-7 we show the gap vs. the error probability. It is seen that the gap Γ is larger

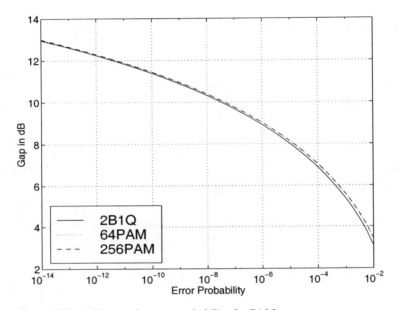

Figure 2-7   Gap vs. the error probability for PAM.

than 9 dB for an error probability less than 10^{-7}. Coding can significantly reduce the gap. For an M-ary QAM, the symbol error probability is given [2] by

$$P_{QAM} \leq 4 \cdot \left(1 - \frac{1}{\sqrt{M}}\right) \cdot Q\left(\sqrt{SNR} \cdot \sqrt{\frac{3}{M-1}}\right) \qquad (2\text{-}15)$$

if $b = \log_2 M$ is even. If b is odd, we have [2]

$$P_{QAM} \leq 4 \cdot \left(1 - \frac{3}{2 \cdot M}\right) \cdot Q\left(\sqrt{SNR} \cdot \sqrt{\frac{6}{2 \cdot M - 1}}\right) \qquad (2\text{-}16)$$

Therefore, we have the SNR in dB given by

2.3 Channel Capacity and Error Probability

$$SNR_{QAM} \geq 20 \cdot \log\left\{Q^{-1}\left(\frac{1}{4\cdot\left(1-\frac{1}{\sqrt{M}}\right)} \cdot P_{QAM}\right)\right\} + 10 \cdot \log\left(\frac{M-1}{3}\right) \quad (2\text{-}17)$$

if $b = \log_2 M$ is even. If b is odd, we have

$$SNR_{QAM} > 20 \cdot \log\left\{Q^{-1}\left(\frac{1}{4\cdot\left(1-\frac{3}{2M}\right)} \cdot P_{QAM}\right)\right\} + 10 \cdot \log\left(\frac{2M-1}{6}\right) \quad (2\text{-}18)$$

We plot the SNR requirement as a function of M for different error probabilities in Fig. 2-8. The gap can be easily found by using equation (2-9) and (2-17). It is given by

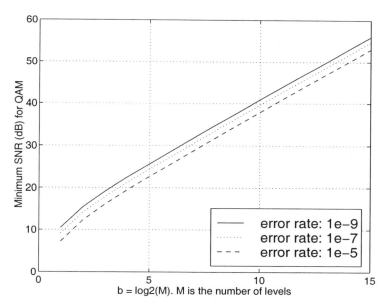

Figure 2-8 Minimum SNR for un-coded M-ary QAM for different error probabilities.

$$\Gamma = 20 \cdot \log\left\{Q^{-1}\left(\frac{1}{4\cdot(1-1/\sqrt{M})} \cdot P_{QAM}\right)\right\} - 4.77 \quad (2\text{-}19)$$

for even $b = \log_2 M$. In Fig. 2-9, we show the gap vs. the error probability. It is seen

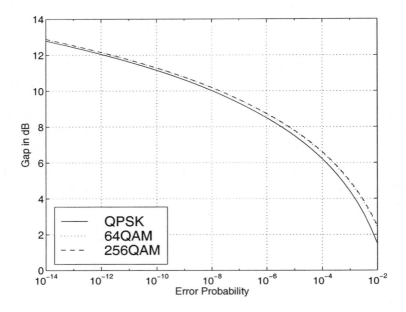

Figure 2-9   Gap vs. the error probability rate for QAM.

that the gap is larger than 9 dB for a symbol error rate less than 10^{-7}. Coding can significantly reduce the gap to 3 to 5 dB. Some extremely powerful coding schemes such as Turbo codes or Concatenated codes can reduce this gap to 1 to 3 dB.

2.4   DUPLEXING METHOD

Duplexing or multiplexing is referred to the exchange of information in both directions of a connect. Do not confuse duplexing with multiple access which is based on insulating signals used in different connections from each other in a wireless radio system.

There are several types of duplexing techniques. The space-division duplexing (SDD) uses physically different media. For instance, the T1/E1 services use different pairs for each direction of transmission. Sometimes, this kind of duplexing is referred to as dual-simplexing in that there are two simplex (unidirectional) transmission channels.

The time-division duplexing (TDD) uses the same physical channel. When one-direction transmission is activated, the other direction is de-activated. One of the preliminary proposals for very high-data-rate DSL (VDSL) uses TDD. Some digital cordless telephones such as CT3 and DECT systems use TDD.

2.4 Duplexing Method

The frequency-division duplexing (FDD) or frequency-division multiplexing (FDM) separates the physical channel into two or more distinguished bands such that one direction of transmission does not coincide in frequency with the other direction of transmission. Most cellular systems such as NMT, GSM, IS-54, etc., use FDD. FDD is also supported as an ANSI and ITU standard for ADSL.

Another technique is called echo-cancellation duplexing (ECD), or full duplexing that is widely used in xDSL. Echo cancellation duplexing systems allow transmission in both directions on the same physical channel at the same time. ECD is based on the concept that the receiver always knows what the transmitter at the same end sends to the other direction. The total signal that the receiver sees is the summation of the receive signal that was sent from the other end of the line and the transmit signal (or the echo signal) that it sends itself. By subtracting the transmit signal from the total signal, the receive signal can be recovered. ECD is used for instance in ISDN, HDSL, and ADSL. ECD usually requires both an analog echo rejection (hybrid) and a digital adaptive filter. In Fig. 2-10 we show the concept of the hybrid. All the impedances are

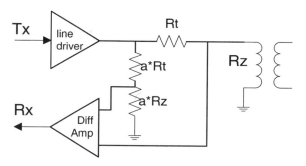

Figure 2-10 Simplified diagram for hybrid.

shown in the figure. Assume that the transmit signal at the line driver output is V_{Tx} and the receive signal at the transformer is V_{Rx}. The signal after the difference amplifier (neglecting the gain in the difference amplifier) is given by

$$V_{out} = \left\{ \frac{R_z}{R_t + R_z} \cdot (V_{Tx} - V_{Rx}) + V_{Rx} \right\} - \frac{a \cdot R_z}{a \cdot R_t + a \cdot R_z} \cdot V_{Tx} =$$

$$= \frac{R_t}{R_t + R_z} \cdot V_{Rx}$$

(2-20)

It is seen that the output does not contain any Tx signal or echo signal. However, the derivation above is based on the assumption that the impedance of the network perfectly matches the line impedance. Due to the vast variation in the line impedance, the echo signal is usually present at the receiver. If the hybrid does not provide a sufficient echo rejection, the data converter requirements increase significantly since the ADC needs to quantize the weak signal together with the strong echo signal. Adaptive hybrids (capable of adjusting the impedance to match the line impedance) can increase the echo rejection if properly designed. All the received echo signal can be eliminated by digital adaptive filters, provided that the Rx path has high enough dynamic range and linearity.

2.5 MULTI-CARRIER SYSTEMS

Multi-carrier systems are systems that have more than one carrier. Orthogonal frequency division multiplexing (OFDM) is used in digital audio broadcasting (DAB) and has been suggested for use in mobile communications. For xDSL, the discrete multi-tone (DMT) modulation has been adopted as an ANSI [9] and ITU standard for ADSL.

DMT is essentially the same as OFDM. Each carrier within DMT or OFDM is usually QAM. With DMT or OFDM, the modulation and demodulation can be achieved in the discrete domain by using IFFT and FFT. One of the drawbacks of multi-carrier systems is the increased PAR which calls for more stringent requirements on the analog building blocks.

Assume that the individual carrier is a QAM signal having a PAR_i and all the carriers are un-correlated, the PAR of a DMT or OFDM consisting of m-carriers is given [7] by

$$PAR_{DMT} = \frac{1}{\sqrt{m}} \cdot \sum_{i=1}^{m} PAR_i \qquad (2\text{-}21)$$

Assuming that all the sub-carriers have the same PAR_{sub}, the PAR of a DMT having m-sub-carrier is given by

$$PAR_{DMT} = \frac{1}{\sqrt{m}} \cdot \sum_{i=1}^{m} PAR_i = \sqrt{m} \cdot PAR_{sub} \qquad (2\text{-}22)$$

With a maximum of 255 sub-carriers in a DMT-ADSL system, the PAR is increased by 16 or 24 dB. This is directly translated into a 24-dB increase in the dynamic range requirement for data converters. Fortunately the peak level does not occur so often.

If the number of sub-carriers is large, the DMT signal approaches a Gaussian distribution, and the probability for a certain peak level to occur can be derived. In Fig. 2-11 we show the clipping probability as a function of the PAR for a DMT signal. Based on this, we can derive the PAR given a clipping probability (i.e., the probability when the peak-to-average ratio is larger than a given PAR). In order to achieve a clipping probability less than 10^{-7}, the PAR of the ADSL DMT signal is ca 5.3. It is obvious that the single-carrier QAM has a smaller PAR than the multi-carrier DMT or OFDM.

2.6 Minimum Data Converter Requirements

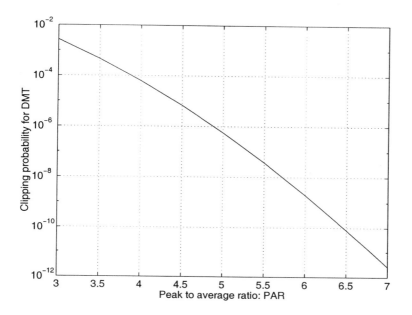

Figure 2-11 Clipping probability as a function of the *PAR* for DMT signals.

2.6 MINIMUM DATA CONVERTER REQUIREMENTS

The minimum data requirements for telecommunications are usually determined by

- the desired error probability or SNR requirement,
- the peak-to-average ratio after digital filtering (i.e, with the consideration of the influence of the excess bandwidth), and
- the duplexing method.

The term minimum refers to the fact that the data converters have to meet these criteria in order to be used in a given communication system.

2.6.1 DAC for Baseband

For baseband transmissions without digital transmit filters, the required DAC resolution is directly determined by the number of levels in the baseband signal. For instance, an $M = 2^b$ PAM needs a b-bit DAC.

If digital transmission filters are used, the DAC requirement for baseband line codes is determined by the required SNR and PAR. According to equations in Chapter 1, the SNR for the DAC (assuming a single sinusoidal input when measuring the SNR) used in a baseband transmission system is therefore given by

$$SNR_{DAC} = SNR_{PAM} + 20 \cdot \log\left(\frac{PAR}{1.4}\right) \quad (2\text{-}23)$$

where the SNR_{PAM} requirement is determined by the error probability requirement and the number of levels in the line code signal, and the PAR is determined by the number of levels in the line code signal and the excess bandwidth.

Notice that the SNR_{PAM} requirement is given at the receive end. There are many factors that contribute to the noise. The DAC quantization noise, the channel additive white Gaussian noise (AWGN), and the ADC quantization are major contributors. If these three sources have equal contributions, the DAC SNR requirement needs to increase by at least 4.78 dB. In this chapter, all the effects are collectively considered as a design margin and are not included in the derivations for simplicity.

The SNR_{PAM} requirement is related to the error rate requirement and the number of levels in the line code by equation (2-13). The peak-to-average ratio is related to the number of levels in the line code by equation (2-2). Therefore, we have

$$\begin{aligned} SNR_{DAC} = {} & 20 \cdot \log\left\{Q^{-1}\left(\frac{1}{2 \cdot \left(1 - \frac{1}{M}\right)} \cdot P_{PAM}\right)\right\} + 10 \cdot \log\left(\frac{M^2 - 1}{3}\right) \\ & + 10 \cdot \log\left(\frac{3 \cdot (M-1)}{M+1}\right) + 20 \cdot \log(\beta) - 3 \\ = {} & 20 \cdot \log\left\{Q^{-1}\left(\frac{1}{2 \cdot \left(1 - \frac{1}{M}\right)} \cdot P_{PAM}\right)\right\} \\ & + 20 \cdot \log(M-1) + 20 \cdot \log(\beta) - 3 \end{aligned} \quad (2\text{-}24)$$

where M is the number of level in the baseband code, P_{PAM} is the error probability, β is related to the excess bandwidth. Notice that the factor β is not the same as the excess bandwidth parameter α. The smaller α is, the larger β is. Deriving the factor β as a function of the excess bandwidth parameter α is tedious, depending on the transmit pulse shapes. The larger the excess bandwidth, the less stringent is the requirement on the DAC.

If the Nyquist bandwidth (half the clock frequency) is higher than the signal bandwidth, the resolution requirement on the DAC is relaxed, i.e.,

2.6 Minimum Data Converter Requirements

$$SNR_{DAC} = 20 \cdot \log \left\{ Q^{-1} \left(\frac{1}{2 \cdot \left(1 - \frac{1}{M}\right)} \cdot P_{PAM} \right) \right\}$$

$$+ 20 \cdot \log(M-1) + 20 \cdot \log\beta - 10 \cdot \log OSR - 3$$

(2-25)

where OSR is the oversampling ratio, defined as the ratio between the Nyquist bandwidth and signal bandwidth.

The number of bits is related to SNR by

$$N_{DAC} = \frac{SNR_{DAC} - 1.76}{6.02}$$

(2-26)

In Fig. 2-12, we plot the DAC requirement as a function of M for PAM. β is assumed

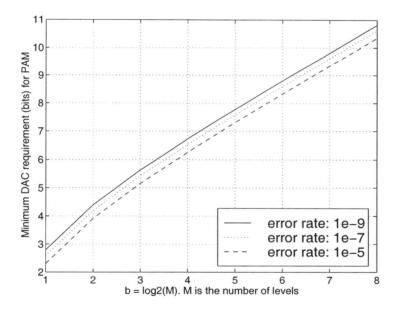

Figure 2-12 DAC requirement for PAM.

to be 2 and no oversampling is used. It is seen that for every doubling in the number of levels M for PAM, the DAC requirement increases by 1 bit. If the error probability is 10^{-7}, and the number of levels is very large, we have the approximation given by

$$N_{DAC} \approx \frac{20 \cdot \log(M) + 20 \cdot \log(\beta) - 10 \cdot \log(OSR) + 9.8}{6.02}$$

$$= b + \frac{20 \cdot \log(\beta) - 10 \cdot \log(OSR) + 9.8}{6.02}$$

(2-27)

2.6.2 DAC for Passband

Following the same derivation, we have the DAC requirement for an M-ary QAM given by

$$SNR_{DAC} = 20 \cdot \log\left\{Q^{-1}\left(\frac{1}{4 \cdot \left(1 - \frac{1}{\sqrt{M}}\right)} \cdot P_{QAM}\right)\right\}$$

$$+ 10 \cdot \log\left\{\frac{(M-1) \cdot (\sqrt{M} - 1)}{(\sqrt{M} + 1)}\right\}$$

$$+ 20 \cdot \log(\beta) - 10 \cdot \log(OSR)$$

(2-28)

where M is the number of levels in the QAM code, P_{QAM} is the error probability, β is determined by the excess bandwidth, and OSR is the oversampling ratio. The number of bits is related to the SNR by equation (2-26).

In Fig. 2-13, we show the DAC requirement as a function of the number of levels M.

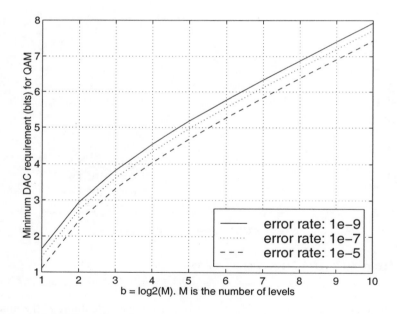

Figure 2-13 DAC requirement for QAM.

β is assumed to be 1.6 and no oversampling is used. It is seen that for every doubling in the number of levels M for QAM, the DAC requirement increases by 0.5 bit.

2.6 Minimum Data Converter Requirements

If the error probability is 10^{-7}, and the number of levels is very large, we have the approximation given by

$$N_{DAC} \approx \frac{10 \cdot \log(M) + 20 \cdot \log(\beta) - 10 \cdot \log(OSR) + 12.6}{6.02}$$

$$= \frac{b}{2} + \frac{20 \cdot \log(\beta) - 10 \cdot \log(OSR) + 12.6}{6.02} \qquad (2\text{-}29)$$

2.6.3 DAC for DMT

One of the disadvantages of DMT is the increased PAR. Suppose that there are m sub-carriers in the DMT and that every sub-carrier has the same number of levels, M. Since the PAR of the DMT increases by $\sqrt{m}$ compared to its individual sub-carrier, by using equation (2-28), we have

$$SNR_{DAC} = 20 \cdot \log \left\{ Q^{-1}\left(\frac{1}{4 \cdot \left(1 - \frac{1}{\sqrt{M}}\right)} \cdot P_{QAM} \right) \right\}$$

$$+ 10 \cdot \log \left\{ \frac{(M-1) \cdot (\sqrt{M} - 1)}{(\sqrt{M} + 1)} \right\} \qquad (2\text{-}30)$$

$$+ 10 \cdot \log(m) - 10 \cdot \log(OSR)$$

where M is the number of levels in the sub-carriers, m is the total number of sub-carriers, and OSR is the oversampling ratio. The modulation and demodulation for DMT (usually using IFFT and FFT) is different from a single-carrier QAM. Therefore, the excess bandwidth parameter does not apply.

Theoretically, with 255 tones, the DAC requirement for the DMT increases by 4 bits compared to its corresponding QAM. Fortunately, when m is large, the DMT tones can be treated as AWGN. With a clipping probability of 10^{-7}, we have the approximation given by

$$SNR_{DAC} \approx 20 \cdot \log\left\{ Q^{-1}\left(\frac{1}{4} \cdot 10^{-7}\right)\right\} + 10 \cdot \log\frac{(M-1)}{3}$$

$$+ 20 \cdot \log(PAR) - 10 \cdot \log(OSR) - 3 \qquad (2\text{-}31)$$

$$\approx 10 \cdot \log(M - 1) - 10 \cdot \log(OSR) + 21$$

Notice that the DAC SNR requirement is considerably less than the transmit MTPR in the ADSL standard [9]. The requirement in the ADSL standard is given by considering other aspects such as design margins and the influence of Tx path on the Rx path, etc.

The number of bits is related to the SNR by equation (2-26) and it is given by

$$N_{DAC} \approx \frac{b}{2} + 3.2 - \log(OSR) \tag{2-32}$$

In Fig. 2-14 we show the DAC requirement as a function of the number of levels in

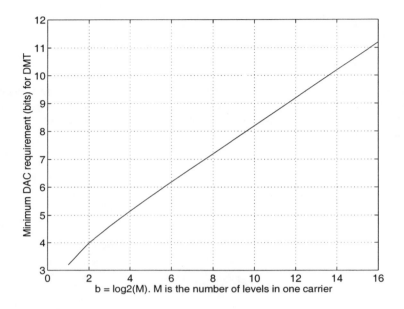

Figure 2-14 DAC requirement for DMT.

one carrier. We assume that all the carriers have been modulated with the same number of levels. In the DMT standard, the maximum number of bits that is allowed to be modulated on a single carrier is 15, i.e., 32768-QAM. This QAM has a PAR of 2.45. The DMT signal that has a PAR of 5.3 increases the DAC requirement by 1.1 bits. Due to the influence of the excess bandwidth in a real implementation, the PAR of a single carrier system, e.g, CAP can be somewhere between 3.6~4. Therefore, the use of DMT usually increase the DAC requirement by ~0.5 bits. This may not seem to be significant for DACs. However, it can increase the requirements on line drivers significantly [10].

2.6.4 ADC Requirements

The transmitted signal is attenuated by the wire or line. However, the line attenuates the noise generated in the transmitter as well. Therefore the SNR will not be degraded until the attenuated transmitter noise is comparable to the background noise on the line. For ADSL systems, the background thermal noise is -140 dBm/Hz or 32 nV/sqrt(Hz).

At the receiver, the SNR after the ADC must meet the error probability requirement for a given modulation scheme. The total noise present at the ADC output includes the

2.6 Minimum Data Converter Requirements

attenuated transmitter noise, the background noise, and the quantization noise generated by the ADC. Therefore, the SNR at the ADC output is given by

$$SNR_{total} = 10 \cdot \log \frac{T^2 \cdot G^2 \cdot P_s}{\sigma^2_{ADC} + T^2 \cdot G^2 \cdot \sigma^2_{DAC} + G^2 \cdot \sigma^2_n}$$

$$= SNR_{DAC} - 10 \cdot \log \left(1 + \frac{\sigma^2_{ADC}}{T^2 \cdot G^2 \cdot \sigma^2_{DAC}} + \frac{\sigma^2_n}{T^2 \cdot \sigma^2_{DAC}} \right) \quad (2\text{-}33)$$

where the SNR_{DAC} is the transmit (or DAC) signal-to-noise ratio, T is the line attenuation, G is the receive gain before the ADC, P_s is the transmit signal power, σ^2_{ADC} is the ADC quantization noise power, σ^2_{DAC} is the DAC noise power, and σ^2_n is the background noise on the line.

It is seen that due to the presence of the background noise and the ADC quantization noise, the DAC SNR requirements must be increased. In a real implementation, the DAC and ADC usually have the same or close to the same signal swing, and therefore we usually set the receive gain to compensate for the attenuation. For most cases, we can assume that the line noise is negligible. (We will discuss later the influence of the line noise). Now we have

$$SNR_{total} = SNR_{DAC} - 10 \cdot \log \left(1 + \frac{\sigma^2_{ADC}}{\sigma^2_{DAC}} \right) \quad (2\text{-}34)$$

Using the equations above, we can find out the DAC and ADC requirements by repeating the above process. It is obvious that the requirements are dependent on the relative resolution of the ADC and DAC.

If the DAC and ADC have the same number of bits, both the DAC and ADC resolution requirements are given by the equations (2-23) to (2-31) plus 0.5 bit accounting for the contribution from both the DAC and the ADC. In a practical implementation, a 1~2 bits margin is usually needed.

2.6.5 Influence of the Duplexing Methods

In the above discussion, we assumed that the signal at the ADC is only the signal sent from the other end and the noise, i.e, the duplexing method is the FDD with an ideal analog band-splitting filter in front of the ADC. If the band-splitting filter is not perfect, the echo signal (the signal being transmitted at the same end) will enter the ADC, calling for a higher dynamic range. (The un-filtered echo signal can be filtered by a digital filter.) For EC based systems, a portion of the echo signal will be at the ADC input depending on the analog network, significantly increasing the ADC requirements. (The echo signal will be cancelled by a digital adaptive filter, or echo canceller.) However, the noise generated by the DAC cannot usually be cancelled. Therefore, the DAC requirements also increase significantly. To derive the actual requirement for EC-based systems, we usually need to find out how much echo signal relative to the receive signal will be present at the ADC. The difference between the total signal at the ADC input and the receive signal at the input determines the increase in the requirements for the data converters.

Assume SNR_o is the data converter requirement without echo signal and SNR_e is the data converter with the presence of the echo signal. They are related by

$$SNR_e = SNR_o + 10 \cdot \log\left(\frac{P_{Rx} + P_{echo}}{P_{Rx}}\right) \quad (2\text{-}35)$$

where P_{Rx} is the receive signal power at the ADC input and P_{echo} is the echo signal at the ADC input.

If the analog band-splitting filter is not ideal for FDM systems, a portion of the Tx or echo signal will be present at the ADC input. It will increase the data converter requirements as given by equation (2-35).

For long reach where the receive signal is very weak compared with the echo signal, the data requirements is increased dramatically if the analog hybrid cannot provide enough echo rejection.

2.7 OPTIMUM DATA CONVERTERS FOR ADSL

In the above discussion, we derived the data converter requirements from the system requirements. For more advanced digital transmission systems, the modulation (including number of modulated bits), the duplexing method, and/or the signal bandwidth change according to the environment. A typical example is the ADSL. For most modern digital transmission systems, there is also a need to minimize the number of analog components. Therefore, higher performance data converters are preferred. The question is how high performance is enough to meet a specific standard for a given practical environment. This section is devoted to answering this question.

2.7.1 Optimum ADCs

Without any analog filter in front of an ADC for either FDM or EC-based system, the ADC is supposed to quantize both the echo signal and the received signal. We need to find out the requirement on the ADC. Since the transformer does not change the SNR, we can assume the transformer ratio to be 1:1 without losing any generality. If the hybrid network and some optional gain stages before the ADC do not introduce appreciable noise, the gain before the ADC does not change the SNR. We can assume a unity gain for simplicity in the derivation.

If the ADC is sampled at f_s, the power spectral density of the background noise including interference on the loop is PSD_n, the transmit power spectral density is PSD_{Tx} (bandwidth from f_{t1} to f_{t2}), the echo rejection is assumed to be $T_e(f)$, the power spectral density of the signal sent from the other side is PSD_{Rx} (bandwidth from f_{r1} to f_{r2}), and the loop attenuation of the received signal is $T_l(f)$.

The echo power at the ADC input in dBm is therefore given by

2.7 Optimum Data Converters for ADSL

$$P_{echo} = 10 \cdot \log \left\{ \int_0^{f_s/2} 10^{\frac{PSD_{Tx}}{10}} \cdot 10^{\frac{-T_e(f)}{10}} \cdot df \right\}$$

$$= 10 \cdot \log \left\{ \int_{f_{t1}}^{f_{t2}} 10^{\frac{PSD_{Tx}}{10}} \cdot 10^{\frac{-T_e(f)}{10}} \cdot df \right\} \quad (2\text{-}36)$$

$$= PSD_{Tx} - T_e + 10 \cdot \log BW_{Tx}$$

where BW_{Tx} is the transmit bandwidth. In the above derivation, we have used the averaged transmit PSD in dBm/Hz and the averaged echo rejection $T_e(f)$ in dB.

The power of the received signal excluding the echo at the ADC input in dBm is given by

$$P_{Rx} = 10 \cdot \log \left\{ \int_0^{f_s/2} 10^{\frac{PSD_{Rx}}{10}} \cdot 10^{\frac{-T_l(f)}{10}} \cdot df \right\}$$

$$= 10 \cdot \log \left\{ \int_{f_{r1}}^{f_{r2}} 10^{\frac{PSD_{Rx}}{10}} \cdot 10^{\frac{-T_l(f)}{10}} \cdot df \right\} \quad (2\text{-}37)$$

$$= PSD_{Rx} - T_l + 10 \cdot \log BW_{Rx}$$

where BW_{Rx} is the received signal bandwidth. In the above derivation, we have used the averaged received PSD in dBm/Hz and the averaged line attenuation $T_l(f)$ in dB.

The total signal power in dBm at the ADC input is therefore given by

$$P_{sig} = 10 \cdot \log \left\{ 10^{\frac{P_{echo}}{10}} + 10^{\frac{P_{Rx}}{10}} \right\} \quad (2\text{-}38)$$

The total line noise power in dBm at the ADC input is given by

$$P_n = 10 \cdot \log \left\{ \int_0^{f_s/2} 10^{\frac{PSD_n}{10}} \cdot df \right\} = PSD_n + 10 \cdot \log \left(\frac{f_s}{2} \right) \quad (2\text{-}39)$$

It appears that the SNR requirement for the ADC is given by the ratio of the signal power vs. noise power at the ADC input. However, the peak-to-average ratio (PAR) significantly increases the requirement due to the fact that the SNR measure of an ADC is based on a single sinusoidal input which has a *PAR* of 1.4. Therefore, we have

$$SNR_{ADC} = P_{sig} - P_n + 20 \cdot \log\left(\frac{PAR}{1.4}\right)$$

$$= 10 \cdot \log\left\{10^{\frac{PSD_{Tx} - T_e + 10 \cdot \log BW_{Tx}}{10}} + 10^{\frac{PSD_{Rx} - T_l + 10 \cdot \log BW_{Rx}}{10}}\right\} \quad (2\text{-}40)$$

$$- PSD_n - 10 \cdot \log\left(\frac{f_s}{2}\right) + 20 \cdot \log\left(\frac{PAR}{1.4}\right)$$

(Strictly speaking, a margin should be included due to the DAC and ADC quantization noise.) Now, we need to find out the SNR requirement needed for the ADC as a function of the echo rejection T_e and the background noise power spectral density PSD_n.

Different lengths of the loop introduce different loop attenuation of the received signal. The smallest attenuation occurs when the loop length is zero. By assuming a zero line attenuation $T_l = 0$ dB, we can find out the maximum SNR needed for the ADC. However, the SNR or carrier to noise ratio (CNR) of each sub-carrier only needs to be large enough to demodulate a maximum 15-bit QAM signal with a given bit error rate for ADSL. By using equation (2-18), the SNR for the QAM is required to be larger than 55 dB for an error probability better than 10^{-7}. Considering a 3-dB coding gain and a 6-dB noise margin, the SNR only needs to be 55-3+6=58 dB to guarantee an error rate less than 10^{-7}. Therefore, the received maximum signal power spectral density only needs to be $PSD_n + 58$ dBm/Hz. If the loop is extremely short, we can reduce the PSD_{Rx} as the standard suggests.

Therefore, we have the SNR requirement for the ADC for ADSL without any analog filter, given by

$$SNR_{ADC} = 10 \cdot \log\left\{10^{\frac{PSD_{Tx} - T_e + 10 \cdot \log BW_{Tx}}{10}} + 10^{\frac{PSD_n + 58 + 10 \cdot \log BW_{Rx}}{10}}\right\} \quad (2\text{-}41)$$

$$- PSD_n - 10 \cdot \log\left(\frac{f_s}{2}\right) + 11.6$$

The SNR is related to the number of bits in an ADC given by

$$N(bits) = \frac{SNR_{ADC} - 1.76}{6.02} \quad (2\text{-}42)$$

Notice that the quantization noise contribution is not factored into the above equations for simplicity. We use the noise margin in the following discussions to account for it as well for other influences.

In the above discussion, we have assumed that the ADC is sampled at f_s and the signal bandwidth is $f_s/2$. If the signal bandwidth is much lower than $f_s/2$, the ADC needs less

2.7 Optimum Data Converters for ADSL

dynamic range. Since the ADC noise floor is governed by the input signal, we can integrate the noise within the receive band to get the total in-band ADC noise. The peak input signal power does not vary, therefore, we have the SNR requirement on an oversampling ADC given by

$$SNR_{ADC} = 10 \cdot \log \left\{ 10^{\frac{PSD_{Tx} - T_e + 10 \cdot \log BW_{Tx}}{10}} + 10^{\frac{PSD_n + 58 + 10 \cdot \log BW_{Rx}}{10}} \right\} \\ - PSD_n - 10 \cdot \log(BW_{Rx}) + 11.6 \qquad (2\text{-}43)$$

If the echo signal dominates, we have

$$\begin{aligned} SNR_{ADC} &\approx PSD_{Tx} - T_e + 10 \cdot \log(BW_{Tx}) \\ &\quad - PSD_n - 10 \cdot \log(BW_{Rx}) + 11.6 \\ &= PSD_{Tx} - T_e - PSD_n + 10 \cdot \log\left(\frac{BW_{Tx}}{BW_{Rx}}\right) + 11.6 \end{aligned} \qquad (2\text{-}44)$$

It is seen that using a receive filter reduces the requirement on the ADC if the transmit bandwidth is larger than the receive bandwidth.

A) Optimum ADC for ADSL-CO

At the ADSL-CO, we have [9]

- PSD_{Tx} = -40 dBm/Hz
- Maximum BW_{Tx} = 1104-26 = 1078 KHz
- Maximum BW_{Rx} = 138-26 = 112 kHz

We plot the SNR and number of bit requirements in Fig. 2-15. It is seen that with a small loop noise floor (-140 dBm/Hz as specified as the background thermal noise floor for ADSL) and poor echo rejection, the ADC requirement is formidably high. However, with good echo rejection which is usually feasible at the CO side, and large noise floor (due to the interferences from other lines and from other services), the ADC is realizable.

B) Optimum ADC for ADSL-CP

At the ADSL-CP, we have [9]

- PSD_{Tx} = -38 dBm/Hz
- Maximum BW_{Rx} = 1104-26 = 1078 KHz
- Maximum BW_{Tx} = 138-26 = 112 kHz

We plot the SNR and number of bits requirements in Fig. 2-16 It is seen that with a small loop noise floor and poor echo rejection, the ADC requirement is formidably

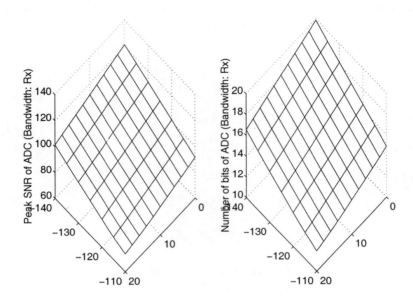

Figure 2-15 ADSL-CO ADC requirements vs. echo rejection and background noise. X-axis: background noise and interference power spectral density in dBm/Hz; Y-axis: echo rejection in dB.

high especially considering the bandwidth.

2.7.2 Optimum DACs

In the above discussions, we assumed that the echo signal will only increase the received signal power. In reality, the noise floor in the echo signal may be another limitation to achieve a high SNR. The noise at the ADC input consists of two parts, one being the contribution due to the line noise and the other being the contribution due to the DAC assuming the noise contributions from the line driver and hybrid network are negligible.

To calculate the noise contribution from the DAC, we need to calculate the voltage gain from the DAC output to the ADC input. The gain consists of the gain of the line driver, echo rejection by the hybrid network and the gain stage before the ADC. In a practical design, ADCs and DACs usually have a comparable signal range and we usually set the gain such that the ADC dynamic range is fully utilized. If the echo signal dominates at the ADC input, the total voltage gain from the DAC output to the ADC input is unity. If the receive signal is not negligible compared to the echo signal, the total gain from the DAC to the ADC is less than unity. Therefore, as long as the DAC has comparable or smaller noise floor than the ADC, the noise floor in the DAC will not degrade the receiver performance significantly (< 3 dB). Since the ADC noise floor is set by the PSD_n, the DAC noise floor should be less than PSD_n.

If we integrate the noise within the Tx band, and compare it to the total received signal

2.7 Optimum Data Converters for ADSL

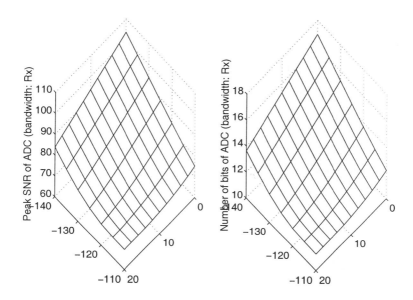

Figure 2-16 ADSL-CP ADC requirements vs. echo rejection and background noise. X-axis: background noise and interference noise spectral density in dBm/Hz; Y-axis: echo rejection in dB.

(since ADC and DAC have the same signal swing) we can derive the SNR requirement for the DAC, i.e.,

$$SNR_{DAC} = P_{sig} - P_n + 20 \cdot \log\left(\frac{PAR}{1.4}\right)$$

$$= 10 \cdot \log\left\{10^{\frac{PSD_{Tx} - T_e + 10 \cdot \log BW_{Tx}}{10}} + 10^{\frac{PSD_{Rx} - T_l + 10 \cdot \log BW_{Rx}}{10}}\right\}$$

$$- PSD_n - 10 \cdot \log(BW_{Tx}) + 20 \cdot \log\left(\frac{PAR}{1.4}\right)$$

$$\approx PSD_{Tx} - PSD_n - T_e + 11.6$$

(2-45)

In the above derivation, we assume that the echo dominates at the ADC input. The SNR is related to the number of bits in a DAC is given by

$$N(bits) = \frac{SNR_{DAC} - 1.76}{6.02}$$

$$\approx \frac{PSD_{Tx} - PSD_n - T_e + 9.84}{6.02}$$

(2-46)

Notice that we only require that the DAC noise within the receive band is governed by equation (2-46). Outside the received band, the noise floor is governed by how many bits are modulated on each sub-carrier according to the ADSL standard.

C) Optimum DAC for ADSL-CO

At the ADSL-CO, we have [9]

- PSD_{Tx} = -40 dBm/Hz
- Maximum BW_{Tx} = 1104-26 = 1078 KHz
- Maximum BW_{Rx} = 138-26 = 112 kHz

The simulated DAC requirement is shown in Fig. 2-17.

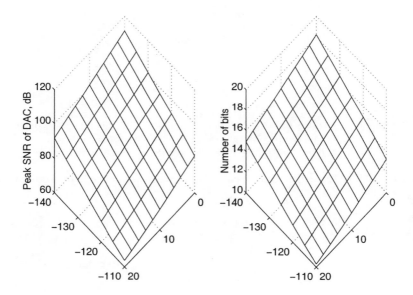

Figure 2-17 ADSL-CO DAC requirements vs. echo rejection and background noise. X-axis: background noise and interference power spectral density in dBm/Hz; Y-axis: echo rejection in dB.

D) Optimum DAC for ADSL-CP

At the ADSL-CP, we have [9]

- PSD_{Tx} = -38 dBm/Hz
- Maximum BW_{Rx} = 1104-26 = 1078 KHz
- Maximum BW_{Tx} = 138-26 = 112 kHz

The simulated DAC requirement is shown in Fig. 2-18.

2.8 ADC Requirements for Wideband Radio

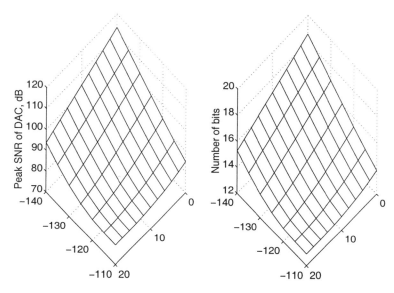

Figure 2-18 ADSL-CP DAC requirements vs. echo rejection and background noise. X-axis: background noise and interference power spectral density in dBm/Hz; Y-axis: echo rejection in dB.

2.8 ADC REQUIREMENTS FOR WIDEBAND RADIO

A digital radio (transceiver) is exactly the same as an analog radio except one difference; some of the analog functions are replaced with their digital equivalent, i.e, data converters are usually used in digital radio in the radio link. Notice that digital radio is not necessarily digital modulation and demodulation. As a matter of fact, digital radio can do an excellent job in receiving analog modulated radio as well.

We further divide digital radio into two categories, the narrow band, and the wideband. By narrowband we mean that there are sufficient analog filtering functions that eliminate or significantly reduce the un-desired signals such that only the signal of interest is present at the ADC. By wideband, we simply means that all the signals including the adjacent channels are present at the ADC without much attenuation by analog circuitry and the filtering is done digitally in the later stages in the DSP. A wideband radio can be used to receive all the channels within an entire band such as cellular or other similar wireless services. The obvious advantage is the sharing of most hardware among all channels. In order to fully utilize the DSP power in a wideband radio, the mixer and the ADC must have a high dynamic range. And they are usually most difficult to design. A typical wideband receiver is shown in Fig. 2-19. The radio signal is received by the antenna. The radio frequency (RF) filter or tuner selects the band of interest (more than one channel is within the band). The received signal is amplified before it is fed to the mixer to generate the intermediate frequency (IF) sig-

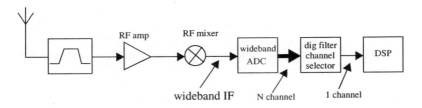

Figure 2-19 A typical wideband radio receiver.

nal. The ADC is used to convert the IF signal into a digital signal. The digital IF signals are processed by a digital decimation filter and digital channel selector before it reaches the DSP block. Depending on the noise level after the RF mixer, some analog filters might be needed for anti-aliasing purposes.

2.8.1 IF Sampling

In order to sample the wideband IF, the wideband ADC must have a high sampling rate governed by the Nyquist theory. However, since the wideband IF is bandpass filtered, the ADC data rate is only required to be twice the bandwidth, not twice the highest IF. When the ADC data rate is less than twice the highest IF, it is usually referred to as undersampling. For undersampling ADCs, the input sample-and-hold must have a bandwidth capable of sampling the highest IF signal.

One of the biggest challenges in designing a radio architecture is the placing of the IF, since the ADC tend to generate unwanted spurs that show up in the digital spectrum of the data conversion. For narrow band radio, there is only the signal of interest. If we know that a certain harmonic (for instance, the third harmonic) dominates, we can place the IF and the ADC sampling rate such that the strong harmonic is out of the signal band of interest. This can be easily achieved by frequency translation. When the IF signal is sampled at f_s, the output spectrum will appear at the following frequencies.

$$\sum_{n=-\infty}^{n=\infty} \pm f_{IF} + n \cdot f_s \tag{2-47}$$

where f_{IF} is the IF (it could be a whole band) and f_s is the ADC sampling rate. The minus sign prior to the f_{IF} indicates that this spectrum is reversed, which means that the higher frequencies are translated to the lower frequencies and vice versa. The k-th harmonic will appear at the following location

$$\sum_{n=-\infty}^{n=\infty} \pm k \cdot f_{IF} + n \cdot f_s \tag{2-48}$$

By properly choosing the IF and the data rate, the harmonics can be placed out of the band of interest. With this technique, the distortion requirement on the ADC is greatly

2.8 ADC Requirements for Wideband Radio

reduced.

For wideband radio, it is more complicated. Besides the distortion generated by each carrier, the inter-modulation may be the dominating factor, making the above-discussed technique less appealing. Therefore, the ADC converters for wideband radio must have superior linearity.

2.8.2 ADC Requirements

Since the ADC in a wideband radio receiver quantizes all the channels within the entire band, no automatic gain control (AGC) can be used to compensate for weak/strong signals. Reducing the gain for strong signals reduces the sensitivity to weak signals, and increasing the gain for weak signals overloads the ADC in presence of strong adjacent signals. In a practical implementation, some kind of AGC (actually attenuation) should be used to avoid overloading the ADC which would cause all calls to be disconnected rather than just dropping the calls with very weak signals. In principle, the ADC requirement in a wideband radio is very similar to the ADC requirement for DMT.

To find the optimum ADC, we need to know the noise floor, the signal power, and the peak-to-average ratio.

The fundamental noise floor is due to the atmospheric noise received by the antenna, its PSD in dBm/Hz is given by

$$PSD_{na} = 10 \cdot \log\left(\frac{B \cdot T}{0.001}\right) \approx -174 dBm/Hz \qquad (2\text{-}49)$$

where B is the Boltzman constant (1.38×10^{-23} J/K) and T is the temperature in K. The above approximation assumed T is equal to 300 K.

The RF signal passes through an RF filter or tuner with a signal bandwidth designated by BW. Therefore, the *rms* noise power within the total signal bandwidth is given by

$$P_{na}(dBm) = PSD_{na} + 10 \cdot \log(BW) \qquad (2\text{-}50)$$

Assume that within the bandwidth BW, there are m channels, each having a maximum power level P_c. For most radio systems, the maximum power level is specified by the standard, e.g, it is -13 dBm for GSM. For most radio systems, constant envelop modulation is used, i.e, the carrier amplitude is not altered according to the digital bit stream. For instance, GSM uses Gaussian minimum-shift keying (GMSK) that basically is a frequency modulated (FM) signal preceded by a Gaussian low-pass filter.

Assuming that all carriers are un-correlated, the maximum signal power within the bandwidth BW is given by

$$P_{sig}(dBm) = P_c + 10 \cdot \log(m) \qquad (2\text{-}51)$$

The number of channels m within the bandwidth BW can be easily found out given the channel bandwidth and the channel spacing (the distance between the centers of two adjacent channels).

For a single carrier or single channel, the peak-to-average ratio is 1.4 since the carrier is usually a sine or cosine. Assuming that the channels are un-correlated, we have the worst case peak-to-average ratio given by

$$PAR = 1.4 \cdot \sqrt{m} \tag{2-52}$$

(If there are many channels, the total signal behaves like AWGN. The PAR is about 5.3 for a clipping factor of 10^{-7}.)

The RF amplifier and mixer will introduce noise. Assume that the total noise figure from the antenna to the ADC input is NF (dB). The total noise at the ADC input is therefore increased by NF (dB). Since the conversion gain amplifies both the signal and noise, it does not have impact on the ADC dynamic range requirement. Assuming the oversampling ratio is OSR (the sampling rate divided by twice the bandwidth BW), we have the ADC peak SNR requirement with the bandwidth BW given by

$$\begin{aligned} SNR_{ADC} &= P_{sig} - (P_{na} + NF) + 20 \cdot \log\left(\frac{PAR}{1.4}\right) - 10 \cdot \log(OSR) \\ &= P_c - (P_{na} + NF) + 20 \cdot \log(m) - 10 \cdot \log(OSR) \\ &= P_c - \left(PSD_{na} + 10 \cdot \log\frac{BW}{m}\right) - NF \\ &\quad + 10 \cdot \log(m) - 10 \cdot \log(OSR) \end{aligned} \tag{2-53}$$

Notice that BW/m is a constant, i.e, the channel spacing. The first line in the equation is the dynamic range for a single carrier. It is seen that the ADC requirement increases with the number of the channels.

To derive the minimum ADC requirement, we only need to replace the noise floor at the input due to the atmospheric noise and the RF noise figure with required minimum carrier power and its carrier-to-noise ratio for a given bit error rate.

2.9 SUMMARY

In this chapter, we have briefed communication-related terminologies and concepts and discussed the impacts of communication systems on data converter requirements. In order to meet a communication system specification, we have derived minimum data converter requirements that are mostly influenced by the error probability and the peak-to-average ratio. We have also discussed optimum data converters for ADSL that are a function of application environments such as the transmit power spectrum, the receive power spectrum, the hybrid rejection, and the background noise. Wideband radio for multi-channel applications are similar to DMT but with much fewer carriers. We have derived optimum A/D converter requirements for wideband radio as a function of the single carrier dynamic range, the number of channels, the noise figure, and the background noise. To apply these knowledge to a specific system, we usually need

2.9 Summary

to add 1~2 bit (or even more) design margin depending on the system trade-offs. Also coding can reduce the data converter requirements by 3~6 dB, which has not been treated in this chapter but can be easily added to the equations. As far as distortions are concerned, they usually follow the SNR requirements without being effected by oversampling.

REFERENCES

[1] Jerry G. Gibson, The mobile communication handbook, CRC press, 1996.

[2] T. Starr, J. Cioffi, and P. Silverman, Understanding digital subscriber line technology Prentice Hall, 1999.

[3] J. G. Proakis, Digital communications, McGraw-Hill, 1997.

[4] IEEE Draft P802.3ab/D.30, Physical layer specification for 1000 Mb/s operation on four pairs of category 5 or better balanced twisted pair cable (1000BASE-T), June 12, 1998.

[5] R. Bates and D. Gregory, Voice and data communications handbook, McGraw-Hill, 1997.

[6] G. H. Im and J. J. Werner, "Bandwidth-efficient digital transmission over unshielded twisted-pair wiring," IEEE J. on Selected areas in communication, Vol. 12, No. 9, Dec 1995, pp. 1643-55.

[7] W. Y. Chen, DSL simulation techniques and standards development for digital subscriber line systems, Macmillan Technical Publishing, 1998.

[8] E.E. Shannon, "A mathematical theory of communication," Bell systems Technical Journal, vol.27, 1948, pp.379-423 (Part I), pp.623-656 (Part II).

[9] T1.413 Issue 2, Sept 26, 1997.

[10] F. Larsen, A. Muralt, and N. Tan, "AFEs for xDSL," Electronics Times 1999 Analog & Mixed-signal Application conferences, Oct 5~7, Santa Clara, CA.

3 OVERVIEW OF HIGH-SPEED A/D CONVERTER ARCHITECTURES

3.1 INTRODUCTION

There are several well-known ADC architectures with different properties making them more or less suitable for a certain specification. To choose the proper architecture, their limitations must be investigated. The oversampling sigma-delta converter is often a suitable architecture when a relatively high resolution over a moderately high bandwidth is required. However, for instance VDSL applications, requires a large resolution (~ 12 bits) over a bandwidth larger than 10 MHz. The high bandwidth makes the sigma-delta converter less attractive for such applications since the oversampling makes the sampling frequency very high. Therefore other architectures must be considered.

In this chapter we summarize some of the most important limitations of different high-speed ADC architectures. The focus is on high performance over large bandwidths. In Sec. 3.2 the influence of the S/H circuit and sampling time uncertainty is discussed. Flash converters are the subject of Sec. 3.3 where we also briefly consider meta-stability in comparators. The semi-flash converter and its limitations are discussed in Sec. 3.4 while folding and interpolating converters are treated in Sec. 3.5. The pipelined ADC is the most popular architecture for high speed, high resolution applications. The properties of the pipelined ADC are reviewed in Sec. 3.6. The discussion in this chapter is focused on high-speed ADCs with a high signal bandwidth, but Sec. 3.7 briefly deals with oversampling sigma-delta converters and section Sec. 3.8 deals with some other types of low speed converters. For very high sampling rates it may not be possible to use a single ADC. In such applications the time-interleaved converter may be an option. The time-interleaved converter is the subject of Sec. 3.9. Finally a survey of the reported performance of high-speed ADCs is presented in Sec. 3.10.

3.2 SAMPLE-AND-HOLD (S/H) CIRCUITS

In all ADCs it is necessary to sample-and-hold (S/H) the signal. In flash converters, this is done in the digital domain after the comparators but an analog S/H circuit is used at the input of most CMOS wide band ADCs. Without a S/H at the input the performance at high signal frequencies of any ADC is usually very poor. This will become clear in the following discussions.

3.2.1 Sampling-Time Uncertainty

The dynamic performance of a S/H circuit is limited by the precision in the sampling instants [1]. Variations in the sampling instants may be caused by clock jitter, switch imperfections and signal dependent delays. The effect of the sampling time uncertainty is illustrated in Fig. 3-1 where the desired sampling time is t but due to circuit

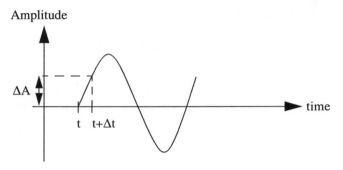

Figure 3-1   Effect of sampling time uncertainty.

imperfections the actual sampling time is $t + \Delta t$. The error ΔA caused by this delay depends on the slope of the input signal and for a sinusoidal input signal the worst case slope is at the zero crossing. The first-order approximation of the peak error is given by

$$\Delta A_{max} = \Delta t \cdot \frac{\partial}{\partial t} V_{in}(t)\Big|_{t=0} = \omega \cdot A \cdot \cos(\omega t) \cdot \Delta t\Big|_{t=0} = \omega \cdot A \cdot \Delta t \quad (3\text{-}1)$$

where A is the amplitude of the sinusoid. The error clearly increases at high signal frequencies but is independent of the sampling frequency. Many of the errors causing sampling time uncertainty can be reduced by using clever circuit techniques as for instance bottom plate sampling. The performance is however fundamentally limited by clock jitter in the switch signals. If the clock jitter is assumed to be random noise with variance σ_t^2 the error power can be approximated as [2]

$$v_{jn}^2 = \sigma_t^2 \cdot \frac{1}{T}\int_0^T \left(\frac{\partial}{\partial t}V_{in}(t)\right)^2 dt \quad (3\text{-}2)$$

where T is the integration time. For periodic signals the integration time can be chosen as the signal period. If the input signal is assumed to be a sinusoidal, the error power can by using (3-2) be calculated as [2]

$$v_{jn}^2 = (2\pi f_{in} A)^2 \sigma_t^2 / 2 \quad (3\text{-}3)$$

where A is the amplitude of the sinusoid and f_{in} the signal frequency.

3.2 Sample-and-Hold (S/H) Circuits

Thus the SNDR is limited by

$$\text{SNDR} = 10 \cdot \log \frac{1}{(2\pi f_{in} \sigma_t)^2} \qquad (3\text{-}4)$$

or by using the ENOB

$$\text{ENOB} = \frac{10 \cdot \log \frac{1}{(2\pi f_{in} \sigma_t)^2} - 1.76}{6.02} \qquad (3\text{-}5)$$

It should be noted that the SNDR is independent of the sampling frequency. The obtainable ENOB as function of the clock jitter, σ_t, is plotted for different input frequencies in Fig. 3-2. The figure shows that the performance degrades rapidly at high

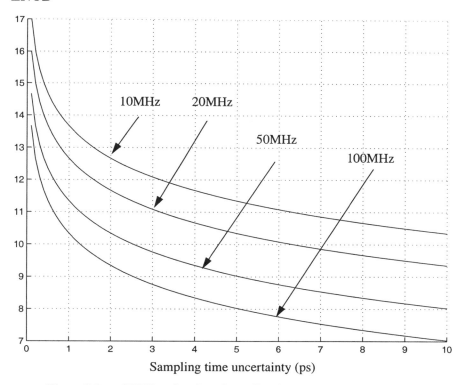

Figure 3-2 ENOB as function of sampling time uncertainty.

signal frequencies. To achieve 10 ENOB at 20 MHz the sampling time uncertainty must be in the order of 6 to 7 ps.

3.2.2 Thermal Noise

The thermal noise is a limiting factor in the S/H circuits. It usually dominates over the flicker noise (the other noise source present in MOS circuits) in wideband circuits. Due to the sampling, the thermal noise is folded back into the signal band, introducing a fundamental limitation.

Most high-speed S/H circuits use passive sampling to get a high bandwidth and a high sampling rate. In this type of sampling circuits the thermal noise power is kT/C [4]. Assuming a sinusoidal input signal the SNDR is given by

$$SNDR = 10 \cdot \log\left(\frac{V_{in}^2}{2 \cdot kT} \cdot C\right) \qquad (3\text{-}6)$$

where V_{in} is the amplitude of the input signal.

Since the voltage swing is limited by the power supply voltage, the only way to increase the SNDR is to increase the capacitor. However, for every doubling of the capacitor, the SNDR only increases by 3 dB and a large capacitance implies a large power consumption. For moderate resolutions, the capacitor size calculated from the thermal noise requirement is very small and a larger capacitor may be chosen for other reasons. For high resolutions a capacitor of several pF is needed to handle the thermal noise. In addition to this the opamp noise may be in the same order as the kT/C noise, leading to further increase of the capacitor.

3.2.3 Nonlinearity

The linearity of the input S/H must be at least as good as the resolution of the ADC. The errors may be caused by nonlinear switch on resistance, clock feedthrough errors, finite amplifier gain, parasitic capacitors or any other error sources present in the circuit. It should be noted that most of the errors increases at high signal and sampling frequencies.

3.3 FLASH CONVERTERS

Flash converters can reach very high sampling rates since the only analog building block is the comparator. The principle of the flash converter is illustrated in Fig. 3-3. There are one comparator for each decision level in the converter. The reference levels for the comparators are usually generated by a resistor string. The signal at the output of the comparators is a thermometer coded representation of the input signal. A decoder is used to generate a more convenient representation at the output.

The main problem with this architecture is that the number of comparators increases exponentially with the number of bits. For N bits, $2^N - 1$ comparators are needed. Due to the large number of comparators the number of bits is usually limited to 10, since the chip area and power consumption would be too large for higher resolutions. Thus the SNDR is limited to 60 dB. One more problem caused by the large number of comparators is the large input capacitance. The circuit driving the ADC must therefore

3.3 Flash Converters

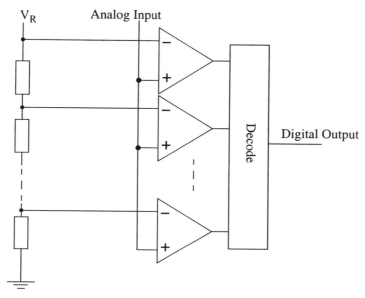

Figure 3-3 The Flash Converter.

handle a large capacitive load. The pure flash ADC is not suitable for telecommunications applications where a high resolution is required. However, a flash converter or at least a comparator is always used as a part of all ADC architectures. It is therefore necessary to know the limitations of the flash converter.

3.3.1 Static Errors

The static errors affect the accuracy in the decision levels of the converter. Any deviation from the ideal levels will caused DNL and INL errors. Errors in the reference levels are caused by resistor mismatch, which usually is in the order of 2 to 0.1% without laser trimming [5], and offsets in the comparators. The errors must be less than [6] (no missing codes are guaranteed for INL < +/- 1/2 LSB)

$$\varepsilon_{ADC} = \pm LSB/2 = \pm FS/2^{N+1} \tag{3-7}$$

where FS is the full scale range of the input signal.

3.3.2 Dynamic Errors

The dynamic errors typically increase at high signal and sampling frequencies. They are mainly caused by timing errors. There are four main sources of timing errors [6] in the flash converter

1. *Skew of the clock and input signal at different places on the chip.*

It gives frequency dependent errors and degrades the performance at high signal frequencies. It can be reduced by using an input S/H circuit.

2. *Limited rise/fall times of the sampling clock.*

They give signal dependent turn off times in the sampling switches. It can be reduced by, for instance, feedback [1] or bottom-plate sampling techniques [3]. The rise and fall times can to a limited extent be reduced by increasing the size of the clock drivers, but this will also increase the power consumption.

3. *Signal-dependent delays.*

Signal dependent delays can cause distortion in e.g. comparators (see below). These errors can be reduced by increasing the bandwidth of the circuit or in some cases by using an input S/H circuit.

4. *Sampling clock jitter.*

Sampling clock jitter is present in all ADCs and will act as a fundamental limitation in the circuit at high signal frequencies.

3.3.3 Signal Dependent Delay in Comparators

In [6] it is shown that the delay in the comparators is signal dependent and that this will cause third-order distortion. Using a single pole model for the comparator the SNDR due to signal dependent delay can be approximated as [6]

$$SNDR = 10 \cdot \log \left(\frac{3\pi \cdot f_b \cdot e^{\frac{V_{lr} f_b}{V_{fs} f_{in}} + 1}}{2 \cdot f_{in}} \right) \qquad (3\text{-}8)$$

where f_{in} is the input frequency, f_b the -3 dB bandwidth of the comparator, V_{lr} the linear range of the comparator and V_{fs} the full scale voltage of the converter.

For a 10-bit accuracy the bandwidth of the comparator should be about 10 times the input signal bandwidth. The timing errors can be reduced by using a S/H circuit at the input of the ADC and thus moving the timing problems to S/H where they are easier to handle. A problem with this solution is that the sampling frequency is reduced since it is difficult to design a high speed S/H, especially if the number of comparators is large.

3.3.4 Meta-Stability

A severe limitation in very high-speed converters is meta-stability. It increases rapidly at high sampling frequencies. Meta-stability is caused by the finite gain in the comparators. Certain input signals generate output signals that can not be detected by the digital circuits. It is common to use comparators with a pre-amplifier followed by a regenerative latch. The metastability error probability for an N-bit converter using this type of comparators is [7]

$$P_E = \frac{2(2^N-1)V_o}{V_R A} \cdot e^{-t_r/\tau} \tag{3-9}$$

where V_R is the analog input range, V_o the output voltage swing required for valid logic levels, A is the combined voltage gain of the pre-amplifier and the latch's gain in the transparent state, τ is the regenerative time constant for the latch and t_r is the resolution time of the latch. It is obvious from (3-9) that the performance deteriorates at high sampling frequencies since the error probability is exponentially related to the resolution time. The effect of these errors in the flash converter depends on the decoding logic and the performance can be improved by using pipelined latches [7].

3.4 SEMI-FLASH CONVERTERS

To reduce the number of comparators the semi-flash converter can be used. In this type of ADC a coarse ADC generates the most significant bits. The residue, the difference between the input signal and the D/A converted signal from the coarse ADC, is converted by a fine A/D converter (see Fig. 3-4). The conversion speed of the semi-flash

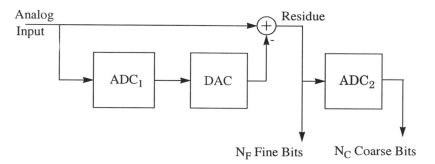

Figure 3-4   The Semi-Flash Converter.

converter is less than half compared to the flash converter but the number of comparators is only

$$2^{N_C} + 2^{N_F} - 2 \tag{3-10}$$

where N_C is the number of bits in the coarse ADC and N_F the number of bits in the fine ADC.

There are two additional building blocks in the semi-flash converter compared with the flash converter, a DAC and a subtractor. It is common to use a residue amplifier after the subtractor as shown in Fig. 3-5 to avoid small residue signals which are sensitive to noise. The gain of the amplifier is usually chosen as

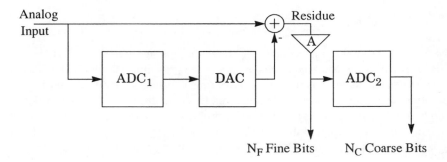

Figure 3-5   The Semi-Flash converter with residue amplifier.

$$A = 2^{N_C} \tag{3-11}$$

to make the signal swing equal for both the ADCs. This simplifies the design since the same reference ladder can be used by both ADCs.

There are in principle three ways of performing the D/A conversion and subtraction, by using voltage, current or charge. All of these circuit techniques have different properties and different accuracy-speed trade-offs [8].

3.4.1   Static Errors

To avoid missing codes and large linearity errors the total error should be less than $\pm LSB/2$. The highest accuracy is required for the ADC and DAC in the first stage and the total error must be within the total resolution of the semi-flash converter, i.e. the error must be [9]

$$\left|\varepsilon_{DAC}\right| = \left|\varepsilon_{ADC_1}\right| \leq \frac{\pm FS}{2^{N_C + N_F + 1}} \tag{3-12}$$

The errors introduced by the second ADC and the residue amplifier only need an accuracy corresponding to the resolution in the second stage. Hence we tolerate larger relative errors in the second stage. If the residue amplifier is assumed to have a gain of 2^{N_C} the errors are limited by [9]

$$\left|\varepsilon_A\right| = \left|\varepsilon_{ADC_2}\right| \leq \frac{\pm FS}{2^{N_F + 1}} \tag{3-13}$$

The influence of some errors in the first stage can be reduced by using digital correction [10]. This technique is based on the fact that the errors in the first stage cause the residue to be out of the input range of the second ADC. By reducing the gain of the residue amplifier this can be avoided. The gain factor of the residue amplifier is usually reduced by a factor 2 to handle the errors in the first stage and to make the decod-

3.4 Semi-Flash Converters

ing of the digital output easy. The gain reduction will introduce redundancy in the converter which also means that the resolution of the converter corresponds to $N_C + N_F - 1$ bits rather than $N_C + N_F$ bits. With digital correction, the nonlinearity errors in the first ADC can be much larger without degrading the performance [9]. We now have that

$$|\varepsilon_{ADC1}| \le \frac{FS}{2^{N_C+1}} \qquad (3\text{-}14)$$

The errors in the DAC must still be within the total resolution according to (3-12) even though the second stage ADC is not overloaded. This is because the actual output levels of the DAC must be known to decode the digital output correctly. The static converter linearity is ultimately limited by the linearity of the DAC and calibration for the DAC is necessary if the resolution is high. The DAC calibration can be performed in both the analog [11] and in the digital domain [12] but requires additional hardware.

3.4.2 Dynamic Errors

The dynamic errors are caused by timing errors in the building blocks. The skew between the two input signals to the subtractor is difficult to match. The effect of the skew error is frequency dependent and the dynamic performance degrades at high signal frequencies. The maximal timing error [13], assuming a full scale sinusoidal input signal, is limited by

$$\Delta t \le \frac{1}{2^{N_C+N_F+1}\pi f_{in}} \qquad (3\text{-}15)$$

where f_{in} is the input frequency. If, however, one bit of digital correction is used, larger timing errors can be tolerated and the maximal timing error is only limited by

$$\Delta t \le \frac{1}{2^{N_C+1}\pi f_{in}} \qquad (3\text{-}16)$$

i.e. the resolution in the first stage. Without digital correction the required resolution is the total resolution of the ADC. To achieve a 10 bit resolution at 20 MHz input frequency the delay skew must be in the order of 8 ps which is very difficult to achieve. If digital correction is used and the resolution in the first stage is 5 bits the maximum delay skew is about 500 ps.

To avoid the delay problems, a S/H circuit can be used at the input. However the sampling frequency may be reduced since it is difficult to design an accurate high-speed S/H circuit especially when the capacitive load at the input is large. There are also semi-flash converters having a S/H circuit both at the input and between the two stages. This type of converter can be viewed as a special case of the pipelined converter discussed Sec. 3.6.

3.5 FOLDING AND INTERPOLATING CONVERTERS

In folding and interpolating A/D converters analog preprocessing is used to reduce the power consumption and the chip area while maintaining a high sampling rate [14]. The principle behind the folding technique is to perform the conversion in two steps just as for the semi-flash converter, but the coarse and fine converters operate in parallel, as illustrated in Fig. 3-6.

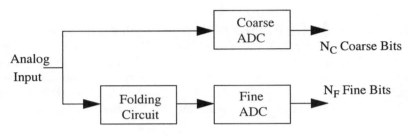

Figure 3-6 The folding ADC.

The analog folding circuit in front of the fine ADC generates the same signal as the subtractor in the semi-flash converter. Since the output from the coarse converter is not used in the fine converter the conversion rate is approximately the same as for the flash converter while the number of comparators is only $2^{N_C} + 2^{N_F} - 2$. The folding signal for $N_c = 2$ is shown in Fig. 3-7. The sawtooth waveform is difficult to gener-

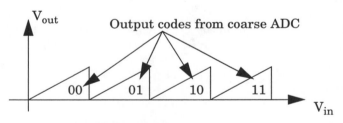

Figure 3-7 The transfer function of the folding circuit for N = 2.

ate and a triangular waveform is preferred in the actual implementation but it is still difficult to get sharp corners in the waveform. Therefore the linearity for large output signals of the folding circuit is poor.

This problem can be circumvented by using several folding circuits with different offsets and thus only utilizing a small linear portion of the input range. Unfortunately the number of folding circuits becomes large and the complexity of the converter may be in the same order as for the flash converter. Therefore some of the folding signals are generated by resistive interpolation. However too many interpolations will create accuracy problems [6] and most of the folding or interpolating ADCs reported so far have a resolution less than 10 bits. The resolution can be increased by performing the folding in several stages. This technique was used in [15], [57] to achieve 12 bits resolution at 60 MS/s. The advantages of the folding-interpolating architecture are normally best utilized in a bipolar or BiCMOS process since the bipolar transistor can make the bandwidth in the folding circuit very large.

The static errors in the folding-interpolating converter are caused by comparator offsets and in-accuracies in the folding and interpolation circuits. The dynamic errors are caused by timing errors and a limiting factor is the delay mismatch between the two sub-converters. The effect of these errors can not be as easily corrected as was the case for the semi-flash converter. Another problem with the folding technique is that the required bandwidth of the folded signal is increased by the folding rate (the number of times the signal is folded) which is difficult to handle especially in CMOS. These problems can be partially solved by using a S/H circuit at the input or a distributed S/H circuit as in [16].

3.6 PIPELINED CONVERTERS

Multi-stage pipelined ADCs [9] are one of the most popular architectures for high-speed applications. They are similar to the semi-flash converters and consists of several cascaded stages each with a low resolution ADC, a S/H-amplifier and a DAC as illustrated in Fig. 3-8. The properties of the pipelined converter is similar to the semi-

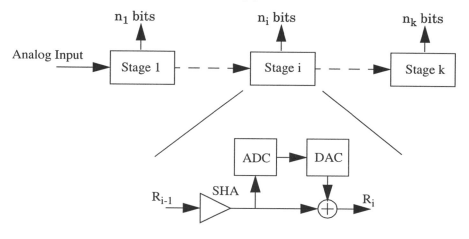

Figure 3-8 The pipelined ADC.

flash converter and will be discussed in more detail in chapter 9. The main difference between the semi-flash ADC and the pipelined ADC is that there is a S/H circuit between the stages which increases the maximum sampling rate of the converter. Due to the pipelining the output word is delayed several clock cycles but in most applications this is not a severe limitation.

3.6.1 Static Errors

The accuracy of the sub-ADCs need not be very high if digital correction is used. The only circuits needing an accuracy corresponding to the full resolution of the converter are the input S/H circuit and the DAC in the first stage. The static performance is mainly limited by the matching errors of the DAC and for a resolution of more than 10 to 12 bits it is usually necessary to use calibration.

3.6.2 Dynamic Errors

Any delay between the signal paths in the first stage will give distortion unless an input S/H is used. The delay skew is limited by the same expressions used for the semi-flash converter in Sec. 3.4. Due to the lower stage resolution, larger delays can normally be handled in the pipelined converter than in the semi-flash converter, assuming digital correction is used.

3.7 OVERSAMPLING SIGMA-DELTA ADCS

Oversampling sigma-delta ADCs [17, 18] are based on the principle that the conversion error can be highpass filtered and later removed by digital filters. The requirements on the analog parts are relaxed and high-resolutions can be achieved. The drawback of this type of converter is that for high resolutions the signal bandwidth is small due to the oversampling. A second-order sigma-delta modulator is shown in Fig. 3-9. The modulator consists of two discrete-time integrators, two DACs and one quan-

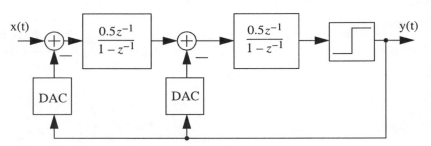

Figure 3-9 Second-order sigma-delta modulator.

tizer. If the quantization error is modelled as white noise the output signal is given by

$$Y(z) = z^{-2}X(z) + (1 - z^{-1})^2 E(z) \qquad (3\text{-}17)$$

where $E(z)$ is the quantization noise in the quantizer and $X(z)$ is the input signal. The quantization noise is highpass filtered while the input signal is only delayed. By removing the high frequency noise with digital filters and decimating the signal, an output signal having a small quantization noise is achieved. Every doubling of the sampling rate will provide $L + 0.5$ extra bits, where L is the order of the modulator.

3.8 Low Speed Converters

To increase the resolution, the order of the modulator or the resolution of the quantizer must be increased. A problem with increasing the resolution of the quantizer is that nonlinearities in the DACs will directly limit the resolution of the ADC. A DAC with only two output levels have only offset and gain errors while a multi-bit DAC also have non-linearities.

There are also problems with increasing the order of a single-stage modulator. Special architectures to avoid instability must be used and a small input swing may be required to avoid saturation of the modulator. This may necessitate large capacitors and a large capacitance spread in the implementation which reduces the speed [19]. By using a multi-stage structure these problems are avoided, but sensitivity to matching errors between the analog and digital parts limits the resolution. A more detailed discussion on sigma-delta converters is given in chapter 11.

There are some alternatives to increasing the bandwidth. In [20] a 16-bit converter with an oversampling ratio of only 8 was designed by combining a sigma-delta modulator with a pipelined converter. The modulator uses a multi-bit quantizer and dynamic element matching must be used to reduce the distortion caused by the DAC. In [67] the modulator is pipelined in such that way the no oversampling is needed. In this case a resolution of 13-bits was achieved.

3.8 LOW SPEED CONVERTERS

There are several types of low speed converters. They are not so common in high-speed communications application since the conversion speed often is too small. They may however be used in time-interleaved converters which is the topic of the next section [22]. Here we only briefly consider some of the most common low speed converters.

The successive approximation converter [21, 22] shown in Fig. 3-10 consists of a

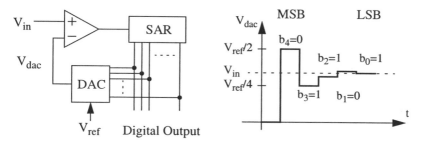

Figure 3-10 Successive approximation A/D converter.

comparator a DAC and a successive approximation register (SAR). Starting with the MSB the comparator determines whether the input signal is larger or smaller then the DAC output. The result is stored in the SAR. The process is repeated for all the bits in the DAC. This is illustrated in Fig. 3-10 for a 5-bit converter. The number if iterations is directly determined by the number of bits in the converter. The speed is therefore reduced when the resolution is high. The static performance is limited by the DAC linearity and for high resolutions calibration is needed. To get a high resolution at high signal frequencies a S/H circuit is needed at the input of the converter.

The algorithmic or cyclic converter shown in Fig. 3-11 is based on the same algorithm

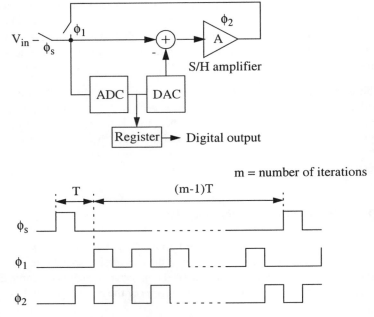

Figure 3-11   Cyclic A/D converter.

as the pipelined converter but the same stage is used to calculate all the bits. Hence the speed is reduced by the required number of iterations. The conversion starts by sampling the input signal and converting the analog input to a coarse digital value by the sub-ADC. The digital value is stored in a register. The residue is calculated by a DAC and a subtractor and stored in the S/H amplifier. In the next clock cycle the amplified residue is fed back to the input of the sub-ADC and the procedure is repeated. The properties of the cyclic converter are very similar to the pipelined converter. Digital correction can be applied by decreasing the residue gain. With digital correction the requirements on the sub ADC are relaxed.

The dual-slope integrating converter [23] shown in Fig. 3-12 is very accurate. The input signal is first integrated for a fix period of time, T_1. Hence, the output of the integrator is proportional to V_{in}. The integrator input is then switched to a reference voltage. The time until the integrator output becomes zero, T_2, is proportional to the input value and is measured by a digital counter. The zero crossing is detected by a comparator. The integrating converter can not be used for high signal bandwidths since the

3.9 Time-Interleaved (Parallel) Converters

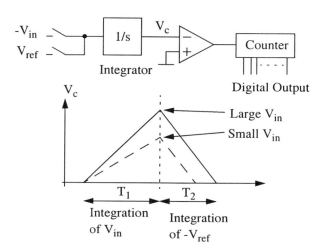

Figure 3-12 Dual-slope A/D converter.

maximum number of clock cycles is 2^N, where N is the number of bits.

3.9 TIME-INTERLEAVED (PARALLEL) CONVERTERS

An attractive way to increase the conversion rate of ADCs is to use time-interleaving techniques where several ADCs operate on different clock phases [24]. This enables a higher total conversion speed but also introduces new problems which are mainly caused by mismatch between the channels. The concept of time-interleaving is illustrated in Fig. 3-13 where the converters can be of any type and are operated at f_s/M. The speed requirements on each converter are relaxed by a factor M but the number of converters is at same time increased by the same factor, which may result in a large chip area and power consumption. The increase in power and area is not necessarily a linear function of M compared to using only one converter, since smaller bias currents can be used due to the lower speed. It is also sometimes possible to share some of the building blocks between the channels [25]. There are three main error sources in time-interleaved systems, phase skew errors, gain errors and offset errors [24, 26, 27].

3.9.1 Offset Errors

Differences in offsets in the channels cause distortion in the output signal. The distortion is not signal dependent and will appear at (see chapter 10)

$$f_s/M \cdot m, \, m = 1, 2, ..., M-1 \tag{3-18}$$

If the offset in channel k is denoted o_k and is assumed to be normally distributed random variables with zero mean and variance σ_o^2, the SNDR can be approximated as

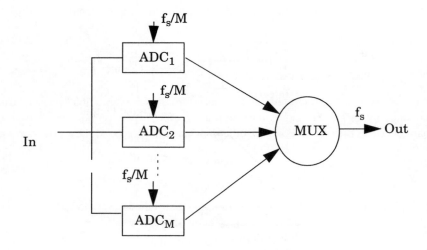

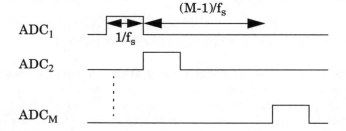

Figure 3-13 Parallel ADCs and corresponding sub-ADC timing.

(see chapter 10)

$$SNDR = 20 \cdot log\left(\frac{1}{\sigma_o}\right) \quad (3\text{-}19)$$

3.9.2 Gain Errors

Different channel gains cause distortion in the output signal. If the gain in channel k is denoted a_k and is assumed to be a normally distributed random variable with mean a and variance σ_a^2, the SNDR can be approximated as (see chapter 10)

$$SNDR = 20 \cdot log\left(\frac{a}{\sigma_a}\right) - 10 \cdot log\left(1 - \frac{1}{M}\right) \quad (3\text{-}20)$$

where M is the number of channels. The distortion tones are located at

$$f_{in} + m \cdot f_s/M, \quad m = 1, ..., M-1 \tag{3-21}$$

3.9.3 Phase Skew Errors

When using parallel converters, as shown in Fig. 3-13, several clock phases, usually generated by an on-chip clock generator, are required. Any phase skew between the channel causes distortion. For sinusoidal input signals the phase skew errors will generate distortion tones at [26]

$$f_{in} + m \cdot f_s/M, \quad m = 1, ..., M-1 \tag{3-22}$$

If the phase skew errors are treated as normally distributed random variables with zero mean and variance σ_t^2 the SNDR can be approximated as [26]

$$\text{SNDR} = 20 \cdot log\left(\frac{1}{\sigma_t 2\pi f_{in}}\right) - 10 \cdot log\left(\frac{(M-1)}{M}\right) \tag{3-23}$$

The last term depends on M (the number of channels) but changes only about 3 dB as M changes from 2 to infinity. Thus the number of channels have quite small effect on the total SNDR. The delay skew σ_t should be less than about 10 ps to get 10 bits resolution at 20 MHz input frequency. This is difficult to achieve in a CMOS process.

3.10 REPORTED PERFORMANCE

To compare the performance using the different ADC architectures, some high-speed ADCs are listed in Table 3-1. In the *Type* column parallel ADCs are specified as e.g. 2 Par-Pipe which means that there are 2 parallel channels and that the sub-ADCs are pipelined converters. The last two columns specify the dynamic performance at high input frequencies for the ADCs. The f_{in} column contains the input frequency and the *SNDR* column the measured SNDR. Some of the ADCs can operate at different sample frequencies and supply voltages. In such a case the values used for the dynamic measurements in the last columns are given as bold numbers. It should be noted that the converters are implemented for different supply voltages using different processes. It is therefore difficult to make a fair comparison. The intention here is only to give an overview of the achieved performance to get a rough indication on which architecture to choose for a certain application.

Table 3-1 ADC Performance

Ref.	Type	Process	Bits	f_s [MHz]	V_{dd} [V]	Power [mW]	Area [mm^2]	f_{in} [MHz]	SNDR [dB]
[28]	Flash	CMOS	6	200	5	400	2.7	low	35

Table 3-1 ADC Performance

Ref.	Type	Process	Bits	f_s [MHz]	V_{dd} [V]	Power [mW]	Area [mm^2]	f_{in} [MHz]	SNDR [dB]
[7]	Flash	CMOS	7	80	-	307	8.2	20	32
[29]	Flash	Bipolar	8	300	5.2	3300	33	100	35
[30]	Flash	CMOS	8	20	5	540	26.5	10	46
[31]	Flash	Bipolar	8	200	10.2	2000	15.2	50	35
[32]	Flash	Bipolar	8	500	5.2	3100	20.3	-	-
[33]	Flash	Bipolar	6	200	9.5	1100	9	50	34
[34]	Flash	BiCMOS	6	80	5	400	6	50	34
[35]	Semi-Flash	Bipolar	10	75	-	-	20	40	51
[36]	Semi-Flash	Bipolar	12	50	5	575	13.7	20	68
[37]	Semi-Flash	CMOS	8	50	5	600	13.8	20	44
[38]	Semi-Flash	CMOS	10	20	2	20	12	10	42
[39]	Semi-Flash	Bipolar	10	75	10	2000	20	20	54
[40]	Semi-Flash	BiCMOS	10	100	5	950	19	50	48
[41]	Semi-Flash	Bipolar	10	75	10	800	16	40	57
[42]]	Semi-Flash	BiCMOS	10	50	5	500	16	20	48
[43]	Semi-Flash	Bipolar	10	50	5	750	11	20	55
[44]	Semi-Flash	Bipolar	12	128	10.2	5700	40	50	58
[45]	Semi-Flash	CMOS	8	20	5	50	4.5	10	41
[46]	Semi-Flash	CMOS	10	20	5	75	1.6	10	58.7
[47]	Semi-Flash	CMOS	10	16-25	3.3	195	0.8	14	56
[48]	Folding	CMOS	8	80-125	3.3-5	150	4	10	32
[49]	Folding	BiCMOS	8	100-200	5	575	4	20	32
[50]	Folding	Bipolar	10	300	5.2	4	39	100	43

3.10 Reported Performance

Table 3-1 ADC Performance

Ref.	Type	Process	Bits	f_s [MHz]	V_{dd} [V]	Power [mW]	Area [mm^2]	f_{in} [MHz]	SNDR [dB]
[51]	Folding	CMOS	8	80	3.3	80	0.3	40	42
[52]	Folding	CMOS	8	70	3.3.-5	110	0.7	10	32
[53]	Folding	Bipolar	8	650	4.5	850	4.2	150	48.6
[54]	Folding	Bipolar	8	100	5.2	800	12	50	45
[55]	Folding	Bipolar	8	55	5	300	6	10	46
[56]	Folding	CMOS	10	50	5	240	2	10	54
[57]	Folding	BiCMOS	12	50	5	300	8	10	<64
[58]	Folding	BiCMOS	10	40	5	65	0.8	4.1	57
[59]	Pipelined	BiCMOS	10	40	5	400	19	20	55
[60]	Pipelined	CMOS	10	20	5	50	13	10	48
[61]	Pipelined	CMOS	10	20	3.3	35	10.5	10	55
[62]	Pipelined	BiCMOS	10	20	10	1000	48	10	50
[63]	Pipelined	CMOS	10	20	5	240	8.7	10	58
[64]	Pipelined	CMOS	10	15-20	2.5	30	6.6	10	<40
[65]	Pipelined	CMOS	10	20	3	135	7	10	45
[66]	Pipelined	CMOS	10	14	1.5	36	5.7	4	54
[67]	Pipelined - SD	CMOS	13	18	5	324	47	8	71
[68]	Pipelined	CMOS	12	40	5	415	-	62	70
[69]	Pipelined	CMOS	12	25	5	280	-	20	68
[70]	Pipelined	CMOS	12	65	5	500	-	32	70
[71]	Pipelined	CMOS	12	20	5	300	-	10	68
[72]	2 Par-Pipe	CMOS	10	32-40	2.7-5.5	85	4	10	45
[73]	14 Par-Succ	CMOS	10	40-70	5	267	8.9	-	-
[74]	2 Par-Pipe	CMOS	10	40-50	5	900	18	20	49
[75]	4 Par-Flash	GaAs, Bipolar	6	1000	-	16000	-	-	-
[24]	4 Par-Succ	CMOS	7	2.5	-	250	0.04	-	-
[3]	4 Par-Pipe	CMOS	10	95-100	5	1100	50	50	50

Table 3-1   ADC Performance

Ref.	Type	Process	Bits	f_s [MHz]	V_{dd} [V]	Power [mW]	Area [mm^2]	f_{in} [MHz]	SNDR [dB]
[25]	2 Par-Pipe	CMOS	8	52	5	250	15	20	46
[76]	4 Par-Pipe	CMOS	8	85	-	1100	25	40	40
[77]	2 Par-Semi	CMOS	12	1	5	25	14.3	1	59

From the discussion in the previous sections it is clear that it is very difficult to get good dynamic performance without a S/H circuit at the input. When a S/H or residue amplifier is needed it is usually the opamp that sets the speed limit. Since the dynamic performance over a large bandwidth is the main concern in this chapter some of the CMOS high-speed ADCs in the table above were plotted to make the comparison easier. The resolution and sampling frequency for these A/D converters is shown in Fig. 3-14. There are several publications with a resolution of 8 to 10 bits in the 10 to 100

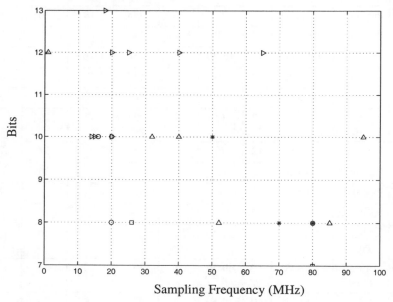

□ Flash ○ Semi_Flash * Folding/Interpol. ▷ Pipelined △ Parallel

Figure 3-14   Resolution and sample frequency for CMOS A/D converters.

MHz range and a few industrial products with a resolution of 12 bits. It should be noted that the fabrication processes used for the industrial products are usually better suited for analog design at high frequencies than the processed used for many of the converters found in publications. Therefore a direct comparison of the performance is

3.11 Summary

not entirely fair. Shown in Fig. 3-15 is the ENOB for high input frequencies. It is seen

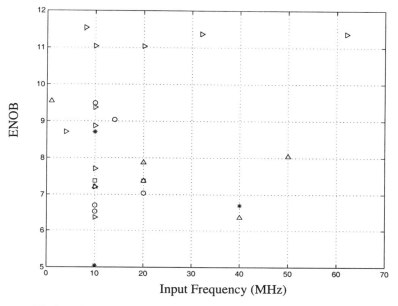

▫ Flash ○ Semi_Flash * Folding/Interpol. ▷ Pipelined △ Parallel

Figure 3-15 ENOB and input frequency for CMOS A/D converters.

that except for the industrial products very few converters have more than 8 ENOB at input signals larger than 10 MHz. The best performance is shown for pipelined and parallel converters with a S/H circuit at the input.

3.11 SUMMARY

In this chapter high-speed ADC architectures for telecommunications were discussed. We were mainly focused on architectures suitable for high signal frequencies and high sampling rates but both static and dynamic errors in the converters were briefly considered. In the following we summarize the most important conclusions.

1. Flash and Folding Converters

Due to large area and large power consumption flash and folding converters are best suited for resolutions < 10 bits at high sampling rates. These types of ADCs does not always have an input S/H circuit since this would limit the conversion speed. Without a S/H circuit at the input the dynamic performance at high input frequencies is poor. Some two-stage folding converters have been reported suitable for high resolutions at high sampling rates. They are however usually implemented in a bipolar or BiCMOS process.

2. Semi-Flash and Pipelined Converters

For higher resolutions pipelined or semi-flash converters are better suited than flash or folding converters. The requirements on the comparators can be relaxed by using digital correction. However, inaccuracies in the DACs necessitates calibration techniques for large resolutions. Due to the pipelining the pipelined converter can reach higher sampling rates than the semi-flash converter.

3. Time-Interleaved Converters

In the time-interleaved converter several ADCs operating at a lower speed are connected in parallel. Mismatch in the channels will cause distortion but can be reduced by calibration techniques and the dynamic performance at high input frequencies is limited by phase-skew errors between the channel. These errors are reduced by an input S/H. However, the input S/H must operate at the full speed of the ADC. Therefore it is difficult to design.

4. Oversampling Sigma-Delta Converters

Oversampling ADCs are relatively insensitive to imperfections in the analog components which makes them suitable for high resolutions. The main drawback is that the oversampling limits the signal bandwidth. The oversampling sigma-delta converter is normally used for signal bandwidths up to a few MHz.

5. Overall

Timing and delay errors are the limiting factor at high signal frequencies in all types of converters and can be reduced by using a S/H circuit at the input. The dynamic performance of the A/D converter at high signal frequencies is, thus, to a large extent determined by the input S/H circuit. A key component in an accurate S/H circuit is a high-gain, high-speed opamp. For high speed the opamp may be very power consuming, especially for high resolutions. The performance of ADCs is fundamentally limited by thermal noise and clock jitter. The effect of these noise sources can not be removed by an input S/H circuit.

REFERENCES

[1] B. Razavi, "Design of Sample-and-Hold Amplifiers for High-Speed Low-Voltage A/D Converters", *IEEE 1997 Custom Integrated Circuits Conf.*, pp. 59-66, 1997.

[2] M. Shinagawa, Y. Akazawa and T. Wakimoto "Jitter Analysis of High-Speed Sampling Systems", *IEEE J. of Solid-State Circuits*, vol. 25, no. 1, pp. 220-4, Feb. 1990.

[3] K. Y. Kim, N. Kusayanagi and A. A. Abidi "A 10-b, 100-MS/s CMOS A/D Converter", *IEEE J. of Solid-State Circuits*, vol. 32, no. 6, pp. 302-11, Dec. 1997.

[4] K R. Laker and W. M. C. Sansen *Design of Analog Integrated Circuits and Systems*, McGraw-Hill, Inc. 1994.

[5] Gray P. R. and Meyer R. G.: *Analysis and Design of Analog Integrated Circuits*, John Wiley & Sons, Inc. Third Edition, 1993.

[6] R. J. van de Plassche, *Integrated Analog-to-Digital and Digital-to-Analog Converters*, Kluwer Academic Publishers, Boston, MA, USA, 1994, ISBN 0-7923-9436-4.

[7] C. L. Portmann. and T. H. Y. Meng, "Power-Efficient Metastability Error Reduction in CMOS Flash A/D Converter", *IEEE J. of Solid-State Circuits*, vol. 31, no. 8, pp. 1133-40, Aug. 1996.

3.11 Summary

[8]   B. Razavi, *Principles of Data Conversion System Design*, IEEE Press, 1995.

[9]   S. H. Lewis, "Optimizing the Stage Resolution in Pipelined, Multistage, Analog-to-Digital Converters for Video-Rate Applications", *IEEE Trans. on Circuits and Systems-II*, vol. 39, no. 8, pp. 516-23, Aug. 1992.

[10] S. H. Lewis and P. R. Gray, "A Pipelined 5-Msample/s 9-bit Analog-to-Digital Converter", *IEEE J. of Solid-State Circuits*, vol. SC-22, no. 6, pp. 954-61, Dec. 1987.

[11] H. S. Lee, D. A. Hodges and P. R. Gray, "A Self-Calibrating 15 Bit CMOS A/D Converter", *IEEE J. of Solid-State Circuits*, vol. SC-19, no. 6, pp. 813-19, Dec. 1984.

[12] S. H. Lee and B. S. Song, "Digital-Domain Calibration of Multistep Analog-to-Digital Converters", *IEEE J. of Solid-State Circuits*, vol. 27, no. 12, pp. 1679-88, Dec. 1992.

[13] G. S. Østrem, *High-Speed Analog-to-Digital Converters for Mixed Signal ASIC Applications*, Ph.D. Dissertation, FYS.EL.-rapport 1996:38, Norwegian University of Science and Technology, 1996.

[14] M. P. Flynn and D. J. Allstot, "CMOS Folding A/D Converters with Current-Mode Interpolation", *IEEE J. of Solid State Circuits*, vol. 31, no. 9, pp. 1248-57, Sept. 1996.

[15] P. Vorenkamp and R. Roovers, "A 12b 50MSample/s Cascaded Folding and Interpolating ADC", *1997 IEEE Intern. Solid-State Circuits Conf.*, pp. 134-5, 1997.

[16] A. G. W. Venes and R. J. van de Plassche, "An 80-MHz, 80-mW, 8-b CMOS Folding A/D Converter with Distributed Track-and-Hold Preprocessing", *IEEE J. of Solid-State Circuits*, vol. 31 no. 12, p. 1846-53, Dec. 1996.

[17] N. Tan, *Oversampling A/D Converters and Current-Mode Techniques*, Ph.D Dissertation, Dept. Electrical Engineering, no. 360, Linköping University, 1994.

[18] J. C. Candy and G. C. Temes, *Oversampling Delta-Sigma Data Converters*, IEEE Press, New York, NY, USA, 1992, ISBN 0-87942-281-5.

[19] N. Tan, *Switched-Current Design and Implementation of Oversampling A/D Converters*, Kluwer Academic Publishers 1997.

[20] T. L. Brooks, D. H. Robertson, D. F. Kelly, A. Del Muro and S. W. Harston, "A Cascaded Sigma-Delta Pipeline A/D Converter with 1.25 MHz Signal Bandwidth and 89 dB SNR", *IEEE J. of Solid-State Circuits*, vol. 32, no. 12, pp. 1896-1906, Dec. 1997.

[21] C. Hammerschmied and Q. Huang,: "A MOSFET-Only, 10b, 200kSample/s A/D Converter Capable of 12b Untrimmed Linearity", *1997 IEEE Intern. Solid-State Circuits Conf.*, pp. 132-3, 1997.

[22] J-E. Eklund, *A/D Conversion for Sensor Systems*, Linköping Studies in Science and Technology, Dissertation no. 491, 1997.

[23] P. E. Allen and D. R. Holberg *CMOS Analog Circuit Design*, Holt, Rinehart and Winston, Inc. 1987.

[24] W. C. Black and D. A Hodghes "Time Interleaved Converter Arrays", *IEEE J. of Solid-State Circuits*, vol. SC-15, no. 6, pp. 1022-29, Dec. 1980.

[25] K. Nagaraj, J. Fetterman, J. Anidjar, S. Lewis and R. G. Renninger "A 250-mW, 8-b, 52-Msamples/s Parallel-Pipelined A/D Converter with Reduced Number of Amplifiers", *IEEE J. of Solid-State Circuits*, vol. 32, no. 3, p.312-20, March 1997.

[26] Y.C. Jenq, "Digital Spectra of Nonuniformly Sampled Signals: Fundamentals and High-Speed Waveform Digitizers", *IEEE Trans. on Instrumentation and Measurement*, vol. 37, no. 2, pp. 245-51, June 1988.

[27] A. Petraglia and S. K. Mitra, "Analysis of Mismatch Effects Among A/D Converters in a Time-Interleaved Waveform Digitizer", *IEEE Trans. on Instrumentation and Measurement*, vol. 40, no. 5, pp. 831-5, Oct. 1991.

[28] J. Spalding, D. Dalton, "A 200M sample/s 6b flash ADC in 0.6 mu CMOS", *1996 IEEE Intern. Solid-State Circuits Conf.*, pp.p.320-1, 1996.

[29]   Y. Nejime, M. Hotta and S. Ueda "An 8-b ADC with Over-Nyquist Input at 300-Ms/s Conversion Rate", *IEEE J. of Solid-State Circuits*, vol. 26, no. 9, pp. 1302-8, Sept. 1991.

[30]   M. J. M. Pelgrom, A. C. J. v. Rens, M. Vertregt and M. B. Dijkstra "A 25 MS/s 8-bit CMOS A/D Converter for Embedded Application", *IEEE J. of Solid-State Circuits*, vol. 29, no. 8, pp. 879-86, Aug. 1994.

[31]   C. W. Mangelsdorf "A 400-MHz Flash Converter with Error Correction" *IEEE J. of Solid-State Circuits*, vol. 25, no. 1, pp. 184-191, Feb. 1990.

[32]   Y. Gendai, Y. Kamatsu, S. Hirase and M. Kawata, "An 8b 500 MHz ADC", *1991 IEEE Intern. Solid-State Circuits Conference*, pp.172-3, 1991.

[33]   H. Reyhani and P. Quinlan "A 5 V, 6-b, 80 Ms/s BiCMOS flash ADC", *IEEE J. of Solid-State Circuits*, vol. 29, no. 8, pp. 873-8, Aug. 1994.

[34]   B. Zojer, R. Petschacher and W. A. Luschnig, "A 6-bit/200-MHz Full Nyquist A/D Converter", *IEEE J. of Solid-State Circuits*, vol. SC-20, no. 3, p.780-5, June 1985.

[35]   B. Zojer, B. Astegher, H. Jessner and R. Petschacher "A 10b 75MHz Subranging A/D Converter", *1990 IEEE Intern. Solid-State Circuits Conf.*, pp. 164-5, 1990.

[36]   F. Murden and R. Gosser, "12b 50MSample/s two-stage A/D converter", *IEEE Intern. Solid-State Circuits Conf.*, pp. 278-9, 1995.

[37]   M. Ishikawa and T. Tsukahara, "An 8-bit 50-MHz CMOS subranging A/D converter with pipelined wide-band S/H", *IEEE J. of Solid-State Circuits*, vol. 24, no. 6, pp. 1485-91, Dec. 1989.

[38]   M. Yotsuyanagi, H. Hasegawa, M. Yamaguchi, M. Ishida and K. Sone, "A 2 V 10 b 20 MSample/s Mixed-Mode Subranging CMOS A/D Converter", *1995 IEEE Intern. Solid-State Circuits Conf.*, pp. 282-3, 1995.

[39]   R. Petschacher, B. Zojer, B. Astegher., H. Jessner and A. Lechner, "A 10-b 75-MSPS Subranging A/D Converter with Integrated Sample and Hold", *IEEE J. of Solid-State Circuits*, vol. 25, no. 6, pp. 1339-46, Dec. 1990.

[40]   K. Sone, Y. Nishida, N. Nakadai, "A 10-b 100Msample/s Pipelined Subranging BiCMOS ADC," *IEEE J. of Solid-State Circuits*, vol. 28, no. 12, pp. 1180-6, Dec. 1993.

[41]   W. T. Colleran and A. A. Abidi, "A 10-b 75-MHz Two-Stage Pipelined Bipolar A/D Converter", *IEEE J. of Solid-State Circuits*, vol. 28, no. 12, pp. 1187-99, Dec. 1993.

[42]   T. Miki, H. Kouno, T. Kumamoto, Y. Kinoshita, T. Igarashi and K. Okada, "A 10-b 50 MS/s 500-mW A/D Converter Using a Differential-Voltage Subconverter", *IEEE J. of Solid-State Circuits*, vol. 29, no. 4, pp. 516-22, April 1994.

[43]   P. Vorenkamp and J. P. M. Verdaasdonk, "A 10b 50MS/s Pipelined ADC", *1992 IEEE Intern. Solid-State Circuits Conference*, pp. 32-3, 1992.

[44]   R. Jewett, K. Poulton, K. C. Hsieh and J.Doernberg, "A 12b 128Msample/s ADC with 0.005LSB DNL", *1997 IEEE Intern. Solid-State Circuits Conf.*, pp. 138-9, 1997.

[45]   S. Hosotani, T. Miki, A. Maeda and N. Yazawa, "An 8-bit 20-MS/s CMOS A/D Converter with 50-mW Power Consumption", *IEEE J. of Solid-State Circuits*, vol. 25, no. 1, pp. 167-72, February 1990.

[46]   B. Brandt and J. Lutsky "A 75mW 10b 20MSample/s CMOS Subranging ADC with 59dB SNDR", *1999 IEEE Intern. Solid-State Circuits Conference*, pp. 322-23, 1999.

[47]   H. van der Ploeg and R. Remmers "A 3.3V 10b 25MSamples/s Two-Step ADC in 0.35um CMOS", *1999 IEEE Intern. Solid-State Circuits Conf.*, pp. 318-19, 1999.

[48]   M. P. Flynn and D. J. Allstot, "CMOS Folding A/D Converters with Current-Mode Interpolation", *IEEE J. of Solid State Circuits*, vol. 31, no. 9, pp. 1248-57, Sept. 1996.

[49]   Ø. Moldsvor and G. Østrem, "An 8-bit, 200 MSPS Folding and Interpolating ADC", *Norchip Conf.*, p. 97-105, 1996.

[50]   H. Kimura, A. Matsuzawa, T. Nakamura and S. Sawada, "A 10-b 300-MHz Interpolated-Parallel A/D Converter", *IEEE J. of Solid-State Circuits*, vol. 28, no. 4, pp. 438-46, April 1993.

3.11 Summary

[51] A. G. W. Venes and R. J. van de Plassche, "An 80-MHz, 80-mW, 8-b CMOS Folding A/D Converter with Distributed Trach-and-Hold Preprocessing", *IEEE J. of Solid-State Circuits*, vol. 31 no. 12, p. 1846-53, Dec. 1996.

[52] B. Nauta and A. G. W. Venes, "A 70-MS/s 110-mW 8-b CMOS folding and interpolating A/D converter", *IEEE J. of Solid-State Circuits*, vol. 30, no. 12, p.1302-8, Dec. 1995.

[53] J. van Valburg and R. J. van de Plassche, "An 8-b 650-MHz Folding ADC", *IEEE J. of Solid-State Circuits*, vol. 26, no. 12, p. 1662-6, Dec. 1992.

[54] R. J. van de Plassche and P. Baltus, "An 8-b 100-MHz Full Nyquist Analog to Digital Converter", *IEEE J. of Solid-State Circuits*, vol. 23, no. 6, pp. 1334-44, Dec. 1988.

[55] R .E. J. van de Grift , I. W. J. M. Rutten and M. van der Veen, "An 8-b Video ADC Incorporating Folding and Interpolation Techniques", *IEEE J. of Solid-State Circuits*, vol. 22, no. 6, pp. 944-53, December 1987.

[56] K. Bult, A. Buchwald and J. Laslowski, "A 170mW 10b 50MSample/s CMOS ADC in 1mm2", *1997 IEEE Intern. Solid-State Circuits Conf.*, pp. 136-7, 1997.

[57] P. Vorenkamp and R. Roovers, "A 12-b, 60-MSample/s Cascaded Folding and Interpolating ADC" *IEEE J. of Solid-State Circuits*, vol. 32, no. 12, pp. 1876-86, Dec. 1997.

[58] G. Hoogzaad and R. Roovers "A 65mW 10b 40MSample/s BiCMOS Nyquist ADC in 0.8mm^2", *1999 IEEE Intern. Solid-State Circuits Conf.*, pp. 320-21, 1999.

[59] T. H. Shu, K. Bacrania and R. Gokhale, "A 10-b 40-Msample/s BiCMOS A/D Converter", *IEEE J. of Solid-State Circuits*, vol. 31, no. 10, p. 1507-10, Oct. 1996.

[60] W. C. Song, H. W. Choi, S. U. Kwak and B. S. Song, "A 10-b 20-Msample/s Low-Power CMOS ADC", *IEEE J. of Solid-State Circuits*, vol. 30, no. 5, p. 514-21, May 1995.

[61] T. B. Cho and P R. Gray, "A 10b, 20 MSamples/s, 35 mW Pipeline A/D Converter", *IEEE J. of Solid-State Circuits*, vol. 30, no. 3, p. 166-72, March 1995.

[62] P. Real, D. H. Robertson, C. W. Mangelsdorf, and T. L. Tewksbury, "A Wide-Band 10-b 20-Ms/s Pipelined Using Current-Mode Signals", *IEEE J. of Solid-State Circuits*, vol. 26, no. 8, pp. 1103-09, Aug. 1991.

[63] S. H. Lewis S.H., H. S. Fetterman, G. F. Gross, R. Ramachandran and T. R. Viswanathan, "A 10-b 20-Msample/s Analog-to-Digital Converter", *IEEE J. of Solid-State Circuits*, vol. 27, no. 3, pp. 351-8, March 1992.

[64] K. Kusumoto, A. Matsuzawa and K. Murata, "A 10-b 20-MHz 30-mW Pipelined Interpolating CMOS ADC", *IEEE J. of Solid-State Circuits*, vol. 28, no. 12, pp. 1200-6, Dec. 1993.

[65] M. Ito, T. Miki, S. Hosotani, T. Kumamoto, Y. Yamashita, M. Kijima and K. Okada, "A 10b 20Ms/s 3V-Supply CMOS A/D Converter for Integration into System VLSIs", *IEEE Intern. Solid-State Circuits Conf.*, p. 48-9, 1994.

[66] A. M. Abo and P. R. Gray, "A 1.5-V, 10-bit, 14-MS/s CMOS Pipeline Analog-to-Digital Converter", *IEEE J. of Solid-State Circuits*, vol. 34 no. 5, pp. 509-606, Aug. 1999.

[67] S Paul, H.-S. Lee, J. Goodrich, T. Alailima and D. Santiago, "A Nyquist-Rate Pipelined Oversampling A/D Converter", *1999 IEEE Intern. Solid-State Circuits Conf.*, pp. 54-5, 1999.

[68] "AD9224", Analog Devices, Inc., http://www.analog.com.

[69] "AD9225", Analog Devices, Inc., http://www.analog.com.

[70] "AD9226", Analog Devices, Inc., http://www.analog.com.

[71] "ADS805", Burr-Brown, Inc., http://www.burr-brown.com.

[72] K. Nakamura, M. Hotta, L. R. Carley and D. J. Allstot, "An 85 mW, 10 b, 40 Msample/s CMOS Parallel-Pipelined ADC", *IEEE J. of Solid-State Circuits*, vol. 30, no. 6, pp. 173-83, Dec. 1995.

[73] J. Yuan and C. Svensson, "A 10-bit 5-MS/s Successive Approximation ADC Cell Used in a 70-MS/s ADC Array in 1.2-um CMOS", *IEEE J. of Solid-State Circuits*, vol. 29 no. 8, pp. 866-872, Aug. 1994.

[74] M. Yotsuyanagi, T. Etoh and K. Hirata, "A 10 b 50 MHz pipelined CMOS A/D converter with S/H", *IEEE J. of Solid-State Circuits*, vol. 28, no. 3, p.292-300, March 1993.

[75] K. Poulton K., J. J. Corcoran and T. Hornak, "A 1-GHz 6-bit ADC System", *IEEE J. of Solid-State Circuits*, vol. SC-22, no. 6, pp. 962-70, Dec. 1987.

[76] S. G. Conroy, D. W. Cline and P. R. Gray, "An 8-b 85-MS/s Parallel Pipeline A/D Converter in 1-um CMOS", *IEEE J. of Solid-State Circuits*, vol. 28, no. 4, p.447-54, April 1993.

[77] M. K. Mayes, S. W Chin and L. L Stoian, "A Low-Power 1 MHz, 25 mW 12-bit Time-Interleaved Analog-to-Digital Converter", *IEEE J. of Solid-State Circuits*, vol. 31, no. 2, p.169-78, Feb. 1996.

4 OVERVIEW OF D/A CONVERTER ARCHITECTURES

4.1 INTRODUCTION

In this chapter we present different techniques for converting a digital signal into an analog signal representation. The approaches differ in speed, chip area, power efficiency, achievable accuracy, etc. It is therefore necessary to understand which converter algorithms or architectures to choose for the specific application. For example, when the conversion bandwidth is relatively small, it could be advantageous to use a higher sampling ratio and some oversampling technique to reduce the noise energy within the signal band. If very high speeds are required, a flash converter architecture should be used. However, the trade-off in converter design is normally between resolution and bandwidth (or the update frequency over signal frequency ratio). The higher bandwidth the lower resolution is generally achievable.

In Sec. 4.2 we discuss the concept of Nyquist-rate conversion, and in Sec. 4.3 to Sec. 4.8 we show different implementations for Nyquist-rate DACs. There is a limited number of DAC architectures that are good candidates for high-speed and high-performance applications. One is the current-steering DAC, which uses a number of binary scaled elements which generate the output value.

In Sec. 4.9, we discuss the concept of oversampling D/A conversion. In this book we refer to an oversampling DAC (OSDAC) to a converter operating at a higher update frequency and also containing a noise shaping loop. Often, the OSDAC utilizes some kind of a Nyquist-rate converter (with a lower nominal resolution) together with digital modulators and analog filters.

To further increase the performance of DACs, some special enhancement techniques can be used, as for example dynamic randomization and dynamic element matching (DEM) techniques. This is discussed in Sec. 4.10.

In Sec. 4.11 we compare a number of DAC architectures reported in the literature and in data sheets from vendors. We present an overview of the performance in terms of linearity vs. update and signal frequencies.

4.2 NYQUIST-RATE D/A CONVERTERS

In general the Nyquist-rate converter is required for wide band applications where oversampling techniques are impossible due to the high clocking frequency. In for example, audio applications an oversampling D/A converter (OSDAC) is preferred. Since a Nyquist-rate converter also is used in most OSDACs we first present some general and common Nyquist-rate DAC architectures. We discuss application areas and highlight some of the advantages and disadvantages with the different types of converter architectures. The architectures that are discussed, are the binary weighted, thermometer coded, and encoded DAC architectures. The different architectures can be combined into hybrid DACs where the advantages of different individual architectures are used.

We are considering DACs for telecommunication applications and therefore we focus on candidates suitable for high speed and high resolution. However, lower-speed DACs such as the algorithmic DAC are also briefly presented. These lower-speed DACs can be used in pipeline DACs where the throughput is increased by pipelining.

To illustrate the principle of the DAC, usually single-ended circuits are used. In a real implementation they are differential circuits to improve linearity and SNR. In this chapter, we consider only binary offset coded inputs. To represent a negative output of the converter, the reference elements of the most significant bit get the opposite sign, i.e., a 2's complement coded input.

We use three modes of circuit technology; *voltage-mode, current-mode*, and *charge-redistribution mode*. We associate *voltage-mode* with a DAC where the element values are given by voltage levels as for example in a resistor-string which divides a voltage reference level into a number of different amplitude levels. With *current-mode* the DAC elements are currents, as for example switched current sources or resistors dividing a major current into weighted subcurrents. Finally, *charge-redistribution* DACs are using the switched-capacitor technique (SC). Note that the SC technique usually is refered to as a voltage-mode technique as well.

4.2.1 Nyquist-Rate DACs

In Nyquist-rate DACs the input signal bandwidth f_B is equal to the Nyquist frequency, $f_B < f_N = f_s/2$, where f_s is the update frequency of the DAC [1, 2]. The entire possible frequency range for fully recoverable signals according to the sampling theorem.

In almost every practical DAC, the output is sample&held. The sampling period is $T_s = 1/f_s$ and, as was shown in Chapter 1, the spectrum is repeated, centered at multiples of the update frequency, f_s. There is also a sinc-weighting of the signal in the frequency domain and in Fig. 4-1 we show the characteristic attenuation A (in dB) of the signal as function of f_{sig}/f_s, where f_{sig} is the output signal frequency. The attenuation of the signal at half the update frequency $f_B = f_N$ (indicated by the dashed line in the figure) is 3.9 dB. To correct for the attenuation, either anti-sinc filters or an oversampling strategy is needed.

As illustrated in Fig. 4-2b) an analog continuous-time low pass (LP) filter with a cut-off frequency f_N is used to remove images. With an anti-sinc filter, the attenuation from the sinc-weighting is reduced. This can be done both in the analog and the digital domain.

4.2 Nyquist-Rate D/A Converters

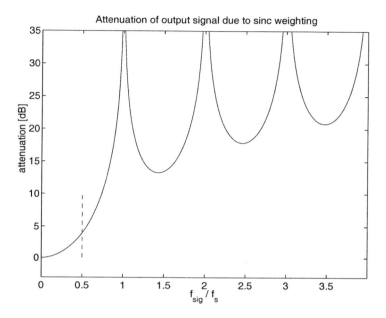

Figure 4-1 Sinc attenuation of the output signal as function of the signal to sampling frequency ratio.

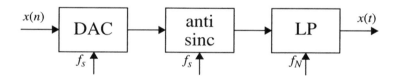

Figure 4-2 DAC with continuous-time filters, such as a low pass filter to remove images and an anti-sinc filter to reduce sinc distortion.

The digital input signal has a resolution of N bits, hence a quantization noise is introduced. This noise can for a larger number of bits be considered as white [2] and its power spectral density (PSD) is constant within the Nyquist frequency range. The influence of the noise, or the signal-to-noise ratio (SNR), was discussed in Chapter 1 and for a full-scale input sinusoid, the SNR is

$$SNR \approx 6.02 \cdot N + 1.76 \text{ dB} \tag{4-1}$$

where N is the nominal number of bits in the converter. For a lower resolution, the quantization error becomes signal correlated and hence distortion is introduced.

4.2.2 Nyquist-Rate DACs with Oversampling

To overcome the difficulties in the realization of anti-sinc and low pass image rejection filters, almost all Nyquist-rate DACs are oversampled, i.e., the signal bandwidth is $f_B < f_N = f_s/2$. We could apply an input signal to the DAC creating an output signal spectrum as illustrated in Fig. 4-3 where we do not use the entire Nyquist range.

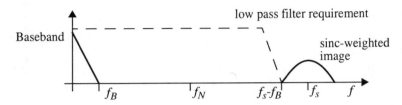

Figure 4-3 DAC input signal when not using the entire frequency range. The low pass filter requirement is illustrated with a dashed line.

If we would use a sampling frequency of $f_s = 8 \cdot f_B$ we would have a 0.22 dB attenuation of the signal at f_B. In general, the attenuation of the signal at f_B is given by

$$A(f_B/f_s) = -20 \cdot \log \left| \frac{f_s}{\pi f_B} \cdot \sin \frac{\pi f_B}{f_s} \right| \qquad (4\text{-}2)$$

Therefore, in most DACs, a higher update frequency than double the bandwidth, $f_N = 2 \cdot f_B$, is used to achieve a lower sinc attenuation and to reduce the image-rejection filter order.

Oversampling can also improve the performance and the improvement by using a lower bandwidth can be found by using the definition of SNR in (4-1). Let the signal power be given by P_s and the quantization noise power spectral density be constant, i.e., $S_q(f) = q^2/\Delta f$. From (4-1) we have that

$$SNR = 10 \cdot \log \frac{P_s}{S_q(f) \cdot f_N} \text{ dB} \qquad (4\text{-}3)$$

Due to the LP filter, we only consider the noise up to the frequency f_B and (4-3) can be written as

$$SNR = 10\log\left(\frac{P_s}{S_q(f) \cdot f_B} \cdot \frac{f_B}{f_N}\right) = 10\log \frac{P_s}{S_q(f) \cdot f_B} - 10\log \frac{f_N}{f_B} \text{ dB} \qquad (4\text{-}4)$$

We have that

$$SNR_B = SNR + 10 \cdot \log \frac{f_N}{f_B} \text{ dB} \qquad (4\text{-}5)$$

where SNR_B is the signal-to-noise ratio within a smaller frequency band. From (4-5) we have the intuitive conclusion that the smaller signal bandwidth compared to the update frequency, the higher SNR. We also have from Fig. 4-3 that the transition band of the following LP filter can be much wider and the filter order will be reduced. Since the sinc-attenuation is low, there is no need for an anti-sinc filter. In the frequency-domain, some of the dynamic errors, such as CFT from switches, are also moved from the signal band to higher frequencies.

Digital filters in the DSP are sometimes updated at the Nyquist rate $(2 \cdot f_B)$ and therefore interpolation is used to increase the update frequency for D/A conversion. We can sample the signal at a higher frequency, but this will introduce signal images within the frequency range f_B to $f_s - f_B$. The analog image rejection filter has to remove these images, meaning that the filter order becomes high (due to the narrow transition band). A better choice is to use digital interpolation filters at the DAC input to remove these images. This is why Nyquist-rate DACs with oversampling also is referred to as interpolating DACs [3, 4, 5].

Note that the maximum achievable resolution within f_B is given by the resolution of the preceding digital circuits. The interpolation filters "compress" the frequency domain and the SNR is still given by this internal resolution. Often this resolution can be considerably larger than that of the DAC itself and hence the DAC introduces quantization noise when reducing the number of bits. Therefore, we have that the maximum achievable resolution according to (4-5) is given by the resolution of the circuits at the input of the DAC.

Higher resolution can be achieved within a small signal bandwidth if we also succeed to spectrally move the noise *introduced by the DAC* to frequencies outside the narrow signal band. This is the concept of oversampling D/A converters (OSDACs) and is further described in Sec. 4.9.

4.3 BINARY WEIGHTED DAC ARCHITECTURE

The binary weighted DAC utilizes a number of reference elements that are binary weighted. The output signal can be written as

$$X_a(nT) = A_{os} + A_0 \cdot (b_0(nT) + 2 \cdot b_1(nT) + \ldots + 2^{N-1} \cdot b_{N-1}(nT)) \quad (4\text{-}6)$$

where A_0 is the reference, A_{os} is the offset, $\{b_i(nT)\}_{i=0}^{N-1}$ are the input bits, and T is the update period of the DAC. In Fig. 4-4 we show the concept of the binary weighted DAC. The total sum of all weights is $(2^N - 1) \cdot A_0$.

One of the drawbacks with this architecture is that for a large number of bits, the difference between the MSB (b_{N-1}) and LSB (b_0) weight is very large and the converter is very sensitive to mismatch errors.

The advantage is that the number of switches and digital encoding circuits become minimal in this architecture. The elements to implement the multiplication and addition operations may be current sources, resistors, or capacitors. We describe the current-steering, R-2R ladder, and charge redistribution DAC architectures below.

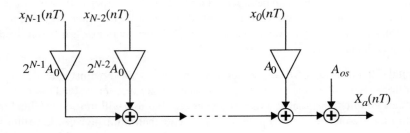

Figure 4-4   Binary weighted DAC with an offset level, A_{os}.

4.3.1   Current-Steering DAC

The switched-current (SI) technique is a natural choice in a CMOS process, since the reference and sum elements as well as switches are relatively easy to implement. The general architecture of a binary weighted current-steering DAC is shown in Fig. 4-5

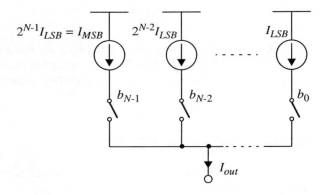

Figure 4-5   An N-bit current-steering binary weighted DAC.

and discussed in [2, 6]. The switches are controlled by the input bits, b_i, where $i = 0, 1, ..., N-1$ and N is the number of bits. b_0 is the LSB and the corresponding current source has the DC value I_{LSB}. The source controlled by bit b_i, i.e., the i-th LSB current source, is preferably formed by connecting 2^i LSB current sources (unit current sources) in parallel, hence the MSB current source has the DC value $I_{MSB} = 2^{N-1} \cdot I_{LSB}$. The use of unit element sources increases the matching of the sources [7, 8, 9]. The output current, I_{out}, of the DAC shown in Fig. 4-5 is given by

$$I_{out}(k) = I_{LSB} \cdot b_0 + 2I_{LSB} \cdot b_1 + ... + 2^{N-1}I_{LSB} \cdot b_{N-1} = I_{LSB} \cdot k \qquad (4\text{-}7)$$

where k is the digital input given by

4.3 Binary Weighted DAC Architecture

$$k = b_0 + 2 \cdot b_1 + \ldots + 2^{N-1} \cdot b_{N-1} = \sum_{l=0}^{N-1} 2^l \cdot b_l \qquad (4\text{-}8)$$

The current-steering DAC has the advantages of being quite small for resolutions below 10 bits and it is very fast. The major disadvantage is its sensitivity to device mismatch, glitches, and current source output impedance for higher number of bits. Another good property of the current-steering DAC is that it has a very high power efficiency since all power is directed to the output. The current-steering is suitable for high-speed high-resolution applications, especially when special care is taken to improve the matching of the converter. The current sources are typically implemented with cascoded NMOS or PMOS transistors.

To achieve monotonicity and reduce the influence of glitches (see Chapter 1), as well as reducing the sensitivity to matching errors, the DAC should be segmented into a coarse and fine part. The coarse part is then thermometer coded and find part is kept binary weighted. This is refered to as a segmented converter and is further discussed in Sec. 4.6.

4.3.2 R-2R Ladder DAC

It is easy to construct a resistor-based binary-weighted DAC. But the resistance spread is very large for a large number of bits N. The better choice is the R-2R ladder architecture. This DAC architecture provides an architecture suitable in processes capable of implementing linear resistances. An N-bit R-2R ladder architecture is shown in Fig. 4-6 and discussed in [1, 2, 10, 11, 12]. The current sources are all equally large,

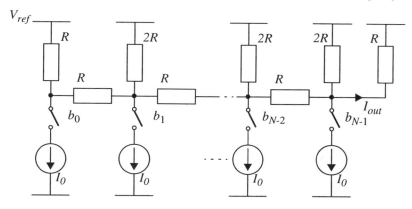

Figure 4-6 An N-bit R-2R ladder DAC.

I_0, and the switches are controlled by the bits b_i. At every node, the impedance is always R. The resistive network divides the currents from the individual current sources to the output and the output current is given by

$$I_{out}(k) = \frac{I_o}{2^{N-1}} \cdot \sum_{l=0}^{N-1} b_l \cdot 2^l = \frac{I_o}{2^{N-1}} \cdot k \qquad (4\text{-}9)$$

This DAC architecture is power inefficient, since there is a current loss through the resistive network.

Due to the fact that all current sources are equally large, the matching can be improved. In a poor process, the resistors can be non-linear and contain signal-dependent capacitive parts yielding distortion. Time-skew between switching instants generates glitches in this architecture.

In this R-2R ladder architecture there is the same amount of current through all switches, which makes the design of the switches simpler. However, the internal voltage nodes are still AC varying and therefore the current sources will have varying terminal voltages.

As a conclusion, one of the major advantages is that we only have a few number of different component sizes to implement, i.e., two resistor sizes, R and $2R$, one current source size, I_0, and one type of current switch. This allows a more regular layout and since the current sources all have the same size. Trimming or calibration of the current sources can also be applied [2, 10]. R-2R ladder DACs are more widely used in bipolar processes where high-quality resistors are available.

4.3.3 Charge Redistribution DAC

The charge redistribution DAC is a switched capacitor (SC) circuit, where the charge stored on a number of binary weighted capacitors is used to perform the conversion. See Fig. 4-7 for an example of an N-bit converter found in [1]. The most significant

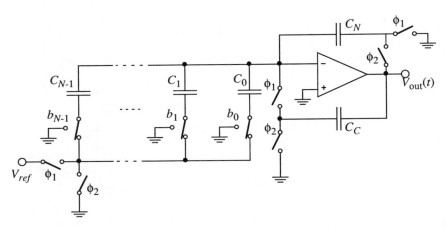

Figure 4-7 An N-bit charge-redistribution DAC.

capacitor C_{N-1} is 2^{N-1} times larger than the least significant capacitor $C_0 = C$, i.e., $C_{N-1} = 2^{N-1} \cdot C$. Typically, the weighted capacitors are created using a number of unit capacitors.

At time nT (on phase ϕ_1) the bits b_i determine which of the binary weighted capacitors that should be charged from the reference voltage, V_{ref}. During this phase, the plates of capacitor C_N are connected to ground and virtual ground at the input of the opamp, i.e., there is no charge on C_N, and $q_N(nT) = 0$. Capacitor C_C is used for offset compensation.

The total charge on the binary weighted capacitors at time point nT is given by

$$q_T(nT) = V_{ref} \sum_{l=0}^{N-1} C_l \cdot b_l = V_{ref} \sum_{l=0}^{N-1} 2^l \cdot C \cdot b_l = V_{ref} \cdot C \cdot k(nT) \qquad (4\text{-}10)$$

At time point $nT + T/2$, on phase ϕ_2, the weighted capacitors are discharged since their plates are connected to DC and virtual grounds. The charge is redistributed to ground and C_N. The charge on C_N at the end of the settling is

$$q_N(nT + \frac{1}{2}T) = C_N \cdot V_{out}(nT + T/2) \qquad (4\text{-}11)$$

Charge conservation gives

$$q_N(nT + \frac{1}{2}T) = -q_T(nT) \qquad (4\text{-}12)$$

Using (4-10) and (4-11) in (4-12) we have

$$V_{out}(nT + \frac{1}{2}T) = \frac{C}{C_N} \cdot k(nT) \cdot V_{ref} \qquad (4\text{-}13)$$

The phases, ϕ_1 and ϕ_2, are non-overlapping. The architecture in Fig. 4-7 is insensitive only to offset voltage and finite gain of the amplifier. The limitations of the converter are the matching of the capacitors, the switch-on resistance, and finite bandwidth of the amplifier.

4.4 THERMOMETER CODED DAC ARCHITECTURE

The thermometer coded DAC architecture utilizes a number of equally weighted elements. The binary input code is encoded into a thermometer code as illustrated in Table 4-1 for a 3-bit input code. Generally, with N binary bits, we have $M = 2^N - 1$ thermometer coded bits. The analog output at the time nT is given by

$$X_a(nT) = A_{os} + A_0 \cdot \sum_{i=1}^{M} c_i(nT) \qquad (4\text{-}14)$$

where $c_i(nT) \in \{0, 1\}$, $1 \leq i \leq M$ are the thermometer coded bits. The signal flow graph (SFG) of the thermometer coded DAC is shown in Fig. 4-8.

Table 4-1   Decimal, binary offset, thermometer, and walking-one code representations.

Decimal	Binary offset	Thermometer	Walking-one
0	000	0000000	0000000
1	001	0000001	0000001
2	010	0000011	0000010
3	011	0000111	0000100
4	100	0001111	0001000
5	101	0011111	0010000
6	110	0111111	0100000
7	111	1111111	1000000

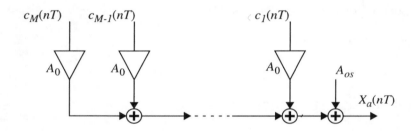

Figure 4-8   Thermometer coded DACs use a number of equally large amplification factors.

In the thermometer coded DAC the reference elements are all equally large and the matching of the individual elements becomes simpler than for the binary case. The total sum of all weights is $2^N - 1$. The transfer function of the thermometer coded converter is monotonic and the DNL and INL is improved compared to the binary version. The requirements on element matching is also relaxed. In fact, if the matching is within a 50% margin, the converter is still monotonic [13].

In Fig. 4-9 we show a current-steering implementation of a thermometer coded DAC with binary-to-thermometer encoder. All current sources are equally large, I_{unit}. For a larger number of bits, the digital circuits converting the binary code (X) into thermometer code (C) and the number of interconnecting wires become large, since the number of outputs is growing exponentially. This implies a more complex circuit layout. The encoding circuit can easily be pipelined and the propagation time through the encoder can be controlled. This is further discussed in Chapter 13.

4.5   ENCODED DAC ARCHITECTURE

An architecture very similar to the thermometer coded DAC is the direct encoded

4.5 Encoded DAC Architecture

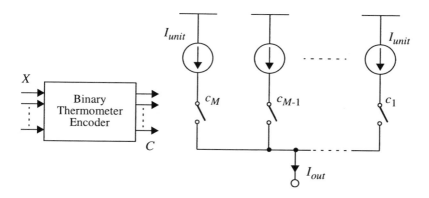

Figure 4-9 A thermometer coded current-steering DAC with encoding circuit.

DAC as the SFG illustrates in Fig. 4-10. In this architecture, linearly weighted refer-

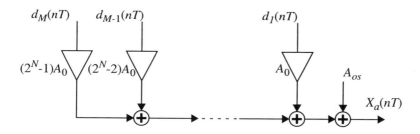

Figure 4-10 Direct encoded DAC uses a number of linearly weighted element values.

ences are used. For an N-bit converter we need $M = 2^N - 1$ references, one for each level. The output value is given by

$$X_a(nT) = A_{os} + A_0 \cdot \sum_{i=1}^{M} A_i \cdot d_i(nT) \tag{4-15}$$

where $d_i(nT) \in \{0, 1\}, 1 \leq i \leq M$ are given by a "walking one" code shown in Table 4-1.

In the resistor-string DAC, the principle of this architecture becomes clear. This architecture, illustrated in Fig. 4-11 is basically a voltage-mode direct encoded DAC architecture. In this DAC, M switches are used, one for each word, i.e., $M = 2^N - 1$. The weighting elements are equally large resistors in a resistor string. If the number of bits is high, the number of resistors becomes very high. The reference voltage V_{ref} is feeding the resistors and from voltage division, we have that the voltage level V_i is given by

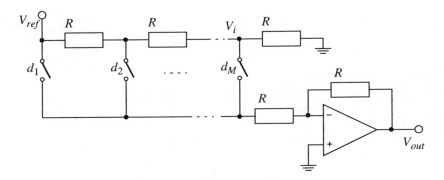

Figure 4-11 An N-bit resistor string DAC where $M=2^N-1$.

$$V_i = \frac{i \cdot R}{R_{tot}} \cdot V_{ref} = \frac{i}{2^N - 1} \cdot V_{ref} \qquad (4\text{-}16)$$

The switches, implemented with for example NMOS transistors or transmission gates, connect the selected voltage V_i to the opamp and the DAC output is given by

$$V_{out}(nT) = -\sum_{i=1}^{M} d_i(nT) \cdot V_i = -\frac{V_{ref}}{M} \cdot \sum_{i=1}^{M} i \cdot d_i(nT) \qquad (4\text{-}17)$$

where $M = 2^N - 1$.

This architecture has some drawbacks; for high speed, the design of the operational amplifier becomes difficult, the matching of the resistors in the resistor string is crucial for the overall accuracy, and the RC-timing through the switches and wires may also limit the bandwidth and linearity of the DAC. The encoder, converting the binary input into a "walking one" code, is a straight-forward design. But still the routing and chip area may become large and complex. The encoding circuit can be pipelined and the propagation time through the encoder is a minor problem.

Mostly, the encoded DAC shows good monotonicity behavior as well as an improved DNL and INL over the binary architecture.

The encoder can be simpler if a tree selection architecture [1] is used. This also reduces the total number of switches. The disadvantage of using such a tree architecture is the additional delay through the increased number of switches' layers.

4.6 HYBRID DAC ARCHITECTURE

As has been addressed previously, different types of DAC architectures have their different advantages and disadvantages. An intuitive conclusion from this, is to use a combination of different architectures to improve the overall performance. This is

4.6 Hybrid DAC Architecture

illustrated in Fig. 4-12 where the different sub-converters can be of completely dif-

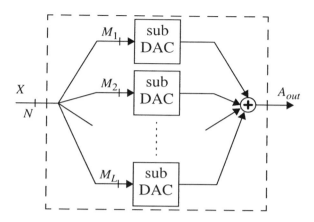

Figure 4-12 Hybrid DACs use a combination of a number of different types of DACs.

ferent types. Each sub-DAC converts M_i bits and in the simplest configuration we have that the total resolution is equal to the sum of the resolutions of the sub-converters, i.e., $N = \sum M_i$. The subconverters have different reference levels.

A popular hybrid converter architecture is the so called segmented architecture as shown in Fig. 4-13. In this N-bit architecture, the M more significant bits are encoded

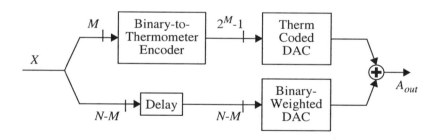

Figure 4-13 An N-bit segmented DAC where M MSBs are thermometer coded.

into a thermometer code and the $M-N$ less significant bits are binary weighted. There is a need of an encoder for the MSBs, which takes an M-bit input signal generates $2^M - 1$ one-bit output signals. When increasing M the size of the encoder grows exponentially and a larger amount of switches and interconnection are needed. However, the advantages are the improved device matching and linearity over binary weighted DACs. One of the key issues is to find the optimum number of MSBs to encode into a thermometer code. Reasonable sizes for M is in the range from 4-8, dependent on the total number of bits N.

In most high-speed high-resolution DACs we find that segmented or multi-segmented architectures have been used (see Sec. 4.11). The multi-segmented architecture uses a thermometer code representation for clusters, e.g., the M_1 MSBs are clustered and encoded, the M_2 next MSBs are clustered and encoded, and the remaining $N - M_1 - M_2$ bits are kept binary weighted. In Chapter 13 we present the design of segmented current-steering DACs, where $N = 14$ and $M = 4$.

4.7 LOW-SPEED DAC ARCHITECTURE

Although we are mainly interested in wide band, high-resolution converters for communications, we also briefly overview some lower-speed DAC architectures, such as the algorithmic (or cyclic) DACs. In lower-speed DACs the internal clock period is smaller than the sample period. These lower-speed architectures can however be pipelined (Sec. 4.8) to increase the throughput to the cost of additional hardware.

4.7.1 Algorithmic DAC Architecture

Assuming a binary offset code, the output of an algorithmic (or cyclic) DAC is

$$A(k) = A_0 \cdot k = A_0 \cdot \sum_{l=0}^{N-1} 2^l \cdot b_l \qquad (4\text{-}18)$$

where 2^l are the weights for the corresponding bit b_l and A_0 is the LSB reference value. The expression in (4-18) can be rewritten as

$$A(k) = A_0 \cdot [b_0 + 2(b_1 + 2(\ldots + 2(b_{N-2} + 2b_{N-1})))] \qquad (4\text{-}19)$$

where b_0 is the LSB and b_{N-1} is the MSB. (4-19) is a recursive formula and if we assume that the digital word is generated bit-wise, hence in serial with the MSB first, we can write

$$\tilde{A}(m) = 2 \cdot \tilde{A}(m-1) + x_{N-m} \qquad (4\text{-}20)$$

where $m = 1, \ldots, N$ and $\tilde{A}$ is an intermediate value with $\tilde{A}(0) = 0$. We need N iterations to generate an analog value, which is multiplied by the reference value and added to an offset, giving the analog output value

$$A(k) = A_0 \cdot \tilde{A}(N) + A_{os} \qquad (4\text{-}21)$$

Therefore, the algorithmic DAC requires a higher internal computational speed since the output value has to be generated within one output update period. We have that $T_s \geq N \cdot T_{alg}$ where N is the number of bits, T_{alg} is the clock period of the internal accumulator, and T_s is the update period of the DAC.

In Fig. 4-14 we show a block-level implementation of an algorithmic DAC with an MSB-first operation. The structure realizes the recursive formula (4-19) and it uses a

4.7 Low-Speed DAC Architecture

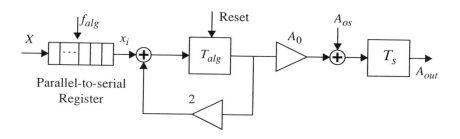

Figure 4-14 Schematic view of an algorithmic DAC.

parallel-to-serial interface where the serial values are stored in a register operating at the higher speed, $f_{alg} = 1/T_{alg}$. The digital bit values $x_i \in \{0, 1\}$ are fed into an accumulator, where the input signal is added to the output signal, which is fed back and amplified by a factor two. The output signal from the accumulator is multiplied with the LSB reference and an offset A_{os} may be added to the value. This output value is sample&held at the lower speed T_s. When each word has been processed, after $N \cdot T_{alg}$, the accumulator is reset.

The accumulator can also be fed with the LSB first and then the gain factor in the feedback becomes 1/2 instead. This gives a lower gain in the feedback loop and is a better choice when implementing the converter.

The advantage of the algorithmic DAC is the low number of circuit components. This reduces the number of interconnections and chip area in general.

4.7.2 Switched-Current Algorithmic DAC

Algorithmic DACs have been successfully implemented in both the switched-capacitor (SC) [14] and switched-current (SI) technique [15, 16]. Since there is an accumulation phase, the algorithmic DAC is not suitable for high-speed, though, the resolution can be quite high at intermediate frequencies [15].

In Fig. 4-15 we show a current-steering algorithmic DAC as proposed in [16]. This DAC utilizes a feedback amplifier of 1/2 instead of a factor 2 as was used in Fig. 4-14 and hence the digital input signal is given by the LSB first. The converter needs three equal subcircuits that are interconnected. Transistor M2 is biased with $I_0/2$ and therefore its size aspect ratio should be half the ratios of M1 and M3.

The circuit is controlled by four different switching signals, ϕ_1 to ϕ_4. On phase ϕ_1 the input bit x_i is determining if the current I_{bit} should be added to the accumulator. The I_{bit} current is equal to the maximum current for all bits i. This current is added to the intermediate accumulator output current from memory cell M3. The sum is stored in memory cell M1. This sum is divided by two in the M2 transistor during phase ϕ_2. Finally, on phase ϕ_4 the stored accumulated current in M3 is switched to the output of the DAC. The switching phase ϕ_3 is used to restore the accumulator and

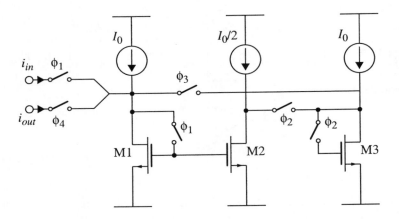

Figure 4-15 A switched-current (SI) implementation of an algorithmic DAC.

hence it should be opened when the first bit of a new word should be fed to the DAC.

To reach a high resolution of the SI algorithmic converter the bias currents must be large and the transistors and switches also have to be large.

4.8 PIPELINED DAC ARCHITECTURE

Low-speed, clocked circuits can be pipelined to increase the throughput of the whole system. This is done to the cost of more hardware and power consumption.

When pipelining algorithmic DACs (previous section) we can break the loop into a pipeline as shown in Fig. 4-16a) [17]. The architecture shown in the figure uses an off-

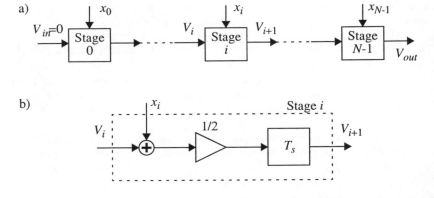

Figure 4-16 a) An N-stage pipelining DAC and b) view of the i-th stage.

4.8 Pipelined DAC Architecture

set binary code, but it can easily be modified to handle 2's complement code as well. Each stage in the pipeline, see Fig. 4-16b), uses an adder, a 1/2 amplifier, and a S/H circuit.

In Fig. 4-17 we show an SC implementation of one of the stages in the pipeline. Two

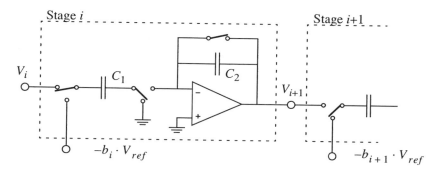

Figure 4-17 An SC implementation of the i th stage in a pipelining DAC.

capacitors and an opamp are used. The circuit is clocked with two non-overlapping clock phases, ϕ_1 and ϕ_2. The transfer function can be derived by studying the charge redistribution. At time instant $t = nT_s$ (during phase ϕ_1) V_i is charging capacitor C_1 with

$$q_1(t) = C_1 \cdot V_i(t) \tag{4-22}$$

The opamp output is auto-zeroed and the C_2 capacitor is discharged and hence $q_2(t) = 0$. At time instant $t + T_s/2$, i.e., during phase ϕ_2. the C_1 capacitor is charged to

$$q_1(t + \tfrac{1}{2}T_s) = C_1 \cdot x_i(t) \cdot (-V_{ref}) \tag{4-23}$$

and C_2 has the charge

$$q_2(t + \tfrac{1}{2}T_s) = C_2 \cdot V_{i+1}(t + \tfrac{1}{2}T_s) \tag{4-24}$$

By charge conservation we have

$$q_1(t) = q_1(t + \tfrac{1}{2}T_s) + q_2(t + \tfrac{1}{2}T_s) \tag{4-25}$$

Using (4-22), (4-23), and (4-24) in (4-25), we get

$$C_1 \cdot V_i(t) = C_2 \cdot V_{i+1}(t + \tfrac{1}{2}T_s) - C_1 \cdot b_i(t) \cdot V_{ref} \tag{4-26}$$

which gives

$$V_{i+1}(t + \frac{1}{2}T_s) = \frac{C_1}{C_2} \cdot [V_i(t) + V_{ref} \cdot b_i(t)] \tag{4-27}$$

The SC S/H should have the gain of 1/2, hence $C_2 = 2C_1 = 2C$. We also find from the configuration, that the next stage cannot sample V_{i+1} during the auto-zeroing phase, therefore the next stage has to have reversed clock phases. The circuits in Fig. 4-16 and Fig. 4-17 are designed for a binary offset code, but can easily be modified to cover 2's complement as well. Due to the non-idealities in the S/H amplifier (as further discussed in Sec. 6.6), this type of converter is not suitable for very high-speed and high-resolution applications.

4.9 OVERSAMPLING D/A CONVERTER (OSDAC)

We found in Sec. 4.2.2 that by increasing the update frequency with respect to the signal bandwidth, we could reduce the requirements on the continuous-time filter and improve the in-band SNR. If the input signal to the DAC is interpolated, the maximum achievable resolution is given by the preceding digital circuits. The Nyquist frequency to the signal bandwidth ratio is denoted oversampling ratio (OSR)

$$OSR = \frac{f_N}{f_B} = \frac{f_s}{2f_B} \tag{4-28}$$

From (4-28) we have that for a Nyquist-rate converter the oversampling ratio is $OSR = 1$. From (4-5) we have that the SNR within the signal bandwidth is expressed as

$$SNR_B = 6.02 \cdot N + 1.76 + 10 \cdot \log OSR \text{ dB} \tag{4-29}$$

We see that the effective number of bits within this frequency range is given by

$$N_B = N + \frac{10}{6.02} \cdot \log OSR \leq N_D \tag{4-30}$$

where N_D is the resolution (bits) given by the word length of the preceding digital circuits. From (4-30) we find that for each doubling of the OSR, the effective number of bits is increased by half a bit. To reach a high resolution we must choose a large OSR. For example, with a nominal DAC resolution of $N = 10$ bits and a desired resolution of $N_D = 16$ bits, an $OSR = 2^{12} = 4096$ would be required. The linearity of the converter must also meet 16 bit accuracy.

It is obvious that this approach is not feasible for high resolutions since the OSR becomes very large. With noise shaping, the quantization noise introduced by the DAC can be spectrally shaped. Hence, we can move noise power outside of the signal band by high pass filtering the noise. This reduces the oversampling ratio. At the output we already have an LP filter with cut-off frequency f_B which will attenuate this noise as well. This is the essence of the noise shaping oversampling D/A converter (OSDAC).

4.9 Oversampling D/A Converter (OSDAC)

Several different OSDAC architectures can be found in the literature [1, 2, 18-23]. The OSDAC consists of an interpolator, a sigma-delta modulator, internal Nyquist-rate DAC, and continuous-time filters as illustrated in Fig. 4-18. Sometimes, discrete-time

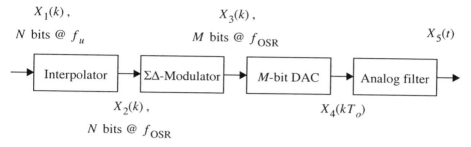

Figure 4-18 General OSDAC structure containing interpolator, sigma-delta modulator, DAC, and analog low pass filter.

filters preceeds the continuous-time filter.

The $(X_1, ..., X_5)$ signals in the OSDAC (Fig. 4-18) are illustrated by the example in Fig. 4-19 where OSR = 4:

The input signal (a) is using the whole bandwidth up to the frequency $f_N = f_u/2$. The interpolator increases the update frequency of the input signal to $f_{OSR} = OSR \cdot f_u$ and introduces a number $(OSR - 1)$ of images in the frequency domain as shown in b). To relax the requirements on the modulator and filters the interpolator also includes filtering, i.e., interpolation filters. In c) we show the result of using such filters.

The modulator truncates the N-bit input signal into an M-bit signal. This operation introduces a large amount of truncation error, but the modulator is also designed to high pass filter this truncation noise and hence the noise power can be moved out of the signal band.

The M-bit DAC (operating at the higher update frequency f_{OSR}) generates the analog output signal which becomes sinc weighted in the frequency domain due to the inherent S/H (e). With a lower number of bits, it may be easier to design a high-resolution, high-linearity DAC, since there is a lower number of analog reference elements that should be matched. In fact, by reducing the number of bits to only $M = 1$, linearity can be guaranteed [1, 23].

The continuous-time, low pass filter attenuates the out of band noise from the modulator and the images from the M-bit DAC (f).

Below, we briefly overview the different components of the OSDAC.

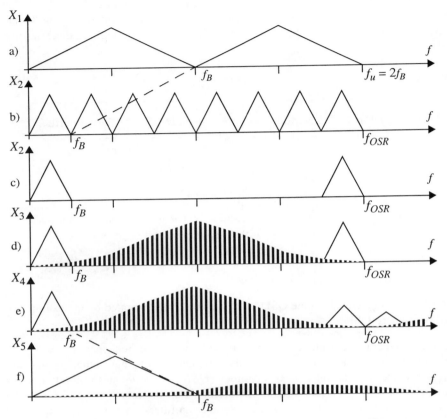

Figure 4-19 Example of the frequency power spectra for individual signals in an OSDAC with OSR=4: a) Original spectrum, b) interpolated spectrum, c) filtered interpolated signal, d) output of the modulator, e) output signal from DAC, and f) final output signal.

4.9.1 Interpolator and Interpolation Filters

The interpolator adds $OSR - 1$ zeros between each sample of its input signal as

$$y(n) = \begin{cases} x(\dfrac{n}{OSR}) & n = m \cdot \text{OSR} \\ 0 & n \neq m \cdot \text{OSR} \end{cases} \quad (4\text{-}31)$$

where x is the input signal, OSR is the oversampling ratio, and m is an integer number. $y(n)$ is updated with the oversampling frequency $f_{OSR} = OSR \cdot f_u = 1/T_o$. The z-transform of the expression in (4-31) is given by

$$Y(z) = X(z^{OSR}) \quad (4\text{-}32)$$

4.9 Oversampling D/A Converter (OSDAC)

where $z = e^{j\omega T_o}$ on the unit circle. From (4-32) we see that the operation in (4-31) also introduces images (Fig. 4-19b). We can relax the requirements on the following modulator and filters by removing or attenuating these images with interpolation filters [24]. The spectrum is shown in Fig. 4-19c). One way to implement the interpolation filters is to use a one-stage interpolation filter, as for example the K-tap FIR filter in Fig. 4-20. The input signal $x(n)$ is updated with the frequency f_u, but the delay

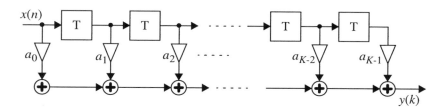

Figure 4-20 General one-stage FIR structure interpolation filter. The delay T is related to the oversampling frequency.

elements (T) are related to the oversampling frequency and hence the output signal $y(k)$ is updated with the frequency f_{OSR}.

If the transition or the signal bands of the interpolation filter have to be narrow or if the attenuation in the stop band has to be very high, the number of taps in the FIR filter is large [24, 25]. A large number of taps increase chip area and power consumption since more additions and multipliers are required. In that case, IIR filters are preferred over FIR filters. However, the design of IIR filters can be more complex and the maximum sample frequency may be lower than for the FIR filter. For IIR filters, the phase is not linear, which may not be acceptable in some applications.

Another way to implement the interpolation filters is to use multi-stage filters [18, 24, 25]. In Fig. 4-21 we show the concept of a multi-rate filter of R stages and with a total

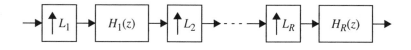

Figure 4-21 Principle description of multi-stage interpolation filtering.

interpolation or oversampling ratio of

$$OSR = L_1 \cdot L_2 \cdot \ldots \cdot L_R \tag{4-33}$$

where L_i are the interpolation factors of the individual interpolators. Using (4-32) we find that the total transfer function of the filter in Fig. 4-21 becomes

$$H_{OSR}(z) = H_1(z^{L_2 \cdot \ldots \cdot L_R}) \cdot H_2(z^{L_3 \cdot \ldots \cdot L_R}) \cdot \ldots \cdot H_R(z) \quad (4\text{-}34)$$

where $H_i(z)$ are the sub-filters of the multi-stage filter. By interpolating in stages and using frequency masking techniques the total filter $H_{OSR}(z)$ can be designed with very narrow pass and transition bands [25].

A simple filtering function is achieved by letting the modulator sample the input signal at the higher frequency f_{OSR}. This implies an interpolating function as

$$y(n) = x(\left\lceil \frac{n}{OSR} \right\rceil) \quad (4\text{-}35)$$

The corresponding weighting of the signal spectrum in the frequency domain becomes in the order of

$$\left| \mathrm{sinc}(\frac{\omega T_O}{2} \cdot OSR) / \mathrm{sinc}(\frac{\omega T_O}{2}) \right| \quad (4\text{-}36)$$

However, with this operation, the signal is slightly attenuated at higher frequencies within the signal band and the attenuation of images is rather poor for a high performance OSDAC. To increase the attenuation ot the out of band images, we can use a higher-order sinc filter.

4.9.2 Sigma-Delta Modulator

The sigma-delta ($\Sigma\Delta$) modulator truncates an N-bit input signal into an M-bit output signal. The truncation noise (error) *introduced by the modulator* is spectrally shaped with a high pass filtering function and the input signal is all pass (or low pass) filtered. The shaped noise is attenuated by the following continuous-time analog low pass filter.

Generally, the modulator contains some kind of a feedback loop, which can be of two kinds; *signal feedback*, where the output signal is fed back, and *error feedback*, where the truncation error is fed back.

Dependent on the order and architecture of the modulator (i.e., the order of the HP filter) the noise shaping function is different. It can be shown [1, 2, 18] that the SNR within the frequency range up to f_B is approximately given by

$$SNR_B = SNR + (20 \cdot K + 10) \cdot \log SNR + C \quad (4\text{-}37)$$

where K is the modulator order, SNR is determined by the resolution at the output of the modulator, and C is a number. (4-37) holds if there are no zeros in the noise shaping function. The improvement is approximately given by $K + 0.5$ bits for each doubling of the OSR [2]. Still the achievable SNR is given by the resolution of the DSP preceding the OSDAC. The gain in using noise shaping is obvious when comparing (4-29) with (4-37).

The signal feedback modulator feeds back the truncated output signal, $y(n)$, [1], as illustrated in Fig. 4-22a) for a first order modulator. $A(z)$ is a discrete-time accumul-

4.9 Oversampling D/A Converter (OSDAC)

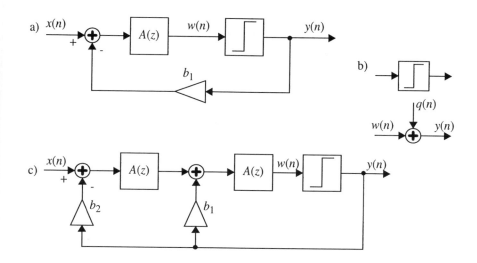

Figure 4-22 a) Principle of the 1st (a) and 2nd (c) order signal feedback sigma-delta modulator and model of the truncation error q(n) (b).

tor with the transfer function given by

$$A(z) = \frac{z^{-1}}{1-z^{-1}} \tag{4-38}$$

The output signal is subtracted from the input signal $x(n)$ and in the frequency domain, we find the transfer function

$$Y(z) = \frac{A(z)}{1+b_1 \cdot A(z)} \cdot X(z) + \frac{1}{1+b_1 \cdot A(z)} \cdot Q(z) \tag{4-39}$$

where $Q(z)$ is the truncation error as modeled in Fig. 4-22b). Using (4-38) in (4-39), we find the signal transfer function, $STF(z)$

$$STF(z) = \frac{z^{-1}}{1+(b_1-1)z^{-1}} \tag{4-40}$$

and the noise transfer function, $STF(z)$

$$STF(z) = \frac{1-z^{-1}}{1+(b_1-1)z^{-1}} \tag{4-41}$$

The signal should be low or all pass filtered and the noise high pass filtered, which states the requirements on the filters $H(z)$ and $G(z)$. For a first order modulator, the STF could be given by a delay only, hence an all pass filter for $b_1 = 1$

$$STF(z) = z^{-1} \tag{4-42}$$

and the first order NTF is described by the high pass filter

$$NTF(z) = 1 - z^{-1} \tag{4-43}$$

For the second order modulator shown in Fig. 4-22c, we can choose the NTF as

$$NTF(z) = (1 - z^{-1})^2 \tag{4-44}$$

which holds for $b_1 = 2$ and $b_2 = 1$. In Fig. 4-23 the NTFs for the first and second

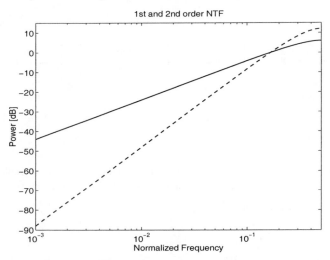

Figure 4-23 NTF of a first order (solid) and second order (dashed) sigma-delta modulator.

order modulator are shown. The second order modulator (dashed) shows a better noise attenuation at lower frequencies than the first order modulator (solid).

For high-performance applications we have to use higher-order modulators. As for oversampling A/D converters (OSADC) the modulator may not be stable for these cases. Therefore, we introduce zeros in the NTF which improves stability, but makes the noise shaping less effective. In Fig. 4-24 an example of an K-th order signal-feedback architecture. The coefficients a_i and b_i determine the NTF and STF. The filters $A(z)$ are given by discrete-time accumulators (4-38). The accumulators do not contain multipliers and they show a good immunity against truncation noise [26]. For a fourth order modulator, the STF is given by

$$STF(z) = \frac{A^4}{1 + b_1 A + (a_1 + a_2 + b_2)A^2 + (a_2 b_1 + b_3)A^3 + (a_1 a_2 + a_2 b_2 + b_4)A^4} \tag{4-45}$$

and the NTF is given by

4.9 Overampling D/A Converter (OSDAC)

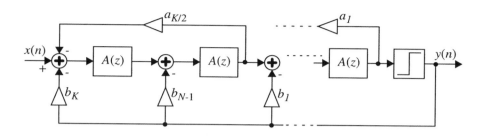

Figure 4-24 General structure of a higher order (N) modulator using signal feedback structure.

$$NTF(z) = \frac{(1 + a_1 A^2)(1 + a_2 A^2)}{1 + b_1 A + (a_1 + a_2 + b_2)A^2 + (a_2 b_1 + b_3)A^3 + (a_1 a_2 + a_2 b_2 + b_4)A^4} \quad (4\text{-}46)$$

where $A = A(z)$ is the accumulator given by (4-38). The NTF zeros are

$$z_{1,2} = 1 \pm j\sqrt{a_1} \text{ and } z_{3,4} = 1 \pm j\sqrt{a_2} \quad (4\text{-}47)$$

The a_i:s are used to move the zeros close to the pass band edge in order to further attenuate the noise in the NTF stop band. The output spectrum of a 5th order, 14-bit input, one-bit output sigma-delta modulator is shown in Fig. 4-25. The input signal is

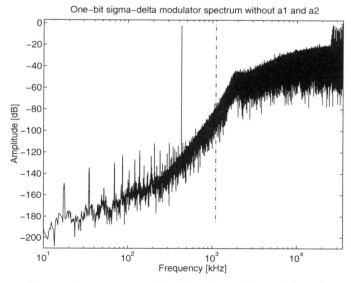

Figure 4-25 Output spectrum of the 5th order modulator without the a_i parameters.

a -6 dBFS sinusoid at the frequency 431.25 kHz, the update frequency is 70.656 MHz and the oversampling ratio $OSR = 32$. The passband edge is indicated with the dashed line. The a_i:s in the NTF are all set to zero. The SNDR within the signal band is approximately 63 dB, corresponding to 10.2 bits. With the zeros, i.e., non-zero a_i:s, included, the $SNDR$ is improved. The zeros attenuate the noise at the signal band edge (dashed line) and a sharper noise shaping transition band is achieved. At lower frequencies the attenuation of the noise is not as strong. The simulation result when applying an –6 dBFS input sinusoid at 431.25 kHz is shown in Fig. 4-26. The $SNDR$

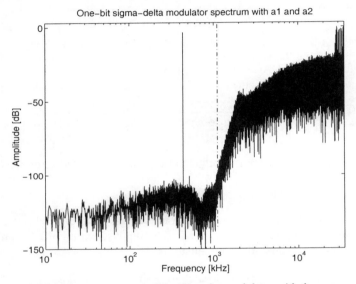

Figure 4-26   Output spectrum of the 5th order modulator with the a_i parameters.

is found to be approximately 79 dB, corresponding to 12.8 bits.

The poles are given by rather complex expressions and the complexity of the filter increases with increased modulator order. Since the output signal is truncated, the stability of the modulator has to be carefully investigated.

Dependent on the number of bits representing the output signal of the modulator, they can be divided into two groups; multi-bit and single-bit modulators. For the single-bit modulator the loop gain is high and therefore the stability of the modulator has to be carefully examined [2, 18]. For a single-bit output, the output signal is equal to the MSB of the quantizer's input.

4.9.3   M-bit DAC

With the M-bit DAC the output signal from the modulator is converted into an analog signal. The M-bit DAC has to operate at the oversampling frequency and is therefore a Nyquist-rate converter with oversampling. Naturally, the DAC has to have the same linearity accuracy as the final output. The design of a DAC with a lower number bits becomes generally less complex, since the number of circuit components are lower,

4.9 Oversampling D/A Converter (OSDAC)

interconnection wires are shorter, etc.

To guarantee a very high linearity, a one-bit DAC can be used. The one-bit DAC has a very high linearity, since the transfer function between its minimum ("0"-input) and maximum ("1"-input) values can be expressed with a completely straight line even though there are matching errors [1]. These matching errors only give rise to gain and offset errors. However, other effects such as linearity of switches, voltage drops, and parasitics, limit the linearity.

Using a one-bit DAC implies that the continuous-time low pass filter has to have a high order due to the high noise power at high frequencies. A way to relax these requirements is to use a so called semi-digital FIR filter [23]. The semi-digital filter is very similar to a regular FIR filter and can be constructed by using a number of cascaded (in time) one-bit DACs having different reference levels as illustrated in Fig. 4-27. The reference levels, i.e., the constants a, are generated with analog components.

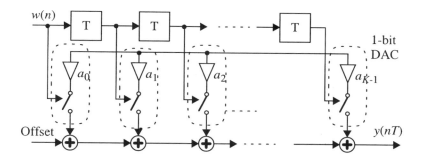

Figure 4-27 Cascaded one-bit DACs forming a K-tap FIR filter structure.

The inaccuracy in these coefficients affect the filter transfer function and does not introduce any other errors. By designing the FIR filter as a low pass filter attenuating the out-of-band modulator noise, the continuous-time filter can be implemented with a lower order.

The output signal of the FIR filter (and the M-bit DAC in general) is piece-wise linear and pulse shaped due to S/H functions, and it is given by

$$y(nT) = a_0 \cdot w(nT) + a_1 \cdot w(nT - T) + a_{K-1} \cdot w(nT - (K-1)T) \quad (4\text{-}48)$$

where K is the number of filter taps and $T = 1/f_{OSR}$ is the update cycle time. The number of coefficients is inversely proportional to the width of the transition band [25]

$$K \sim \frac{1}{\Delta \omega T} = \frac{1}{\omega_a T - \omega_c T} \quad (4\text{-}49)$$

where $\omega_c T$ is the normalized pass band frequency and $\omega_a T$ is the normalized stop band frequency. Hence, for a narrow transition band filter, which must be used for a low oversampling ratio, a long FIR filter is required.

4.9.4 Continuous-Time Filter

Due to the S/H elements in the M-bit DAC, the output contains frequency-domain images centered at multiples of the oversampling frequency. The analog continuous-time low pass filter attenuates these images and the modulator noise at higher frequencies are attenuated. In addition, especially if a one-bit DAC is used, an additional discrete-time filter e.g., switched-capacitor filter, is needed before the continuous-time filter to further attenuate the shaped noise.

4.10 SPECIAL IMPROVEMENT TECHNIQUES

Some of the architectures discussed in the previous sections prove to be very suitable for high-resolution and high-speed. However, to reach very high performance some additional techniques can or must be used. These techniques can be of several different kinds, as for example calibration, trimming of internal reference values, randomization, etc. [2, 10, 18, 27 - 31].

We find three different approaches to improve performance; first we can with for example laser trimming improve the matching between circuit components during the fabrication phase, secondly we can measure the DAC output with an A/D and use a feedback loop for error correction during operation, and third, we can use techniques that during operation improve the performance by smoothening distortion terms into noise without any prior knowledge about the errors.

Consider the second case; if we know the input signal and the non-linear transfer function we can in some sense predict the distortion terms in the output. In the reality, this will be very hard, since the non-linear transfer function is most likely not known. If we are able to observe and analyze the DAC output we may be able to feed back enough information to reduce an amount of the non-linearity (or errors). A compensating DAC could be used in parallel as illustrated in Fig. 4-28 to add an inverse func-

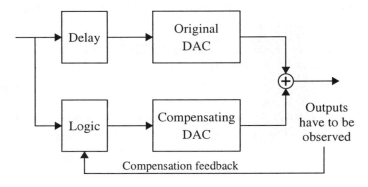

Figure 4-28 Use of an additional DAC to compensate and calibrate errors in the transfer function of the original DAC.

tion to the original DAC output. Another approach is to use an inverse non-linear function in the digital domain, hence changing the digital input data to the DAC. This is often referred to as intentional pre-distortion [29]. A third approach is instead of compensating the output signal to change the internal parameters (references) of the DAC using the fed back information. For example, the individual sizes of the weights in the DAC could be modified in order to minimize the distortion instead of compensating the output signal [32].

The drawback with these methods is obvious. An observer that is performing an analog-to-digital conversion is needed as well as an automatic compensating routine that operates at the same speed as the original converter. It should also be noted that a compensating DAC must not only compensate for static errors, since the dynamic performance determines the quality of the original DAC at higher frequencies.

Therefore, we emphasize methods to reduce non-linearity by not knowing the actual transfer function. Especially we refer to randomization techniques to smoothen distortion into noise as discussed in Sec. 4.10.1. These methods are advantegeous in applications where linearity requirements are more important than the noise requirements. The use of internal reference sources to calibrate DAC weights is also similar to using a calibration routine for internal DAC references which is further described in Sec. 13.3.4.

4.10.1 Dynamic Randomization and Element Matching

In a segmented converter the specifications on performance may still not be met although an optimal number of bits has been chosen for segmentation. This is due to the influence of matching errors between reference levels in the DAC. In a conventional thermometer coded DAC, (see Fig. 4-8), the input number, $k = 1$, is represented by using one fixed reference element in the DAC, e.g., current source or capacitor, $k = 2$ is represented by using two fixed references, etc. Using this approach, the matching errors become strongly signal-dependent, since a reference is associated with specific input codes, and we will have distortion at the output.

To reduce the distortion, the binary to thermometer encoder can be designed so that, at different times, different references are chosen to represent $k = 1$, i.e., we do not fix a reference to a certain code. If we choose the references in a way that is uncorrelated with the signal, the matching errors will no longer be signal dependent and hence the error will become noise. This is achieved by using a randomizer or scrambler as illustrated in Fig. 4-29, In a real implementation the randomization can be implemented with a pseudo-random binary sequence (PRBS) generator. We can also choose to assign the references in a cyclic way, which does not give a completely uncorrelated error signal, but the improvement can be significant.

The advantage can be realized by considering the output current for a thermometer coded current-steering DAC. Assume that there are M equally large current sources where all sources, I_i, $i = 1, ..., M$, have the same nominal size, I_{src}. For each current source a relative mismatch error, δ_i, is associated, hence

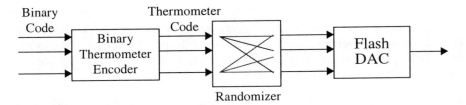

Figure 4-29 Randomization of bits in a thermometer coded DAC. The matching error becomes uncorrelated with the signal.

$$I_i = I_{src} \cdot (1 + \delta_i) \qquad (4\text{-}50)$$

The errors are fixed for the specific chip and without randomization, the output current for code $k = 00011\ldots11$ (k number of 1's) is

$$I_{out}(k) = \sum_{i=1}^{k} I_i = \sum_{i=1}^{k} I_{src} \cdot (1 + \delta_i) = I_{src} \cdot k \cdot \left(1 + \frac{1}{k}\sum_{i=1}^{k}\delta_i\right) \qquad (4\text{-}51)$$

From (4-51) we find that with each code k we can associate a specific relative current error of magnitude

$$\delta(k) = \frac{1}{k}\sum_{i=1}^{k}\delta_i \qquad (4\text{-}52)$$

which introduces distortion.

Assume that for $I_{out}(k)$ we could use any k current sources out of the set of M sources with equal probability and independent of k. This will in mean give a relative current error which is equal for all input codes

$$\overline{\delta(k)} = \frac{1}{M}\sum_{i=1}^{M}\delta_i \qquad (4\text{-}53)$$

An equally large relative error for all k will only give rise to a gain error at the output. The total error power (for a full-scale signal) is approximately kept constant inspite of the randomization process since the error power still is dependent on the size of the matching errors, δ_k. In the frequency domain the distortion terms are reduced to the cost of a higher noise floor. To illustrate dynamic randomization, we show in Fig. 4-30. a Matlab simulation result where a full-scale sinusoid has been applied to an 8-bit thermometer coded DAC. To all current sources in the DAC a Gaussian distributed random error with standard deviation equal to 12% is applied. In Fig. 4-30a) we find clear distortion terms and in b) the distortion has been reduced and the noise floor has increased. The SNDR is in both cases approximately 41 dB.

When shifting the assignment of the sources in a cyclic way, the errors may still be

4.10 Special Improvement Techniques

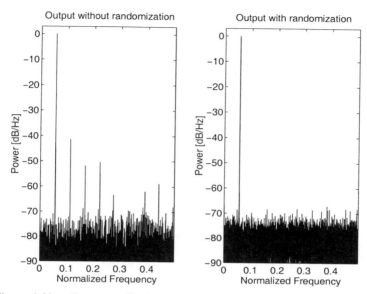

Figure 4-30 Output amplitude spectrum from a thermometer coded 8-bit DAC a) without randomization and b) with randomization.

slightly signal-dependent, but the linearity errors are reduced. It is hard to predict and find how much the result is improved. The cyclic shift has the advantage of not needing any PRBS generator and the circuit architecture can be implemented with less complexity. As previously, the distortion is reduced to the cost of a higher noise floor.

Dynamic element matching (DEM) techniques can be of several kinds [1, 2, 30, 31, 33] and in this book we understand a DEM technique where the principle is to divide the converter into a number of subconverters with a lower resolution, as proposed in [30, 31].

Basically, the method is similar to the randomization technique. In Fig. 4-31 we sketch a concept found in [30, 31]. We have an N-bit binary input and a binary switching tree of N layers. In the k-th layer we have 2^{N-k} switching blocks, $S_{k,j}$. Each of those switching block has one $(k+1)$-bit input and two k-bit outputs. For the single switching block in the N-th layer, $S_{N,1}$ the number of input bits is increased to $N+1$ by adding a '0' as new LSB.

The objective of each switching blocks, is to direct the input signal through the tree. Dependent on implementation, a control bit for each switching block, $c_{i,j}$, determines how the input should be directed. Actually, the switching tree performs the binary to thermometer encoding. The thermometer code is fed to the $M = 2^N$ one-bit DACs terminating the tree, i.e., an N-bit thermometer coded DAC as illustrated with dashed line in Fig. 4-33.

If all control bits $c_{i,j}$ are fixed and if there are matching errors in the DACs, there will

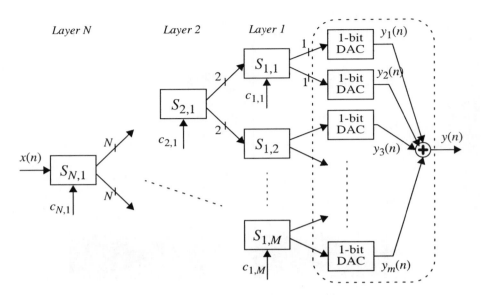

Figure 4-31 An N-bit DAC using dynamic element matching and a bank of one-bit DACs.

be distortion. If we let $c_{i,j}$ be given by a PRBS generator, the signal will pass differently through the tree and distortion is smoothened into signal-independent error as for the dynamic randomization. We can however add the possibility to spectrally shape the circuit errors to higher frequencies, outside the signal band, by letting the control bits $c_{i,j}$ be given by a $\Sigma\Delta$ modulator with a noise shaping function. This action prevents us from using the entire Nyquist range, and therefore we can, at the input of the DAC, add a lower-order $\Sigma\Delta$ modulator which also shapes the input quantization noise to higher frequencies, further increasing the dynamic range within the signal band. The out-of-band noise is attenuated by a continuous-time filter at the output of the converter.

The one-bit DACs are linear and matching errors will only give rise to linear gain and offset errors. (If we do not consider second order effects, such as signal-dependent reference element values). This property makes it possible to shape the errors spectrally and move the error power outside the signal band.

If the number of input bits is high, the tree width and routing will become very large. For a 14-bit DAC we will have 16384 one-bit DACs at the output. Therefore, to reduce the design complexity, the tree can be shortened and terminated at the L-th layer and terminated by 2^{N-L+1} L-bit DACs. Multi-bit DACs will however introduce a non-linear gain due to the matching errors, and the shaping of the error does not become as powerful. The methods show an improved performance to a modest increase of hardware [34] over the regular dynamic randomization technique.

The dynamic matching techniques increase the switching noise in the DAC. The design has to be done carefully so that the gain in dynamic range from less distortion

4.11 DAC Comparison

not is lower than the loss of dynamic range from the increase of the glitch energy. Due to the complexity (chip area, delay) of the randomizer, the technique is mostly preferred when the number of bits is small.

4.11 DAC COMPARISON

In this section we conclude the properties of the DACs discussed in this chapter. We show in tables and figures the properties of different DAC types and architectures as well as reported performance in the literature.

In Fig. 4-32 and Fig. 4-33 and we show a summary of the measured and reported performance from international transactions, journals, text books, etc., as well as information from commercial circuit providers. There are different means of performance and resolution and in most publications we find reported performance on DNL, INL, and settling time. These are insufficient for communications and instead we use measures such as SFDR and SNDR to characterize the converters. In Fig. 4-32 we use the SFDR or SNDR on the y-axis and the *sample frequency* on the x-axis. For the same converters, we use in Fig. 4-33 the SFDR and SNDR on the y-axis and the *signal frequency* on the x-axis.

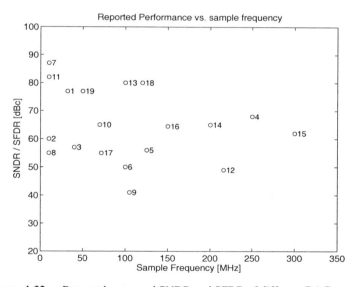

Figure 4-32   Reported measured SNDR and SFDR of different DACs as a function of the sample frequency.

In Table 4-2, we show a list of the DACs used for the summary together with a brief description of the DAC architecture, circuit technique, etc. In the literature we find mostly current-steering converters as suitable candidates for high speed and high resolution. Concluding the summaries in the figures and the table we find a rather clear trade-off between signal frequency and linearity as illustrated with the dashed line in Fig. 4-33.

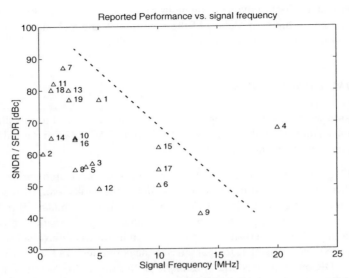

Figure 4-33   Reported measured SNDR and SFDR of different DACs as a function of the signal frequency.

Table 4-2   List of reported DAC performance.

No	Ref	Implementation	Res. [bits]	Supply [V]	Process	Fsignal [MHz]	Fsample [MS/s]	SFDR [dBc]
1	[3]	Current-steering. 4x Interpolation.	14	5	CMOS	5.01	32	77
2	[6]	Current-steering. Segmented.	10	3.3	CMOS	0.3	10	60
3	[7]	Current-steering. Binary weighted.	10	5	CMOS	4.43	40	57
4	[13]	Current-steering. Segmented.	10	3.3	CMOS	20	250	68
5	[27]	Hybrid current-steering. Thermometer coded.	10	5	CMOS	3.9	125	56
6	[35]	Current-steering. Thermometer coded.	10	5	BiCMOS	10	100	50
7	[36]	Hybrid Current-steering and R-2R ladder. Thermometer coded MSBs	14	5/-5.2	BiCMOS	2.03	10	87
8	[44]	Current-steering. Segmented.	10	1.5	CMOS	3	10	55

Table 4-2   List of reported DAC performance.

No	Ref	Implementation	Res. [bits]	Supply [V]	Process	Fsignal [MHz]	Fsample [MS/s]	SFDR [dBc]
9	[50]	Current-steering. Segmented.	8	5	CMOS	13.5	105	41
10	[51]	Current-steering. Segmented.	10	3.3	CMOS	3	70	65
11	[52]	Hybrid Current-steering and R-2R ladder. Thermometer coded MSBs	16	5	BiCMOS	1.23	10	82
12	[53]	Oversampling. Noise shaping.	8	3.3	CMOS	5	216	49 SNDR
13	[54]	Current-steering. Segmented.	14	5	CMOS	2.48	100	80
14	[55]	Current-steering. Segmented.	12	2.7	CMOS	1	200	65
15	[56]	Current-steering. Multi-segmented.	12	3.3	CMOS	10	300	62
16	[57]	Current-steering. Segmented.	14	2.7	CMOS	3	150	64.5
17	[58]	Hybrid. Segmented R-2R ladder.	12	-5.2	Bipolar	10	72	55
18	[59]	Oversampling. Noise shaping.	14	2.5	CMOS	1	120	80 SNDR
19	[60]	Current-Steering. Segmented. 5 V supply. 14-bit.	14	5	CMOS	2.51	50	77

4.12 SUMMARY

In this chapter we have presented some of the basic DAC architectures that are suitable for high-speed and high-resolution applications. We have also outlined some possible techniques for implementation, such as the charge-redistribution, current-steering, etc. As a guidance, we have summarized a number of reported performances of DACs that are suitable for higher-speed applications in an overviewing figure and table.

REFERENCES

[1]   D.A. Johns and K. Martin, *Analog Integrated Circuit Design*, John Wiley & Sons, New York, NY, USA, 1997, ISBN 0-471-14448-7

[2]   R.J. van de Plassche, *Integrated Analog-to-Digital and Digital-to-Analog Converters*, Kluwer Academic Publishers, Boston, MA, USA, 1994, ISBN 0-7923-9436-4

[3]   Analog Devices, Data Sheet, AD9774, Interpolating DAC, 1999.
[4]   L. Maliniak, "CMOS Data Converters Usher In High Performance At Low Cost," *Electronic Design*, pp. 69-70, 72, 74, 76, Feb., 1998.
[5]   A. Bindra. "Wideband DAC Fosters Multicarrier, Multimode Transmission," *Electronic Design*, pp. 37-8, 40, June 1999.
[6]   N. Tan, E. Cijvat, and H. Tenhunen, "Design and implementation of high-performance CMOS D/A converter," in *Proc. of the IEEE 1997 International Symposium on Circuits and Systems, ISCAS'97*, Hong Kong, June, 1997.
[7]   C.A.A. Bastiaansen, D. Wouter, J. Groenewald, H.J. Schouwenaars, and H.A.H. Termeer, "A 10-b 40-MHz 0.8-mm CMOS Current-Output D/A Converter," *IEEE Journal of Solid-State Circuits*, vol. 26, no. 7, pp 917-21, July 1991
[8]   J. Bastos, M. Steyaert, A. Pergoot, and W. Sansen, "Mismatch characterization of submicron MOS transistors," *Analog Integrated Circuits and Signal Processing*, vol. 12, no. 2, pp. 95-106, Feb. 1997
[9]   M.J.M Pelgrom, A.C.J. Duinmaijer, and A.P.G. Welbers, "Matching Properties of MOS Transistors," *IEEE Journal of Solid-State Circuits*, vol. 24, no. 5, pp. 1433-9, Oct. 1989
[10] A. Biman and D.G. Nairn, "Trimming of current mode DACs by adjusting V_t," in *Proc. of the IEEE 1996 International symposium on Circuits and Systems, ISCAS'96*, Atlanta, GA, U.S.A., June, 1996.
[11] P. E. de Haan, S. van den Elshout, E. A. M. Klumperink, and K. Bult, "Analysis of a current mode MOST-only D-A converter,", in *Proc. of the 1994 European Solid-State Circuits Conference, ESSCIRC'94*, Gif-sur-Yvette, France, 1994. p. 188-91
[12] N. Tan, "Design and Implementation of a High-Speed Bipolar DAC," in *Proc. of the IEEE 1997 International Solid-State Circuits Conference, ISSCC'97*, San Fransico, CA, U.S.A., June, 1997
[13] C.H. Lin and K. Bult, "A 10-b, 500-MSample/s CMOS DAC in 0.6 mm^2," *IEEE Journal of Solid-State Circuits*, vol. 33, No. 12, Dec. 1998, p.1948-58.
[14] M.J.M. Pelgrom and M. Roorda, "An algorithmic 15-bit CMOS digital-to-analog converter," *IEEE Journal of Solid-State Circuits*, vol. 23, No. 6; Dec. 1988, p.1402-5
[15] P. Riffaud-Desgreys, E. Garnier, Ph. Roux, and Ph. Marchegay, "New structure of algorithmic DAC in switched current technique," in *Proc. of 3rd international conference on Advanced A/D and D/A Conversion Techniques and their Applications*, Glasgow, Scotland, 26-28 July, 1999, pp. 111-4,
[16] H. Träff, T. Holmberg, and S. Eriksson, "Application of Switched Current Technique to Algorithmic DA- and AD-Converters," In *Proc. of IEEE International Symposium on Circuits and Systems, ISCAS'91*, June 1991, pp. 1549-52
[17] R. Unbehauen and A. Cichocki, *MOS switched-capacitor and continuous-time integrated circuits and systems*, Springer-Verlag 1989.
[18] J.C. Candy and G. C. Temes, *Oversampling Delta-Sigma Data Converters*, IEEE Press, New York, NY, USA, 1992, ISBN 0-87942-281-5
[19] J.W. Kim, B.M. Min, J.S. Yoo, and S.W. Kim, "An Area-Efficient Sigma-Delta DAC with a Current-Mode Semidigital IFIR Reconstruction Filter," in *Proc. of the IEEE 1998 International Symposium on Circuits and Systems, ISCAS'98*, Monterey, CA, USA, May 31 - June 3, 1998
[20] H. Lin, J. Barreiro da Silva, B. Zhang, and R. Schreier, "Multi-Bit DAC with Noise-Shaped Element Mismatch," in *Proc. of the IEEE 1996 International Symposium on Circuits and Systems, ISCAS'96*, vol. 1, pp. 235-8, Atlanta, GA, USA, May 12-15, 1996
[21] T. Ritoniemi, V. Eerola, T. Karema, and H. Tenhunen, "Oversampled A/D and D/A Converters for VLSI System Integration," in *Proc. of the 3rd Annual IEEE ASIC Seminar and Exhibit*, pp. P8/7.1-12, New York, NY, USA, 1990
[22] T. Ritoniemi, T. Karema, and H. Tenhunen, "Design of Stable High Order 1-bit Sigma-Delta Modulators," in *Proc. of the IEEE 1990 International Symp. on Circuits and Systems, ISCAS'90*, Vol. 4, pp. 3267-70, New Orleans, LA, USA, May 1-3, 1990

4.12 Summary

[23] D. K. Su and B. A. Wooley, "A CMOS oversampling D/A Converter with a Current-Mode Semi-Digital Reconstruction Filter," in *Proc. of the IEEE 1993 International Solid-State Circuits Conference, ISSCC'93*, San Fransisco, CA, USA, 1993, pp. 230-1

[24] N. J. Fliege, *Multirate Digital Signal Processing: Multirate Systems, Filter Banks, Wavelets*, John Wiley & Sons, Chichester, UK, 1994, ISBN 0-471-93976-5

[25] H. Johansson, *Synthesis and Realization of High-Speed Recursive Digital Filters*, Linköping Studies in Science and Technology, Dissertation no. 534, 1998, ISBN 91-7219-207-0

[26] S. Chu and C.S. Burrus, "Multirate Filter Design Using Comb Filters," *Transactions on Circuits and Systems I*, vol. 31, No. 11, Nov. 198, pp. 913-24.

[27] S.-Y. Chin and C.-Y. Wu, "A 10-b 125-MHz CMOS Digital-to-Analog Converter (DAC) with Threshold-Voltage Compensated Current Sources," *IEEE Journal of Solid-state Circuits*, vol. 29, no. 11, pp. 1374-80, Nov. 1994

[28] S. Boiocchi, S. Brigati, G. Caiulo, and F. Maloberti, "Self-Calibration in High Speed Current Steering CMOS D/A Converters," in *Proc. of the 2nd International Conferenece on 'Advanced A-D and D-A Conversion Techniques and their Applications'*, London, UK, 1994, pp. 148-52

[29] A. Netterstrom, "Using Digital Pre-Distortion to Compensate for Analog Signal Processing Errors," in *Proc. of the IEEE 1990 International Radar Conference*, pp. 243-8, Arlington, VA, USA, May 7-10, 1990

[30] I. Galton, "Spectral Shaping of Circuit Errors in Digital-to-Analog Converters," *IEEE Trans. on Circuits and Systems II: Analog and Digital Signal Processing*, vol. 44, no. 10, pp. 808-17, Oct. 1997

[31] H.T. Jensen and I. Galton, "A Low-Complexity Dynamic Element Matching DAC for Direct Digital Synthesis," *Transactions on Circuits and Systems II: Analog and Digital Signal Processing*, vol. 45, no. 1, pp. 13-27, Jan. 1998

[32] G. Stehr, F. Szidarovsky, and O.A. Paulusinski, and D. Andersson, "Performance Optimization of binary weighted current-steering D/A converters," *Applied Mathematics and Computation*, to be published, 1999.

[33] R.K. Henderson and O.J.A.P. Nys, "Dynamic element matching techniques with arbitrary noise shaping function," in *Proc. of the IEEE 1996 International Symposium on Circuits and Systems. Circuits and Systems, ISCAS'96*, Atlanta, GA, U.S.A. 1996, Vol. 1, p.293-6

[34] N.U. Andersson and J.J. Wikner, "A comparison of DEM techniques in current-steering DACs," in *Proc. of the 17th Norchip Conf., NORCHIP'99*, Oslo, Norway, Nov. 7-8, 1999

[35] I.H.H. Jørgensen and S.A. Tunheim, "A 10-bit 100MSamples/s BiCMOS D/A Converter," *Analog Integrated Circuits and Signal Processing*, vol. 12, pp. 15-28, 1997

[36] B.J. Tesch and J.C. Garcia, "A Low Glitch 14-b 100MHz D/A Converter," *Journal of Solid-State Circuits*, vol. 32, no. 9, pp. 1465-9, Sept. 1997

[37] C. Toumazou, J.B. Hughes. and N.C. Battersby, *Switched-Currents : An Analogue Technique for Digital Technology*, IEE, Peter Peregrinus Ltd. 1993.

[38] B.G. Henriques and J.E. Franca, "High-Speed D/A Conversion with Linear Phase sinx/x Compensation," in *Proc. of the IEEE 1993 International Symp. on Circuits and Systems, ISCAS'93*, vol. 2, pp. 1204-7, Chicago, IL, USA, May 3-6, 1993

[39] B.G. Henriques, K. Kananen, J.E. Franca, and J. Rapeli, "A 10 Bit Low-Power CMOS D/A Converter with On-Chip Gain Error Compensation," in *Proc. of the IEEE 1995 Custom Integrated Circuits Conference, CICC'95*, pp. 215-8, Santa Clara, CA, USA, May 1995

[40] B. Leung, "BiCMOS Current Cell and Switch for Digital-to-Analog Converters", *IEEE Journal of Solid-State Circuits*, vol. 28, no. 1, pp. 68-71, Jan. 1993

[41] K. Maio, S. I. Hayashi, M. Hotta, T. Watanabe, S. Ueda, and N. Yokozama, "A 500-MHz 8-bit D/A Converter," *IEEE Journal Solid-State Circuits*, vol. 20, no. 6, pp. 1133-7, Dec. 1985

[42] Y. Nakamura, T. Miki, A. Maeda, H. Kondoh, and N. Yazawa, "A 10-b 70-MS/s CMOS D/A Converter," *IEEE Journal of Solid-State Circuits*, vol. 26, no. 4, pp. 637-42, Apr. 1991

[43] R.J. Romanczyk and B.H. Leung, "BiCMOS Circuits for High Speed Current Mode D/A Converters," *IEEE Journal of Solid-State Circuits*, vol. 30, no. 8, pp. 923-34, Aug. 1995

[44] N. Tan, "A 1.5-V 3-mW 10-bit 50 MS/s CMOS DAC with Low Distortion and Low Intermodulation in Standard Digital CMOS Process," in *Proc. of the IEEE 1997 Custom Integrated Circuits Conference, CICC'97*, pp. 599 - 602, Santa Clara, CA, USA, May 1997

[45] N. Tan and J.J. Wikner, "A CMOS Digital-to-Analog Converter Chipset for Telecommunication," *IEEE Magazine on Circuits and Devices*, vol. 13, no. 5, pp. 11-16, Sept. 1998

[46] X. Xu, G. C. Temes, and R. Schreier, "The Implementation of Dual-Truncation Sigma Delta D/A Converters," in *Proc. of the IEEE 1992 International Symposium on Circuits and Systems, ISCAS'92*, vol. 2, pp. 597-600, San Diego, CA, USA, May 1992

[47] P. Ju, K. Suyama, P. Ferguson Jr., and W. Lee, "A Highly Linear Switched-Capacitor DAC for Multi-Bit Sigma-Delta D/A Applications," in *Proc. of the 1995 IEEE International Symposium on Circuits and Systems, ISCAS'95*, vol. 1, pp. 9-12, Seattle, WA, USA, April 29 - May 3, 1995

[48] H. Lin, J. Barreiro da Silva, B. Zhang, and R. Schreier, "Multi-Bit DAC with Noise-Shaped Element Mismatch," in *Proc. of the IEEE 1996 International Symposium on Circuits and Systems, ISCAS'96*, vol. 1, pp. 235-8, Atlanta, GA, USA, May 12-15, 1996

[49] R. W. Stewart and E. Pfann, "Oversampling and Sigma-Delta Strategies for Data Conversion," *Electronics & Communication Engineering Journal*, vol. 10, no. 1, pp. 37-47, Feb. 1998

[50] J.M. Fournier and P. Senn, "A 130-MHz 8-b CMOS Video DAC for HDTV Applications," *Journal of Solid-State Circuits*, vol. 26, no. 7, pp 1073-1076, July 1991

[51] A. Fraval and F. Dell'ova, "A 10-bit 70MHz 3.3V CMOS 0.5um D/A Converter for Video Applications," *Transactions on Consumer Electronics*, vol. 41, no. 3, Aug. 1995, pp. 875-9

[52] D.A. Mercer, "Two Approaches To Increasing Spurious Free Dynamic Range in High Speed DACs," in *Proc. of the IEEE 1993 Bipolar/BiCMOS Circuits and Technology Meeting*, 1993, pp. 80-83.

[53] D. Birru and E. Roza, "Video-Rate D/A Converter Using Reduced Rate Sigma-Delta Modulation," in *Proc. of the IEEE 1998 Custom Integrated Circuits Conference, CICC'98*, 1998, pp. 241-4

[54] B.J. Tesch, P.M. Pratt, K. Bacrania, and M. Sanchez, "A 14-b 125 MSPs Digital-to-Analog Converter and Bandgap Voltage Reference in 0.5um CMOS," in *Proc. of the IEEE 1999 International Symposium on Circuits and Systems, ISCAS'99*, Orlando, FL, U.S.A., June 1998, pp. 452-5

[55] A. Van den Bosch, M. Borremans, J. Vandenbussche, G. Van der Plas, A. Marques, J. Bastos, M. Steyaert, G. Gielen, and W. Sansen, "A 12 bit 200 MHz Low Glitch CMOS D/A Converter," in *Proc. of the IEEE 1998 Custom Integrated Circuits Conference, CICC'98*, San Fransisco, CA, 1998, pp. 249-52

[56] A. Marques, J. Bastos, A. Van den Bosch, J. Vandenbussche, M. Steyaert, and W. Sansen, "A 12b Accuracy 300MSample/s Update Rate CMOS DAC," in *Proc. of the IEEE 1998 International Solid-State Circuits Conference, ISSCC'98*, San Fransisco, CA, 1998, pp. 216-7, 440

[57] J. Vandenbussche, G. Van der Plas, A. Van den Bosch, W. Daerns, G. Gielen, M. Steyaert, and W. Sansen, "A 14b 150MSample/s Update Rate Q^2 Random Walk CMOS DAC," in *Proc. of the IEEE 1999 International Solid-State Circuits Conference, ISSCC'99*, San Fransisco, CA, 1999.

[58] C.G. Martinez and S. Simpkins, "A Monolithic 12-Bit Multiplying DAC for NTSC and HDTV Applications, " in *Proc. of the IEEE 1989 Bipolar Circuits and Technology Meeting*, 1989, pp. 52-5.

[59] K. Falaksahi, C.-K. Ken Yang, and B.A. Wooley, "A 14-bit, 5-MHz Digital-to-Analog Converter Using Multi-bit Sigma-Delta Modulation," in *Proc. of the IEEE 1998 Symposium on VLSI Circuits*, 1998, pp. 164-5.

[60] Analog Devices, Data Sheet, AD9764, TxDAC series, 1999.

5 OVERVIEW OF CIRCUIT TECHNIQUES

5.1 INTRODUCTION

Both ADCs and DACs are sampled data systems, therefore they require the use of circuit techniques that can handle discrete-time analog signals. Depending on whether the analog signals are voltages or currents the circuits are often divided into voltage-mode or current-mode circuits. In CMOS the switched-capacitor (SC) technique is usually used for voltage-mode circuits while the switched-current (SI) technique is used for current-mode circuits.

A S/H circuit is needed in all sampled data systems. In Sec. 5.2 a general overview of errors in S/H circuits is given. Limitations of the SI technique is discussed in Sec. 5.3 while the SC technique is treated in Sec. 5.4.

5.2 SAMPLE-AND-HOLD CIRCUITS

The S/H circuit is a crucial part of most discrete-time systems. The S/H circuit samples the input signal when a hold command is given and stores the signal value for a certain amount of time. There are a number of effects that will degrade the performance of an S/H circuit. Referring to Fig. 5-1, the aperture time is the delay time from when the hold command is given until the sampling switch is actually turned off. Variations in the aperture time will cause distortion and limit the maximum sampling frequency. The sample-to-hold settling time is the time it takes for the signal to settle within a certain accuracy. The pedestal error is an offset generated by clock feedthrough from the sampling switch. This error is usually signal dependent and causes distortion. The droop rate specifies how fast the held value will decrease due to leakage currents and therefore imposes a lower limit on the sample frequency. Another source of error in the S/H circuit which is not illustrated in the figure is the feedthrough error from the input signal caused by capacitive coupling from the input to the sampling capacitor.

The acquisition time is the time from the end of the hold phase until the output signal of the S/H tracks the input signal within a certain accuracy. For high-speed operation the acquisition time must be small and the bandwidth of the S/H circuit must be large.

The errors described above limits the performance of any S/H circuit except the aperture error which is mainly a problem for S/H circuits with continuous-time input signals. A S/H circuits with a continuous-time inputs signal is sometimes referred to as a

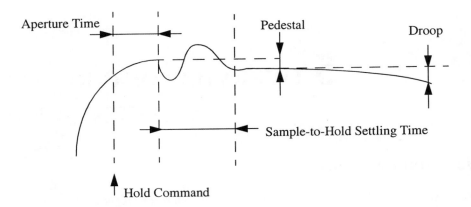

Figure 5-1   Errors in the S/H circuit.

track-and-hold (T/H) circuit and is slightly more difficult to design due to the aperture error.

5.3   SWITCHED-CURRENT TECHNIQUES

The basic SI-memory cell [1] is simple and requires no high-gain amplifiers and no linear capacitors. Therefore this technique is suitable for implementation in a digital CMOS process with a low supply voltage. The basic operation of SI circuits is illustrated in Fig. 5-2.

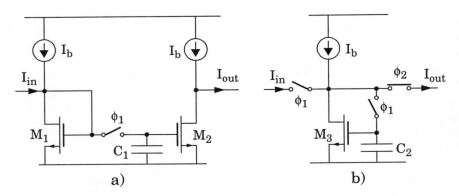

Figure 5-2   The SI-memory cell a) first generation b) second generation.

There are in principle two types of circuits referred to as the first (FG) and second generation (SG) memory cell. In the FG memory cell the input current is converted to a voltage by the diode connected transistor M_1 which is sampled-and-held on the capacitor C_1. This capacitor can be only the gate-source capacitance of M_2 or an additional physical capacitor. The voltage is then converted to a current by transistor M_2 and the output current is a sampled-and-held copy of the input current. The principle of the SG cell is the same as for the FG cell but the same transistor M_3 is used both for the current to voltage conversion and for generating the output current. It is therefore necessary to use two additional switches to control the direction of the current in M_3. Hence the FG cell can provide an output current on both clock phases while the SG cell can only provide an output current on one clock phase. It is clear the both the FG and SG cell can perform a S/H or a T/H function.

In the following we assume that there are no additional capacitors to store the gate-source voltage, since this give the highest speed of the memory cell. It has also been assumed that all the bias currents are equal, i.e. $I_{b1} = I_{b2} = I_{b3}$, and that all transistors have the same size.

5.3.1 Mismatch

Since the SG cell uses only one transistor the only source of mismatch is variations in the drain-source voltage. The two transistors in the FG cell must be matched. A first order approximation of the relative error in output current caused by mismatch is given [2] by

$$\frac{\Delta I}{I} = \frac{\Delta \beta}{\beta} + \frac{2\Delta V_{GS}}{V_{GS} - V_T} - \frac{2\Delta V_T}{V_{GS} - V_T} + \frac{\Delta \lambda}{\frac{1}{V_{DS}} + \lambda} + \frac{\Delta V_{DS}}{V_{DS} + \frac{1}{\lambda}} \tag{5-1}$$

where β is the transconductance parameter, V_{GS} is the gate-source voltage, V_{DS} the drain-source voltage, λ the channel length modulation factor and V_T is the threshold voltage of the transistor. The transconductance parameter is given by

$$\beta = \mu C_{ox} \cdot W/L \tag{5-2}$$

where C_{ox} is the gate capacitance per square, μ is the mobility, W is the width and L is the length of the transistor.

To avoid large V_{GS} mismatch due to voltage drops in the power supply wires, the power supply wires should be made very wide. Variations in V_{DS} cause errors in the output signal, but these errors can be reduced by circuit techniques, as will be explained in the following.

There is an important difference between the two types of SI circuits in that the output current can be easily amplified in a FG cell by simply increasing the size of M_2. This can not be done in the SG cell. To amplify the current in the SG cell we can either use an additional transistor to generate an amplified current in the output phase or use an additional memory cell operated on a third clock phase ϕ_3 to get a mismatch free amplification. The drawback of the second approach is that the extra clock phase decreases the speed and increases the complexity of the circuit. Another problem is that only integer gain factors can be generated by the second approach. Therefore the first approach is often used and there may be process dependent mismatch even if the

SG memory cell is used.

V_T and β mismatch are usually the dominant sources of error. It has been shown that the correlation between these mismatch errors are small and that the variance is inversely proportional to the area of the transistor [3]. In the CMOS process used in [3], the current mismatch error for a transistor pair with $WL = 1000$ μm^2 and $V_{GS} - V_T = 1$ V is 0.14% which is comparable to capacitor matching. However, for low voltage circuits it may be necessary to decrease V_{GS} and the effect of V_T mismatch will increase. For high-speed circuits smaller devices are normally used which implies that the matching can not be as good as for low-speed circuits.

5.3.2 Speed

The -3 dB bandwidth of the memory cell is determined by [1] (if the effect of the switches are ignored)

$$\omega_{-3dB} = \frac{g_m}{C_{tot}} \tag{5-3}$$

where C_{tot} is the total input capacitance at the input node. For the FG cell $C_{tot} = 2C_{ox}WL$ if both transistors are equal and for the SG cell $C_{tot} = C_{ox}WL$. The transconductance is determined [1] by

$$g_m = \mu C_{ox} \frac{W}{L}(V_{GS} - V_T) = \sqrt{2\mu C_{ox} \frac{W}{L} I_D} \tag{5-4}$$

and the pole is determined by (SG case)

$$|\omega_{-3dB}| = \frac{\sqrt{2\mu I_D}}{\sqrt{C_{ox}WL^{3/2}}} = \mu \frac{(V_{GS} - V_T)}{L^2} \tag{5-5}$$

It should be noted that the bandwidth of the FG cell is a factor 2 smaller due to the larger capacitance. In (5-5) it is seen that to increase the speed with a fixed V_{GS} the length must be reduced. To get the same speed for the FG and SG memory cell, either the length of the transistors in the FG circuit must be decreased or the bias current increased.

5.3.3 Transmission Errors

Due to the finite output impedance of the memory transistor there will be a transmission error and the low frequency transfer function is given by [2]

$$\frac{i_{out}}{i_{in}} = \frac{-1}{1 + \frac{2r_{in}}{r_{out}}} = \frac{-1}{1 + \frac{2g_{ds}}{g_m}} \tag{5-6}$$

where $g_{ds} = I_b \lambda$. The transmission error can be reduced by increasing the output

5.3 Switched-Current Techniques

impedance or decreasing the input impedance of the memory cell. The output impedance can be increased by using a cascode transistor or the regulated cascode technique as shown in Fig. 5-3 for the SG case. The output impedance is for the cascoded mem-

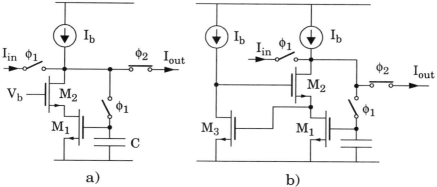

Figure 5-3 a) The cascoded memory cell b) regulated cascode memory cell.

ory cell

$$r_{out} = r_{ds1} \cdot r_{ds2} \cdot g_{m2} \tag{5-7}$$

and for the regulated cascode cell

$$r_{out} = r_{ds1} \cdot r_{ds2} \cdot g_{m2} \cdot r_{ds3} \cdot g_{m3} \tag{5-8}$$

where r_{dsi} is the output resistance of transistor M_i and g_{mi} the transconductance of transistor M_i.

The input impedance of the memory cell can be decreased by inserting an amplifier as shown in Fig. 5-4. The input impedance is now reduced to

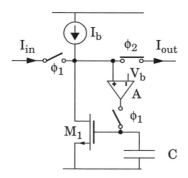

Figure 5-4 Memory cell with feedback.

$$r_{in} = \frac{1}{A \cdot g_m} \qquad (5\text{-}9)$$

where A is the gain of the amplifier and g_m the transconductance of the memory transistor.

A third approach is to use the S^2I technique [4] where the input current is sampled twice as illustrated in Fig. 5-5. The input current is first sampled by a coarse memory

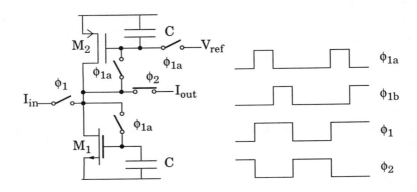

Figure 5-5 The S^2I memory cell.

(M_1) and then the remaining error is sampled by the fine memory (M_2). Since the error current sampled by M_2 is small the gate-source voltage will be almost constant. The input of this memory cell will thus act as a virtual ground. The transmission error, ε, of the S^2I cell is [2]

$$\varepsilon = \varepsilon_c \cdot \varepsilon_f \qquad (5\text{-}10)$$

where ε_c is the transmission error of the coarse memory cell and ε_f is the transmission error of the coarse memory cell. This technique can easily be extended to a S^nI technique where n memory cells are used to further reduce the transmission error.

5.3.4 Noise

In MOS circuits there are in principle two types of noise, 1/f noise and thermal noise. The SG cell has an advantage over the FG cell since the inherent correlated double sampling will cancel the low frequency 1/f noise [1]. For high frequency circuits the thermal noise is the dominant noise source and the contribution from the 1/f noise is relatively small.

The output referred current noise spectral density of the memory cell is [1]

$$S_i(f) = \frac{8}{3} \cdot kT \cdot g_m \cdot (1 + k_{bias}) \qquad (5\text{-}11)$$

5.3 Switched-Current Techniques

where k is the Boltzmann's constant, T the absolute temperature, g_m the transconductance of the memory transistor and k_{bias} a factor taking the noise contribution from the current source into account. The total noise power can be calculated as the spectral density times the noise bandwidth, i.e.

$$i_n^2 = S_i(f) \cdot BW_n = S_i(f) \cdot \frac{\omega_{-3dB}}{4} = \frac{8}{3} \cdot kT \cdot (1 + k_{bias}) \cdot \frac{g_m^2}{4 \cdot C_{tot}} \quad (5\text{-}12)$$

where BW_n is the noise bandwidth and C_{tot} the total capacitance at the gate of the memory transistor. Assuming a sinusoidal input signal the signal power can be calculated as

$$i_{in,rms}^2 = \frac{m_i^2 I_b^2}{2} \quad (5\text{-}13)$$

where I_b is the bias current in the memory cell and m_i is the modulation index (the ratio of the input current amplitude and the bias current).

The dynamic range (DR) of the memory cell can now be calculated as [1]

$$DR = \frac{i_{in,rms}^2}{i_n^2} = \frac{3 \cdot m_i^2 I_b^2 \cdot C_{tot}}{4 \cdot kT \cdot g_m^2 \cdot (1 + k_{bias})} \quad (5\text{-}14)$$

The DR can be increased without affecting the speed of the circuit. This is achieved by connecting several memory cells in parallel (W is increased by a factor K and I_b is increased by a factor K). This means that for every doubling of the bias current the DR is increased by 3 dB. To make the comparison to SC circuits easier the noise current can be expressed as an input referred noise voltage with power

$$v_n^2 = \frac{i_n^2}{g_m^2} = \frac{2 \cdot kT \cdot (1 + k_{bias})}{3 \cdot C_{tot}} \quad (5\text{-}15)$$

The dynamic range can be expressed in the voltage domain as

$$DR = \frac{V_{in}^2}{8/3 \cdot kT} C_{tot} \quad (5\text{-}16)$$

where V_{in} is the input voltage swing. For current-mode circuits the input voltage swing is usually quite small (a few hundred mV) compared to voltage-mode circuits and SI-circuits are therefore relatively noisy.

5.3.5 Clock Feedthrough Errors

When the sampling switch is turned off a portion of the channel charge is transferred to the sampling capacitor. Due to capacitive coupling through the overlap capacitors the switch control signal will introduce additional charge on the sampling capacitor. The additional charge gives a pedestal error in the output signal. The error current caused by the clock feedthrough errors can to the first order be approximated as [2]

$$\Delta i \approx \omega_{-3dB} \cdot (k_i + k_d \cdot i/I_b) = g_m/C_{tot} \cdot (k_i + k_d \cdot i/I_b) \qquad (5\text{-}17)$$

where i is the input current, I_b the bias current in the memory cell, g_m the transconductance of the memory transistor k_i and k_d are coefficients determined by process parameters, overlap and channel capacitors and switching voltages. Since both g_m, k_i and k_d are signal dependent the clock feedthrough will cause both offset, gain errors and distortion. These errors will degrade the performance especially for high bandwidths. To reduce the signal dependent error the voltage swing at the sampling switch should be minimized but this conflicts with the high voltage swing required for improving the dynamic range.

5.3.6 Aperture Errors

The aperture error in the S/H is caused by signal dependent delays and signal dependent switch turn-off times. Both the source voltage and the threshold voltage of the sampling switch is signal dependent in SI circuits. Therefore aperture time errors will result and thereby distortion. Both aperture and clock feedthrough errors can be reduced by keeping the voltage swing at the sampling switch small. There are circuits techniques using feedback to improve performance ([5], [6]) but they are more difficult to design since the stability of the feedback loop must be considered. They also require the use of floating capacitors.

5.4 SWITCHED-CAPACITOR TECHNIQUES

There are different ways of realizing a S/H circuit using the SC techniques. A possible solution is the circuit shown in Fig. 5-6. The circuit is parasitic insensitive which is

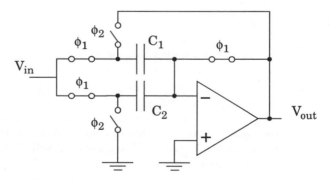

Figure 5-6 Switched-capacitor S/H circuit.

preferable for high accuracy circuits. The circuit uses two non-overlapping clock phases, ϕ_1 and ϕ_2. On phase ϕ_1 the opamp is auto-zeroed while the input voltage is sampled by C_1 and C_2. On phase ϕ_1 C_2 is grounded while C_1 is connected to the

5.4 Switched-Capacitor Techniques

output of the opamp. The output voltage on phase ϕ_2 is given by

$$V_{out} = \frac{C_1 + C_2}{C_1} \cdot V_{in} \qquad (5\text{-}18)$$

5.4.1 Mismatch

The gain of a SC circuits is usually determined by a capacitor ratio. There are some exceptions, e.g. if $C_2 = 0$ in the S/H circuit shown Fig. 5-6, the gain is always unity regardless of the capacitance value. However, most SC circuits rely on accurate capacitor matching. The value of integrated circuit capacitors is given by [7]

$$C = \frac{\varepsilon_{ox}}{t_{ox}} \cdot A = C_{ox} \cdot A \qquad (5\text{-}19)$$

where ε_{ox} is the dielectric constant of the silicon dioxide, t_{ox} the thickness of the oxide, A the area of the capacitor and C_{ox} the capacitance per area. Mismatch is caused by variations in C_{ox} and A. The relative capacitance error is given by [2]

$$\frac{\Delta C}{C} = \frac{\Delta C_{ox}}{C_{ox}} + \frac{\Delta A}{A} \qquad (5\text{-}20)$$

where $\Delta C_{ox}/C_{ox}$ is the relative error in C_{ox} and $\Delta A/A$ is the relative error in area. The accuracy of the capacitance value is usually not very good but for SC circuits we are more interested in the accuracy of a capacitor ratio. The relative error of a capacitor ratio is approximately [7]

$$\frac{\Delta \frac{C_2}{C_1}}{C_2/C_1} \approx \frac{\Delta C_2}{C_2} - \frac{\Delta C_1}{C_1} \qquad (5\text{-}21)$$

Hence the accuracy of a capacitor ratio is approximately determined by the difference between the individual errors of the capacitors. It is therefore important to try to make the errors as equal as possible and to reduce the influence of systematic errors due to e.g. overetching. It is a general practice to use unit capacitors [8] to avoid systematic errors. This means that the capacitors are composed of an array of small equal unit capacitors. The capacitor matching is often expressed in SD%[FS] which is the standard deviation of the error in a capacitor ratio expressed as percent of full-scale (total capacitance in the unit-capacitor array). Experimental results in [9] indicate that the matching is inversely proportional to the square root of the unit capacitor area and directly proportional to the perimeter-to-area ratio. A quite weak dependence on the capacitor ratio was found which disagrees with the results in [10] showing a square-root dependence and the results in [11] showing virtually no dependence at all. The difference is probably explained by differences in the processes and layout techniques that were used in the measurements.

We conclude that a large area and a large perimeter-to-area ratio improves the matching. It should, however, be noted that matching is usually improved only to a certain limit. If the capacitors are larger than a certain value the matching does not improve any more. Matching errors in the range of 0.1% or even better is achievable. This is better than transistors matching especially for low-supply voltages.

5.4.2 Speed

The speed of the circuit depends on the opamp architecture. Assuming a single stage opamp the unity gain frequency is determined by [8]

$$\omega_u = \frac{g_m}{C_L} \tag{5-22}$$

where g_m is the transconductance of the input transistor of the opamp and C_L the total load capacitance at the output of the opamp. For the two-stage opamp the load capacitor C_L should be replaced by the compensation capacitor C_c in the expression for the unity-gain frequency. The load capacitance is given by

$$C_L = C_{next} + \frac{C_1(C_2 + C_p)}{C_1 + C_2 + C_p} \tag{5-23}$$

where C_{next} is the input capacitance of the following SC circuit and C_p is the parasitic capacitor at the input of the opamp.

The -3dB frequency of the S/H circuit is determined by

$$|\omega_{-3dB}| = \beta \cdot \omega_u \tag{5-24}$$

where β is the feedback factor given by

$$\beta = \frac{C_1}{C_1 + C_2 + C_p} \tag{5-25}$$

Hence the -3dB frequency is given by

$$|\omega_{-3dB}| = \frac{g_m}{C_{next} + \frac{C_1(C_2 + C_p)}{C_1 + C_2 + C_p}} \cdot \frac{C_1}{C_1 + C_2 + C_p} \tag{5-26}$$

where C_p is the parasitic capacitor at the input of the opamp. The parasitic capacitor consists of the C_{gs} capacitor of the input transistor and the bottom plate-to-ground capacitors of C_1 and C_2. If the gain is unity, i.e. $C_2 = 0$, and the parasitic capacitor C_p is small compared to C_1, the -3dB frequency can be approximated as

$$|\omega_{-3dB}| \approx \frac{g_m}{C_{next}} \tag{5-27}$$

5.4 Switched-Capacitor Techniques

The expression for the speed of the SC circuit is thus in principle similar to the speed of SI circuits. The circuit in Fig. 5-6 is also similar to the SI circuits in that the speed is reduced when the gain is increased. If the two-stage opamp is used we have

$$|\omega_{-3dB}| = \frac{g_m}{C_c} \cdot \frac{C_1}{C_1 + C_2 + C_p} \tag{5-28}$$

The speed is thus independent of the load capacitance (this is only true when ignoring the high frequency poles). In theory the speed could be very high by choosing a small C_c but in practice this is not possible since the circuit will be unstable.

The expressions above were derived on clock phase ϕ_2 since it is most likely that the speed is lowest in this clock phase especially for large gain factors. When designing SC circuits it is important to consider both clock phases with respect to speed and stability. In clock phase ϕ_1 the total load capacitance at the output is

$$C_{L,1} = C_1 + C_2 + C_p + C_{next,1} \tag{5-29}$$

where $C_{next,1}$ is the input capacitance of the following stage in clock phase ϕ_1. The feedback factor is $\beta = 1$.

The speed of SC circuits as well as SI circuits are obviously dependent on the opamp and S/H configuration. From the simple models used here we do not see any strong indication on a significant difference in speed between the SC and SI techniques.

5.4.3 Transmission Errors

Finite opamp gain will cause additional gain errors in the total transfer function. The transfer function including the effect of finite opamp gain is given by

$$V_{out} = \frac{C_1 + C_2}{C_1 + \frac{C_1 + C_2}{A}} \cdot V_{in} \tag{5-30}$$

where A is the gain of the opamp. The effect of the finite gain is thus similar to the effect of finite output resistance in SI circuits.

5.4.4 Noise

A fundamental limitation on SC circuits is imposed by the thermal noise in the circuit. The flicker noise can in principle be made negligible small compared to the thermal noise by using circuit techniques such as correlated double sampling and chopper stabilization [12]. In the following analysis we neglect the noise from the opamp. In a more detailed analysis the opamp noise should be taken into account since it may be as large as or even larger than the switch noise. The noise voltage power spectral density of a switch is

$$S_v(f) = 4kT \cdot R_{on} \tag{5-31}$$

where R_{on} is the switch on-resistance, k is the Bolzmann's constant and T is the absolute temperature. The total noise power can be calculated as the spectral density times the noise bandwidth, i.e.

$$v_n^2 = S_v(f) \cdot BW_n = S_i(f) \cdot \frac{\omega_{-3B}}{4} = 4kT \cdot R_{on} \cdot \frac{R_{on}}{4 \cdot C_s} = \frac{kT}{C_s} \quad (5\text{-}32)$$

where BW_n is the noise bandwidth and C_s the sampling capacitor. Assuming a sinusoidal input signal the signal power can be calculated as [8]

$$v_{in,rms}^2 = \frac{v_a^2}{2} \quad (5\text{-}33)$$

where v_a is the amplitude of the sinusoid.

The dynamic range (DR) of the memory cell can now be calculated as

$$DR = \frac{v_a^2}{2kT} \cdot C \quad (5\text{-}34)$$

In conclusion, the voltage noise levels in SI and SC for the same sampling capacitor are approximately the same while the voltage swing in SC circuits usually is larger than SI circuits. Therefore SI circuits are more noisy than SC circuits [13], especially for low supply voltages.

5.4.5 Aperture and Clock Feedthrough Errors

Aperture errors and signal dependent clock feedthrough errors are caused by variations in the source and threshold voltage of the sampling switch. The signal dependent part of these errors can be reduced by using bottom-plate sampling [14]. This technique is illustrated in Fig. 5-7. There are two control signals for each sampling capac-

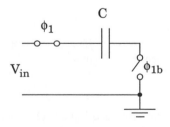

Figure 5-7   Bottom-plate sampling.

itor where ϕ_{1b} is opened slightly before ϕ_1. When the switch controlled by ϕ_{1b} is opened the charge on C can not change. Therefore, any error caused by the switch controlled by ϕ_1 will have no impact on the stored value. Since the switch controlled by ϕ_{1b} is always connected to ground or virtual ground the only remaining errors is an offset from the clock feedthrough. This technique is not easily applied to the basic SI memory cell.

5.5 SUMMARY

We now summarize some of the main features of the two techniques discussed in this chapter [2].

SI Circuits

- Require no high-gain amplifiers and no linear capacitors which makes the technique suitable for implementation in a low-voltage digital CMOS process
- Due to their simplicity and the slightly smaller capacitive loads it is easier to design high bandwidth circuits using the SI technique.
- Due to the simple and straight forward structure of the circuits it is easier to design SI circuits and they can easily be made modular.

SC Circuits

- Since the voltage swing in SC circuits is larger than in SI, SC circuits have higher dynamic range.
- When using bottom-plate sampling, SC circuits are less sensitive to clock feedthrough errors and aperture errors. Therefore SC circuits have smaller distortion.
- The matching accuracy in SC circuits is usually better than for SI circuits.

For high performance analog circuits the SC technique is the clear choice. The SI technique could be a good candidate for medium performance analog circuits in a low voltage CMOS process.

REFERENCES

[1] C. Toumazou, J. B. Hughes and N. C. Battersby, Eds.: *Switched-Currents : An Analogue Technique for Digital Technology*, Peter Peregrinus Ltd. 1993.

[2] N. Tan: *Switched-Current Design and Implementation of Oversampling A/D Converters*, Kluwer Academic Publishers 1997.

[3] M. J. M. Pelgrom, A. C. J. Duinmaijer and A. P. G. Welbers, "Matching Properties of MOS Transistors", *IEEE J. of Solid-State Circuits*, vol. SC-24, pp. 1433-39, Oct. 1989.

[4] J. B. Hughes and K. W. Moulding, "S^2I: A two-step Approach to Switched-Currents", *IEEE Intern. Symp. on Circuits and Systems*, ISCAS-93, vol. 2, pp. 1235-8, 1993.

[5] D. G. Nairn, "Zero-Voltage Switching in Switched Current Circuits", *Intern. Symp. on Circuits and Systems*, ISCAS-94, vol. 5, pp. 289-92, 1994.

[6] P. Shah and C. Toumazou, "A New High Speed Low Distortion Switched-Current Cell" *Intern. Symp. on Circuits and Systems*, ISCAS-96, vol. 1, pp. 421-24, 1996.

[7] P. E. Allen and D. R. Holberg, *CMOS Analog Circuit Design*, Holt, Rinehart and Winston, Inc., 1987.

[8] D. A. Johns and K. Martin, *Analog Integrated Circuit Design*, John Wiley & Sons, Inc. 1997.

[9] R. Singh and A. B. Bhattacharyya, "Matching Properties of Linear MOS Capacitor", *Solid-State Electronics*, vol. 32, no. 4, pp. 299-306, 1989.

[10] J. L. McCreary, "Matching Properties, and Voltage and Temperature Dependece of MOS Capacitors", *IEEE J. of Solid-State Circuits*, vol. SC-16, no. 6, pp. 608-16, Dec. 1981.

[11] J. B. Shyu, G. C. Temes and F. Krummenacher, "Random error Effects in Matched MOS Capacitors and Current Sources", *IEEE J. of Solid-State Circuits*, vol. SC-19, no. 6, pp. 948-55, Dec. 1984.

[12] R. Gregorian and G. C. Temes,: *Analog MOS Integrated Circuits*, John Wiley & Sons, Inc., 1986.

[13] R. Castello, F. Montecchi, F. Rezzi and A. Baschirotto, "Low-Voltage Analog Filters", *IEEE Trans. on Circuits and Systems-I*, vol. 42, no. 1, pp. 827-40, Nov. 1995.

[14] K. Y. Kim, N. Kusayanagi and A. A. Abidi "A 10-b, 100-MS/s CMOS A/D Converter", *IEEE J. of Solid-State Circuits*, vol. 32, no. 6, pp. 302-11, Dec. 1997.

6 ANALOG FUNCTIONAL BLOCKS

6.1 INTRODUCTION

In chapter 5 the SC and SI circuit techniques and their properties were briefly reviewed. In this chapter we show in more detail how these circuit techniques can be used to make some important analog functional blocks such as integrators and multiplying DACs (MDAC). They are key components in sigma-delta and pipelined ADCs respectively. The S/H circuit used at the input of most ADCs can be regarded as a special case of the MDAC and is not treated separately. We also show how to derive the maximum speed and how to calculate the thermal noise in the circuits. A more detailed investigation on how these circuit limitations affects the performance of the ADC is given in chapter 9 for pipelined converters and in chapter 11 for oversampled sigma-delta converters. In Sec. 6.2 we introduce the MDAC which is an important part of the pipelined converter. SI MDACs for high speed are considered in Sec. 6.3 while the limitations on speed and the thermal noise are derived in Sec. 6.4. The discussion on SI MDACs is concluded by making a comparison on different SI architectures in Sec. 6.5. In Sec. 6.6 we consider the SC MDAC and investigate the effect of non-ideal opamps while the thermal noise in SC MDACs is discussed in Sec. 6.7. Finally the speed and thermal noise of integrators are derived in Sec. 6.8 for the SI technique and in Sec. 6.9 for the SC technique.

6.2 THE MULTIPLYING DAC

The pipelined converter contains several stages each with a low resolution ADC and an analog circuit calculating the analog output signal to the following stage. The analog circuit contains a DAC, a subtractor and a S/H amplifier and is sometimes referred to as a multiplying DAC (MDAC) [1], see Fig. 6-1. The output signal of the MDAC is

$$out_i = G_i \cdot (in_i - DAC_i) \qquad (6\text{-}1)$$

where G_i is the gain of the MDAC, in_i the analog input signal and DAC_i the output signal of the DAC. If $DAC_i = 0$ the MDAC behaves as a S/H circuit while if $in_i = 0$ it behaves as a DAC. Hence the S/H circuit and the DAC can regarded as special cases of the MDAC and are not treated separately.

The limitations of the MDAC using both the SI and the SC techniques are investigated

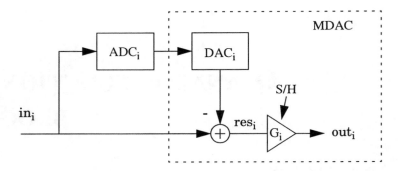

Figure 6-1 The MDAC in one stage of a pipelined ADC.

in the following sections.

6.3 MDACS IN SI PIPELINED ADCS

There are relatively few publications on SI ADCs especially for high-speed applications. Some of the publications and their performance are listed in Table 6-1. All the

Table 6-1 SI ADCs

Ref.	Process	Bits	Sampling freq. MHz	V_{dd} V	Power mW	Area mm^2	Signal freq. MHz	SNDR dB
[2]	Bi-CMOS	10	20	10	1000	48	10	50
[3]	CMOS	8	15	5	350	2.4	14	<40
[4]	CMOS	8	4.5	5	128	5.8	1	47
[5]	CMOS	10	3	3	82.5	2.7	20	40
[6]	CMOS	8	10	2.5	60	7	0.5	41

listed ADCs are pipelined converters. In [2] both current-mode and voltage-mode signals are used. The flash converters use voltage-mode signals while the signal in the main path is currents. The ADC has three-stages with one bit of digital correction in each stage. The architecture is shown in Fig. 6-2. In the figure voltage-mode signals are indicated with a V while current-mode signals are indicated with an I. The residue gain amplification is performed in two stages using two S/H amplifiers with gains of 2 and 4 respectively. The residue in the second stage is generated by only one S/H amplifier with a gain of 8. Two clock phases, ϕ_1 and ϕ_2 are used as shown in the Table 6-2. The circuit is designed in a BiCMOS process and both high-gain amplifiers and linear capacitors are used for the V/I conversion and S/H circuit.

6.3 MDACs in SI Pipelined ADCs

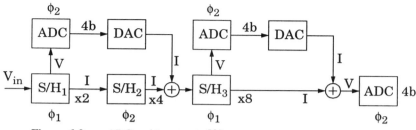

Figure 6-2 ADC architecture in [2].

Table 6-2 The operation of the stages in [2].

Stage	Block	ϕ_1	ϕ_2
1	S/H$_1$	sample	hold
1	S/H$_2$, ADC	hold	sample
1	DAC	settling	-
2	S/H$_3$	sample	hold
2	ADC	sample	hold
2	DAC	-	settling
3	ADC	hold	sample

In [3] the S^{2}I technique [7] is used for the memory cells. A high clock frequency is achieved by using unity-gain MDACs. This means that the amplification factor of the residue amplifier is one. The reference currents of the DACs must therefore be reduced by a factor two for each stage in the pipeline as illustrated in Fig. 6-3.

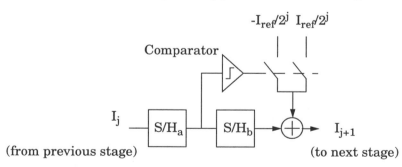

Figure 6-3 ADC architecture in [3].

There are two approaches to implementing the S/H circuits. One option is to use current mirrors to copy and amplify the currents which requires two clock phases. The second approach is to use ratio independent switching which requires more than two clock phases. In [3] the second approach was adopted and the total number of clock phases is four. The sampling rate is thus 1/4 of the clock frequency which in this case means $f_{clk} = 60$MHz and $f_s = 15$MS/s. The clocking of one stage in the pipeline is illustrated in Table 6-3. All the stages are clocked in the same way.

Table 6-3 The operation of the stages in [2].

Stage	Block	ϕ_1	ϕ_2	ϕ_3	ϕ_4
j	S/H$_a$	sample	hold to S/H$_b$	hold to Comp.	-
j	S/H$_b$	hold to I$_{j+1}$	sample	hold	hold
j	Comp.	-	reset	sample	hold to DAC
j	DAC	settle	-	-	-

In [4] a residue amplifiers with a gain of two is used which means that the reference currents can be equal in all the stages, see Fig. 6-4. The operation of two consecutive

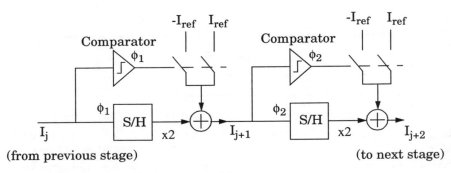

Figure 6-4 ADC architecture in [4].

stages is illustrated in Table 6-4.

Table 6-4 The operation of the stages in [4].

Stage	Block	ϕ_1	ϕ_2
j	S/H	sample	hold
j	Comp.	sample	hold
j	DAC	-	settling
j+1	S/H	hold	sample
j+1	Comp	hold	sample
j+1	DAC	settling	-

In this case the amplification by two is implemented by using current mirrors and thus only two clock phases are needed. Both the comparator and the S/H circuit are clocked on the same clock phase. The architecture in [5] is similar to the architecture in [4] but

6.4 Performance Limitations in SI MDACs

in this case the redundant-signed-digit (RSD) algorithm [8] is used to enable digital correction. The RSD algorithm is discussed in chapter 9. The architecture in [6] is shown in Fig. 6-5. The clocking of two consecutive stages is shown in Table 6-5. In

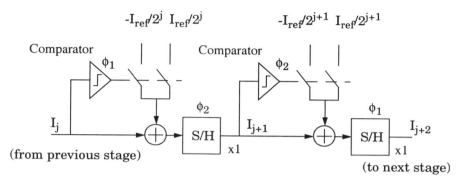

Figure 6-5 ADC architecture in [6].

Table 6-5 The operation of the stages in [6].

Stage	Block	ϕ_1	ϕ_2
j	Comp.	sample	hold
j	S/H	hold	sample
j	DAC	-	settling
j+1	Comp.	hold	sample
j+1	S/H	sample	hold
j+1	DAC	settling	-

this architecture the comparator and the S/H of one stage are clocked on different clock phases. It is thereby possible to perform the D/A conversion before the S/H, which may be advantageous as will be explained in the following sections.

6.4 PERFORMANCE LIMITATIONS IN SI MDACS

When considering the speed and the noise in the circuits we will neglect the effect of the switches. The circuits are usually designed in such a manner that the switches have a small influence, but in a more accurate analysis their effect should be taken into account.

6.4.1 Speed

The speed of the circuits is determined by the -3 db bandwidth of the memory cell and the required number of clock phases. The 3-dB cut-off frequency of a memory cell is

$$\omega_{-3dB} = \frac{g_m}{C_{tot}} \tag{6-2}$$

where C_{tot} is the total capacitance at the gate of the memory transistor and g_m the transconductance of the memory transistor. Assuming a single pole system the relative settling error is determined by

$$\varepsilon_r = e^{-\frac{t_s}{\tau}} \tag{6-3}$$

where $\tau = 1/\omega_{-3dB}$ is the time constant and t_s the available settling time. The available settling time is given by

$$t_s = \frac{T_s}{N_p} \tag{6-4}$$

where T_s is the total sampling period of the circuit and N_p is the required number of sampling phases within one sample period. We assume that all sampling phases have equal duration. The maximum sampling frequency is given by

$$f_s = \frac{1}{T_s} = \frac{1}{t_s \cdot N_p} = \frac{\omega_{-3dB}}{\ln\left(\frac{1}{\varepsilon_r}\right) \cdot N_p} \tag{6-5}$$

6.4.2 Residue Amplification

There are two approaches to amplifying a current, using current mirrors or ratio-independent SI circuits. The two approaches for a gain of two are shown in Fig. 6-6. The amplifier in Fig. 6-6a) works as a current mirror where the gate voltage of transistors M_2 and M_3 is sampled and held by a switch. The number of transistors connected in parallel at the output of the current mirror determines the gain of the amplifier. If all transistor are assumed to be equal, the bandwidth of this circuit is

$$\omega_{in,a} = \frac{g_m}{C_{gs}(1+G)} \tag{6-6}$$

where G is the residue gain, g_m the transconductance and C_{gs} the gate-source capacitor of the transistors. Two sampling phases is needed regardless of the gain factor, one sampling phase for the current stage, ϕ_1 and one for the following stage, ϕ_2.

The ratio-independent amplifier in Fig. 6-6a) first stores the input voltage on the gate of transistor M_4 (ϕ_1) and then stores the same voltage on the gate of transistor M_5 (ϕ_2). On clock phase ϕ_3 both the current from M_4 and M_5 are sent to the output and it is sampled by the following stage. If all the stages are equal the number of sampling phases is given by

$$N_{p,b} = 2G \tag{6-7}$$

6.4 Performance Limitations in SI MDACs

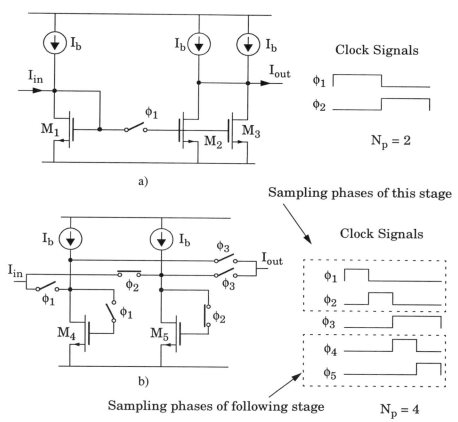

Figure 6-6 Current amplifier with gain = 2. a) current mirror b) ratio-independent.

The input bandwidth of the memory transistors is

$$\omega_{in,b} = \frac{g_m}{C_{gs}} \qquad (6\text{-}8)$$

The main advantage of the ratio-independent approach is that the gain factor is not dependent on transistor matching. A disadvantages of the ratio-independent cell is that there are many clock signal which makes the circuits more complex. Another limitation is that only integer gain factors can be realized. The current mirror approach is thus more flexible. It is obvious from (6-6) and (6-8) that the speed is reduced in both cases if the residue gain factor is increased. Therefore unity gain MDACs are preferable in high speed applications.

6.4.3 Distributing Currents to Several Outputs

The output current of the MDAC must be sent to both the S/H circuit and the ADC of the following stage. Two circuits performing this function are shown in Fig. 6-7. In the first case, Fig. 6-7a), the two output currents are simply generated by using two

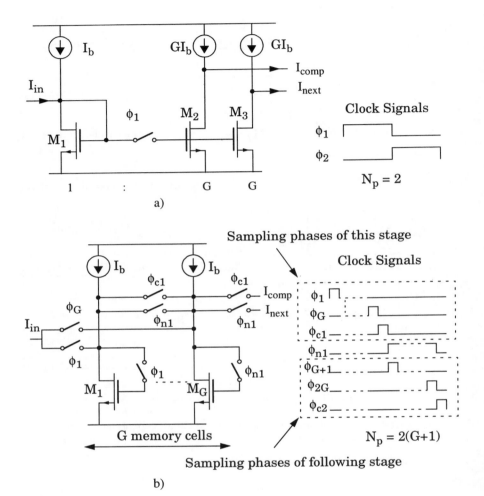

Figure 6-7 Copying currents a) current-mirror b) ratio-independent.

output transistors. The input bandwidth of the circuit is given by

$$\omega_{in1} = \frac{g_m}{(1+2G)C_{gs}} \quad (6\text{-}9)$$

where G is the residue gain. Compared to the result in (6-6) the speed is reduced almost by a factor two if G is large. In the circuit in Fig. 6-7b) an additional sampling phase is needed to send the output current to the comparator of the next stage. The input bandwidth is the same as in (6-8) but the number of clock phases is now $2(G+1)$. The maximum speed for the two circuits in Fig. 6-7 is

6.4 Performance Limitations in SI MDACs

$$f_{s,a} = \frac{g_m}{C_{gs} \cdot \ln(1/\varepsilon_r) \cdot 2 \cdot (1 + 2G)} \qquad (6\text{-}10)$$

and

$$f_{s,b} = \frac{g_m}{C_{gs} \cdot \ln(1/\varepsilon_r) \cdot 2 \cdot (1 + G)} \qquad (6\text{-}11)$$

respectively. Hence the circuit in Fig. 6-7b) is almost twice as fast as the circuit in Fig. 6-7a).

6.4.4 Time-Interleaved Stage

From the discussion above it is clear that a small gain factor will increase the speed and if possible unity gain MDACs should be used. Two attractive features of the current mirror approach is that only two clock phases are needed and that non integer gain factors can be realized. Non-integer gain factors are needed when using the calibration technique in [9]. A third advantage of the current mirror cell is that the voltage glitches at the switching instants are usually smaller than for the ratio-independent circuits. The main disadvantages of the current mirror circuit are the higher power consumption, the larger chip area and the distortion caused by transistor mismatch. In Fig. 6-8

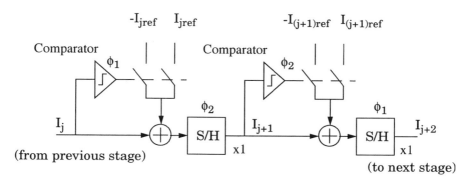

Figure 6-8   ADC architecture in [6].

we show a current mirror based stage with reduced area and power consumption [6]. The clocking of two consecutive stages is shown in Table 6-6. In this architecture the

Table 6-6   The operation of the stages in [6].

Stage	Block	ϕ_1	ϕ_2
j	Comp.	sample	hold
j	S/H	hold	sample
j	DAC	-	settling

Table 6-6   The operation of the stages in [6].

Stage	Block	φ₁	φ₂
j+1	Comp.	hold	sample
j+1	S/H	sample	hold
j+1	DAC	settling	-

comparator and S/H of one stage are clocked on different clock phases. Unity-gain MDACs are used and therefore the reference currents are scaled down by a factor of two for each stage in the pipeline i.e.

$$I_{(j+1)ref} = \frac{I_{jref}}{2} \tag{6-12}$$

where I_{jref} is the reference current in the j-th stage and $I_{(j+1)ref}$ is the reference current in the next stage. The output current of a stage is the summation of the output current of the internal D/A converter and the output current of the S/H circuit. It is given by

$$I_{j+1} = I_{in} - \sum_{k=1}^{j} b_k \cdot \frac{I_{ref}}{2^k} \tag{6-13}$$

where I_{in} is the input current, I_{j+1} the output current from stage j and b_k is ±1 depending on the signal from the comparator in stage k. Due to the unity-gain MDACs the power consumption and area will be smaller in the LSB stages since the current swing gets smaller and smaller in the pipeline. The drawback with the smaller currents is that the signal gets more sensitive to noise and other error sources. However the required accuracy of the MDACs are lower for the LSB stages.

Since the comparator and the S/H circuit are time interleaved and clocked on different clock phases the memory cell can be simplified as shown in Fig. 6-9. The capacitive

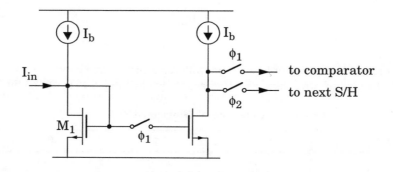

Figure 6-9   Simplified memory cell.

6.4 Performance Limitations in SI MDACs

load is decreased and both the power consumption and area are reduced compared to the solution in Fig. 6-7a). It should be noted that when calculating the speed of the circuit we must consider that the signal now must settle through both the S/H and the comparator on the same clock phase. However, the settling time of the comparator is usually small compared to the settling time of the S/H circuit. If the speed reduction caused by the comparator is ignored the maximum speed of the MDAC is given by

$$f_{max,b} = \frac{g_m}{C_{gs} \cdot \ln(1/\varepsilon_r) \cdot 2 \cdot (1+G)} \quad (6\text{-}14)$$

which is comparable to speed of the circuit in Fig. 6-7b).

6.4.5 D/A Conversion

We have not yet considered the clocking of the DACs and the generation of the residue signal. In Fig. 6-10, two circuits to generate the residue are shown. The first circuit

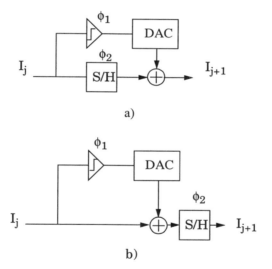

Figure 6-10 Residue generation with subtraction a) before and b) after S/H.

adds the DAC current after the S/H circuit while the second circuit adds the DAC current before the S/H circuit. The advantage of the second approach is that the bias current and transistor sizes in the S/H can be reduced by a factor 2 due to the smaller signal swing. This means that noise, area and power is a factor two smaller.

6.4.6 Noise

The thermal noise in the circuits limits the resolution of the converter. The noise current power spectral density for the thermal noise of an MOS transistor is given by [10]

$$S(f) = \frac{8}{3} \cdot kT \cdot g_m \qquad (6\text{-}15)$$

where k is the Boltzmann's constant, T the absolute temperature and g_m the transconductance of the memory transistor. Assuming a single pole system and white noise the total noise power is given by [10]

$$i_n^2 = S(f) \cdot BW_N \qquad (6\text{-}16)$$

where BW_N (frequency in Hz) is the noise bandwidth related to the -3 dB bandwidth as

$$BW_N = \frac{\pi}{2} f_{-3dB} = \frac{\omega_{-3dB}}{4} \qquad (6\text{-}17)$$

When this noise is sampled all the high frequency noise is folded back to the signal band and the total sampled noise power is the same. By using (6-16) we have

$$i_n^2 = \frac{2}{3} \cdot kT \cdot g_m \cdot \omega_{-3dB} \qquad (6\text{-}18)$$

First we calculate the noise in the circuit shown in Fig. 6-11. The gain of the circuit is

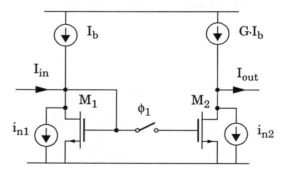

Figure 6-11   First generation memory cell with noise sources.

G. The bias current and W/L ratio of transistor M_2 is G times larger than for transistor M_1. We neglect the noise from the bias current sources. The noise contribution from the current sources adds to the total noise in exactly the same way as the noise from M_1 and M_2. Their effect can be included by adding a factor $1 + k_{bias}$ in the expression for the total noise [11]. k_{bias} is the relative size of the transconductance in the current source compared to the transconductance of the current mirror transistor. There are two noise sources present in the circuit which both contribute to the total noise. First the noise contribution from M_1 is sampled on the gate capacitor of M_2. The voltage noise power is given by

6.4 Performance Limitations in SI MDACs

$$v_{n1}^2 = \frac{i_{n1}^2}{g_{m,in}^2} = \frac{\frac{2}{3} \cdot kT \cdot g_{m,in} \cdot \omega_{in}}{g_{m,in}^2} \tag{6-19}$$

where $g_{m,in}$ is the transconductance of M_1, and ω_{in} is the input bandwidth, i.e.

$$\omega_{in} = \frac{g_{m,in}}{C_{gs1} + C_{gs2}} = \frac{g_{m,in}}{C_{gs1} \cdot (1+G)} \tag{6-20}$$

In the hold phase the sampled noise is transferred to the output and we have

$$i_{n1,out}^2 = v_{n1}^2 \cdot g_{m,out}^2 = \frac{g_{m,out}^2}{g_{m,in}^2} \cdot \frac{2}{3} \cdot kT \cdot g_{m,in} \cdot \omega_{in} \tag{6-21}$$

where $g_{m,out}$ is the transconductance of M_2. In the hold phase the noise from M_2 will also contribute. The noise power is given by

$$i_{n2,out}^2 = \frac{2}{3} \cdot kT \cdot g_{m,out} \cdot \omega_{out} \tag{6-22}$$

where $g_{m,out}$ is the transconductance of M_2 and ω_{out} is the bandwidth of the following stage. Assuming uncorrelated noise sources the total noise at the output is

$$\begin{aligned} i_{n,tot}^2 &= i_{n1,out}^2 + i_{n2,out}^2 \\ &= \frac{2}{3} \cdot kT \cdot \left(\frac{g_{m,out}^2}{g_{m,in}^2} \cdot g_{m,in} \cdot \omega_{in} + g_{m,out} \cdot \omega_{out} \right) \end{aligned} \tag{6-23}$$

If we for simplicity assume that $\omega_{out} = \omega_{in}$ have that

$$g_{m,out} = G \cdot g_{m,in} \text{ and } \omega_{in} = \omega_{out} = \frac{g_{m,in}}{(1+G) \cdot C_{gs}} \tag{6-24}$$

Hence the total noise is

$$i_{n,tot}^2 = \frac{2}{3} \cdot kT \cdot G \cdot \frac{g_{m,in}^2}{C_{gs}} \tag{6-25}$$

When using the second generation approach, see Fig. 6-12, G transistors are needed to get a gain of G. The noise for each transistor can be calculated using (6-23) and assuming $g_{m,in} = g_{m,out}$, i.e.

$$i_{n,i}^2 = \frac{2}{3} \cdot kT \cdot (g_{m,in} \cdot \omega_{in} + g_{m,in} \cdot \omega_{out}), \ i = 1, ..., G \tag{6-26}$$

The noise from each memory transistor adds to the total noise at the output. For simplicity we assume $\omega_{in} = \omega_{out} = g_{m,in}/C_{gs}$, which yields

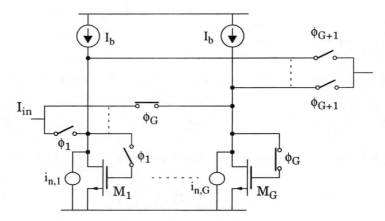

Figure 6-12 Second generation current amplifier with noise sources.

$$i_{n,tot}^2 = \sum_{i=1}^{G} i_{n,i}^2 = \frac{2}{3} \cdot kT \cdot 2G \cdot \frac{g_{m,in}^2}{C_{gs}} \qquad (6\text{-}27)$$

Hence the noise in the ratio independent case is twice as large as in the current mirror case. Sometimes it is more convenient to calculate the input referred voltage noise since it can be directly compared to the input voltage swing. For the circuit in Fig. 6-11 we have

$$v_{n,in}^2 = \frac{i_{n,tot}^2}{g_{m,out}^2} = \frac{i_{n,tot}^2}{(g_{m,in} \cdot G)^2} = \frac{2}{3} \cdot kT \cdot \frac{1}{C_{gs} \cdot G} \qquad (6\text{-}28)$$

It is interesting to note that the voltage noise only depends on the residue gain and the size of the sampling capacitor. Since the voltage swing is limited by the supply voltage the only way to improve the SNR is to increase the sampling capacitor.

6.4.7   Dynamic Range

The dynamic range at the output is determined by the signal power at the output and the total noise at the output. If the rms value of the input signal is $i_{rms,in}$ the dynamic range is given by

$$DR = \frac{i_{rms,out}^2}{i_{n,tot}^2} \qquad (6\text{-}29)$$

or expressed in dB

$$DR = 20 \cdot \log\left(\frac{i_{rms,out}}{i_{n,tot}}\right) \qquad (6\text{-}30)$$

6.5 Comparisons of SI MDACS

It may be more convenient to refer the noise to the input as a voltage. i.e.

$$DR = \frac{v_{rms,in}^2}{v_{n,in}^2} \tag{6-31}$$

where $v_{rms,in}^2$ is the power of the input signal. The DR of the circuit in Fig. 6-11 can for instance be calculated as

$$DR = \frac{v_{rms,in}^2}{v_{n,in}^2} = \frac{v_{rms,in}^2}{\frac{2}{3} \cdot kT \cdot \frac{1}{C_{gs} \cdot G}} = \frac{v_{rms,in}^2 \cdot C_{gs} \cdot G}{\frac{2}{3} \cdot kT} \tag{6-32}$$

To improve the DR we must clearly increase the voltage swing or the sampling capacitor.

6.5 COMPARISONS OF SI MDACS

To conclude the discussion on SI MDACs in this chapter we compare the speed, DR, area and power of the different architectures. The three types considered here are shown in Fig. 6-13. The first type, shown in Fig. 6-13a), is similar to the architecture in [4] where both the comparator and the S/H circuit use the same clock phase which requires that first generation SI circuits are used. The circuit in Fig. 6-13b) corresponds to the time-interleaved stage in Sec. 6.4.4 and uses current mirror circuits with the comparator and S/H operated on different clock phases. The architecture in Fig. 6-13c) is based on ratio-independent circuits where several clock phases are required. In all cases we assume that the bandwidth of the comparator is large enough not to influence the speed. We also assume that all unit transistors are of equal size and that the transconductance is g_m and that the gate-source capacitance is C_{gs}. The different architectures are in the following referred to as Type 1, Type 2 and Type 3. The residue gain is denoted G. The speed of the circuits is determined by the number of clock phases and the bandwidth, ω_{in}. All these results have been derived in the previous sections and are summarized in Table 6-7. The speed of the Type 2 MDAC is optimis-

Table 6-7   Speed of the different MDACs.

Type	Clock Phases	ω_{in}	f_s
1	2	$\dfrac{g_m}{(1+2G)C_{gs}}$	$\dfrac{g_m}{2(1+2G)\ln(1/\varepsilon_r)C_{gs}}$
2	2	$\dfrac{g_m}{(1+G)C_{gs}}$	$\dfrac{g_m}{2(1+G)\ln(1/\varepsilon_r)C_{gs}}$
3	2(1+G)	$\dfrac{g_m}{C_{gs}}$	$\dfrac{g_m}{2(1+G)\ln(1/\varepsilon_r)C_{gs}}$

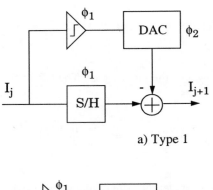

a) Type 1

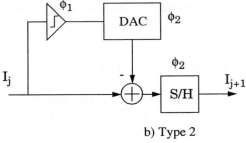

b) Type 2

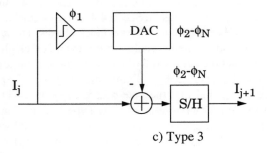

c) Type 3

Figure 6-13 The different MDAC architectures.

tic in the sense that the speed of the comparator has been assumed to be very high not to influence the total speed of the circuit. The effect of the comparator can be included by approximating the bandwidth of the MDAC as

$$\omega_{tot} = \frac{1}{\frac{1}{\omega_{SH}} + \frac{1}{\omega_{comp}}} \tag{6-33}$$

where ω_{SH} is the bandwidth of the memory cell and ω_{comp} is the bandwidth of the comparator.

The noise in the MDACs can be derived by using (6-23). We assume $\omega_{out} = \omega_{in}$. For

6.5 Comparisons of SI MDACS

the converter of type 1 we have

$$i^2_{n,tot} = \frac{2}{3} \cdot kT \cdot \left(\frac{g_{m,out}}{g_{m,in}} + 1\right) \cdot \omega_{in} \cdot g_{m,out} =$$

$$= \frac{2}{3} \cdot kT \cdot (G+1) \cdot \frac{g_m}{(1+2G)C_{gs}} \cdot g_m \cdot G \qquad (6\text{-}34)$$

For the converter of type 2 we have

$$i^2_{n,tot} = \frac{2}{3} \cdot kT \cdot \left(\frac{g_{m,out}}{g_{m,in}} + 1\right) \cdot \omega_{in} \cdot g_{m,out} =$$

$$= \frac{2}{3} \cdot kT \cdot (G+1) \cdot \frac{g_m}{(1+G)C_{gs}} \cdot g_m \cdot G = \frac{2}{3} \cdot kT \cdot \frac{g_m^2}{C_{gs}} \cdot G \qquad (6\text{-}35)$$

Finally for the converter of type 3 we have

$$i^2_{n,tot} = \left(\frac{2}{3} \cdot kT \cdot \left(\frac{g_{m,out}}{g_{m,in}} + 1\right) \cdot \omega_{in} \cdot g_{m,out}\right) \cdot G =$$

$$= \left(\frac{2}{3} \cdot kT \cdot 2 \cdot \frac{g_m}{C_{gs}} \cdot g_m\right) \cdot G = \frac{2}{3} \cdot kT \cdot \frac{g_m^2}{C_{gs}} \cdot 2G \qquad (6\text{-}36)$$

The noise performance is summarized in Table 6-8.

Table 6-8 Noise in the different MDACs.

Type	Output referred noise
1	$\frac{2}{3} \cdot kT \cdot \frac{(1+G) \cdot G}{(1+2G)} \cdot \frac{g_m^2}{C_{gs}}$
2	$\frac{2}{3} \cdot kT \cdot G \cdot \frac{g_m^2}{C_{gs}}$
3	$\frac{2}{3} \cdot kT \cdot 2G \cdot \frac{g_m^2}{C_{gs}}$

It is very difficult to accurately estimate the chip area of the complete ADC since e.g. switches, wires and the digital circuits will consume a portion of the area. In the following the total area of the memory transistor (assumed to have width W and length L) is used to compare the relative area of the circuits. This is not a very accurate estimate but can give a rough indication on whether the MDAC has a large area or not. The total DC current of the S/H circuit is used as measure of the power consumption of the MDAC. Each unit transistor is assumed to use the current I_b. The area and power is summarized in Table 6-9.

Table 6-9 Noise in the different MDACs.

Type	Area	Power
1	$(1+2G)WL$	$(1+2G)I_b$
2	$(1+G)WL$	$(1+G)I_b$
3	GWL	GI_b

A few observations can be made about the results above.

- For all types of MDACs the speed is increased if the currents are scaled down in the pipeline. This can be achieved by reducing both the residue gain factor and the reference currents of the following stage. For a 1 bit/stage converter unity gain MDACs can be used. To avoid too small currents in the later stages of the pipeline, some of the LSB stages may use a larger residue gain factor [6].

- When comparing the speed of the circuits, Type 3 is the best choice. Type 2 is the second best choice but for high gain factors the difference is small. It should also be noted that non-overlapping clock phases must be used for the voltage switches, meaning that a portion of the sampling period can not be used for settling. Hence, increasing the number of clock phases, which is necessary in MDAC of Type 3, is likely to reduce the total available settling time.

- MDAC of Type 1 has the lowest noise but also the lowest speed, largest area and power consumption. To get the same DR for all the circuits the bias current for Type 2 and 3 can be increased. This will increase both the power and area to make them comparable to Type 1. However, an advantage of Type 2 and 3 is that the D/A conversion can be performed before the S/H circuit without decreasing the speed. This means that if unity gain MDACs are used, the bias currents can be decreased in the S/H circuits which improves the DR.

- It should also be noted that expressions above are based on very simple models and several limiting factors, such as clock feedthrough, have been neglected. The equations can be regarded as the fundamental limitation of the relative performance.

6.6 SC MDACS

In this section we discuss SC MDACs. A circuit commonly used [12] to implement the MDAC in pipelined ADCs is shown in Fig. 6-14. The circuit is single-ended but the actual implementation would normally be fully differential. The MDAC in the figure has N input bits $b_{N-1}, \ldots, b_0$ where b_{N-1} is the MSB and b_0 the LSB. One bit of digital correction is used and therefore the residue gain is decreased by a factor 2. We will use this circuit in the following to illustrate how important parameters as noise and bandwidth can be calculated for SC MDACs. We assume that the DC gain of the opamp is A_0, the parasitic capacitance at the input node of the opamp is C_p. On the

6.6 SC MDACs

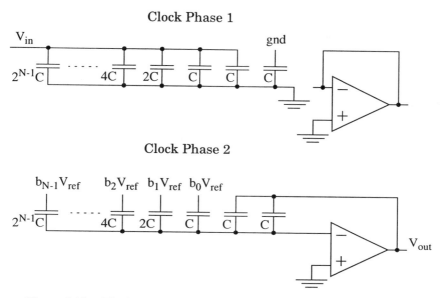

Figure 6-14 The SC MDAC of a pipelined converter.

first clock phase all capacitors but one are connected to the input voltage of the circuit, making the total charge on the top plates of the capacitors

$$q_1 = V_{in} \cdot (2^{N-1}C + \ldots + C + C) \tag{6-37}$$

On the next clock phase two of the capacitors are connected to the output while the other capacitors are connected to $\pm V_{ref}$ depending on the output from the sub-ADC. The total charge on the top plates of the capacitors on this clock phase is

$$q_2 = \left(V_{out} + \frac{V_{out}}{A}\right) \cdot 2C + V_{ref} \cdot (2^{N-1} \cdot b_{N-1} \cdot C + \ldots + b_0 \cdot C) \\ + \frac{V_{out}}{A} \cdot (2^{N-1}C + \ldots + C) + \frac{V_{out}}{A} \cdot C_p \tag{6-38}$$

where $b_{N-1}, \ldots, b_0 = \pm 1$ are determined by the digital output bits from the sub-ADC. The total charge is conserved and the output signal can be calculated as

$q_1 = q_2 \Rightarrow$

$$V_{out} = \frac{V_{in} \cdot 2^{N-1} - \frac{V_{ref}}{2} \cdot (2^{N-1} \cdot b_{N-1} + \ldots + b_0)}{1 + \frac{C_{tot}}{2A_0 C}} \tag{6-39}$$

where $C_{tot} = 2^{N-1}C + \ldots + C + 2C + C_p$ is the total capacitance connected to the input node of the opamp. The reference voltages in the DACs are chosen as

$$V_{ref-} = -\frac{FS}{2} \text{ and } V_{ref+} = \frac{FS}{2} \tag{6-40}$$

where FS is the full-scale input swing of the converter.

6.6.1 Effect of Finite Opamp Gain and Parasitic Capacitors

Due to the finite opamp gain and parasitic capacitors there will be a gain error in the MDAC as can be seen in (6-39). By using the first order Taylor expansion we get the following approximation

$$\frac{1}{1+\varepsilon} \approx 1 - \varepsilon \Rightarrow \frac{1}{1 + \frac{C_{tot}}{2A_0 C}} \approx 1 - \frac{C_{tot}}{2A_0 C} \tag{6-41}$$

Hence the relative gain error can be approximated as

$$\varepsilon_G = \frac{C_{tot}}{2A_0 C} = \frac{2^N + 1 + \frac{C_p}{C}}{2A_0} \approx \frac{2^N - 1}{A_0} \tag{6-42}$$

where the approximation is valid for large N. The relative gain error is the same for both the input signal and the DAC signal. This means that the gain error can be modelled as an error at the output of the subtractor.

6.6.2 Speed

The relative settling error assuming linear settling and a single pole opamp is determined by

$$\varepsilon_r = e^{-t_s \cdot \omega_{-3dB}} \tag{6-43}$$

where ω_{-3dB} is the -3dB bandwidth and t_s is the settling time. The settling error causes a gain error at the output of the MDAC. When using two clock phases with 50% duty cycle the settling time is, limited by

$$t_s < \frac{1}{2f_s} \tag{6-44}$$

where f_s is the sampling frequency. The unity-gain bandwidth of the opamp is related to the bandwidth as

$$\omega_u = \omega_{-3dB}/\beta \tag{6-45}$$

where β is the feedback factor. The feedback factor in the evaluation phase is

$$\beta = \frac{2C}{(2^N+1)C + C_p} = \frac{2}{(2^N+1) + C_p/C} \tag{6-46}$$

The unity gain bandwidth of a single stage opamp is determined by

$$\omega_u = \frac{g_m}{C_{load}} \tag{6-47}$$

where g_m is the transconductance of the input device of the opamp and C_{load} is the total load capacitance at the output of the opamp. The load capacitance is determined by

$$C_{load} = \beta \cdot ((2^N-1)C + C_p) + C_{next} \tag{6-48}$$

where C_{next} is the capacitive load of the following stage. If the following stage is identical the load capacitance is given by

$$C_{load} = \beta \cdot ((2^N-1)C + C_p) + 2^N C \approx 2^N C \tag{6-49}$$

where the last approximation is valid for large N. When the circuit is designed, both clock phases must be considered. The feedback factor of the opamp in clock phase 1 is $\beta = 1$ while the load is only C_p. To avoid that the opamp get unstable it may be necessary to connect an additional load capacitor to the output in this clock phase.

The maximum speed of the MDAC can now be calculated as

$$f_s = \frac{1}{2t_s} = \frac{\omega_{-3dB}}{\ln(1/\varepsilon_r) \cdot 2} = \frac{\omega_u \cdot \beta}{\ln(1/\varepsilon_r) \cdot 2} = \frac{\frac{g_m}{2^N C} \cdot \frac{2}{(2^N+1) + C_p/C}}{\ln(1/\varepsilon_r) \cdot 2} \approx \frac{g_m}{\ln(1/\varepsilon_r) \cdot 2^{2N} C} \tag{6-50}$$

The assumption of linear settling is not valid when the gain is small, i.e. when the feed back factor is large (se chapter 7). The settling will be slew rate limited and the settling time is increased compared to the above expressions.

6.7 NOISE IN SC MDACS

There are in principle two types of noise in MOS circuits, thermal noise and 1/f noise. The thermal noise is white while 1/f noise is frequency dependent. 1/f noise dominates at low frequencies while the thermal noise is usually dominant for wide-band circuits. The 1/f can also be decreased by circuits techniques such as correlated double sampling. Therefore the 1/f noise is neglected in the following.

6.7.1 kT/C Noise

When a signal is sampled on a capacitor, the thermal noise generated by the switch will alias since the inherent bandwidth is wider than half of the sampling frequency. The total power is equal to kT/C, independent of the switch-on resistance [10]. The noise in the signal bandwidth, BW_s is

$$v_{n,C}^2 = \frac{k \cdot T}{C} \cdot \frac{BW_s}{f_s/2} = \frac{k \cdot T}{C} \cdot \frac{1}{OSR} \qquad (6\text{-}51)$$

where k is the Boltzmann's constant, T is the absolute temperature, C the sampling capacitor, f_s the sampling frequency and OSR the oversampling ratio. If the circuit is fully differential and the sampling capacitance is C in both the positive and negative signal path the kT/C noise within the signal bandwidth is twice that of (6-51). However the effective signal swing is also doubled and therefore the DR is improved by 3 dB.

6.7.2 Thermal Noise in Opamp

The input referred noise power spectral density for a single-stage opamp is given by (see chapter 7)

$$S_{n,amp}(f) = 2 \cdot \frac{8}{3} \cdot \frac{k \cdot T}{g_m} \cdot (1 + n_t) \qquad (6\text{-}52)$$

where g_m is the transconductance of the input transistor and n_t is the noise contribution factor due to the other noise sources. From equation (6-52) it is seen that we can increase the input transconductance or decrease the noise contribution factor of the other noise sources in order to decrease the input-referred noise. We will see shortly that increasing the input transconductance is hardly an option due to the increase in noise bandwidth. As we will see in chapter 7 the noise contribution factor is small for the telescopic opamp. The telescopic opamp is therefore the best choice from a noise point of view. The noise contribution from the opamp is normally the same regardless if the circuit is single-ended or fully differential. Thus the dynamic range is increased by 6 dB when using fully differential circuits if only the opamp noise is considered.

6.7.3 Input Referred Thermal Noise of SC Amplifier I

In Fig. 6-15, we show an SC amplifier configuration. The parasitic capacitance at the input is pC. The capacitor bC and cC represent the load capacitance switched to the opamp output on clock phase ϕ_2 and ϕ_1 respectively. For the moment we neglect the noise contribution from the switch on-resistance. The feedback factor on clock phase ϕ_1 is

$$\beta_1 = 1 \qquad (6\text{-}53)$$

while on ϕ_2 it is

6.7 Noise in SC MDACs

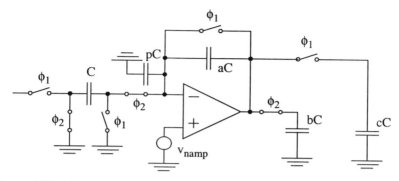

Figure 6-15 S/H amplifier I.

$$\beta_2 = \frac{a \cdot C}{a \cdot C + C + p \cdot C} = \frac{a}{a + 1 + p} \tag{6-54}$$

The corresponding load capacitors for the opamp are

$$C_{L,1} = C \cdot p + C \cdot c \tag{6-55}$$

and

$$C_{L,2} = b \cdot C + \beta_2 \cdot (C + p \cdot C) = (b + \beta_2 \cdot (1 + p)) \cdot C \tag{6-56}$$

Assuming a single-stage opamp the unity gain bandwidth is

$$\omega_u = \frac{g_m}{C_L} \tag{6-57}$$

and the -3 dB bandwidth

$$\omega_{-3dB} = \omega_u \cdot \beta \tag{6-58}$$

The noise bandwidth s given by [10]

$$BW_n = \omega_{-3dB}/4 \tag{6-59}$$

The total noise power in the two clock phases can now be calculated by using $v_n^2 = S(f) \cdot BW_n$ [10]

$$v_{namp,1}^2 = \frac{4}{3} \cdot \frac{k \cdot T}{C_{L,1}} \cdot \beta_1 \cdot (1 + n_t) = \frac{4}{3} \cdot \frac{k \cdot T}{C \cdot (p + c)} \cdot (1 + n_t) \tag{6-60}$$

and

$$v_{namp,2}^2 = \frac{4}{3} \cdot \frac{k \cdot T}{C_{L,2}} \cdot \beta_2 \cdot (1+n_t)$$
$$= \frac{4}{3} \cdot \frac{k \cdot T}{(b \cdot (a+1+p)+a \cdot (1+p)) \cdot C} \cdot a \cdot (1+n_t) \quad (6\text{-}61)$$

It is seen from equation (6-60) and (6-61) that the input transconductance has no effect on the input referred noise. This is because when we increase the input transconductance to decrease the input-refer noise spectral density, we increase the noise bandwidth, making the noise power constant. In order to evaluate the impact of the thermal noise generated in the opamp, we must find the input-referred noise of the SC amplifier. The z-domain transfer function from the opamp input can be derived by observing that the total charge on capacitors pC, C and aC is conserved from clock phase ϕ_1 to ϕ_2. The output signal is given by

$$V_{out}(z) = \frac{1+a+p}{a} \cdot V_{namp,2}(z) - \frac{p}{a} \cdot V_{namp,1}(z) \cdot z^{-1/2} \quad (6\text{-}62)$$

where the first term corresponds to the output referred opamp noise generated during clock phase ϕ_2 and the second term to the output referred opamp noise generated during clock phase ϕ_1. Since all the samples of the thermal noise are uncorrelated the two noise terms in (6-62) will add. However if the noise samples are correlated as for 1/f noise, low frequency noise would be slightly attenuated due to the minus sign on the second term of (6-62). Since the gain of the SC circuit is approximately $1/a$ the total input referred noise caused by the opamp can be calculated as

$$v_{n,in}^2 = (1+a+p)^2 \cdot v_{n,amp2}^2 + p^2 \cdot v_{n,amp1}^2 + \frac{kT}{C}$$
$$= \frac{4}{3} \cdot kT \cdot \frac{(1+n_t)}{C} \left(\frac{(1+a+p)^2 \cdot a}{b \cdot (a+1+p)+a \cdot (1+p)} + \frac{p^2}{c+p} \right) + \frac{kT}{C} \quad (6\text{-}63)$$

In the above expression we have also included the kT/C-noise stored on the sampling capacitor. We see that if the noise contribution factor, n_t, is small and the capacitors at the output are large the kT/C-noise dominates.

6.7.4 Input Referred Thermal Noise of SC Amplifier II

Another popular SC amplifier configuration is shown in Fig. 6-16. The difference is that the feedback capacitor is also used to sample the input. The noise can in this case be derived in similar way as was done for S/H amplifier I in the previous section. Both the feedback factors and the z-domain transfer function are the same. The only difference is that in this case the gain of the S/H amplifier is $1+1/a$. Hence the input-referred noise is

6.7 Noise in SC MDACs

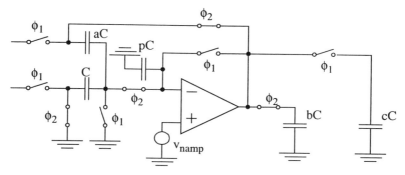

Figure 6-16   S/H amplifier II.

$$v_{n,in}^2 = \frac{(1+a+p)^2}{(1+a)^2} \cdot v_{n,amp2}^2 + \frac{p^2}{(1+a)^2} \cdot v_{n,amp1}^2 + \frac{kT}{(a+1)C}$$

$$= \frac{4kT}{3} \cdot \frac{(1+n_t)}{C} \left(\frac{\left(1+\frac{p}{(1+a)}\right)^2 \cdot a}{b(a+1+p)+a(1+p)} + \frac{p^2}{(1+a)^2(c+p)} \right) \quad (6\text{-}64)$$

$$+ \frac{kT}{(a+1)C}$$

We see that if the noise contribution factor, n_t, is small and the capacitors at the output are large the kT/C-noise dominates.

6.7.5   Thermal Noise in MDAC

In this section we calculate the thermal noise in the MDAC in Fig. 6-14. We consider the thermal noise from the switches and the opamp. To simplify the analysis the parasitic capacitor at the input of the opamp is ignored and the opamp gain is assumed to be infinite. On clock phase ϕ_1 only the noise from the switches contributes. The noise voltage stored on the capacitors is [10]

$$v_n^2 = \frac{kT}{(2^N+1) \cdot C} = \frac{kT}{C_{tot}} \quad (6\text{-}65)$$

where C_{tot} is the sum of all capacitors in the MDAC. The total noise charge can be calculated as

$$q^2 = \frac{kT}{C_{tot}} \cdot C_{tot}^2 = kT \cdot (2^n+1) \cdot C \quad (6\text{-}66)$$

On clock phase ϕ_2 this charge will be converted to a voltage. Hence the total noise generated during ϕ_1 referred to the output of the MDAC is

$$v_{n,\phi_1}^2 = \frac{q^2}{(2C)^2} = \frac{kT \cdot (2^N + 1)}{4C} \approx \frac{kT}{C} \cdot 2^{N-2} \quad (6\text{-}67)$$

where the last approximation is valid for large N. The thermal noise sources on clock phase ϕ_2 are shown in Fig. 6-17. Noise sources v_{n1} and v_{n2} represent the thermal

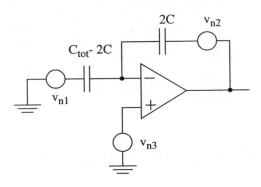

Figure 6-17 Noise sources in clock phase 2.

noise from the switch on-resistance while v_{n3} represents the thermal noise from the opamp. On clock phase ϕ_2 all three noise sources will contribute. The noise contribution from a noise source referred to the output of the circuit can be calculated as

$$v_{out,k}^2 = S_k(f) \cdot |H_k|^2 \cdot BW_{N,k}, \quad k = 1, 2, 3 \quad (6\text{-}68)$$

where $S_k(f)$ is the spectral density of the noise source k, $|H_k|$ the gain from noise source k to the output of the MDAC and $BW_{N,k}$ the noise bandwidth. The noise bandwidth is for all the noise sources

$$BW_{N,k} = \frac{\beta \cdot \omega_u}{4} \quad (6\text{-}69)$$

where β is the feedback factor and ω_u the unity gain bandwidth of the opamp. The factors $|H_k|$ can be shown to be

$$|H_1| = \frac{C_{tot} - 2C}{2C}, \quad |H_2| = 1, \quad |H_3| = \frac{C_{tot}}{2C} \quad (6\text{-}70)$$

and the spectral densities are

$$S_1(f) = 4kT \cdot R_1, \quad S_2(f) = 4kT \cdot R_2, \quad S_3(f) = 2 \cdot \frac{8kT}{3g_m} \cdot (1 + n_t) \quad (6\text{-}71)$$

where g_m is the transconductance of the input device of the opamp, n_t the noise contribution factor, R_1 the switch resistance of the switch at the input and R_2 the switch resistance of the switch in the feedback loop of the opamp. It is common that $1/g_m$ is larger than the on-resistance of the switches. If this is the case the contribution from the opamp will dominate and we have

$$v_{n,\phi_2}^2 = \frac{16kT}{3g_m} \cdot \left(\frac{C_{tot}}{2C}\right)^2 \cdot \frac{\beta \cdot \omega_u}{4} \cdot (1 + n_t) \tag{6-72}$$

The unity-gain bandwidth, ω_u, is determined by the transconductance, g_m, and the total capacitive load on the opamp output, C_{load}. Hence we have

$$v_{n,\phi_2}^2 = \frac{16kT}{3g_m} \cdot \left(\frac{C_{tot}}{2C}\right)^2 \cdot \frac{\beta \cdot \frac{g_m}{C_{load}}}{4} \cdot (1 + n_t) =$$

$$= \frac{16}{3}kT \cdot \left(\frac{2^N + 1}{2}\right)^2 \cdot \frac{2^N + 1}{4C_{load}} \cdot (1 + n_t) = \frac{2kT}{3C_{load}} \cdot (2^N + 1) \cdot (1 + n_t) \tag{6-73}$$

The total noise at the output is the sum of the noise contributions in both the clock phases, i.e.

$$v_{tot}^2 = v_{n,\phi_1}^2 + v_{n,\phi_2}^2 \approx \left(\frac{kT}{C} \cdot 2^{N-2} + \frac{2kT}{3C_{load}} \cdot (2^N + 1) \cdot (1 + n_t)\right) \tag{6-74}$$

If the load capacitance is in the same order as the sampling capacitor, i.e. $C_{load} \approx 2^N C$ the noise is, at least for large N, dominated by v_{n,ϕ_1}^2, i.e. the kT/C-noise in the sampling capacitor. If this is the case the input referred noise of the MDAC can be calculated by dividing the output referred noise by the MDAC gain squared, i.e.

$$v_{tot,in}^2 \approx \frac{kT}{C} \cdot \frac{2^{N-2}}{(2^N-1)^2} = \frac{kT}{2^N C} \approx \frac{kT}{C_{tot}} \tag{6-75}$$

It is interesting to note that the total noise only depends on the capacitors in the circuit, not g_m. It should also be noted that the above expressions represent the total noise power. If the output of MDAC is sampled by another SC circuit all the noise power is folded into the Nyquist band. However if the circuit is used as a DAC and the analog wave form is not sampled, the noise power generated during clock phase ϕ_2 is evenly distributed up to the noise bandwidth. Since the noise bandwidth usually is much larger than the sampling frequency, only a small part of the opamp noise power falls in the signal band.

6.8 SI INTEGRATOR

The integrator is an important part of oversampled sigma-delta converters. In this section we briefly investigate fundamental limitations on speed and noise of SI integrators. A more detailed analysis on SI oversampled ADCs can be found in [13]. A lossless integrator using the SI technique is shown in Fig. 6-18. The z-domain transfer

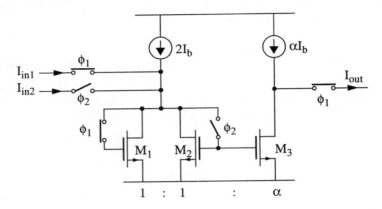

Figure 6-18 Noise sources in clock phase 2.

function of the integrator is given by [13]

$$I_{out}(z) = \frac{\alpha}{1-z^{-1}} \cdot (z^{-1} \cdot I_{in1} - z^{-1/2} \cdot I_{in2}) \tag{6-76}$$

The gain of the integrator, α, is controlled by the dimension ratio of transistor M_2 and M_3 while the sign is determined by how the input current is switched.

6.8.1 Speed

The bandwidth of the circuit in Fig. 6-18 is different in the two clock phases. On clock phase ϕ_1 the bandwidth is determined by transistor M_1, i.e.

$$\omega_{-3dB} = \frac{g_m}{C_{gs}} \tag{6-77}$$

where g_m is the transconductance of transistor M_1 and C_{gs} the gate-source capacitance of M_1. On clock phase ϕ_2 the bandwidth is lower since transistor M_3 adds extra capacitive load. Assuming M_3 to be α times larger than M_1 and M_2 we have

$$\omega_{-3dB} = \frac{g_m}{C_{gs}(1+\alpha)} \tag{6-78}$$

Thus the speed will be limited by (6-78). A small gain factor gives a higher bandwidth.

6.8.2 Noise

There are three noise sources due to transistors M_1, M_2 and M_3 in the integrator. These noise sources will appear between the source and drain of the transistors as shown in Fig. 6-19. To simplify the analysis the two noise sources for M_1 and M_2 are added to

6.8 SI Integrator

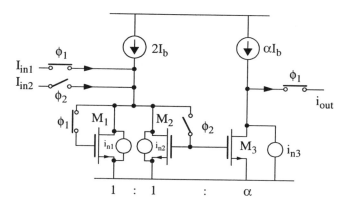

Figure 6-19 Noise sources in SI integrator.

one noise source i_n, where $i_n = i_{n1} + i_{n2}$. The effect of the noise source can now be derived as follows [11]. We assume that the input current is zero. The signal current through M_1 is denoted i_1 and the signal current through M_2 is denoted i_2. The sampling period is T. The noise sources are here treated as sampled noise sources but they are actually continuously changing their output current. The noise currents are not sampled until at the end of each clock phase when the switches are opened. Strictly this would make the delay of the noise signal one half clock period less [11]. However, it has no influence on the results of the noise analysis. We start the analysis at time $t = (n-1)T$ and assume that the switches controlled by ϕ_1 are closed. Transistor M_1 is now diode-connected and its current is

$$i_1(n-1) = i_n(n-1) - i_2(n-1) \tag{6-79}$$

where $i_x(n-1)$ denotes a current at time $t = (n-1)T$.

During the next clock phase $t = (n-1/2)T$ and the above current is held in transistor M_1, i.e.

$$i_1(n-1/2) = i_1(n-1) \tag{6-80}$$

while the current in transistor M_2, now diode-connected, is

$$\begin{aligned} i_2(n-1/2) &= i_n(n-1/2) - i_1(n-1/2) = \\ &= i_n(n-1/2) - i_n(n-1) + i_2(n-1) \end{aligned} \tag{6-81}$$

In the following clock phase the current in M2 is held and we have

$$i_2(n) = i_2(n-1/2) \tag{6-82}$$

This current is mirrored to M_3 and sent to the following stage and we have the total noise current at the output as

$$i_{out}(n) = \alpha \cdot i_2(n) + i_3(n) \tag{6-83}$$

By using the z-transform on the above equations we get the z-domain output current as

$$i_{out}(z) = \frac{\alpha \cdot i_n(z) \cdot (z^{-1/2} - z^{-1})}{1 - z^{-1}} + i_3(z) \tag{6-84}$$

The contribution from i_n has a zero at $z = 1$. This means that low frequency noise as 1/f noise from the transistors is attenuated in the circuit. If the noise is AWGN all samples are independent and the noise contributions from the two clock phases are added as powers. We can now write the output noise as

$$i_{out}(z) = \frac{\alpha \cdot i_{n,tot}(z)}{1 - z^{-1}} + i_3(z) \tag{6-85}$$

where $i_{n,tot}(z)$ is the total noise contribution from M_1 and M_2. If the noise is AWGN the total noise power of $i_{n,tot}(z)$ is

$$i_{n,tot}^2 = i_{n,\phi_1}^2 + i_{n,\phi_2}^2 \tag{6-86}$$

i_{n,ϕ_1}^2 is the noise power in clock phase ϕ_1 and i_{n,ϕ_2}^2 is the noise power in clock phase ϕ_2. These noise power can be calculated as the product of the spectral density and the noise bandwidth, i.e.

$$i_{n1}^2 = \frac{8}{3} \cdot kT \cdot g_m \cdot BW_{noise} = \frac{8}{3} \cdot kT \cdot g_m \cdot \frac{\omega_{-3dB}}{4} = \frac{2}{3} \cdot kT \cdot \frac{g_m^2}{C_{gs}} \tag{6-87}$$

and

$$i_{n2}^2 = \frac{2}{3} \cdot kT \cdot \frac{g_m^2}{(1+\alpha)C_{gs}} \tag{6-88}$$

Hence the noise power is

$$i_{n,tot}^2 = \frac{2}{3} \cdot kT \cdot \frac{g_m^2}{C_{gs}} \cdot \left(1 + \frac{1}{\alpha}\right) \tag{6-89}$$

The noise power generated by M_3 depends on the bandwidth of the load connected to the output. It is given by

$$i_{n3}^2 = \frac{2}{3} \cdot kT \cdot g_m \cdot \alpha \cdot BW_n \tag{6-90}$$

where BW_n is the noise bandwidth of the load at the output of the integrator. It is convenient to refer the noise to the input. This is achieved by dividing the expression in (6-85) with the z-domain transfer function of the integrator (see (6-76)). The input current noise would be

6.9 SC Integrator

$$i_{in}(z) = i_{n,tot}(z) + \frac{i_3(z)}{\alpha} \cdot (1-z^{-1}) \tag{6-91}$$

The noise from M_3 is high-pass filtered. If the integrator is used in an oversampled data converter the noise from M_3 can be neglected. It is interesting to calculate the corresponding input referred noise voltage power since it can be directly compared to the maximum voltage swing to get an upper limit in the SNR. We get the noise voltage by dividing the current noise by the g_m^2. If the contribution from M_3 is neglected we get

$$v_{n,in}^2 = \frac{i_{n,tot}^2}{g_m^2} = \frac{2}{3} \cdot kT \cdot \frac{1}{C_{gs}} \cdot \left(1 + \frac{1}{\alpha}\right) \tag{6-92}$$

Again we see that the voltage noise power is only dependent on the sampling capacitor and not the transconductance of the memory transistors.

6.9   SC INTEGRATOR

In Fig. 6-20. we show a lossless SC integrator configuration that is widely used in

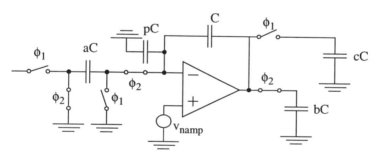

Figure 6-20   SC Integrator.

oversampling ADCs. The transfer function of this integrator is (assuming infinite opamp gain)

$$H(z) = \frac{a \cdot z^{-1}}{1 - z^{-1}} \tag{6-93}$$

It should be noted that the sampling capacitor of the following stage, here represented by bC and cC can be clocked on either ϕ_1 or ϕ_2. By clocking on ϕ_1 the speed of the circuit is slightly larger since the load in the integration phase is smaller.

6.9.1   Speed

The speed of the circuit is limited by the bandwidth in both clock phases. Normally the integration phase is most critical. The -3dB bandwidth is determined by the unity gain frequency of the opamp and the feedback factor. The feedback factor on ϕ_1 is given by

$$\beta_1 = \frac{C}{pC+C} = \frac{1}{p+1} \tag{6-94}$$

while on ϕ_2 it is given by

$$\beta_2 = \frac{C}{pC+aC+C} = \frac{1}{a+p+1} \tag{6-95}$$

For a single-stage opamp the unity-gain frequency is

$$\omega_u = \frac{g_m}{C_{load}} \tag{6-96}$$

The load capacitance on ϕ_1 is given by

$$C_{load,1} = cC + \beta_1 \cdot p \cdot C \tag{6-97}$$

while on ϕ_2 it is

$$C_{load,2} = bC + \beta_2 \cdot (a+p) \cdot C \tag{6-98}$$

Thus the -3dB bandwidth on ϕ_1 is

$$\omega_{-3dB,1} = \omega_u \cdot \beta_1 = \frac{g_m}{(c \cdot (1+a)+p) \cdot C} \tag{6-99}$$

and on ϕ_2

$$\omega_{-3dB,2} = \omega_u \cdot \beta_2 = \frac{g_m}{(b \cdot (a+p+1)+(a+p)) \cdot C} \tag{6-100}$$

We can now use (6-3) to find the relative settling errors in the two clock phases as

$$\varepsilon_{r,1} = e^{-\omega_{-3dB,1} \cdot t_s} = e^{-\frac{g_m}{(c \cdot (1+a)+p) \cdot C} \cdot t_s} \tag{6-101}$$

and

$$\varepsilon_{r,2} = e^{-\omega_{-3dB,2} \cdot t_s} = e^{-\frac{g_m}{(b \cdot (a+p+1)+(a+p)) \cdot C} \cdot t_s} \tag{6-102}$$

where t_s is the available settling time and ε_r is the relative settling error. The settling in the integration phase is usually most critical. Ideally all the charge on the input capacitor aC should be transferred to the integration capacitor C. Due to the finite bandwidth only a part of the charge will be transferred. The remaining portion of the

6.9 SC Integrator

charge on aC is determined by the relative settling error. Hence the transfer function of the integrator is

$$H(z) = \frac{(1 - \varepsilon_{r,2}) \cdot a \cdot z^{-1}}{1 - z^{-1}} \tag{6-103}$$

Thus we can conclude that the finite bandwidth introduces a gain error in the integrator.

6.9.2 Finite Opamp Gain

We assume that the DC gain of the opamp is A_0, the parasitic capacitance at the input node of the opamp is pC. In the first clock phase ($t = (n-1)T$) capacitor aC is connected to the input voltage of the circuit, making the total charge on the top plates of the capacitors

$$q_1 = V_{in}(n-1) \cdot aC + V_{out}(n-1) \cdot C + \frac{V_{out}(n-1)}{A_0} \cdot (pC + C) \tag{6-104}$$

where $V_x(n-1)$ denotes a voltage at time $t = (n-1)T$. In the following clock phase ($t = (n-1/2)T$) the total charge is

$$q_2 = V_{out}\left(n - \frac{1}{2}\right) \cdot C + \frac{V_{out}\left(n - \frac{1}{2}\right)}{A_0} \cdot (pC + C + aC) \tag{6-105}$$

The total charge is conserved and the output signal at $t = (n-1/2)T$ can be calculated as

$$q_1 = q_2 \Rightarrow$$

$$V_{out}\left(n - \frac{1}{2}\right) = \frac{V_{in}(n-1) \cdot aC + \left(\frac{pC + C}{A_0} + 1\right) \cdot V_{out}(n-1)}{C + \frac{pC + C + aC}{A_0}} \tag{6-106}$$

In the following clock phase the output is held by capacitor C, i.e.

$$V_{out}(n) = V_{out}\left(n - \frac{1}{2}\right) \tag{6-107}$$

By using (6-106) and (6-107) the z-domain transfer function can be calculated to be

$$H(z) = \frac{z^{-1} \cdot a}{\left(1 + \frac{p + 1 + a}{A_0} - \left(\frac{p + 1}{A_0} + 1\right) \cdot z^{-1}\right)} \tag{6-108}$$

The expression can be rewritten as [14]

$$H(z) = \frac{r_2 \cdot z^{-1} \cdot a}{1 - \frac{r_2}{r_1} \cdot z^{-1}} \tag{6-109}$$

where

$$r_1 = \frac{\beta_1 \cdot A_0}{1 + \beta_1 \cdot A_0} = \frac{1}{\left(\frac{p+1}{A_0} + 1\right)}$$

$$r_2 = \frac{\beta_2 \cdot A_0}{1 + \beta_2 \cdot A_0} = \frac{1}{\left(\frac{p+1+a}{A_0} + 1\right)} \tag{6-110}$$

Hence the finite opamp gain will cause a gain error and introduce leakage in the integrator.

6.9.3 Effect of Finite Bandwidth and Finite DC Gain

In the previous two sections the transfer function of the integrator was derived considering finite opamp gain and finite bandwidth separately. If both these are taken into account simultaneously the derivation get slightly more complicated. Using a single pole model of the integrator the transfer function is given by [14]

$$H(z) = \frac{r_2 \cdot (1 - \varepsilon_{r,2}) \cdot z^{-1}}{1 - \frac{r_2}{r_1} \cdot \left(1 - \varepsilon_{r,2} \cdot \left(1 - \frac{r_1}{r_2}\right)\right) \cdot z^{-1}} \tag{6-111}$$

where $\varepsilon_{r,2}$ is the relative settling error on clock phase ϕ_2. According to Sec. 6.6.2 $\varepsilon_{r,2}$ is given by

$$\varepsilon_{r,2} = e^{-\omega_{-3dB,2} \cdot t_s} = e^{-\frac{g_m}{(b \cdot (a+p+1) + (a+p)) \cdot C} \cdot t_s} \tag{6-112}$$

We conclude that when the finite opamp gain is included the finite bandwidth will introduce both gain errors and integrator leakage.

6.9.4 Noise

In Sec. 6.7 the noise of S/H amplifiers were derived. The noise in the integrator can be calculated in a similar way. The z-domain output noise from the opamp noise source (v_{namp}, see Fig. 6-20) can be derived based on charge conservation. We assume in the following the opamp gain is infinite and we neglect the influence of finite bandwidth in the opamp on the transfer function. On clock phase ϕ_1 ($t = (n-1)T$) we have the total charge

$$q_1 = V_{out}(n-1) \cdot C - v_{namp}(n-1) \cdot (pC + C) \tag{6-113}$$

6.9 SC Integrator

where $V_x(n-1)$ denotes a voltage at time $t = (n-1)T$. In the following clock phase ($t = (n-1/2)T$) the total charge is

$$q_2 = V_{out}\left(n-\frac{1}{2}\right) \cdot C - v_{namp}\left(n-\frac{1}{2}\right) \cdot (pC + C + aC) \tag{6-114}$$

The total charge in the two phases is conserved and the output signal at $t = (n-1/2)T$ can be calculated as

$$q_1 = q_2 \Rightarrow$$

$$V_{out}\left(n-\frac{1}{2}\right) = V_{out}(n-1) - v_{namp}(n-1) \cdot (p+1) +$$

$$+ v_{namp}\left(n-\frac{1}{2}\right) \cdot (p+1+a) \tag{6-115}$$

In the following clock phase the output is held by capacitor C, but we also have a noise contribution from the noise source. The total charge on C and pC is conserved, i.e.

$$V_{out}\left(n-\frac{1}{2}\right) - v_{namp}\left(n-\frac{1}{2}\right) \cdot (p+1) = V_{out}(n) - v_{namp}(n) \cdot (p+1) \tag{6-116}$$

We can now get an expression for the output signal at $t = nT$ as

$$V_{out}(n) = V_{out}\left(n-\frac{1}{2}\right) + v_{namp}(n) \cdot (p+1) - v_{namp}\left(n-\frac{1}{2}\right) \cdot (p+1)$$

$$= V_{out}(n-1) + (v_{namp}(n) - v_{namp}(n-1)) \cdot (p+1) + \tag{6-117}$$

$$+ a \cdot v_{namp}\left(n-\frac{1}{2}\right)$$

The z-domain transfer function is

$$V_{out}(z) = z^{-1} \cdot v_{namp,1}(z) \cdot (1+p) + v_{namp,2}(z) \cdot \frac{z^{-1/2}}{1-z^{-1}} \cdot a \tag{6-118}$$

where $v_{namp,1}(z)$ is the noise contribution on ϕ_1 and $v_{namp,2}(z)$ is the contribution on ϕ_2. The noise can be referred to the input by dividing the output noise by the transfer function of the integrator, (6-93), i.e.

$$v_{nin}(z) = v_{namp,1}(z) \cdot \frac{(p+1)}{a} \cdot \frac{(1-z^{-1})}{z^{-1}} + v_{namp,2}(z) \cdot z^{1/2} \tag{6-119}$$

If the signal frequency is much lower than the sampling frequency, which is the case in an oversampled sigma-delta modulator, the second term in (6-119) will dominate since the first term is high-pass filtered. Assuming a single-stage opamp the input referred noise is (see Sec. 6.7)

$$v_{nin}^2 = \frac{4}{3} \cdot \frac{kT}{C_{load}} \cdot \beta \cdot (1+n_t)$$

$$= \frac{4}{3} \cdot \frac{kT}{bC + \frac{(a+p)}{a+p+1} \cdot C} \cdot \frac{1}{a+p+1} \cdot (1+n_t) \qquad (6\text{-}120)$$

$$= \frac{4}{3} \cdot \frac{kT \cdot (1+n_t)}{(b(a+p+1)+a+p)C}$$

where n_t is a factor taking the noise contributions from the current sources in the opamp into account. In the above analysis only the opamp noise was considered. The kT/C noise from the passive sampling should also be considered. The total noise would then be

$$v_{nin,tot}^2 = \frac{4}{3} \cdot \frac{kT \cdot (1+n_t)}{(b(a+p+1)+a+p)C} + \frac{kT}{C} \qquad (6\text{-}121)$$

Again it is interesting to note that the noise only depends on the capacitor sizes and not the transconductance of the input device. If the noise contribution factor, n_t, is small and the load capacitance at the output is large, the kT/C-noise from the switch resistance will dominate.

6.10 SUMMARY

In this chapter we have discussed two important analog functional blocks in data converters, the MDAC and the integrator. Both SC and SI circuits were considered. We showed how to calculate the noise and maximum speed of the circuits based on first order models. The maximum speed of the circuits is determined by the bandwidth and the required number of clock phases. The maximum speed normally decreases as the gain of the circuit increases. The thermal noise of the circuits can be referred to the input as a noise voltage. We showed that the noise voltage power for all the circuits is only determined by the capacitors in the circuit.

The results in this chapter will be used in chapter 9 and chapter 11 to investigate the effect of circuit imperfections on pipelined and oversampled sigma-delta converters respectively.

REFERENCES

[1] B. S. Song, S. H. Lee and M. F. Tompsett, "A 10-b 15-MHz CMOS Recycling Two-Step A/D Converter", *IEEE J. of Solid-State Circuits*, vol. 25, no. 6, pp. 1328-38, Dec. 1990.

[2] P. Real, D. H. Robertson, C. W. Mangelsdorf, and T. L. Tewksbury, "A Wide-Band 10-b 20-Ms/s Pipelined Using Current-Mode Signals", *IEEE J. of Solid-State Circuits*, vol. 26, no. 8, pp. 1103-09, Aug. 1991.

[3]   M. Bracey, W. Redman-White, J. Richardson and J. B. Hughes, "A Full Nyquist 15 MS/s 8-b Differential Switched-Current A/D Converter", *IEEE J. of Solid-State Circuits*, vol. 31, no. 7, pp. 945-51, July 1996.

[4]   C. Y. Wu, C. C. Chen, J. J. Cho, "A CMOS Transistor-Only 8-b 4.5-Ms/s Pipelined Analog-to-Digital Converter Using Fully-Differential Current-Mode Circuit techniques", *IEEE J. of Solid-State Circuits*, vol. 30, no. 5, May 1995.

[5]   B. E. Jonsson and H. Tenhunen, "Low-Voltage, 10bit Switched-Current ADC with 20MHz Input Bandwidth", *Electronics Letters*, vol. 34, no. 20, pp. 1904-5, 1st October 1998.

[6]   M. Gustavsson and N. Tan, "New Current-Mode Pipeline A/D Converter Architectures", *IEEE Intern. Symp. on Circuits and Systems*, ISCAS-97, vol. 1, pp. 417-20, 1997.

[7]   J. B. Hughes and K. W. Moulding, "S^2I: A two-step Approach to Switched-Currents", *IEEE Intern. Symp. on Circuits and Systems*, ISCAS-93, vol. 2, pp. 1235-8, 1993.

[8]   B. Ginetti, P .G. A. Jespers and A. Vandemeulebroecke, "A CMOS 13-b Cyclic RSD A/D Converter", *IEEE J. of Solid-State Circuits*, vol. 27, no. 7, p. 957-64, July 1992.

[9]   A. N. Karanicolas, H. S. Lee and K. L. Bacrania, "A 15-b 1 Msample/s Digitally Self-Calibrated Pipeline ADC", *IEEE J. of Solid-State Circuits*, vol. 28, no. 12, pp. 1207-15, Dec. 1993.

[10] D. A. Johns and K. Martin, *Analog Integrated Circuit Design*, John Wiley & Sons, Inc. 1997.

[11] C. Toumazou, J. B. Hughes and N. C. Battersby, Eds.: *Switched-Currents : An Analogue Technique for Digital Technology*, Peter Peregrinus Ltd. 1993.

[12] S. U. Kwak, B. S. Song and K. Bacrania, "A 15-b, 5-Msample/s Low-Spurious CMOS ADC", *IEEE J. of Solid-State Circuits*, vol. 32, no. 12, pp. 1866-75, Dec. 1997.

[13] N. Tan, *Switched-Current Design and Implementation of Oversampling A/D Converters*, Kluwer Academic Publishers 1997.

[14] Y. Geerts, A. M. Merques, M. Steyaert and W. Sansen "A 3.3-V, 15-bit, Delta-Sigma ADC with a Signal bandwidth of 1.1 MHz for ADSL Applications", *IEEE J. of Solid-State Circuits*, vol. 34, no. 7, pp. 927-36, July 1999.

7 BASIC ANALOG CIRCUIT DESIGN

7.1 INTRODUCTION

For SC data converters, opamps and voltage comparators are the fundamental building blocks. For SI data converters, SI memory circuits and current comparators are the fundamental building blocks. We have already touched upon SI memory circuits in the previous chapters. In this chapter we will concentrate on opamps and comparators. For most data converter products, references are usually required to be generated on the same chip. We will also discuss voltage and current references in this chapter.

Following this introduction, we will discuss opamps in Sec. 7.2, with emphasize on high-speed design. Only fully differential architectures will be covered since high performance SC data converters usually use fully differential opamps. In Sec. 7.3, we will discuss voltage comparators. Current comparators are briefed in Sec. 7.4. References will be discussed in Sec. 7.5.

7.2 OPERATIONAL AMPLIFIERS

We can categorize opamps into single-stage and multi-stage architectures. Multi-stage opamp architectures are not well-suited for high-speed high-accuracy SC applications due to the compensation needed to improve the phase margin. Non-accurate compensation may introduce doublets with frequencies lower than the closed-loop -3-dB bandwidth. Low-frequency doublets are hazard for achieving the high-accuracy and high-speed settling. Therefore, we focus us on single-stage opamp architectures. However with a reduced supply voltage, we have to resort to multi-stage architecture due to the limited voltage head room. Two-stage opamps will be covered in the following chapter for low-voltage operation.

There are three popular architectures for single-stage opamps. They are the current-mirror based, folded-cascode, and telescopic architectures. In comparing the architectures, we only focus on speed, noise, and power. We do not consider the DC gain since it can be effectively improved by using the gain boosting technique which will be discussed in this chapter as well.

7.2.1 Slew Rate vs. Unity-Gain Bandwidth

If a single-stage opamp with only one dominant pole is used in a close loop, the output setting responding to a step input is given [2] by

$$V_{out}(t) = V_{step} \cdot (1 - e^{-t \cdot \beta \cdot \omega_u}) \quad (7\text{-}1)$$

where V_{step} is the input step size, β is the feedback factor of the closed loop, and ω_u is the unity-gain bandwidth of the opamp.

To avoid slewing, the slew rate must be larger than the maximum slope at the output, i.e,.

$$SR \geq \frac{d}{dt} V_{out}(t) \bigg|_{max} = V_{step} \cdot \beta \cdot \omega_u \quad (7\text{-}2)$$

Without slewing in the opamp, the non-complete settling only introduces a linear error. However, for most practical opamps such as the ones to be discussed below, the slew rate and the unity-gain bandwidth are closely related. We cannot simply increase the slew rate and the unity-gain bandwidth independently. Higher slew rate usually makes the unity-gain bandwidth higher. For SC data converter applications, the step size, i.e, the difference between two successive values can be very high. Therefore, it is almost practically impossible to avoid slewing in the opamp unless the gate voltage of the input transistor is excessively large or the feedback factor is very small. (This will become evident in the following discussions.) Therefore, we have to guarantee the settling accuracy. Any non-complete settling will introduce distortion, not just a linear error in some practical SC data converters.

7.2.2 Current-Mirror Based Opamp

In Fig. 7-1 we show the schematic of the current-mirror based single-stage opamp [1].

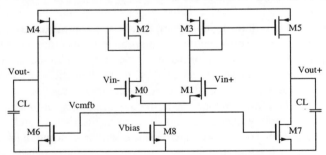

Figure 7-1 Current-mirror based single-stage opamp.

The unity gain bandwidth is given by

$$\omega_u = \frac{n \cdot g_{m0}}{C_L} \quad (7\text{-}3)$$

where g_{m0} is the transconductance of the input device M0 (M1), C_L is the capacitive load and n is the dimension ratio of the current mirrors M2 and M4 (M3 and M5).

The first non-dominant pole is due to the current mirror, and is given by

$$\omega_n \approx \frac{g_{m2}}{C_{gs2} + C_{gs4} + C_{gd4} + C_{gd0} + C_{b2}} \tag{7-4}$$

where g_{m2} is the transconductance of transistor M2 (M3), C_{gs2} is the gate-source parasitic capacitance of transistor M2 (M3). C_{gs4} is the gate-source parasitic capacitance of transistor M4 (M5), C_{gd0} is the gate-drain parasitic capacitance of transistor M0 (M1), and C_{b2} is the total substrate capacitance at the drain of transistor M2 (M3). (Notice that the Miller effect associated with transistor M4 (M5) is not considered here in that practical opamps use a cascode output stage to increase the DC gain.)

The slew rate is given by

$$SR = \frac{2 \cdot I_{bias6}}{C_L} = \frac{n \cdot I_{bias8}}{C_L} \tag{7-5}$$

where I_{bias6} is the quiescent bias current in transistor M6 (M7), and I_{bias8} is the bias current in transistor M8. The factor of 2 is due to the fully differential structure. The maximum slewing current is I_{bias6} for one side while the maximum slewing current is $I_{bias5} = I_{bias6}$ for the other side.

The transconductance of an MOS transistor in the saturation region is given [2] by

$$g_m = \sqrt{2 \cdot \mu \cdot C_{ox} \cdot \frac{W}{L} \cdot I_D} = \frac{2 \cdot I_D}{V_{eff}} \tag{7-6}$$

where μ is the charge mobility, C_{ox} is the gate capacitance per unit area, W/L is the transistor dimension, I_D is the drain current, and V_{eff} is the effective gate voltage (the difference between the gate-source voltage and the threshold voltage). To avoid slewing governed by equation (7-2), we have

$$\frac{V_{eff}}{V_{step}} \geq \beta \tag{7-7}$$

This condition cannot be met in data converters such as oversampling ADCs and pipeline ADCs with low gain stages. For instance, V_{step} (differential) can be as high as 4~6 V, with a feedback factor of 0.5, the effective gate voltage of the input devices must be in the neighborhood of 2~3V. This is too high in most practical designs where the effective gate voltage is designed to be ca 200~500 mV within the whole operation range and for all the process corners. Therefore, slewing will occur in data converter applications. We have to guarantee the settling accuracy in order not to introduce distortion. Both the slew rate and the unity-gain bandwidth must be high to increase the settling accuracy. Increasing n (without increasing the input bias current) seems to increase both the slew rate and the unity gain bandwidth, but it decreases the non-dominant pole frequency, reducing the phase margin. We usually have to increase both the input and output currents to improve the settling.

To have a large phase margin for monotonic settling, we require that the non-dominant pole must be much larger than the unity-gain bandwidth. Since the gate-source capacitance is much larger than the gate-drain capacitance in a saturated MOS transistor, the non-dominant pole is considerably lower than the non-dominant pole in the folded-cascode opamp to be discussed. The lower non-dominant pole is one of the reasons for the rejection of the current mirror based opamp for high-speed applications.

For high performance data converters, the thermal noise generated by the opamp could be another limiting factor besides the kT/C noise. The input referred noise power spectral density of an MOS transistor in the saturation region is given [2] by

$$S(f)|_{MOS} = \frac{8}{3} \cdot k \cdot T \cdot \frac{1}{g_m} \tag{7-8}$$

where k is Boltzmann constant, T is the Kevin temperature, and g_m is the transconductance.

In the current-mirror based opamp, all the transistors except the tail bias current transistor M8 contribute to the total noise. The noise introduced by transistor M8 is only a common-mode noise seen by the input pair, having no impact on the differential signals. Therefore, the input referred noise spectral density of the current-mirror based opamp is given by

$$\begin{aligned} S(f) &= \frac{16}{3} \cdot k \cdot T \cdot \frac{1}{g_{m0}} \cdot \left\{ 1 + \frac{g_{m2}}{g_{m0}} + \frac{g_{m4} + g_{m6}}{g_{m0}} \cdot \frac{1}{n^2} \right\} \\ &\approx \frac{16}{3} \cdot k \cdot T \cdot \frac{1}{g_{m0}} \cdot \left\{ 1 + \frac{g_{m2}}{g_{m0}} \cdot \left(1 + \frac{1}{n}\right) + \frac{g_{m6}}{g_{m0}} \cdot \frac{1}{n^2} \right\} \end{aligned} \tag{7-9}$$

where g_{m0}, g_{m2}, g_{m4}, and g_{m6} are the transconductance of transistor M0, M2, M4, and M6, respectively, and n is the current mirror ratio. It is seen that by increasing the current mirror ratio, we can reduce the noise contribution from the output devices M4 and M6. It also increases the unity gain bandwidth according to equation (7-3), but it decreases the non-dominant pole frequency according to equation (7-4). The phase margin, and therefore the settling time degrades significantly with a large current mirror ratio n.

7.2.3 Folded-Cascode Opamp

In Fig. 7-2 we show the folded-cascode opamp [1]. The unity-gain bandwidth is given by

$$\omega_u = \frac{g_{m0}}{C_L} \tag{7-10}$$

7.2 Operational Amplifiers

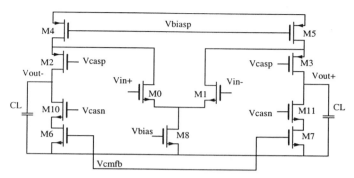

Figure 7-2 Folded-cascode single-stage opamp.

where g_{m0} is the transconductance of the input device M0 (M1), C_L is the capacitive load.

The first non-dominant pole is due to the folded cascode transistors, and is given by

$$\omega_n \approx \frac{g_{m2}}{C_{gs2} + C_{gd4} + C_{gd0} + C_{b2}} \tag{7-11}$$

where g_{m2} is the transconductance of transistor M2 (M3), C_{gs2} is the gate-source parasitic capacitance of transistor M2 (M3), C_{gd4} is the gate-drain parasitic capacitance of transistor M4 (M5), C_{gd0} is the gate-drain parasitic capacitance of transistor M0 (M1), and C_{b2} is the total substrate capacitance at the source of transistor M2 (M3).

Comparing equations (7-4) and (7-11) we realize that the non-dominant pole in a folded-cascode opamp is about 2 times higher than that in a current-mirror based opamp (the current mirror ratio n is assumed to be 1) since there is only one dominating gate-source capacitor in the folded-cascode opamp. Therefore folded-cascode opamps are preferred over current-mirror based opamps for high-speed applications.

The quiescent current in M4(M5) is usually set to $0.5 \cdot I_{bias8} + I_{bias6}$ to avoid systematic offset, where I_{bias8} is the bias current in M8 and I_{bias6} is the bias current in M6(M7). When the input is hard driven, all the tail current I_{bias8} will flow through one transistor, say M0, and no current will flow through transistor M1. Therefore, the maximum charging current at Vout+ is the difference in the quiescent currents of M5 and M7, i.e, $0.5 \cdot I_{bias8}$. The maximum charging current at Vout− becomes $min(0.5 \cdot I_{bias8}, I_{bias6})$ since no current can flow through the PMOS transistor M2 from the drain to the source. To avoid hard turning off of transistor M2 or M3 and unsymmetrical slewing, we usually require $I_{bias6} \geq 0.5 \cdot I_{bias8}$. Therefore, the slew rate is given by

$$SR = \frac{I_{bias8}}{C_L} \tag{7-12}$$

where I_{bias8} is the quiescent bias current in transistor M8. To avoid slewing governed by equation (7-2), we have

$$\frac{V_{eff}}{V_{step}} \geq \beta \tag{7-13}$$

This condition can hardly be met in some data converters such as oversampling ADCs and pipelined ADCs with low gain stages. Therefore, we have to guarantee the settling accuracy in order not to introduce distortion due to the inevitable slewing in these applications. For high-speed applications, we need both a high unity-gain bandwidth and a high slew rate. This means that we need to bias both the input and output devices at a high current and therefore the architecture is not power-efficient just as the current-mirror based opamp.

Since the impedance at the sources of cascode transistors is low, the noise contributions from the cascode transistors can be neglected. Therefore, the input referred noise spectral density of the folded-cascode opamp is given by

$$S(f) = \frac{16}{3} \cdot k \cdot T \cdot \frac{1}{g_{m0}} \cdot \left\{ 1 + \frac{g_{m2} + g_{m6}}{g_{m0}} \right\} \tag{7-14}$$

where g_{m0} is the transconductance of the input device M0 (M1), g_{m4} is the transconductance of transistor M4 (M5), g_{m6} is the transconductance of transistor M6 (M7). In order to reduce the noise, we have to minimize the noise contribution from the output devices M4(M6).

7.2.4 Telescopic Opamp

In order to get rid of PMOS transistors in the signal path, the telescopic opamp [3] shown in Fig. 7-3 can be used. The unity gain bandwidth is given by

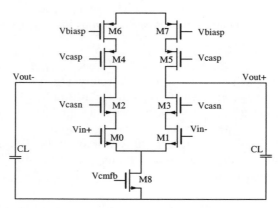

Figure 7-3 Telescopic single-stage opamp.

7.2 Operational Amplifiers

$$\omega_u = \frac{g_{m0}}{C_L} \qquad (7\text{-}15)$$

where g_{m0} is the transconductance of the input device M0 (M1), C_L is the capacitive load.

The first non-dominant pole is created by the cascode transistor M2 (M3), and is given by

$$\omega_n \approx \frac{g_{m2}}{C_{gs2} + C_{gd0} + C_{b2}} \qquad (7\text{-}16)$$

where g_{m2} is the transconductance of transistor M2 (M3), C_{gs2} is the gate-source parasitic capacitance of transistor M2 (M3), C_{gd0} is the gate-drain parasitic capacitance of transistor M0 (M1), and C_{b2} is the total substrate capacitance at the source of transistor M2 (M3).

Due to the use of NMOS transistors in the signal path and less parasitic capacitance at the source of the cascode transistors, the non-dominant pole is higher than that in the current-mirror based opamp and in the folded-cascode opamp.

The slew rate is given by

$$SR = \frac{I_{bias8}}{C_L} \qquad (7\text{-}17)$$

where I_{bias8} is the quiescent bias current in transistor M8. To avoid slewing governed by equation (7-2), we have

$$\frac{V_{eff}}{V_{step}} \geq \beta \qquad (7\text{-}18)$$

This condition hardly be met in some data converters. For instance, V_{step} (differential) can be as high as 4V, with a feedback factor of 0.5, the effective gate voltage of the input devices must be in the neighborhood of 2V. This is too high in most practical designs where the effective gate voltage is designed to be ca 200~500 mV within the whole operation range and for all the process corners. Therefore, the slewing will occur in some applications. We have to guarantee the settling accuracy in order not to introduce distortion. By increasing the bias current, we can increase both the slew rate and the unity-gain bandwidth.

The input referred noise spectral density is given by

$$S(f) = \frac{16}{3} \cdot k \cdot T \cdot \frac{1}{g_{m0}} \cdot \left\{ 1 + \frac{g_{m6}}{g_{m0}} \right\} \qquad (7\text{-}19)$$

where g_{m0} is the transconductance of the input device M0(M1), and g_{m6} is the transconductance of transistor M6 (M7).

Besides the advantages of having a higher non-dominant pole, the telescopic opamp is more power efficient, since no extra currents are needed other than the bias current for the input devices to provide a high slew rate and a high unity-gain bandwidth. The large bias current improves both the slew rate and the unity-gain bandwidth as well reduces the thermal noise.

There is a misconception that there is a drawback due to the three NMOS transistors in series. This makes the output signal swing small. However, for most applications, we only care about the SNR. With the reduced signal swing, we can increase the capacitors to maintain the SNR.

Assume that the average of the maximum source-drain voltage for every transistor is ca 400 mV across process corner and temperature range. We also assume that the supply voltage is 5 V. For the current-mirror based (usually have cascode transistors in the output stage) or the regulated cascode opamps, the fully differential output signal swing is 2*(5-4*0.4)=6.8 Vp-p. For the telescopic opamp, there is one more cascode transistors. Therefore the output signal swing becomes 2*(5-5*0.4)=6Vp-p. By using the telescopic opamp, the signal swing is reduced by 6.8/6. In order to keep the same SNR in a kT/C limited environment, we have to increase the capacitor to (6.8/6)^2, i.e, by 28%. In order to keep the speed, we have to increase the input transconductance g_m and slew rate by 28% and therefore we need to increase the bias current by 28% to 64%. (This large range is due to the fact that we can increase the bias current and the transistor size at the same time or only increase the bias current.) From a pure power dissipation's view point, the telescopic opamp still consumes 18% to 36% less power than other types of opamps. (It is usually a common practice for high speed design that both input and output have comparable bias currents for the current-mirror based and the regulated cascode opamps.) Besides the power advantage, the telescopic opamp is usually faster due to the higher slew rate and the higher non-dominant pole and it is less noisy due to the fewer noise-contributing transistors. Therefore, for SC applications, telescopic opamps should be the top choice.

To utilize the advantages of the telescopic opamp, we need to use different common-mode voltages for the input and output. This can be easily achieved by using a level shifter, either SC or continuous-time.

7.2.5 Gain Boosting Technique

As discussed in previous chapters, we need a very high DC gain in the opamp for high accuracy applications. For example, for a 12-bit settling in an SC amplifier with a gain factor of 2, we need a DC gain over 85 dB. For high-speed applications, short channel transistors are used in opamps which result in a DC gain less than 60 dB. Gain boosting techniques must be employed.

The most popular gain boosting technique is to use a cascode output stage and the gain of the cascode transistors is further increased by employing a feedback [4]. The gain boosting technique applied to the folded-cascode opamp is shown in Fig. 7-4. It can also be readily applied to the current-mirror based opamps and telescopic opamps

7.2 Operational Amplifiers

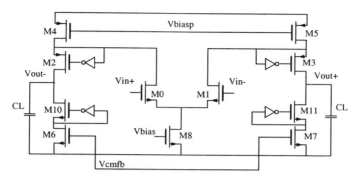

Figure 7-4 Folded-cascode opamp with gain boosting amplifiers.

without much modification. The DC gain is given by

$$A = \frac{g_{m0}}{\left(\frac{g_{ds0} \| g_{ds4}}{A_{M2} \cdot A_p}\right) \| \left(\frac{g_{ds6}}{A_{M10} \cdot A_n}\right)} \qquad (7\text{-}20)$$

where g_{m0} is the transconductance of the input transistor M0 (M1), g_{ds0} is the output conductance of the input transistor M0 (M1), g_{ds4} is the output conductance of the current source transistor M4 (M5), g_{ds6} is the output conductance of the current source transistor M6 (M7), A_{M2} is the gain of the cascode transistor M2 (M3), A_{M10} is the gain of the cascode transistor M10 (M11), A_P is the gain of the gain boosting amplifier for the P-type cascode transistor M2 (M3), and A_N is the gain of the gain boosting amplifier for the N-type transistor M10 (M11).

With a gain for every transistor around 30-40 dB, it is fairly easy to achieve a DC gain of over 90 dB with the architecture shown in Fig. 7-4.

However, due to the gain boosting, a doublet is introduced. In order to have a single-pole settling, the doublet's frequency should be larger than the -3-dB bandwidth of the closed loop. For the stability concern, this doublet's frequency must be less than the non-dominant pole of the main opamp [1] [4]. Therefore, we have

$$\beta \cdot \omega_u < \omega_{gu} < \omega_n \qquad (7\text{-}21)$$

where β is the feedback factor, ω_u is the unity-gain bandwidth of the opamp, ω_{gu} is the unity-gain bandwidth of the gain-boosting amplifier, and ω_n is the non-dominant pole frequency of the opamp. This equation can be easily guaranteed since the load capacitance of the gain-boosting amplifier is considerably smaller than the load capacitance of the main amplifier. Also due to the fact that much smaller capacitance loads the gain-boosting amplifiers, considerably smaller currents and smaller transistors can be used for the gain-boosting amplifiers.

7.2.6 Common-Mode Feedback

Fully differential circuits process signals fully-differentially without any control over the common-mode potential. It is the common-mode feedback circuit that keeps the common-mode potential stable. The generalized fully differential opamp with a CMFB [5] is shown in Fig. 7-5

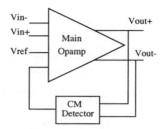

Figure 7-5 Fully differential opamp with CMFB.

The main opamp processes the fully differential input signals. The common-mode (CM) detector detects the common-mode voltage in the outputs. This detected voltage is compared to a desired common-mode reference voltage. Through the feedback action, the detected common-mode voltage is kept close to the common-mode reference voltage. There are several critical factors in designing CMFB circuits. The CM detector should not load the main opamp. The CM detector should have a larger input signal range. The interaction between the CMFB loop and the fully-differential signal processing should be minimized. The CMFB loop should be fast with a moderate to high gain. Therefore it is preferable to feed the CMFB signal to the output devices directly as indicated in Fig. 7-1 and Fig. 7-2

There are two main categories of CMFB circuits. One is the continuous-time and the other is the SC. In continuous-time CMFB circuits [2][5], current mirrors are needed. The extra poles introduced in the current mirrors (and other parts) in the continuous-time CMFB decrease the bandwidth and increase the settling time. Even when NMOS current-mirrors are used, the settling time still degrades if very large bias currents are not used to push the pole frequencies high. Besides the slow speed in continuous-time CMFB circuit, the input signal range of the CM detector is limited and the interference of the CMFB with the fully-differential signal processing is also high. It is therefore a much better choice to use an SC CMFB for high-speed high-accuracy SC circuits. Such an SC CMFB [6] is shown in Fig. 7-6.

Capacitors C_{cm} generate the average of the output voltages (V_{out+} and V_{out-}) of the opamp. The generated signal V_{cmfb} is used to control the current sources in the opamps (e.g., see Fig. 7-1 - Fig. 7-4). The DC voltage across C_{cm} is determined by capacitor C_S which is switched between a bias voltage V_b and the desired common-mode level (V_{cm}). This circuit acts as a simple SC low-pass filter having a DC output signal.

The use of SC CMFB only results in a little bit more capacitive load for the opamp.

7.3 Voltage Comparators

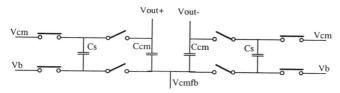

Figure 7-6 SC CMFB.

The bandwidth is determined by the transconductance of the current sources in the opamp which are controlled by V_{cmfb} and the capacitive load. It is therefore a good choice to use NMOS current source controlled by V_{cmfb} as shown in Fig. 7-1-Fig. 7-4 to increase the bandwidth of the CMFB.

Another advantage of using the SC CMFB is the freedom to choose a different common-mode voltage for the input and for the output. This is very important in order to fully utilize the advantages of the telescopic opamp that usually has a smaller input common-mode voltage than the output common-mode voltage.

7.3 VOLTAGE COMPARATORS

High-performance comparators need to amplify a small input voltage (or the difference between the input voltage and a reference voltage) to a level large enough to be detected by digital logic circuits within a very short time. Therefore, we need both high gain and high bandwidth in the comparators. Increase in the gain and in the bandwidth is usually contradictory. High gain yields low bandwidth, and vice versa.

The input to a comparator can also be very large and it overdrives the comparator. If the recovery time is too long, the accuracy of the next comparison will be influenced. The large positive gate voltage of an MOS transistor can also introduce the charge trapping mechanism where electrons are trapped via tunneling (electrons tunnel to oxide traps close to the conduction band). The release of them takes an excess time. Therefore, the voltages within a comparator needs to be clipped for high-speed comparison.

7.3.1 Amplifier-Type Comparator

The natural choice to design a comparator is to use an amplifier. A small voltage at the input is then amplified to a value large enough to be detected by the following digital logic circuits.

Suppose the amplifier consists of n stages each stage being a single-pole system as shown in Fig. 7-7.

The total gain is therefore given by

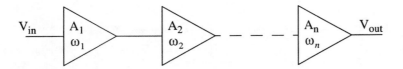

Figure 7-7 An amplifier type comparator.

$$A = A_1 \cdot A_2 \cdot \ldots \cdot A_n = \prod_{i=1}^{n} A_i \tag{7-22}$$

where A_i is the gain of the i-th stage.

In a single-pole system, the settling time constant is related to the -3-dB bandwidth by

$$\tau_i = \frac{1}{\omega_i} \tag{7-23}$$

where τ_i is the settling-time constant of the i-th stage and ω_i is the -3-dB bandwidth of the i-th stage. And the open-loop -3-dB bandwidth is related to the unity gain bandwidth by

$$\omega_i = \frac{\omega_{ui}}{A_i} \tag{7-24}$$

where ω_{ui} is the unity-gain bandwidth of the i-th stage.

Therefore, the settling time constant of the cascaded amplifier can be approximated by

$$\tau = \sum_{i=1}^{n} \tau_i = \sum_{i=1}^{n} \frac{1}{\omega_i} = \sum_{i=1}^{n} \frac{A_i}{\omega_{ui}} \tag{7-25}$$

From equations (7-22) and (7-25) it is clear that it is better to cascade more low-gain stages than to use fewer high-gain stages in order to reduce the settling time.

Suppose that each stage has the same DC gain A_0 and the same unity-gain bandwidth ω_{u0}, equations (7-22) and (7-25) become

$$A = \prod_{i=1}^{n} A_i = A_0^n \tag{7-26}$$

and

7.3 Voltage Comparators

$$\tau = \sum_{i=1}^{n} \frac{A_i}{\omega_{ui}} = \frac{n \cdot A_0}{\omega_{u0}} \qquad (7\text{-}27)$$

If a single stage is used to achieve the same gain, the settling time constant would become A_0^n/ω_{u0}, which is much larger than the settling time given by (7-27). This is why under no circumstance should an opamp be used as a high-performance comparator.

Suppose we need to design a comparator for a 12-bit accuracy, we need to amplify an LSB-voltage to a voltage that can be detected by digital logic circuits. If the LSB-voltage is assumed to be 0.1 mV and the voltage detectable by digital logic circuits is assumed to be 2 V, the small signal gain of the comparator must be larger than 86 dB and the comparator must accomplish the amplification within a given time interval. Recent study has shown that the comparator gain must be much larger than the above given value if metastability is considered in a pipeline A/D converter [7]. If we consider bit error rate, the gain has to be much larger if no positive feedback latch is used. Depending on the use of the comparator in the whole A/D system, the settling of the comparator might be coupled with the settling of the preceding gain amplifier (as in certain type of pipeline A/D converters). Therefore, we cannot assume that we can allocate half of the clock period for the comparator to settle as if it was driven by a voltage source. The settling time must be much shorter than half of the clock period. Even with a multiple stage design, it is difficult to achieve the performance needed for a 12-bit 50-MS/s A/D converter. (Notice that if the settling of comparison is not coupled with the settling of the preceding gain amplifier, the settling of auto-zeroing will be coupled in an SC pipeline A/D converter.)

7.3.2 Latch-Type Comparator

To have a fast comparator, we can use latch-type comparators. It consists of two phases. During the first phase (track mode), the output is reset and the input voltage is tracked, and during the second phase (latch mode), the output is toggled by using a positive feedback (we will show the circuit realization later on). Thanks to the positive feedback, a very fast comparison can be achieved. The time constant of the latch in the latch mode can be analyzed by using two back-to-back gain amplifiers as shown in Fig. 7-8.

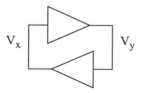

Figure 7-8 An latch type comparator in its latch mode.

Assume that the two amplifiers (single pole) are the same having an input transcon-

ductance g_m, output impedance R_{out}, and load capacitance C_L.
A linearized model gives that

$$g_m \cdot V_x + \frac{V_y}{R_{out}} = -C_L \cdot \frac{dV_y}{dt} \tag{7-28}$$

and

$$g_m \cdot V_y + \frac{V_x}{R_{out}} = -C_L \cdot \frac{dV_x}{dt} \tag{7-29}$$

Rearranging equations (7-28) and (7-29), we have

$$A \cdot V_x + V_y = -\tau \cdot \frac{dV_y}{dt} \tag{7-30}$$

and

$$A \cdot V_y + V_x = -\tau \cdot \frac{dV_x}{dt} \tag{7-31}$$

where A is the DC gain given by $A = g_m \cdot R_{out}$, and τ is the settling time constant of the amplifier given by $\tau = C_L \cdot R_{out} = A/\omega_u = 1/\omega_{-3dB}$, (where ω_u is the unity gain bandwidth and ω_{-3dB} is the -3-dB bandwidth).

Subtracting equation (7-30) from (7-31), we have

$$\Delta V = \frac{\tau}{A-1} \cdot \frac{d}{dt}\Delta V \approx \frac{\tau}{A} \cdot \frac{d}{dt}\Delta V = \omega_u \cdot \frac{d}{dt}\Delta V \tag{7-32}$$

where $\Delta V = V_x - V_y$.

Solving equation (7-32), we have

$$\Delta V = \Delta V_0 \cdot e^{\omega_u \cdot t} = \Delta V_0 \cdot e^{\frac{t}{\tau_l}} \tag{7-33}$$

where ΔV_0 is the initial voltage and τ_l is the settling time constant of the latch given by

$$\tau_l = \frac{1}{\omega_u} = \frac{\tau}{A} = \frac{C_L}{g_m} \tag{7-34}$$

Now it is seen that the settling time constant of the positive feedback is the settling time constant of the individual amplifier divided by the gain of the amplifier. By minimizing the load capacitance and increasing the input transconductance, a very fast toggling can be achieved.

7.3 Voltage Comparators

In order to generate a voltage difference ΔV_{logic}, necessary to be processed by the following digital circuits, we find the time given by

$$T_{\text{latch}} = \tau_l \cdot \ln\left(\frac{\Delta V_{\text{logic}}}{\Delta V_0}\right) \tag{7-35}$$

If the time is larger than the allowed time for the latch phase, metastability occurs. Even the noise at the input can cause metastability [7].

7.3.3 Metal Stability and Error Probability

A comparator has a limited time to settle. If at the end of the available time, the output of the comparator has a voltage that the following logic circuit cannot detect, metastability occurs. When the metastability occurs, the digital output is neither 1 nor 0. This introduces errors in the communication system. The error probability can be approximated by the LSB voltage and the minimum input voltage that the comparator can generate an output that the following digital circuit can generate a deterministic value 1 or 0.

By using equation (7-33), the minimum voltage needed to generate a voltage ΔV_{logic} to toggle the following digital circuit is given by

$$V_{\text{min}} = \Delta V_{\text{logic}} \cdot e^{-\frac{t_0}{\tau_l}} \tag{7-36}$$

where t_0 is the maximum time settling time for the latch-type comparator and τ_l is the settling time constant given by equation (7-34).

If there is a pre-amplifier preceding the latch-type comparator, the error probability in a comparator due to metastability is given by

$$P(e) = \frac{V_{\text{min}}}{V_{LSB}} \cdot \frac{1}{A} = \frac{\Delta V_{\text{logic}}}{V_{LSB}} \cdot \frac{1}{A} \cdot e^{-\frac{t_0}{\tau_l}}$$

$$\approx 2^n \cdot \frac{1}{A} \cdot e^{-\frac{t_0}{\tau_l}} \tag{7-37}$$

where V_{min} is the minimum voltage at the comparator input, V_{LSB} is the LSB voltage, A is the pre-amplifier gain, n is the number of bits of the ADC, t_0 is the maximum time settling time for the latch-type comparator and τ_l is the settling time constant given by equation (7-34). If a two-phase clock is used, the settling time for the latch can be approximated by half of the clock period. We have the following approximation

$$P(e) = 2^n \cdot \frac{1}{A} \cdot e^{-\pi \cdot \frac{BW}{f_s}} \tag{7-38}$$

where BW is the -3dB bandwidth of the latch-type comparator during the latch phase and f_s is the sampling frequency of the ADC. The same equation holds true for an amplifier-type comparator. However its -3dB bandwidth is much smaller than the -3dB bandwidth of a latch-type comparator as can be seen by comparing equations (7-25) and (7-34). The -3-dB bandwidth of a latch-type comparator is equal to the unity-gain bandwidth of the back-to-back gain amplifiers, while the -3-dB bandwidth of an amplifier-type comparator is equal to its unity gain bandwidth divided by its DC gain.

In Fig. 7-9, we show the simulated error probability as a function of the ratio of the -

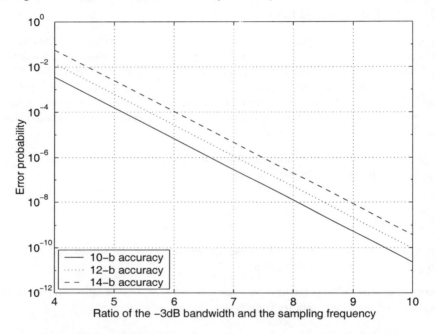

Figure 7-9   Simulated error probability vs. the ratio of the -3-dB bandwidth and the sampling frequency. The gain is assumed to be 1.

3-dB bandwidth and the sampling frequency. The pre-amplifier gain is assumed to be unity. It is seen that by increasing the bandwidth of the latch-type comparator, we can reduce the error probability drastically due to the exponential relationship.

In Fig. 7-10, we show the simulated error probability vs. the comparator gain. The -3-dB bandwidth is assumed to be 4 times larger than the sampling frequency. It is seen that it is not effective to reduce the error probability by increasing the gain due to the linear relationship. Therefore for high-speed ADCs, latch-type comparators are much preferred.

Notice that the error due to the metastability in a comparator does not necessarily introduce an error in the ADC, depending on coding and error correction. Also notice that an error in the ADC does not necessarily imply an error in the whole communication system, i.e, the bit error rate in the ADC does not necessarily impose a lower limit

7.3 Voltage Comparators

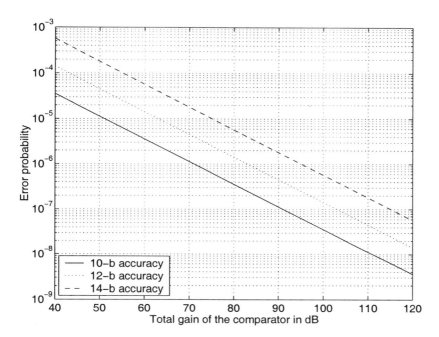

Figure 7-10 Simulated error probability vs. the comparator gain. The -3dB bandwidth is assumed to be 4 times larger than the sampling frequency.

on the bit error rate of the communications system, depending on the ADC architecture and the specific nature of the communication system. As a general rule, however, we should make the error probability in an comparator several order of magnitude lower than the required system bit error rate. For latch-type comparator, this can be easily achieved without much penalty in power dissipation due to the exponential relationship between the error probability and the -3-dB bandwidth.

7.3.4 Offset Cancellation Techniques

Depending on the A/D converter architecture, the offset within a comparator may limit the A/D converter performance. To reduce the offset, offset cancellation or auto-zeroing techniques need to be used. There are two major offset cancellation techniques [8]. One is based on input offset storage, and the other is based on output offset storage as shown in Fig. 7-11.

Both techniques need two non-overlapping clock phases as indicated in the figure. Each topology consists of a gain stage (pre-amplifier), an offset storage capacitor, and a second-stage comparator that could be a latch. With the input offset storage technique, the cancellation is performed by closing a unity-gain loop around the pre-amplifier and storing the offset on the input coupling capacitor. With the output offset storage technique, the offset is cancelled by short-circuiting the pre-amplifier inputs and storing the amplified offset on the output coupling capacitor.

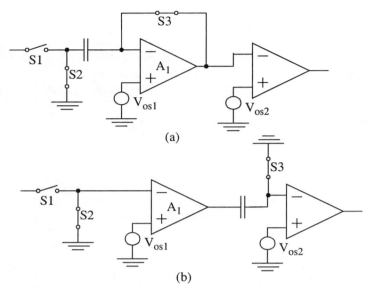

Figure 7-11 Offset cancellation techniques: (a) input offset storage and (b) output offset storage.

With the input offset storage technique, the input referred offset is given by

$$V_{os} = \frac{V_{os1}}{1+A_1} + \frac{\Delta Q}{C} + \frac{V_{os2}}{A_1} \qquad (7\text{-}39)$$

where V_{os1} and A_1 are the input offset and the gain of the pre-amplifier, respectively, ΔQ is the charge injection from switch S3 to the capacitor, and V_{os2} is the offset in the second-stage comparator.

With the output offset storage technique, the input referred offset is given by

$$V_{os} = \frac{\Delta Q}{A_1 \cdot C} + \frac{V_{os2}}{A_1} \qquad (7\text{-}40)$$

where V_{os1} and A_1 are the input offset and the gain of the pre-amplifier, respectively, ΔQ is the charge injection from switch S3 to the capacitor, and V_{os2} is the offset in the second-stage comparator.

Comparing equations (7-39) and (7-40), we see that for similar pre-amplifiers, the input referred offset using the output offset storage is smaller than that using the input offset storage. However, there are some fundamental differences between these two techniques which may make it less attractive to use the output offset storage technique.

With the input offset storage technique, the input to the whole comparator is AC coupled and therefore the input common-mode range is not of a main concern. Since the pre-amplifier is configured in a closed loop fashion, a high gain in the pre-amplifier can be employed to reduce the offset, though it will slow down the comparison as dis-

7.3 Voltage Comparators

cussed in the previous section.

With the output offset storage, the input to the whole comparator is DC coupled and therefore the input common-mode range is limited. Since the pre-amplifier is configured as a open loop operation, the gain has to be limited (to usually less than 20 dB) to ensure operation in the active region across the process corner and temperature range.

Both techniques can be mixed and used repeatedly for multi-stage comparators.

7.3.5 Circuit Example

All the techniques discussed above can be combined to design high-speed, high resolution comparators. An example is shown in Fig. 7-12. (If the offset in the comparator is of less concern as in A/D converters with digital correction, the latch or the latch with a pre-amplifier to prevent kickback can be used.)

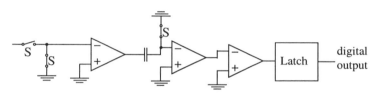

Figure 7-12 A high-speed high-accuracy comparator architecture.

The comparator consists of three pre-amplifiers followed by a positive feedback latch. The input of the first pre-amplifier is DC coupled while the coupling between the first and second pre-amplifier is AC coupled.

In order to have a high-accuracy comparison, the offset needs to be compensated. We use the output offset storage technique to cancel the offset. If the gain of the pre-amplifier is not too large so that the gain of the pre-amplifier remains the same on both phases and the pre-amplifier is not saturated on the amplification phase, this arrangement eliminates the offset error [9] [10]. For a comparator, the offset is only of concern when the input signal is very small, and when the input signal is very small, the pre-amplifier usually does not saturate and it makes this technique an attractive option.

The clock feedthrough on the AC coupling capacitors between the first and the second pre-amplifier is also reduced by the gain of the first pre-amplifier since the input signal has already been amplified by the first pre-amplifier. In order to reduce the metastability of the positive-feedback latch, two pre-amplifiers proceed the latch.

Notice that if the pre-amplifiers are overdriven, it may take an excess time to recover during the reset phase. Clipping is sometimes needed. However, proper designed pre-amplifiers can automatically clip the output voltages.

When the noise is of concern, extra capacitors are needed to be placed at the input of the comparator. Since the comparator is usually preceded by a sample-and-hold amplifier, the noise bandwidth due to the sample-and-hold amplifier is considerably smaller than that given by the capacitance and switch-on resistance. Therefore, the total noise power is less than kT/C and we can use capacitors in the gain amplifier.

Due to the use of the output offset storage, the pre-amplifier must have a low gain. In order to have a wide bandwidth, a fully differential pair with a diode load is one of the choices. Such an example is show in Fig. 7-13.

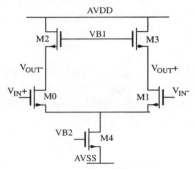

Figure 7-13 The pre-amplifier circuit.

The pre-amplifier consists of the fully differential pair M0 and M1. The load is an NMOS pair contrary to the general practice of using PMOS transistors. The use of the NMOS pair as the load eliminates the need of a level-shifting circuit in order to drive another pre-amplifier. The output voltage is automatically clipped below the positive supply voltage AVDD. The output voltage is also automatically clipped above the negative supply voltage AVSS due to the fact that one of the fully differential pair conducts all the current provided by the current source transistor M4 and therefore the output voltage cannot be smaller than the drain potential of M4. The current source transistor M4 is controlled by an SC common-mode feedback circuit (not shown here) in order for the pre-amplifier to be able to work within a wide common-mode range. If the common-mode feedback is not used, transistor M4 can be biased at a proper current.

The gain of the pre-amplifier is given by

$$A_{pre} \approx \frac{g_{m0}}{g_{m2}} \approx \sqrt{\frac{(W/L)_{M0}}{(W/L)_{M2}}} \qquad (7\text{-}41)$$

where g_{m0} and g_{m2} are the transconductance of transistor M0 (M1) and M2 (M3), $(W/L)_{M0}$ and $(W/L)_{M2}$ are the dimension ratios of transistor M0 (M1) and M2 (M3), respectively. The practical value of the gain is usually around 4 to 10.

The pole frequency at the source of transistor M2 (M3) determines the -3-dB frequency of the pre-amplifier. It is given by

$$\omega_{-3dB}\big|_{pre} = \frac{g_{m2}}{C_L} \qquad (7\text{-}42)$$

where g_{m2} is the transconductance of transistor M2 (M3) and C_L is the total load capacitance. To have a high bandwidth with some gain given by equation (7-41), a large bias current is needed.

One realization of the positive feedback latches [11] is shown in Fig. 7-14. It only con-

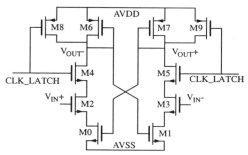

Figure 7-14   The positive-feedback latch circuit.

sumes power during the transition.

The latch basically consists of a back-to-back inverter pair consisting of transistors M0 and M6, and transistors M1 and M7, respectively. When the latch signal clk_latch is high, transistors M4 and M5 open, the potential difference at the outputs due to the input voltage is amplified by the positive feedback. When the latch signal clk_latch goes low, transistors M4 and M5 cut off. To reduce hysteresis, transistors M8 and M9 are used to set the two outputs high.

It is also possible to connect the input transistors M2 and M3 in parallel with transistors M0 and M1, respectively [12], the latch then will consume power in the latch mode.

As discussed in section 7.3.2 and given by equation (7-34), the speed of the latch is determined by the transconductance and the capacitive load. By properly dimensioning the transistor size, the latch is capable of flipping in less than 0.65 ns for a 0.1 mV input in a typical CMOS process. Though the designed positive-feedback latch is fast enough for most high speed A/D converters, without pre-amplifiers the latch will influence the proceeding track-and-hold circuit due to the charge transfer into or out of the latch input when the latch goes from the track mode to the latch mode (a mechanism called kickback). Also the offset voltage in a well-laid out differential pair is still in the tens and hundreds of mV range, far larger than the minimum voltage needed to toggle the latch, the offset compensation in the pre-amplifiers is necessary for A/D converters without digital correction. With digital correction, however, relatively large offset errors in comparators are tolerable, making the use of only the latch such as shown in Fig. 7-14 acceptable.

7.4   CURRENT COMPARATORS

Current comparators or quantizers are used in current-mode circuits. The input to current quantizers is usually a current rather than a voltage. For current inputs, we can directly connect the inputs together to realize summation or subtraction of currents. Different current quantizer structures exist. In Fig. 7-15, we show the widely used current quantizers [13].

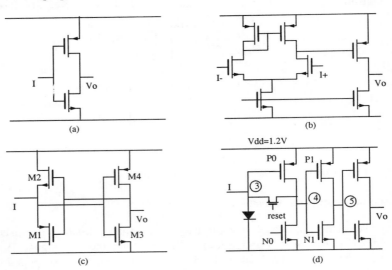

Figure 7-15 Current comparators. (a) inverter as a current quantizer; (b) fully differential voltage comparator as a current quantizer; (c) low-impedance current quantizer; and (d) ultra low-voltage current quantizer.

For the current quantizer of Fig. 7-15 (a), the input current charges or discharges the input parasitic capacitance and causes the potential at the inverter input to change. For positive input current I, i.e., the current is forced into the inverter, the voltage at the inverter input ramps up until it hits the upper voltage rail V_{dd} causing the inverter output to go low. For negative input current I, the current is pulled to ground and the inverter output goes high. For the current quantizer of Fig. 7-15 (b), fully differential SI circuits can directly drive it. The currents from the preceding SI circuits charge or discharge the gate parasitic capacitance of the input transistors of the comparator depending on the direction of the currents. When one branch of the differential outputs from the preceding SI circuits charges the gate parasitic capacitance of one input transistor, the other discharges the gate parasitic capacitance of the other input transistor. The voltages at the differential nodes ramp in opposite direction towards V_{dd} or zero, respectively. The differential voltage at the comparator inputs thus changes according to the direction of the current flows and the output is produced according to the differential voltage. The current quantizers of Fig. 7-15 (a) and (b) are not true current quantizer in the sense that the input impedance is high. The potential change at the current quantizer inputs is large. For high speed applications, low-impedance current quantizers are needed.

The low-impedance current quantizer is shown in Fig. 7-15 (c). The drawback is the large quiescent power consumption. When the input current I flows into the current quantizer of Fig. 7-15 (c), all the current must be sunk by the p-transistor M1, making its gate-source voltage less than the value of its threshold voltage. When the input current I flows from the current quantizer, all the current must be provided by the n-transistor M2, making its gate-source voltage larger than its threshold voltage. Therefore, the gates of M1 and M2 experience a large potential change when the input current changes directions. Due to the low input impedance, the potential change at the input is small. Transistors M3 and M4 act as a voltage amplifier. Small voltage variation at the input due to the input current is amplified. The speed of the current quantizer is determined by the input impedance, the input parasitic capacitance, and the response time of the voltage amplifier consisting of M3 and M4. After optimization, a settling time of 10 ns is achieved with a current step input of 0.5 μA. To use the current quantizer of Fig. 7-15 (c), a fully differential to single ended converter is needed before driving the current quantizer.

To achieve a high speed and high resolution current comparison with a ultra low supply voltage (e.g., 1.5 V), we can use a resettable current quantizer shown in Fig. 7-15 (d). When Reset is high, transistor P0 is diode connected and transistor N0 provides the bias current. We design the DC voltage at node 4 close to the corresponding DC voltage of the driving circuit. Transistors N1 and P1 form an inverter and are dimensioned to have high speed. A minimum sized inverter is used as the load. Due to the very low supply voltage, only the branch consisting of transistors N0 and P0 conducts DC current. This DC current can be designed very small (< 10 uA). Therefore, the current quantizer does not consume much static power. When Reset is high and there is a current I_{in} flowing into the current quantizer, the potential of nodes 3 and 4 changes. When Reset goes low, the gate of P0 is isolated from its drain and the gate voltage changes the potential at node 4. At the same time, the gate voltage keeps changing due to the input current. This makes the potential change at node 4 even faster. Therefore, a fast and accurate comparison can be accomplished. To reduce the settling time when Reset goes high, we use a diode at the input to limit the potential. With an input current of 50 nA, a comparison time of less than 0.25 us is achieved. When the input current increases to 1 uA, the comparison time is less than 15 ns. And the power consumption is less than 2.4 uW for the nominal process parameters at the room temperature. To use the current quantizer of Fig. 7-15 (d), a fully differential to single ended converter is needed before driving the current quantizer.

As discussed above, a latch-type voltage comparator usually follows the current comparator to reduce the error probability.

7.5 VOLTAGE AND CURRENT REFERENCES

7.5.1 Voltage Reference Generation

For high-performance data converters, high-quality reference voltages or currents are needed. Bandgap reference circuits can generate a supply-voltage independent, process independent, and temperature independent voltage [14]. The principle behind a bandgap reference is to utilize the negative temperature coefficient of the base-emitter voltage of a bipolar transistor and the positive temperature coefficient of the thermal voltage (kT/q). Therefore, for high-performance data converters, we must have access to bipolar transistors.

In CMOS process, we can resort to parasitic PNP transistors. A bandgap circuit in CMOS process in shown in Fig. 7-16.

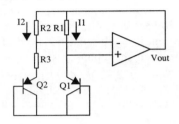

Figure 7-16 Bandgap reference in CMOS process.

Due to the high DC gain in the amplifier, the voltage across the resistors R1 and R2 have equal voltage, forcing the currents ratioed by the ratio of R1 and R2. Therefore the difference between the base-emitter voltages is given by

$$\Delta V_{BE} = V_T \cdot \ln\left(\frac{I_1}{I_2} \cdot \frac{A_2}{A_1}\right) = V_T \cdot \ln\left(\frac{R_2}{R_1} \cdot \frac{A_2}{A_1}\right) \qquad (7\text{-}43)$$

where V_T is the thermal voltage (kT/q, ~26 mV at 300 K), A_1 and A_2 are the emitter area of transistors Q1 and Q2, respectively. This voltage difference is also the voltage across resistor R3. Since R2 and R3 have the same current, the voltage across R2 is given by

$$V_{R2} = \frac{R_2}{R_3} \cdot V_{BE} = \frac{R_2}{R_3} \cdot V_T \cdot \ln\left(\frac{R_2}{R_1} \cdot \frac{A_2}{A_1}\right) \qquad (7\text{-}44)$$

Since resistors R1 and R2 have the same voltage, we have

$$V_{out} = V_{BE1} + \frac{R_2}{R_3} \cdot V_T \cdot \ln\left(\frac{R_2}{R_1} \cdot \frac{A_2}{A_1}\right) \qquad (7\text{-}45)$$

It is seen that the output voltage is determined by the base-emitter voltage that has a negative temperature coefficient and the thermal voltage that have a positive temperature coefficient. By choosing the constant that multiplies with the thermal voltage, we can have a temperature coefficient of the output voltage close to zero at a given temperature. With one accurate temperature independent voltage source, we can generate other voltage references by using a gain amplifier. Since the offset in the amplifier introduces an error in the output voltage, large input devices should be used to minimize the offset.

Since the parasitic PNP transistors in CMOS have a gain of 2~4, the Darlington configuration can be used to provide a higher gain. It is also critical to bias the PNP transistors in the high current gain region.

Bandgap circuits usually have two stable operation points. A start-up circuit is needed. Less known for high performance A/D converters is the noise filtering. All the references should be low-pass filtered by R-C filters with R being poly resistor and C being MOS capacitor.

7.5.2 Current Reference Generation

Accurate currents can not be generated solely on chip. We have to resort to precision external resistors. To generate an accurate current reference, we first need to generate an accurate voltage on chip using the bandgap circuit. By adding an external precision resistor with a low temperature coefficient, we can generate an accurate current reference. An example is shown in Fig. 7-17

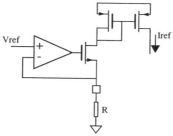

Figure 7-17 Current reference generation.

Assume that the PMOS current mirror transistors are matched and that the gain in the amplifier is infinite. The reference current is then given by

$$I_{ref} = \frac{V_{ref}}{R} \qquad (7\text{-}46)$$

7.5.3 Reference Buffer

The voltage references in an SC data converter need to charge and discharge a capacitor. Assuming that the capacitor is charged to V_{ref} and then completely discharged, the capacitive load has an equivalent impedance given by

$$R_{eq} = \frac{T}{C} \qquad (7\text{-}47)$$

where T is the clock period and C is the capacitance. Notice that if the capacitor is charged to V_{ref} and then charged to $-V_{ref}$ in more advanced configurations discussed in this book, there is a factor of 0.5.

If the clock period is small and the capacitive load is large, the equivalent impedance is low. Driving a low impedance needs a low-impedance voltage sources. Therefore, buffering the reference voltage is a must. In Fig. 7-18, we show a block diagram of the reference buffer.

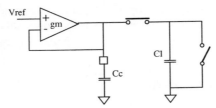

Figure 7-18   Voltage reference buffer.

The output impedance of the buffer is determined by the transconductance of the input devices. If the charging of the load is data independent (all high-performance data converters should have this kind of configuration.), the finite output impedance of the buffer does not introduce distortion but an offset voltage. And this offset voltage can be easily compensate by providing a DC current to the output. It is also possible to reduce the output impedance by using a two-stage amplifier with compensation.

Besides the equivalent low impedance of the SC load, discharging and discharging of the load call for instantaneous action, i.e, large instantaneous currents are needed. To address this issue, the most effective way is to provide a large compensation capacitor Cc as a charge reserve. This large compensation capacitor also limits the noise bandwidth. It is common to use an external capacitor to achieve the best performance. When external capacitors are used, damping of the resonance due to the bond wire and packaging inductance is necessary.

7.6   SUMMARY

In this chapter, we have discussed the basic building blocks for data converters including opamps, voltage and current comparators, reference generators, and reference buffers. For high-performance SC data converters, the telescopic opamp with the SC CMFB is the recommended top choice since it is most power-efficient, least noisy, and fastest. To boost the gain, the gain-boosting technique is recommended. For voltage comparators, the best choice is the positive-feedback latch preceded by low-gain preamplifier(s). When a low-offset is needed, auto-zeroing is recommended. The key to high-performance current comparator is the low-input impedance such that current sources having high output impedance can directly drive it. A complete data converter needs on-chip band-gap reference using the parasitic PNP transistors. With a high-accuracy voltage generated on-chip, we can derive an accurate current by resorting to an external precision resistor. To buffer the voltage reference for high performance data converters, we recommend a large external capacitor.

REFERENCES

[1] J. Huijsing, R. Plassche, and W. Sansen, *Analog Circuit Design*, Kluwer Academic Publishers, 1993.

[2] D. Johns and K. Martin, *Analog Integrated Circuit Design*, John Wiley & Sons, Inc., 1997.

[3] G. Nicollini, P. Confalonieri, and D. Senderowicz, "A fully differential sampled-and-hold circuit for high speed applications," *IEEE J. Solid-State Circuits*, pp. 1461-1465, Oct. 1989.

[4] K. Bult and G. Geelen, "A fast-settling CMOS opamp for SC circuits with 90-dB DC gain, *IEEE J. Solid-State Circuits*, pp. 1397-1384, Dec. 1990.

[5] J. E. Duque-Carrillo, "Control of the common-mode components in CMOS continuous-time fully differential signal processing," *Kluwer J. Analog Integrated Circuits and Signal Processing*, pp.131-139, 1993.

[6] D. Senderowicz, S. Dreyer, J. Huggins, C. Rahim, and C. Laber, "A family of differential NMOS analog circuits for a PCM codec filter chip," *IEEE J. Solid-State Circuit*, pp. 1014-1023, Dec. 1982.

[7] J.-E. Eklund, *"A/D conversion systems for sensor systems,"* Ph.D. dissertation, Linkoping University, 1997.

[8] B. Razavi and B. Wooley, "Design techniques for high-speed high resolution comparators," *IEEE J. Solid-State Circuits*, pp. 1916-1926, Dec. 1992.

[9] E. A. Vittoz, "Dynamic analog techniques," in *Design of MOS VLSI circuits for telecommunication*, eds. Y. Ysividis and P. Antogenetti, Prentice Hall, Englewood Cliffs, NJ, 1985.

[10] R. Poujois and J. Borel, "A low drift fully integrated MOSFET operational amplifier, " *IEEE J. Solid-State Circuits*, pp. 514-521, May 1995.

[11] W.-C. Song, H.-W. Choi, S.-U.Kwak, and B.-S. Song, "A 10-b 20-Msamples/s low power CMOS ADC," *IEEE J. Solid-State Circuits*, pp. 513-521, May 1995.

[12] A. Yukawa, "A CMOS 8 bit high-speed A/D converter IC," *IEEE J. Solid-State Circuits*, pp. 775-779, June 1985.

[13] N. Tan, *Switched-Current Design and Implementation of Oversampling A/D Converters*, Kluwer Academic Publishers, Boston/Dordrecht/London, 1997.

[14] P. Gray and R. Meyer, *Analog and Design of Analog Integrated Circuits*, 3rd edition, John Wiley & Sons, Inc., 1993.

8 LOW-VOLTAGE ANALOG TECHNIQUES

8.1 INTRODUCTION

The general trend in CMOS fabrication is to make the devices smaller and smaller to increase the density and speed of digital circuits. It is also common to decrease the thickness of the gate oxide to increase the driving capability of the transistor which implies that the supply voltage must be decreased to avoid that the electric field in the devices gets too high. The reduced power supply voltage is normally not an advantage for analog design and a low supply voltage may require some special circuit techniques. Such circuit techniques are discussed in this chapter. In Sec. 8.2 we briefly consider the impact of scaling of CMOS processes. opamp suitable for low supply voltages are considered in Sec. 8.3 and techniques to improve the performance of switches are the subject of Sec. 8.4. Finally the switched opamp technique is reviewed in Sec. 8.5.

8.2 IMPACT OF SCALING

The speed of an MOS process is indicated by the cut-off frequency of the MOS transistor. This is the frequency where the current gain is unity. This frequency can be considered to be an upper speed limit for the process and can be approximated as [1]

$$f_c = \frac{\mu \cdot (V_{GS} - V_T)}{2\pi L^2} \qquad (8\text{-}1)$$

where μ is the channel mobility, V_{GS} the gate-source voltage, V_T the threshold voltage and L is the channel length of the transistor. It is seen from (8-1) that the speed is increased by making the MOS transistor length smaller. It should be noted that for very small lengths the speed is proportional to $1/L$ rather than $1/L^2$.

In digital circuits where the minimum length is used for the transistors (this is not always the case in analog circuits) the speed is increased and the density is increased if the transistor dimensions are scaled down. Therefore the trend is to make the MOS transistors smaller and smaller.

As the dimensions of the CMOS process is scaled down to increase speed and circuit density, the supply voltage must also decrease since the gate oxide is made thinner.

This makes the design of high-performance analog circuits in such a process more difficult. There are also a number of short channel effects that will degrade the intrinsic gain of the transistor, again making the analog design more difficult for small dimensions [2, 3].

For digital circuits the dynamic power consumption can be approximated as [4]

$$P = \alpha f_{clk} C_L V_{dd}^2 \tag{8-2}$$

where α is the activity factor, f_{clk} the clock frequency, C_L the total load capacitance and V_{dd} the power supply voltage. In digital circuits reducing the power supply voltage is a way of trading speed for power which in principle can be compensated for by increasing the parallelism. For analog circuits the dynamic range (DR) must be considered. A fundamental limit on the DR in discrete-time circuits is imposed by the thermal noise in the circuits. The thermal noise voltage power for discrete-time circuits is given by [2]

$$v_n^2 = \frac{kT}{C} \tag{8-3}$$

where k is the Boltzmann's constant, T the absolute temperature and C the sampling capacitor. The dynamic range with a sinusoidal input signal with peak-to-peak amplitude V_{dd} and rms power $V_{dd}^2/8$ is given by

$$DR = \frac{V_{dd}^2 C}{8kT} \tag{8-4}$$

If the power supply voltage is reduced the capacitor must be increased to keep the DR and it is necessary to increase the bias currents to maintain the speed of the circuits. The number of transistors that can be stacked on top of each other will be reduced due to the reduced supply voltage. Therefore it may be necessary to use several cascaded gain stages to get a high DC gain which further increase the power consumption. The power consumption in analog circuits may therefore increase when the power supply is reduced. The reduction in feature size is thus not advantageous for analog circuits. But using low supply voltages is necessary if analog and digital circuits are to be fabricated on the same chip or if they share the same supply to reduce system cost. Therefore, low voltage analog techniques are of interest, although we prefer to use a 5-V supply for high performance analog circuits.

8.3 LOW VOLTAGE OPAMPS

High accuracy SC circuits usually require a high-gain opamp and a simple differential stage is rarely sufficient. For low supply voltages the design of the op amp becomes troublesome. The gain can easily be increased without increasing the power consumption by using cascode transistors as in the telescopic opamp [5], but the voltage swing, especially for low supply voltages, is small. To achieve a large gain at low supply voltages we must resort to two-stage opamps architectures. For very low supply voltages

8.3 Low Voltage Opamps

even a three stage configuration may be necessary. The highest output swing is achieved if the cascodes are used in the first stage as shown in Fig. 8-1. The voltages

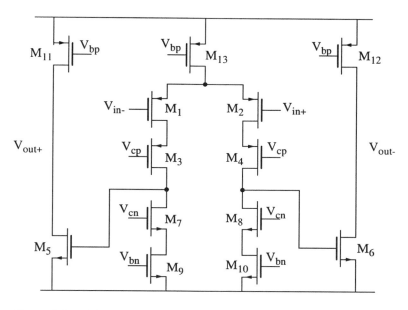

Figure 8-1   Two-stage opamp with cascodes in the first stage.

V_{bp}, V_{cp}, V_{cn} and V_{bn} are bias voltages generated by a bias circuit not shown in the figure. The voltage swing at the output of the first stage is relatively small which makes it possible to use supply voltages as low as ~2 V. By using a folded cascode in the first stage as low as 1.5 V supply voltage can be used. However the power consumption and the noise is large for the folded cascode case. The drawback with the two stage solution is that there are two poles in the transfer function significantly below the unity-gain frequency, which makes frequency compensation mandatory. In the following we consider some compensation techniques which can be applied to the opamp architecture shown in Fig. 8-1. When analyzing the frequency characteristics and the pole positions of the opamp, several approximations are usually required to get useful expressions. These simplified expressions does not accurately model the behavior when the second order-parasitic elements are taken into account. An analytical analysis is very difficult and therefore the effect of second order parasitic elements are investigated by simulations.

8.3.1   Compensation Techniques

A suitable small signal model of the opamp in Fig. 8-1 is shown in Fig. 8-2. If the bulk of all N-transistors are connected to V_{ss} and the bulk of all P-transistors to V_{dd} the component values in the small signal model are given by

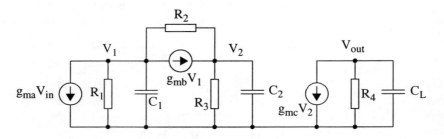

Figure 8-2   Small signal model of the opamp.

$$g_{ma} = g_{m1}, \quad g_{mb} = g_{m3} + g_{mb3}, \quad g_{mc} = g_{m5}$$
$$R_1 = r_{ds1}, \quad R_2 = r_{ds3}, \quad R_3 \approx r_{ds7} r_{ds9} g_{m7}$$
$$R_4 = r_{ds5} \parallel r_{ds11} \tag{8-5}$$
$$C_1 = C_{gs3} + C_{bs3} + C_{bd1},$$
$$C_2 = C_{gs5} + C_{gd7} + C_{gd3} + C_{bd7} + C_{bd3}$$

where g_{mi} is the transconductance of transistor M_i, r_{dsi} is the output resistance of transistor M_i, C_{gsi} is the gate-source capacitor of transistor M_i and C_{bdi} is the bulk-drain capacitance of transistor M_i.

In the small signal model the C_{gd} capacitors of transistors M_1 and M_5 have been neglected as well as the parasitic capacitors at the drain of M_9. The effect of transistor M_{13} has also been neglected. The total capacitive load at the output of the opamp is modeled by C_L. The transfer function of the small signal model have three poles, two in the first stage and one in the second stage.

If $g_m \gg g_{ds}$ and R_3 (output resistance of PMOS current sources in the first stage) is assumed to be very large the poles are approximately given by [2]

$$p_1 \approx \frac{-1}{C_2 \cdot g_{mb} \cdot R_1 \cdot R_2}$$
$$p_2 \approx \frac{-1}{R_4 C_L} \tag{8-6}$$
$$p_3 \approx \frac{-g_{mb}}{C_1}$$

If the effect of R_3 is neglected the DC gain of the opamp is

$$A_0 = g_{ma} g_{mb} g_{mc} R_1 R_2 R_4 \tag{8-7}$$

There are no zeros in the transfer function. If the C_{gd} capacitors of M_1 and M_5 are taken into account, two zeros will appear at

8.3 Low Voltage Opamps

$$z_1 = \frac{g_{ma}}{C_{gd1}}$$

$$z_2 = \frac{g_{mc}}{C_{gd3}} \tag{8-8}$$

The zeros are located at high frequencies and can usually be neglected. It is obvious from (8-6) that there are two poles at quite low frequencies which results in a poor phase margin. Hence frequency compensation is necessary if the opamp is to be used in a feedback configuration with a large feedback factor. Compensation can be performed by using the Miller compensation technique in the second stage [5]. This is done by connecting a compensation capacitor between the input and output of the second stage and results in the small signal model shown in Fig. 8-3. The dominant pole

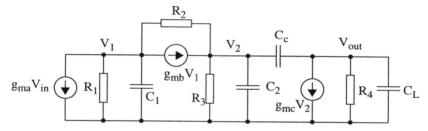

Figure 8-3 Small signal model with Miller compensation.

is now determined by the output resistance of the first stage, $r_{out1} = R_1 R_2 g_{mb}$, and the equivalent Miller capacitor, $C_M \approx C_c R_4 g_{mc}$. Due to the Miller capacitor p_2 has been move to a high frequency and there is now a right half-plane-zero (RHP) [5]. Assuming R_3 to be very large and $g_{ds} \ll g_m$ the poles and zeros with Miller compensation are approximately given by [2]

$$p_1 \approx \frac{-1}{C_c R_4 R_2 R_1 g_{mc} g_{mb}}$$

$$p_2 \approx \frac{-C_c g_{mc}}{C_2 C_c + C_2 C_L + C_c C_L} \approx \frac{-g_{mc}}{C_L}$$

$$p_3 \approx \frac{-g_{mb}}{C_1} \tag{8-9}$$

$$z_1 = \frac{g_{mc}}{C_c}$$

The RHP zero may be located at a low frequency if g_{mc} is small, giving a small phase margin. By adding a resistor in series with the compensation capacitor the zero can be moved to the left half plane to cancel p_2 [2]. If the pole-zero doublet cancel each other perfectly the frequency behavior is similar to that of the single-stage cascoded opamp.

When the pole-zero doublet is not perfectly cancelled a slow settling component appears for the closed loop system. Due to the doublet the relative settling error as function of time in a unity gain configuration is given by [6]

$$\varepsilon_d = \frac{\Delta f_{pz}}{\omega_u} \cdot \exp\left(-\frac{t}{\tau_{pz}}\right) \qquad (8\text{-}10)$$

where Δf_{pz} is the difference between the pole and zero frequency, ω_u the unity gain frequency and $\tau_{pz} = 1/2\pi f_{pz}$ corresponds to the time constant of the doublet. If the doublet is located below the unity-gain frequency the settling behavior may be sensitive to mismatch in the doublet.

A drawback with the Miller compensation technique is that for high frequencies the gate and drain of the second stage gain transistor will track one another due to the compensation capacitor C_c. This can give a poor power supply rejection ratio (PSRR) of single ended opamps [5]. Therefore compensation techniques that improves the PSRR have been developed. These techniques may also improve the settling of the opamp compared to Miller compensation. One way of improving the PSRR is to connect the compensation capacitor to the output of the second stage and the source of the cascode transistor in the first stage [6] which results in the small signal model shown in Fig. 8-4. Unless g_{mb} is very large two of the poles are in many cases complex which is not

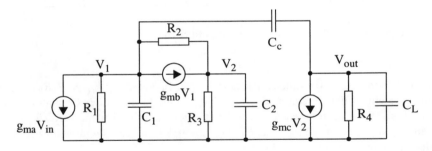

Figure 8-4 Small signal model with the compensation capacitor connected to the cascode of the first stage.

necessarily a disadvantage. This is illustrated in Fig. 8-5. and Fig. 8-6., where the simulated performance of a linear system with one left-half-plane (LHP) zero and three poles is shown. The open loop DC gain in this example was chosen to 1000 and the dominant pole was placed at 1 kHz. Hence the ideal (ignoring parasitic poles and zeros) unity-gain frequency is 1 MHz and the settling time for a relative settling error of 0.1% is

$$t_s = \frac{-\ln(0.001)}{2\pi \cdot 10^6} = 1.1\mu s \qquad (8\text{-}11)$$

The real part of all parasitic poles and zeros in the example were chosen to be three times the unity-gain frequency, i.e.

8.3 Low Voltage Opamps

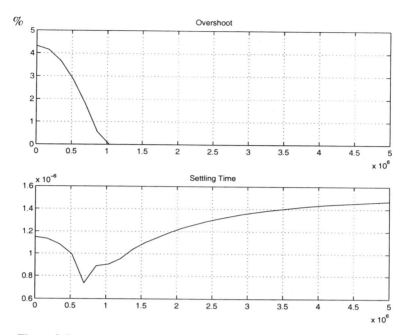

Figure 8-5 Simulated settling time and overshoot as function of the imaginary part of the complex pole pair.

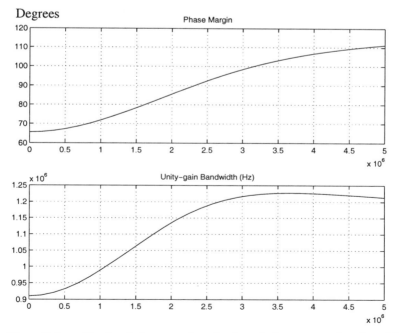

Figure 8-6 Simulated phase margin and unity-gain frequency as function of the imaginary part of the complex pole pair.

$$p_1 = -2\pi 1000$$
$$p_{2,3} = -2\pi \cdot 3 \cdot 10^6 \pm jx \qquad (8\text{-}12)$$
$$z_1 = -2\pi \cdot 3 \cdot 10^6$$

The imaginary part of the poles, x, is swept from 0 to $2\pi \cdot 5 \cdot 10^6$. The system was simulated in a unity-gain configuration using MATLAB and the overshoot and settling time are shown in Fig. 8-5. There is no overshoot in the step response when the imaginary part x is larger than 1 MHz and in the range 0.7 to 1.5 MHz the settling time is slightly smaller than in the single-pole case. The corresponding phase margin and unity-frequency are shown in Fig. 8-6. It should be noted that for large x, both the phase margin and unity-gain frequency is large, but the settling time is not improved. Thus the phase margin and unity-gain frequency are not always good measures of how the amplifier behaves in a closed loop configuration.

The effect of making the poles complex is not always easy to predict especially when there is a large number of parasitic poles and zeros. The slew-rate requirements may increase and the step response may not increase monotonously even though there is no overshoot. However the example shows that from a settling-time point of view complex poles in the open-loop transfer function are not a drawback. The compensation technique proposed in [7] is also a good candidate when a high PSRR is desirable. The principle behind this technique is to feed the current through the compensation capacitor back to the first stage. This can be done by connecting the compensation capacitor between the output of the second stage and the gate of the cascode transistor in the current source in the first stage, i.e. the gates of transistors M_7 and M_8 in Fig. 8-1. If the transconductances of these transistors are large we obtain the small signal model in Fig. 8-7. When using this technique the RHP zero that appeared in the Miller compen-

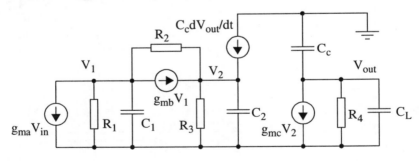

Figure 8-7 Small signal model when using the compensation technique in [7].

sated amplifier is removed and there are three poles approximately determined by [7]

8.3 Low Voltage Opamps

$$p_1 \approx \frac{-1}{C_c R_4 R_2 R_1 g_{mb} g_{mc}}$$

$$p_2 \approx \frac{-g_{mb}}{C_1} \qquad (8\text{-}13)$$

$$p_3 \approx \frac{-g_{mc} C_c}{C_2 \cdot (C_L + C_c)}$$

By keeping g_{mb}, g_{mc} large and C_1, C_2 small the parasitic poles can be moved to high frequencies. If the transconductances of the cascode transistors M_7 and M_8 are in the same order as g_{mb} and g_{mc}, one LHP zero and one more pole will appear in the transfer function. Two of the poles are in many practical cases complex.

In some cases, especially when three or more gain stages are used in the opamp, it may be necessary to use several compensation capacitors to get high performance, such as the nested Miller compensation technique [8] or the compensation technique in [9]. The opamp in Fig. 8-1 is usually considered as a two-stage opamp but there are some advantages with using several compensation capacitors also for this architecture [8]. The compensation technique considered here is illustrated in Fig. 8-8, where a com-

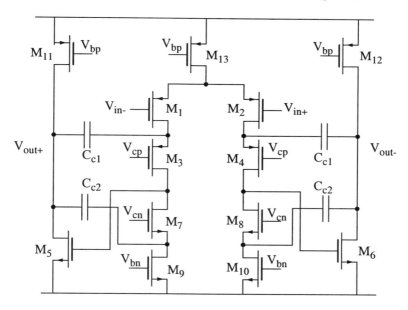

Figure 8-8 Nested compensation of the opamp in Fig. 8-1.

pensation capacitor is connected to both the cascode transistors of the first stage [8]. The corresponding small signal model is shown in Fig. 8-9, where the effect of finite transconductance in M_7 and M_8 has been neglected. The poles and zeros are approximately given by

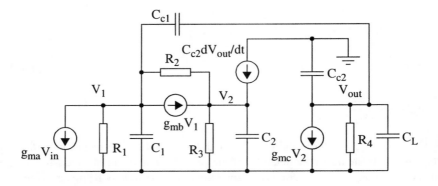

Figure 8-9 Small signal model with the nested compensation technique.

$$p_1 = \frac{-1}{(C_{c1}+C_{c2})R_4R_2R_1g_{mb}g_{mc}}$$

$$p_{2,3} = -\frac{C_{c2}g_{mc}}{2C_2(C_{c2}+C_L)} \pm$$

$$\pm \sqrt{\frac{C_{c2}^2 C_{c1}^2 g_{mc}^2}{(2C_2(C_{c2}+C_L))^2} - \frac{C_{c1}(C_{c2}+C_{c1})g_{mb}g_{mc}}{C_2(C_{c2}+C_L)}} \quad (8\text{-}14)$$

$$z_{1,2} = \pm\sqrt{\frac{g_{mb}g_{mc}}{C_2 C_{c1}}}$$

We assumed that $g_m \gg g_{ds}$ and that the compensation capacitors C_{c1} and C_{c2} are large compared to the parasitic capacitors C_1 and C_2.

For a large phase margin the transconductances g_{mb} and g_{mc} should be large while C_2 should be small, since this moves both the parasitic poles as well as the parasitic zeros to higher frequencies. It should be noted that the expressions in (8-14) are simplified expressions that are useful to get some basic knowledge on how different parameter influence the poles and zeros. For more accurate results all the parasitics, such as the finite transconductance of transistors M_7 and M_8, in the circuit should be taken into account. When these parasitics are included the transfer function will have at least one more pole and one more zero. Thus it is very difficult to analyze the circuit using analytical methods and numerical simulations are more convenient. In the following section we show by an example that the above compensation method can be useful.

8.3.2 Simulations

To verify the effectiveness of the compensation techniques in the previous section we first add a few more parasitic elements to the small signal model as shown in Fig. 8-10.

8.3 Low Voltage Opamps

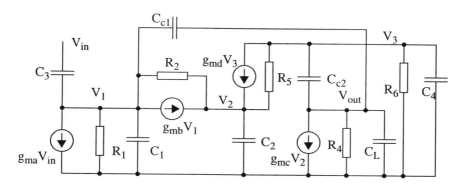

Figure 8-10 Small signal model including parasitic elements.

The new parasitic elements are determined by (referred to the transistor numbering in Fig. 8-8)

$$g_{md} = g_{m7}, \quad R_5 = r_{ds7}, \quad R_6 = r_{ds9}$$
$$C_3 = C_{gd1}$$
$$C_4 = C_{db9} + C_{gs7} + C_{sb7} \quad (8\text{-}15)$$
$$C_5 = C_{gd5}$$

The small signal parameters were taken from a design in a 0.6 μm CMOS process. They are shown in Table 8-1.

Table 8-1 Small Signal Parameters

Param.	Value	Param.	Value
g_{ma}	2.6 mA/V	R_5	40 kΩ
g_{mb}	3 mA/V	R_6	40 kΩ
g_{mc}	1.7 mA/V	C_1	376 fF
g_{md}	3.8 mA/V	C_2	239 fF
R_1	11.5 kΩ	C_3	82 fF
R_2	10.5 kΩ	C_4	547 fF
R_3	∞	C_5	4 fF
R_4	47 kΩ	C_L	4 pF

Ignoring all high frequency poles the unity-gain bandwidth is given by

$$\omega_u = \frac{g_{ma}}{C_{c1} + C_{c2}} \qquad (8\text{-}16)$$

In a unity-gain configuration and assuming 0.025% settling accuracy (approximately 12 bits accuracy) the settling time is

$$t_s = \frac{-\ln(0.0025)(C_{c1} + C_{c2})}{10^{-3}} \qquad (8\text{-}17)$$

In Fig. 8-11 is shown the simulated settling time (using MATLAB) as function of the

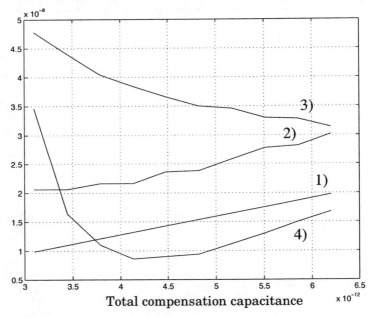

Figure 8-11 Settling times with different compensation techniques.

total compensation capacitance for four different cases:

1. Only the dominant pole is used and the settling time is given by (8-17).
2. Parasitics taken into account and $C_{c1} = 0$ while C_{c2} is changed.
3. Parasitics taken into account and $C_{c2} = 0$ while C_{c1} is changed.
4. Parasitics taken into account and $C_{c1} = 3$ pF while C_{c2} is changed.

The nested compensation technique can in this case give a smaller settling time compared to the other alternatives. By choosing $C_{c2} = 2$ pF the settling time is approximately 10 ns. If C_{c2} is decreased a faster settling time is achieved but the step response will then exhibit some overshoot and therefore C_{c2} should not be too small. The step responses for case 1) and case 4) are shown in Fig. 8-12a. The settling time

8.3 Low Voltage Opamps

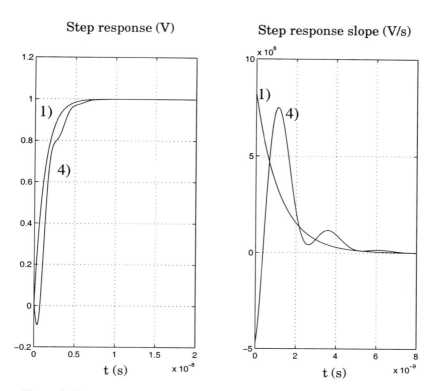

Figure 8-12 a) Step responses and b) step response slope.

is in both cases approximately 10 ns. Typically the slope of the step response of case 4) is larger than for case 1), increasing the slew rate requirements. However in this particular case the maximum slopes are approximately the same as can be seen in Fig. 8-12b, showing the slope of the step responses as a function of time. For case 2) and 3) there was no improvement in the settling time when the Miller capacitor C_5 was increased. The DC gain of the opamp is 97dB, the unity gain frequency is 84 MHz and the phase margin 72°. The poles and zeros of the transfer function are shown in Fig. 8-13a. Compared to the simplified model in Fig. 8-9 there are two more zeros and one more pole in the transfer function. The high frequency RHP zero is caused by C_3 and is approximately given by

$$z_4 \approx \frac{g_{ma}}{C_3} \qquad (8\text{-}18)$$

The finite value on g_{md} (the transconductance of the NMOS cascodes, M_7 and M_8) introduces a pole-zero pair in the left half plane that almost cancel each other. The complex poles are moved compared to the model in Fig. 8-9. Hence the simplified model in Fig. 8-9 does not accurately model the behavior of the circuit when g_{md} is small.

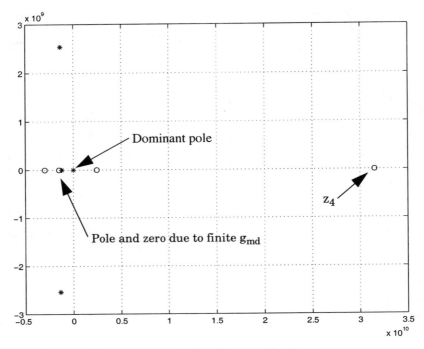

Figure 8-13 Poles and zeros of open loop transfer function.

8.3.3 Common-Mode Feedback

When the opamp is used in a fully differential circuit it is important to have a circuit to regulate the common-mode voltages in the circuit. The most important aspects concerning common-mode feedback (CMFB) is the speed and linearity, while moderate gain is usually acceptable. There are in principle two CMFB techniques. One is the continuous-time CMFB and the other is SC CMFB. Continuous-time CMFB usually suffers from strong nonlineraities. It is usually preferable to use SC CMFB for SC circuits, even though it usually introduces extra capacitive load for the main amplifier. For a two-stage fully-differential opamp, the common-mode voltages in both stages must be controlled. It is possible to sense the common-mode voltage in the second-stage, i.e. at the output, and feed the control signal to the first stage. This kind of CMFB is slow since it has to go through both stages of the main amplifier. A better choice is to use of two CMFBs applied to the individual stages since the individual CMFBs have a much wider bandwidth. For the opamp in figure Fig. 8-8, the effectiveness of the compensation technique is sensitive to the capacitance at the input of the second stage. It is therefore not a good choice to use SC CMFB for the first stage. Continuous-time CMFB is a better choice. The nonlinearity does not present any problem in the two-stage opamp since the second stage has a high gain, making the signal swing at the first stage output small. For the second stage, an SC CMFB is applied. The opamp with compensation capacitors is shown in Fig. 8-14. The CMFB circuit of the first stage consists of transistors M_{c1} and M_{c2}. The transistors are operated in the

8.3 Low Voltage Opamps

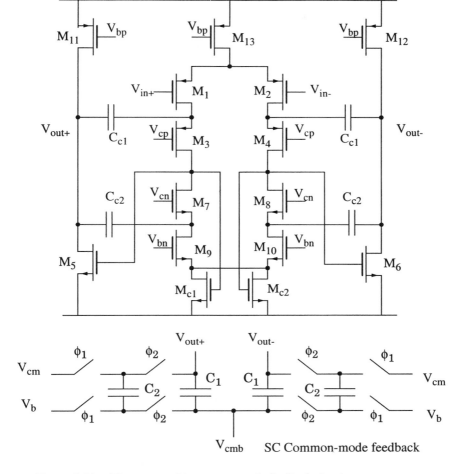

Figure 8-14 The opamp with common-mode feedback circuits.

linear region. The SC CMFB circuit in the second stage consists of four capacitors and eight switches [10] and generates the bias voltage V_{cmb} for the load transistors, M_{11} and M_{12}, in the second stage. V_{cm} is the desired output common-mode voltage and V_b is the desired control voltage in this case the same voltage as is used to bias the current source in the first stage, M_{13}.

8.4 CLOCK VOLTAGE DOUBLERS

For very low supply voltages the switches in a discrete-time circuit becomes a problem. The switch resistance is simply too large due to the low V_{gs} of the switch transistors. Therefore it is necessary to generate switch control signal with a larger swing than V_{dd}. In Fig. 8-15. we show two voltage doubler circuits that can be used for this

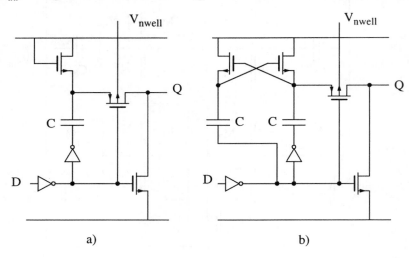

Figure 8-15 Clock doubler circuits. a) Traditional and b) cross-coupled.

purpose. Shown in Fig. 8-15a) is the conventional clock voltage doubler circuit. The drawback of this circuit is the voltage drop across the NMOS transistor. Shown in Fig. 8-15b) is a clock doubler circuit using a cross-coupled pair. Due to the cross-coupling, the sources of the cross-couples transistor pair are charged to the supply voltage without any voltage drop. By applying a square wave input signal with voltage swing V_{dd} at the input D, the two capacitors are self-charged to the supply voltage through the cross-coupled NMOS transistors [11,12,13], and a square wave is generated at the output Q. The high voltage of the square wave output signal is given by

$$V_Q = 2V_{dd} \cdot \frac{C}{C_{load} + C + C_p} \qquad (8\text{-}19)$$

where C is the capacitor connected to the cross-coupled pair, C_{load} the load capacitance at the output, i.e. the gate capacitance of the switch, and C_p is the parasitic capacitance at the output.

In Fig. 8-16 we show the SPECTRE simulation results, of the two clock doubler circuits with a supply voltage of 1.2 V. In the simulation, both load and parasitic capacitances are considered. It is seen that the cross-coupled circuit pumps the voltage to around 2.1 V while the conventional circuit only increases the voltage to about 1.6 V.

8.4 Clock Voltage Doublers

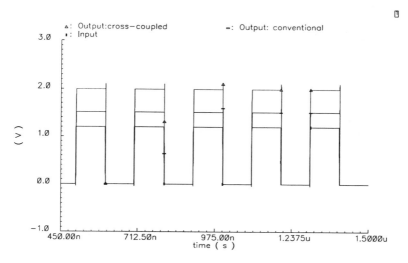

Figure 8-16 Simulation results of the voltage doubler circuits.

8.4.1 High Voltage Generator

The bias voltage for the nwells of the PMOS transistors in the clock double circuits in the previous section must be higher that the supply voltage. This voltage can be supplied by the high voltage generator [11,12] in Fig. 8-17. It needs two square input sig-

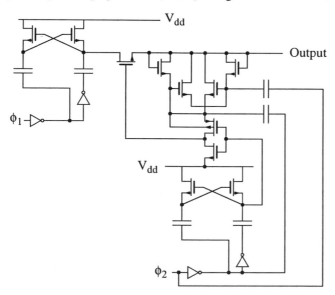

Figure 8-17 High Voltage generator.

nals. The key feature of the high voltage generator is to use feedback to pump the voltage across the capacitors to V_{dd} in the same way as in the cross-coupled clock doubler circuit of Fig. 8-15. Controlled by non-overlapping clock phases (ϕ_1 and ϕ_2), this high voltage generator is capable of generating a voltage of over 2 V if the clock phases have a 1.2 V swing. This voltage can be used to bias the nwell for the PMOS transistors in the clock double circuits of Fig. 8-15 to avoid latch up. Since the current to the nwell is very small, except during start up, the power consumption of the circuit is small. With a 1.2 V supply voltage the output is pumped to about 2.2 V. The simulated output voltage during start-up is shown in Fig. 8-18.

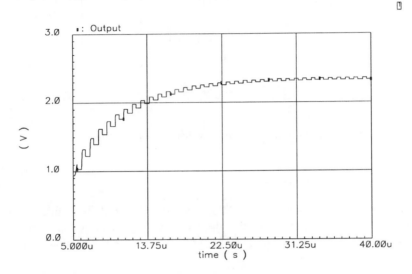

Figure 8-18   Simulation result of high voltage generator.

8.4.2   Bootstrap Switching

The voltage doublers in Fig. 8-15 gives a constant output swing. For switches that need to handle a large voltage swing it would be better if the V_{gs} voltage of the switch tracks the signal. This can be achieved by using a bootstrap switching circuit. Fig. 8-19 shows a bootstrap switching circuit [14]. The switch is represented by transistor M_{11} in this figure. The circuit is controlled by one clock signal ϕ and its inverse. In the off phase when ϕ is low the switch clock voltage, V_g, is low since the gate of M_{11} is discharged through M_7 and M_{10}. Capacitor C_3 is in this clock phase charged to V_{dd}. When ϕ goes high M_9 opens and the lower node of C_3 will track the source voltage of the switch transistor M_{11}. Since ideally no charge can leave capacitor C_3 the gate voltage of the switch transistor is its source voltage plus V_{dd}. Hence the gate voltage tracks the source voltage. This is illustrated in Fig. 8-20. Due to parasitic capacitors in the circuit the voltage across C_3 will be smaller than V_{dd}. The gate voltage can be calculated as [14]

8.4 Clock Voltage Doublers

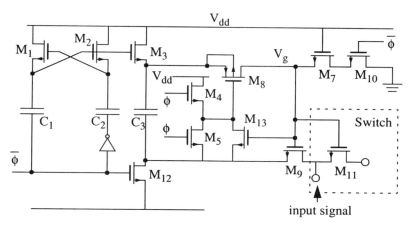

Figure 8-19 Bootstrap switching circuit.

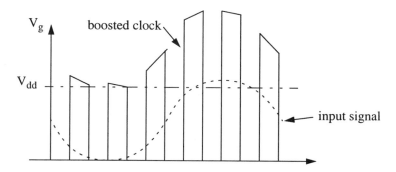

Figure 8-20 Boostrap circuit output.

$$V_g = V_s + \frac{C_3}{C_3 + C_p} \cdot V_{dd} \tag{8-20}$$

where V_s is the source voltage of the switch and C_p is the parasitic load at the gate of the switch transistor when ϕ is high.

8.4.3 Level Conversion Circuit

It is sometimes necessary to convert low voltage digital signals to a higher voltage to interface with other digital circuits using a higher supply voltage. In Fig. 8-21 we show a level-conversion circuit [15]. The input digital signal has a voltage swing of 1.2 V and the output digital signal has a voltage swing equal to V_{ddH}. The PMOS cross-coupled pair (P_0 and P_1) are connected to the high supply voltage, driven differentially (via NMOS pair N_0 and N_1) by the low-voltage swing signal. A minimum-size inverter is used as load. The pull-down NMOS devices, N_0 and N_1, are DC ratioed against the cross-coupled pull-up PMOS devices, so that a low swing input guarantees a correct output transition.

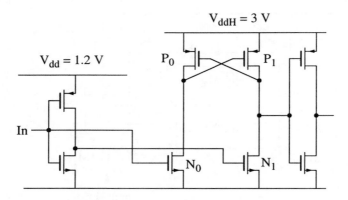

Figure 8-21 Digital level-conversion circuit.

In Fig. 8-22 we show the simulation result of the level-conversion circuit. It is seen

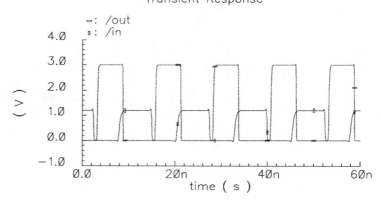

Figure 8-22 Simulation result of digital level conversion circuit.

that we can generate a 3 V square wave output with a 1.2 V square wave input. This level conversion circuit consumes power only during transitions and it consumes no DC power. This level-conversion circuit only works with nwell or twin-well CMOS processes since it requires isolated wells for the PMOS devices that are connected to the high voltage supply and the ones that are connected to the low voltage supply.

8.5 SWITCHED OPAMP TECHNIQUE

The high voltage generators from the previous section can be avoided by using the switched opamp technique [16]. The most difficult switches to realize in a SC circuit are the switches that need to handle a large voltage swing. In the switched opamp technique these switches are removed and we can disconnect the capacitor from the opamp

8.5 Switched Opamp Technique

output by turning off the bias current in the opamp. In [16] the switched opamp technique is used to implement a SC biquad lowpass filter. A standard solution for a SC biquad is shown in Fig. 8-23. In Fig. 8-24 the corresponding switched opamp biquad

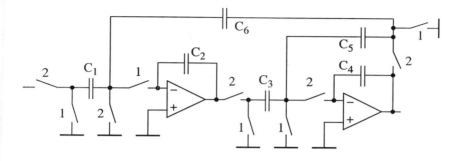

Figure 8-23   Standard SC lowpass biquad.

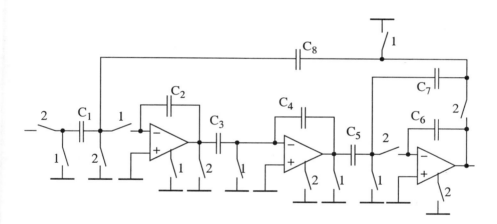

Figure 8-24   Switched opamp biquad.

is shown. With the switched opamp technique the required number of opamps is increased since the output of the integrator can not be used while the opamp is turned off.

The switched opamp in [16] is shown in Fig. 8-25. It is a two stage opamp with Miller compensation. When the clock signal goes low the currents in both the output devices M_5 and M_6 are turned off by when the switch transistors M_9 and M_{10} are opened. The switched opamp technique may not be suitable for very high speed since even though the switches with low voltage swing has been eliminated the switches at the opamp input still have a large resistance due to the small V_{gs}. Since the opamp bias currents can not be turned on instantly it takes a while until it operates properly which further limits the maximum speed of the circuit.

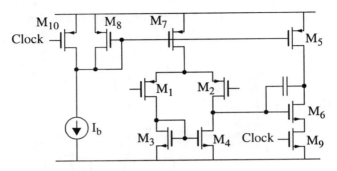

Figure 8-25   Switched opamp.

8.6   SUMMARY

Driven by the need for high-speed and high density in digital circuits the feature size in modern CMOS processes are scaled down. To avoid too large electric fields in the devices which causes reliability problems, the supply voltage must be decreased. This is usually not advantageous for the design of analog circuits. In this chapter we have considered some circuits suitable for low supply voltages. A two stage opamp with a telescopic first stage and a simple common-source output stage was shown to be a good choice for supply voltages down to about 2V. By replacing the first stage with a folded cascode gain stage the opamp can be made to work for supply voltages as low as 1.5V. Different frequency compensation techniques were considered and a nested compensation using two compensation capacitors were shown to have some advantages over single capacitor compensation techniques. For a fully differential implementation the common-mode voltage must be controlled by a common-mode feedback (CMFB) circuit. To avoid a slow common-mode feedback circuit the opamp uses a continuous-time CMFB in the first stage and an SC CMFB circuit in the second stage.

In discrete-time circuits the switches becomes a problem if the supply voltage is low. The effective clock swing can be increased by clock doubler circuits, and thereby reduce the resistance. Some different implementations of such circuits were considered.

An alternative to increasing the swing of the clock signals is to use the switched opamp technique. With this technique switches that need to handle a large voltage swing are removed. Instead the capacitors are disconnected from the opamp output by turning off the bias current in the opamp. A drawback with using the switched opamp technique is that the number of opamps will normally increase since it can only be used on one clock phase. It may also be difficult to use this technique for high speed applications. Therefore, voltage doubler circuits are usually preferred.

REFERENCES

[1] E. S. Yang *Microelectronic Devices*, McGraw-Hill, 1988.

[2] D. A. Johns and K. Martin, *Analog Integrated Circuit Design*, John Wiley & Sons, Inc. 1997.

[3] S. Wong and C. A. T. Salama, "Impact of Scaling on MOS Analogt Performance", *IEEE J. of Solid-State Circuits*, vol. SC-18, no. 1, pp. 106-14, Feb. 1983.

[4] J. M. Rabaey, *Digital Integrated Circuits*, Prentice Hall Electronics and VLSI Series, 1996.

[5] P. E. Allen and D. R. Holberg, *CMOS Analog Circuit Design*, Holt, Rinehart and Winston, Inc. 1987.

[6] K. R. Laker and W. M. C. Sansen, *Design of Analog Integrated Circuits and Systems*, McGraw-Hill, Inc. 1994.

[7] B. K. Ahuja, "An Improved Frequency Compensation Technique for CMOS Operational Amplifiers", *IEEE J. of Solid-State Circuits*, Vol. SC-18, No. 6, pp. 629-33, Dec. 1983.

[8] R. G. H. Eschauzier and J. H. Huijsing, *Frequency Compensation Techniques for Low-Power Operational Amplifiers*, Kluwer Academic publishers, 1995.

[9] G. S. Østrem, *High-Speed Analog-to-Digital Converters for Mixed Signal ASIC Applications*, Ph.D. Dissertation, FYS.EL.-rapport 1996:38, Norwegian University of Science and Technology, 1996.

[10] D. Senderowicz, S. F. Dreyer, J. H. Huggins, C. F. Rahim and C. A. Laber, "A Family of Defferential NMOS Analog Circuits for PCM Codec Filter Chip", *IEEE J. of Solid-State Circuits*, Vol. SC-17, No. 6, pp. 1014-23, Dec. 1982.

[11] Y. Nakagome, et. al., "Experimental 1.5-V 64-Mb DRAM", *IEEE J. of Solid-State Circuits*, vol. 26, pp. 465-472, April 1991.

[12] N. Tan, "Design and Implementation of Digital-to-Analog Converters for Telecommunication", ERICSSON Report, KI/EKA/MERC-96:078, 1996-12-05.

[13] N. Tan, "A 1.5-V 3-mW 10-b 50-Ms/s CMOS DAC with Low Distortion and Low Intermodulation in Standard Digital CMOS Process", *Proc. IEEE Custom Integrated Circuit Conference (CICC)*, California, USA, May, 1997.

[14] A. M. Abo and P. R. Gray, "A 1.5-V, 10-bit, 14.3-MS/s CMOS Pipeline Analog-to-Digital Converter", *IEEE J. of Solid-State Circuits*, vol. 34, no. 5, pp. 599-606, May 1999.

[15] N. Tan: *Switched-Current Design and Implementation of Oversampling A/D Converters*, Kluwer Academic Publishers 1997.

[16] J. Crols and M. Steyaert, "Switched-Opamp: An Approach to Realize Full CMOS Switched-Capacitor Circuits ar very Low Power Supply Voltages", *IEEE J. of Solid-State Circuits*, vol. 29, no. 8, pp. 936-942, Aug. 1994.

9 PIPELINED A/D CONVERTERS

9.1 INTRODUCTION

The pipelined ADC is suitable for applications where a relatively high bandwidth and a high resolution are required. In Sec. 9.2 we review the fundamentals of pipelined ADCs and show how the total resolution of the ADC is determined and we also show how the digital output of the converter is calculated. Digital correction is used to relax the requirements on the sub-ADCs, The principle behind digital correction is explained in Sec. 9.3. For very high resolutions it may be necessary to use calibration to achieve the desired performance. Digital self calibration algorithms are reviewed in Sec. 9.4. Errors in pipelined converter are treated in general terms in Sec. 9.5, while a more detailed investigation of errors sources in an SC MDAC is considered in Sec. 9.6.

9.2 THE PIPELINED CONVERTER

The pipelined ADC [1] consists of several pipelined stages, each containing a sub ADC, a DAC, a subtractor and a residue gain amplifier as shown in Fig. 9-1. The last

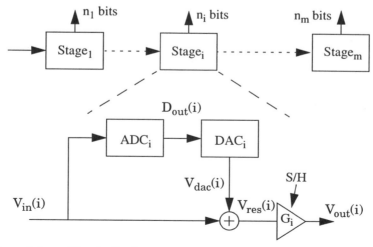

Figure 9-1   The pipelined converter.

stage needs only an sub ADC. Sometimes an additional S/H circuit is used at the input to avoid delay skew errors in the two input signals to the subtractor. The symbols listed in Table 9-1 are used in this chapter.

Table 9-1   Symbols

Symbol	Description
ADC_i	The sub ADC in stage i
DAC_i	The sub DAC in stage i
$V_{res}(i)$	Residue in stage i (before the amplifier)
$V_{in}(i)$	Analog input to stage i
$V_{out}(i)$	Analog output of stage i. Hence $V_{out}(i) = V_{in}(i+1)$
$D_{out}(i)$	The digital output of stage i
D_{out}	Total digital output of entire ADC
$V_{dac}(i)$	The output of the DAC in stage i
FS	Full scale input range of ADC_i (assumed to be equal for all stages)
n_i	Number of output bits in stage i
n_{tot}	Total number of bits in the pipelined converter
N_i	Number of output codes in ADC_i
N_{tot}	Total number of output codes of the pipelined ADC
G_i	Residue gain in stage i
m	Number of stages

The number of output codes of a stage is usually a power of two, i.e. $N_i = 2^{n_i}$ where n_i is the number of bits in stage i but any number of codes can be used. In for example redundant-signed-digit (RSD) converters the number of output codes is 3.

9.2.1 Sub ADC

The sub ADC in stage i is assumed to have N_i output codes denoted $D_{out}(i)$. $D_{out}(i)$ is an integer number ranging from 0 to $N_i - 1$. In the following we assume that an offset binary code is used. Thus the digital output of stage i can be calculated as

$$D_{out}(i) = \sum_{l=0}^{n_i-1} b_{i,l} \cdot 2^l \tag{9-1}$$

where $b_{i,l}$ is the l-th bit in the binary output of stage i. We assume that the input sig-

9.2 The Pipelined Converter

nal range from $-FS/2$ to $+FS/2$ and that the decision levels in ADC_i are equally spaced over the entire input range. This is illustrated in Fig. 9-2 where the input range

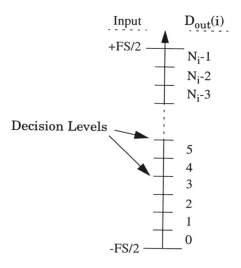

Figure 9-2   Input range and output codes of stage ADC.

has been divided into N_i segments each corresponding to one output code. The analog value corresponding to one LSB is determined by

$$LSB_i = \frac{FS}{N_i} \qquad (9\text{-}2)$$

9.2.2   Sub DAC

The analog output of the sub DAC for a certain code is determined by the expression

$$V_{dac}(i) = \left(D_{out}(i) - \frac{N_i - 1}{2}\right) \cdot \frac{FS}{N_i} \qquad (9\text{-}3)$$

9.2.3   Residue and S/H Amplifier

The residue is the difference between the analog input of the pipelined stage and the DAC output, i.e.

$$V_{res}(i) = V_{in}(i) - V_{dac}(i) = V_{in}(i) - \left(D_{out}(i) - \frac{N_i - 1}{2}\right) \cdot \frac{FS}{N_i} \qquad (9\text{-}4)$$

The transfer function from the input of the stage to the output of the subtractor is sawtooth shaped with the amplitude $FS/2N_i$. This is illustrated in Fig. 9-3. The swing of the residue is thus N_i times smaller than the swing of the input signal to the stage. Here

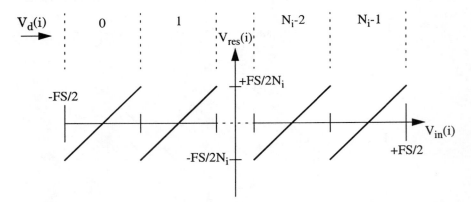

Figure 9-3 Transfer function from input to output of subtractor of one stage in the pipeline.

we have assumed that the full scale range of all the stages are the same. It is therefore necessary to amplify the residue in order to utilize the entire swing of the following stage. Choosing the gain as N_i makes the swing equal to the FS of the following stage. The analog output signal of one stage in the pipeline can be expressed as

$$V_{out}(i) = V_{res}(i) \cdot G_i = \left(V_{in}(i) - \left(D_{out}(i) - \frac{N_i - 1}{2}\right) \cdot \frac{FS}{N_i}\right) \cdot G_i \qquad (9\text{-}5)$$

or if the signal swing of all the stages are made equal the output signal can be written as

$$V_{out}(i) = \left(V_{in}(i) - \left(D_{out}(i) - \frac{N_i - 1}{2}\right) \cdot \frac{FS}{N_i}\right) \cdot N_i \qquad (9\text{-}6)$$

9.2.4 Digital Output

The digital outputs of the stages, $D_{out}(i)$, must be combined to generate the total output code of the pipelined converter. The output signal of the first stage is the input signal to the second stage, i.e. $V_{out}(i) = V_{in}(i+1)$. Hence each segment of the sawtooth shaped output signal of the first stage will be quantized by the second stage. This is illustrated by an example in Fig. 9-4, for a two stage converter where both stages have a resolution of 2 bits. The input range is clearly divided into 16 sections each corresponding to a unique code combination $\{D_{out}(1), D_{out}(2)\}$. The number of codes in a pipelined ADC where $G_i = N_i$ can be calculated as

9.2 The Pipelined Converter

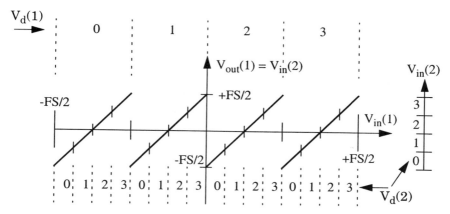

Figure 9-4 Analog output of first stage and corresponding digital output codes of a two stage ADC.

$$N_{tot} = \prod_{i=1}^{m} N_i \qquad (9\text{-}7)$$

or if all the resolutions are chosen as $N_i = 2^{n_i}$ the total number of bits is given by

$$n_{tot} = \sum_{i=1}^{m} n_i \qquad (9\text{-}8)$$

By using that

$$V_{out}(i) = V_{res}(i) \cdot G_i = (V_{in}(i) - V_{dac}(i)) \cdot G_i \qquad (9\text{-}9)$$

the input signal to the first stage of the pipeline can be written as

$$V_{in}(1) = V_{dac}(1) + \frac{V_{out}(1)}{G_1} \qquad (9\text{-}10)$$

Since $V_{in}(2) = V_{out}(1)$ we can use (9-9) one more time to get

$$V_{in}(1) = V_{dac}(1) + \frac{V_{dac}(2)}{G_1} + \frac{V_{out}(2)}{G_1 \cdot G_2} \qquad (9\text{-}11)$$

By repeatedly using (9-9) for all the stages we finally arrive at the expression

$$V_{in}(1) = V_{dac}(1) + \frac{V_{dac}(2)}{G_1} + \dots + \frac{V_{dac}(i)}{G_1 \dots G_{i-1}} + \dots$$
$$\dots + \frac{V_{dac}(m)}{G_1 \dots G_{m-1}} + V_{res}(m) \tag{9-12}$$

where the residue of the last stage, $V_{res}(m)$, corresponds to the quantization error of the converter. It should be noted that the residue of the last stage need not be generated in an implementation of a pipelined converter. By using (9-3) and (9-12) we get

$$V_{in}(1) = \sum_{i=1}^{m} \left(\left(D_{out}(i) - \frac{N_i - 1}{2} \right) \cdot \frac{FS}{N_i} \cdot \frac{1}{\prod_{k=1}^{i-1} G_k} \right) + V_{res}(m) \tag{9-13}$$

We are now interested in finding an expression to calculate the total output code of the converter. The total output should start at 0 and increase by 1 as the digital code in the last stage change by 1. We can rewrite (9-13) as

$$V_{in}(1) = \sum_{i=1}^{m} \left(D_{out}(i) \cdot \frac{N_m}{N_i} \cdot \prod_{k=i}^{m-1} G_k \right) \cdot \frac{\frac{FS}{N_m}}{\prod_{k=1}^{m-1} G_k} + V_{res}(m) + \alpha \tag{9-14}$$

where α is a constant given by

$$\alpha = \sum_{i=1}^{m} \left(-\frac{N_i - 1}{2} \cdot \frac{FS}{N_i} \cdot \frac{1}{\prod_{k=1}^{i-1} G_k} \right) \tag{9-15}$$

The first term in (9-14) is clearly 0 if all the digital outputs, $D_{out}(i)$, are 0, which is what we desired for the total output code. We also see that the last term in the sum, i.e. when $i = m$, is equal to $D_{out}(m)$. Hence, the sum changes by 1 as the code in the last stage changes by one. Therefore, we can use the expression

$$D_{out} = \sum_{i=1}^{m} \left(D_{out}(i) \cdot \frac{N_m}{N_i} \cdot \prod_{k=i}^{m-1} G_k \right) \tag{9-16}$$

to calculate the total output code of the converter.

9.3 Digital Correction

When choosing the residue gains as $G_k = N_i$ the expression simplify to

$$D_{out} = \sum_{i=1}^{m} \left(D_{out}(i) \cdot \prod_{k=i+1}^{m} N_k \right) \qquad (9\text{-}17)$$

We will illustrate how to use (9-16) and (9-17) by an example. Assume that we have a three stage pipelined converter with the stage resolutions 4, 3 and 2 bits. This means that $N_1 = G_1 = 16$, $N_2 = G_2 = 8$ and $N_3 = G_3 = 4$ respectively. According to (9-17) the output code of the pipelined converter can now be calculated as

$$\begin{aligned} D_{out} &= N_2 \cdot N_3 \cdot D_{out}(1) + N_3 \cdot D_{out}(2) + D_{out}(3) \\ &= 32 \cdot D_{out}(1) + 4 \cdot D_{out}(2) + D_{out}(3) \end{aligned} \qquad (9\text{-}18)$$

Since multiplication by 2 is easily performed in the digital domain by shifting, the total output code in this example can be calculated by shifting $D_{out}(1)$ 5 steps, shifting $D_{out}(2)$ 2 steps, not shifting $D_{out}(3)$ and adding all three together. This is illustrated in Fig. 9-5.

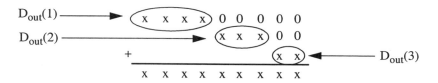

Figure 9-5   Calculation of the output code of a pipelined ADC.

We see by this simple example that no digital processing except shifting is necessary to get the total output code. It is therefore very easy to generate the output code for this type of converter.

9.3 DIGITAL CORRECTION

In a pipelined converter where $G_i = N_i$, the decision levels in the sub ADCs, and thereby also the comparators, must be very accurate if the total resolution of the converter is high. If a decision level in a sub-ADC is moved, the transfer function of the stage will be affected as shown by an example in Fig. 9-6. We see that the output signal swing now is larger than FS. If the input signal to a sub-ADC is larger than +FS/2 or smaller than -FS/2 the output will saturate as shown in Fig. 9-7. Hence there is a large conversion error and the effective resolution of the converter is reduced. To relax the requirements on the comparators, digital correction can be used [2].

In a pipelined ADC with digital correction the residue gain is reduced to introduce redundancy. This is illustrated in Fig. 9-8 for an ADC with two stages where the first

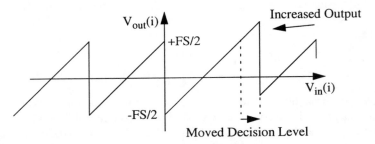

Figure 9-6 Moved ADC decision level increases signal swing.

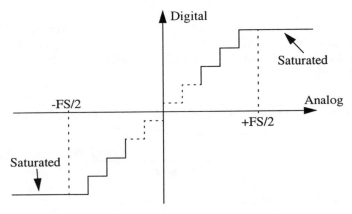

Figure 9-7 The output of an ADC saturates for large inputs.

stage has 4 codes and the second stage has 8 codes. The residue gain has been reduced from 4 to 2. The output signal swing of the first stage is now only half the input range of the following stage. This means that the codes **0,1,6** and **7** will never be used. The gain is usually reduced by a factor 2 but can in principle be chosen arbitrarily. However, it is usually desirable to have the same step size for all the codes. The gain is then restricted to values that give the correct step size at the decision levels in the first stage. This is illustrated in Fig. 9-9 showing the output when one of the decision lines has been moved. In this figure it is seen that moving the decision level will not cause saturation in the following stage since there are now the redundant codes **0, 1, 6,** and **7** in stage 2. The digital correction can correct errors in the comparators as long as the residue is within the FS range of the following stage. The smaller the residue gain, the larger errors can be accepted. The maximum decision line deviation is given by

$$\Delta V = \pm \frac{FS}{2}\left(\frac{1}{G_i} - \frac{1}{N_i}\right) \tag{9-19}$$

If the gain factor is reduced by a factor 2 we get

9.3 Digital Correction

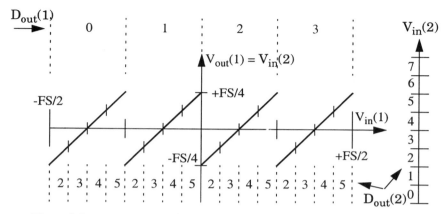

Figure 9-8 Analog output of first stage and corresponding digital output codes for a two stage ADC with reduced residue gain.

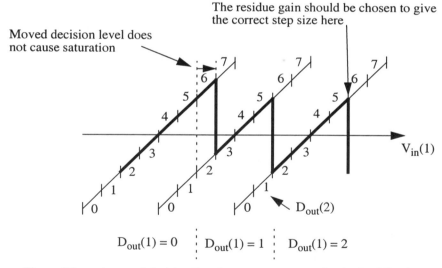

Figure 9-9 A moved decision line does not cause saturation in the following stage when the residue gain is reduced.

$$\Delta V = \pm \frac{FS}{2N_i} \tag{9-20}$$

Hence, the error in ADC_i can be $\pm LSB_i/2$, without causing a large conversion error. A drawback of the digital correction is, as was illustrated by the example above, that several code combinations give the same total output code. The total resolution of the converter is thus decreased when digital correction is introduced unless more stage are added.

9.3.1 Output Codes and Decoding

Due to the reduced residue gain factor in converters with digital correction the total number of codes is reduced compared to a converter without digital correction. All the codes in the first stage are used but only $N_2 \cdot G_1/N_1$ codes are used in the second stage and only $N_i \cdot G_{i-1}/N_{i-1}$ are used in the i-th stage. The total number of codes in the converter can thus be calculated as

$$N_{tot} = N_1 \cdot \prod_{i=2}^{m} N_i \cdot \frac{G_{i-1}}{N_{i-1}} = \prod_{i=1}^{m-1} G_i \cdot N_m \tag{9-21}$$

With digital correction the digital output code can still be calculated using (9-16). For example, assume that we have a three stage converter with the stage resolutions 3, 4 and 5 bits and that the gain in the two first stages is reduced by a factor two to introduce digital correction. This means that we now have $N_1 = 8$, $G_1 = 4$, $N_2 = 16$, $G_2 = 8$, $N_3 = 32$ and $G_3 = 16$. By using (9-16) the output code can now be calculated as

$$\begin{aligned} D_{out} &= \frac{N_3}{N_1} \cdot G_1 \cdot G_2 \cdot D_{out}(1) + \frac{N_3}{N_2} \cdot G_2 \cdot D_{out}(2) + D_{out}(3) \\ &= 128 \cdot D_{out}(1) + 16 \cdot D_{out}(2) + D_{out}(3) \end{aligned} \tag{9-22}$$

The total output code in this example can be calculated by shifting $D_{out}(1)$ 7 steps, shifting $D_{out}(2)$ 4 steps, not shifting $D_{out}(3)$ and adding all three together. This is illustrated in Fig. 9-10. In this case the digital codes from the stages overlap when they

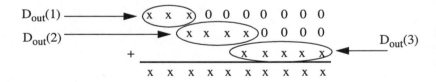

Figure 9-10 Calculation of the output code of a pipelined ADC.

are added. This means that the generation of the digital output code is slightly more complicated since the digital logic must handle carry propagation in the addition. This is not a severe drawback and most high resolution pipelined ADCs use some form of digital correction. There may however be a small problem with directly using (9-16) to calculate the output code, as will be discussed in the following section.

9.3.2 Modified Converter

In the two stage converter used as an example in Fig. 9-8, four codes in the second stage were used for digital correction. This means that for the smallest possible input signal, -FS/2, the output codes from the two stages will be **0** and **2** respectively. By using (9-16) the total output code of the converter is

$$D_{out} = \sum_{i=1}^{m} D_{out}(i) \cdot \frac{N_m}{N_i} \cdot \prod_{k=i}^{m-1} G_k = \frac{8}{4} 2 \cdot D_{out}(1) + D_{out}(2) \quad (9\text{-}23)$$
$$= 4 \cdot 0 + 2 = 2$$

The largest possible input, +FS/2, gives the output codes **3** in the first stage and **5** in the second stage. By again using (9-16) the total output code is 17. Hence there are 16 output codes which corresponds to a 4 bit converter, but since the smallest code is 2 there is a digital offset. Clearly we must remove the digital offset, since codes 16 and 17 can not be represented using 4 bits. One way of doing this would be to change the coding of the second stage as shown in Fig. 9-11.

Figure 9-11 Changed coding of the second stage ADC.

If there are no errors in the decision levels of the first stage the coding is very simple. According to (9-23) the code in stage 1 should be multiplied by 4 (corresponds to two binary shifts) and added to the code in the stage 2. Since the codes in stage 2 without errors now range from **0** to **3** there will be no carry propagation in the addition. However if a decision level in the first stage is moved up significantly the codes **4** and **5** are used for some input signals and a carry bit must be propagated. On the other hand if a decision level is moved down, subtraction must be performed since we have the negative codes **-1** and **-2**. Thus there are three cases, do nothing, addition of carry and subtraction of carry. This makes the testing of the correction logic more difficult [3].

One way to simplify the testing is to introduce offsets in the stages as shown in Fig. 9-12 ([3]). Due to the offsets the transfer function of the stage is shifted to the right as

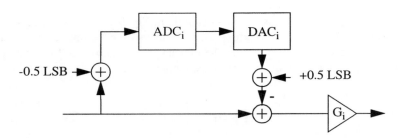

Figure 9-12   One stage with offset.

shown in Fig. 9-13. If we keep the old coding in the second stage with codes ranging

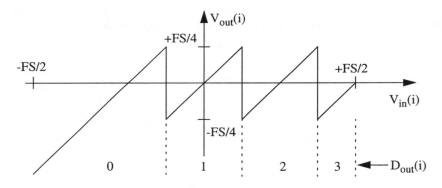

Figure 9-13   Output of a stage with offsets ($N_i = 4$).

from **0** to **7**, the total output code will now range from **0** to **15** which can be represented with four output bits. It is also easier to test the correction logic since subtraction is no longer needed and carry addition will occur even if there are no errors in the decision levels. There is one more advantage of introducing the offsets in that the decision level in the zero crossing has now been moved, improving the linearity for small input signals [3]. The transfer function of the stage with offsets is no longer symmetric. A symmetric transfer function can be achieved by removing the rightmost decision line in Fig. 9-13. The resulting transfer function is shown in Fig. 9-14. There are now only 3 codes in the stage but the output range has been increased, making the total number of codes the same as before. A converter containing stages with the output shown in Fig. 9-14 is usually referred to as a redundant signed digit (RSD) converter [4]. The same technique with introducing offsets and removing one comparator can be used for any number of bits in the stages. The digital output can still be calculated using (9-16) but we must remember to use the original number of codes for N_i before one comparator was removed. There is one small difference depending on the last stage of the pipeline. The last stage contains only an sub ADC since the residue does not need to be generated. It is therefore not necessary to add the offset to the last sub ADC to get a correct

9.3 Digital Correction

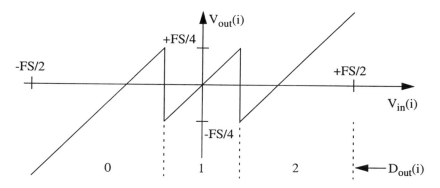

Figure 9-14 Symmetric output with one comparator removed.

coding. The maximum output code in this case is

$$D_{out, max1} = \sum_{i=1}^{m-1} (N_i - 2) \cdot \frac{N_m}{N_i} \cdot \prod_{k=i}^{m-1} G_k + N_m - 1 \qquad (9\text{-}24)$$

Since the minimum code is **0**, the total number of codes is $D_{out, max1} + 1$. However if the offset is introduced also in the last stage (this would save one comparator), only $N_m - 1$ codes are available in the last stage making the maximum code in this case

$$D_{out, max2} = \sum_{i=1}^{m-1} (N_i - 2) \cdot \frac{N_m}{N_i} \cdot \prod_{k=i}^{m-1} G_k + N_m - 2 \qquad (9\text{-}25)$$

There is thus one code less in this case. If the total resolution of the converter is large this will not affect the resolution of the converter.

As an example we calculate the total number of codes in an RSD converter where all the stages are equal as

$$\begin{aligned} N_{tot} = D_{out, max2} + 1 &= \sum_{i=1}^{m-1} (N_i - 2) \cdot \frac{N_m}{N_i} \cdot \prod_{k=i}^{m-1} G_k + N_m - 1 \\ &= \sum_{i=1}^{m-1} 2 \cdot 1 \cdot \prod_{k=i}^{m-1} 2 + 2 - 1 = 2^{m+1} - 1 \end{aligned} \qquad (9\text{-}26)$$

Hence, to get n_{tot} bits in a RSD converter $n_{tot} - 1$ stages are needed.

9.4 DIGITAL CALIBRATION

By using digital correction, large errors in the sub ADCs can be corrected. But the digital correction only works if there are no errors in the residues of the stages. The DAC of the first stage must therefore have a linearity which corresponds to the full resolution of the pipelined ADC. The linearity of the DAC is limited to about 10 to 12 bits by component matching and for high resolutions some calibration algorithm is needed. There are both analog [5] and digital [6] calibration techniques to reduce the effect of DAC errors. The pure digital techniques are usually preferred since almost no additional analog hardware is needed and the speed of the converter is therefore not sacrificed because of the calibration. The idea behind the digital calibration is to estimate the DAC errors in one stage by using the other stages of the converter.

In e.g. [7] the stages in the converter are connected in a circular structure to enable calibration of all the stages, starting with the LSB stage. The actual residue gains and DAC outputs of each stage are in this manner approximated by the other stages of the converter. By repeating this procedure the calibration algorithm will finally converge to a solution where the desired parameters have been approximated with sufficient accuracy to compensate for the errors. It is however not always necessary to calibrate all the stages in the pipeline. It may be sufficient to calibrate only a few of the MSB stages, since the MSB stages are most sensitive to DAC errors. In the following we will briefly describe this calibration technique ([6], [8], [9]).

Assume first that there are errors in the first stage DAC of a pipelined ADC while all the following stages are error-free. When using (9-16) to calculate the output code of the converter it is assumed that the DAC output will always change its value by a certain amount when the input code is changes by 1 and that the residue gain is exactly the nominal value. Any errors in the DAC or residue gain amplifier will cause the correction logic to calculate the wrong output code and the transfer function of the pipelined ADC will have discontinuities as illustrated in Fig. 9-15. The errors causing the non-linearities, $\delta(k)$, can now be corrected in the digital domain if we know their sizes. Their sizes can be estimated by using the following stages in the converter. When the digital code to the DAC in the first stage is changed from $k-1$ to k, the digital output of stages 2 to m will change by an amount here denoted $\Delta(k)$. The errors can now be calculated as

$$\delta(k) = \Delta(k) - \Delta_{ideal} \tag{9-27}$$

where Δ_{ideal} is the ideal change in the digital code when the DAC code in stage 1 is changed from $k-1$ to k. The ADC can be linearized by adding correction terms in the digital domain. These correction terms can be calculated as

$$CORR(k) = \sum_{l=1}^{k} \delta(k) \tag{9-28}$$

and they are stored in a memory from which they can later be retrieved during normal operation. We illustrate this with an example. Assume we have an ADC where the first

9.4 Digital Calibration

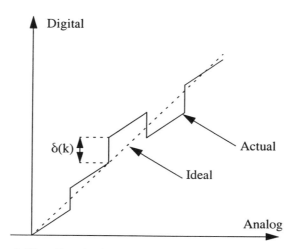

Figure 9-15 Transfer function of pipelined ADC with errors in the first stage.

stage needs calibration, see Fig. 9-16. The first stage has 2 output bits but offsets have

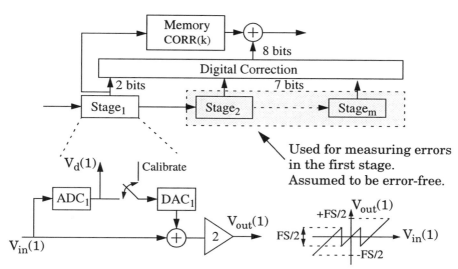

Figure 9-16 Calibration of first stage in a pipelined converter.

been introduced as was described in the previous section such that one comparator can be removed (Fig. 9-12). The number of codes in the first stage ADC is thus only 3 and the analog output signal is described by Fig. 9-14 also shown in Fig. 9-16. The residue gain factor is 2 to enable digital correction. The remaining stages of the pipeline has a total resolution of 7 bits. In Fig. 9-16 we see that increasing the digital code by 1 will decrease the analog value by $FS/2$. This corresponds to a change of 2^6 codes in the 7 bit converter following stage 1, i.e.

$$\Delta_{ideal} = 2^6 \tag{9-29}$$

The calibration now proceed as follows.

1. *Turn the input of the DAC to calibrate (see Fig. 9-16) and force the input code to 0. Apply an input offset such that the output ends up in the upper part of the range (exactly how this is done depends on the implementation). Measure the digital output code from stage 2 to m.*

2. *Change the input code to the DAC to 1. Measure the corresponding digital output and calculate the difference from the previous measurement. This difference is* $\Delta(1)$.

3. *Repeat the above procedure when changing the code from 1 to 2. This gives* $\Delta(2)$. *If more codes are used in the first stage the measurements must be repeated for all the codes to obtain all the* $\Delta(k)$.

4. *Assume* $\delta(0) = 0$ *and* $CORR(0) = 0$. *Calculate* $CORR(1)$ *and* $CORR(2)$ *by using (9-27) and (9-28). Store the correction terms in a memory.*

During normal operation the correction terms are added as illustrated in Fig. 9-17. In

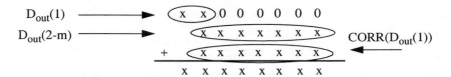

Figure 9-17 Calculation of the output code of a pipelined ADC with calibration.

this way DAC errors and linear gain errors in the first stage can be corrected. The purpose of the above example is to show the principle behind the calibration. The exact procedure to measure the errors depends on the circuit architecture and which type of errors that needs to be calibrated (see e.g. [6], [10]). If more than one stage needs to be calibrated the LSB stages must be calibrated before the MSB stages. The number of stages that needs to be calibrated depends on the total resolution of the ADC and the size of the DAC errors. It may be necessary to add a few more bits at the end of the pipeline to avoid digital truncation errors when the correction terms are measured. The final result can easily be truncated to the desired number of bits.

9.4.1 Other Calibration Techniques

Digital calibration can also be applied if the stages have only one bit per stage [11]. To avoid that the residue gets too large when errors are introduced a non-integer residue gain factor must be used. In [11] the gain is 1.96.

A drawback with the calibration algorithms considered so far is that the ADC can not be used during calibration. In [12] an interpolation technique is used to calculate some of the output samples. The interpolated sampled need not be calculated by the ADC and it is therefore possible to perform the calibration while the ADC is running. In [13] DAC errors are measured by using an additional sigma-delta converter. Due to the circuit architecture the calibration can be performed in the background. Another background calibration algorithm is presented in [14].

There are also some analog circuit techniques that do not really calibrate the ADC but they can still improve the linearity, e.g. capacitor error-averaging [15] and the commutated feedback-capacitor switching (CFCS) technique [16].

9.5 ERRORS IN PIPELINED ADCS

In the following we will discuss how different error sources affect the performance of the pipelined converter [1]. The stage resolution $n_i = \log_2(N_i)$ may be any number of bits but for simplicity we assume that all stage resolutions are equal, i.e.

$$n = n_i, \ 1 \leq i \leq m \tag{9-30}$$

We assume that x bits in every stage is used for digital correction and will not contribute to the final resolution of the ADC. If we assume that all the stages except the last use x bits for correction the resolution of the total ADC is [1]

$$n_{tot} = (m-1)(n-x) + n \tag{9-31}$$

To get the same full-scale conversion range for all the stages the residue amplifier gain should be chosen as

$$G = 2^{n-x} \tag{9-32}$$

It is convenient to introduce r_i as the resolution remaining to be determined from stage i to stage m

$$r_i = n_{tot} - (i-1)(n-x) \tag{9-33}$$

If the error in stage i is denoted e_i, the input referred error e_{in} can be expressed as

$$e_{in} = e_1 + \sum_{i=1}^{m-1} \frac{e_{i+1}}{G^i} \tag{9-34}$$

To keep the errors smaller than LSB/2 (guarantees no missing codes) we have

$$e_i \leq \frac{FS}{2^{n_{tot}+1}} G^{i-1} \tag{9-35}$$

where FS is the full-scale conversion range of the ADC. This means that larger relative errors are acceptable for the LSB stages of the converter. Assuming equal errors e in all the stages gives

$$e_{in} = e\left(1 + \sum_{i=1}^{m-1} \frac{1}{A^i}\right) = F \cdot e \qquad (9\text{-}36)$$

The factor F is minimized for large residue gain factors, G, which indicates that large resolutions in the stages are preferable. However sometimes the errors increase as the stage resolution increases, e.g. it may be more difficult to design a high-speed S/H amplifier with a large gain factor.

The errors in one stage of the pipelined converter can in principle appear at four different places as shown in Fig. 9-18. The error e_1 appearing at the input of the subtrac-

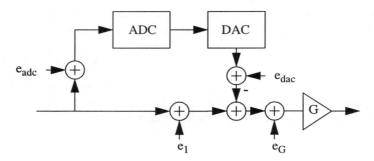

Figure 9-18   Error sources in one stage.

tor will have the same effect as a DAC error and we need not consider this error source separately.

9.5.1   ADC Errors

For most practical situations $x = 1$ is sufficient to correct errors in the sub-ADCs and errors as large as $\pm LSB/2$ at an n-bit level are then acceptable. Therefore, any static errors in the sub-ADCs, e_{adc}, in this type of pipelined converters are of minor concern.

9.5.2   S/H Amplifier Errors

The relative gain errors, e_G, in the S/H amplifiers (SHA) must be smaller than [1]

$$e_{Gi} \le \frac{1}{2^{r_{i+1}}} \qquad (9\text{-}37)$$

where r_{i+1} is the resolution remaining in the stages after stage i. This shows again that the MSB stages must have a more accurate gain compared to the LSB stages. The first stage is the most critical and we have that

9.5 Errors in Pipelined ADCs

$$e_{G1} \leq \frac{1}{2^{n_{tot}-(n-x)}} \tag{9-38}$$

If an input S/H circuit is used it must as accurate as the total resolution of the converter, i.e.

$$e_0 < \frac{1}{2^{n_{tot}}} \tag{9-39}$$

However, gain errors in the input S/H circuit is normally acceptable and it is only the distortion that needs to meet the relation in (9-39).

9.5.3 DAC Errors

There are three types of errors in the DACs that should be considered separately, linearity, offset and gain errors.

1. Offset Errors

Offset errors in the DACs can be replaced by offsets in the sub-ADCs and an offset at the input of the ADC. Due to the digital correction this does not limit the resolution of the ADC.

2. Gain Errors

Gain errors in the DACs can be replaced by gain errors in the sub ADCs and gain errors in the S/H amplifiers. We only consider the first stage gain errors here since the first stage is most critical. Gain errors in ADC_1 are corrected by the digital correction and is not a problem. Gain error at the input SHA will only change the gain of the ADC and will not introduce distortion. The gain error at the output of the first stage is equivalent to a gain error in the S/H amplifier of the first stage and must meet the relation in (9-38).

3. Linearity Errors

Linearity errors in the DACs must be small enough to meet the relation [1]

$$e_{DACi} \leq \frac{FS}{2^{r_i+1}} \tag{9-40}$$

where FS is the full-scale range of the converter. The highest requirements are on the first stage DAC, and we have

$$e_{DAC1} \leq \frac{FS}{2^{n_{tot}+1}} \tag{9-41}$$

It is seen that the linearity must be at least as good as the total ADC linearity. However exactly how component mismatch influence the performance of the converter depends on the implementation. As will be shown in Sec. 9.6.5 a high stage resolution may be better. For high resolutions it may be necessary to use calibration as discussed earlier in this chapter.

From the above discussion it is clear that it is preferable to have only gain and offset errors in the DACs to having linearity errors. A 1-bit DAC has no non-linearity errors, but a low resolution DAC is in conflict with the high stage resolution needed to reduce the effect of SHA gain errors. It is therefore difficult to find the optimum resolution of the stages since it depends on both the specification and the circuit technique used for the implementation. In the following we investigate how circuit limitations in SC circuits influence the performance of the ADC.

9.6 IMPACT OF NON-IDEAL SC CIRCUITS

In this section we discuss the effect of performance limitations in SC pipelined converters. The circuit performing the function of the DAC and the S/H amplifier is sometimes referred to as a multiplying DAC (MDAC). We will in the following use the SC MDAC from chapter 6 as an example. The SC MDAC is shown in Fig. 9-19 and it has

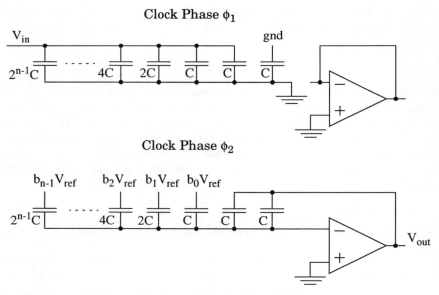

Figure 9-19 The SC MDAC of a pipelined converter.

a gain corresponding to one bit of digital correction.

9.6.1 Finite Opamp Gain

From chapter 6 we have that the transfer function with finite opamp gain is

9.6 Impact of Non-Ideal SC Circuits

$$V_{out} = \frac{V_{in} \cdot 2^{N-1} - \frac{V_{ref}}{2} \cdot (2^{N-1} \cdot b_N + \ldots + b_1)}{1 + \frac{C_{tot}}{2A_0 C}} \quad (9\text{-}42)$$

where A_0 is the DC-gain and C_{tot} the total capacitance at the input of the opamp on ϕ_2. Hence a finite DC-gain will introduce a gain error. According to chapter 6 the relative gain-error can be approximated as

$$\varepsilon_G \approx \frac{C_{tot}}{2A_0 C} = \frac{2^n + 1 + \frac{C_p}{C}}{2A_0} \quad (9\text{-}43)$$

The error must be smaller than LSB/2 referred to the required resolution of the remaining stages in the pipeline. The highest DC-gain is thus needed for the first stage. Due to the digital correction the remaining resolution after the first stage is $n_{tot} - n_1 + 1$ bits, where n_{tot} is the total number of bits in the converter and n_1 the number of bits in the first stage. Hence the required opamp gain in the first stage is determined by

$$\varepsilon_G < \frac{1}{2^{n_{tot} - n_1 + 1} \cdot 2} \Rightarrow$$

$$A_0 > 2^{n_{tot} - n_1 + 1} \left(2^{n_1} + 1 + \frac{C_p}{C} \right) = 2^{n_{tot} + 1} \left(1 + \frac{1 + \frac{C_p}{C}}{2^{n_1}} \right) \quad (9\text{-}44)$$

For a fixed C_p/C ratio the required opamp gain is reduced if the resolution in the first stage is increased. However the unit-capacitor, C, is normally smaller for higher stage resolutions, making the C_p/C ratio larger. Therefore the opamp gain requirement is weak function of the stage resolution.

9.6.2 Noise

In chapter 6 we showed that the performance is fundamentally limited by the kT/C noise in the sampling capacitor. There is also a noise contribution from the opamp but here we ignore this noise source. This a usually a fair assumption when the stage resolution is large. The requirements on the first stage is most critical. The input referred noise of the first stage MDAC can be approximated as

$$v_{n,1}^2 \approx \frac{kT}{2^{n_1} C_1} \quad (9\text{-}45)$$

where C_1 is the unit capacitor size in stage 1, n_1 the number of bits in the first stage and. To avoid that the thermal noise has a large impact on the resolution the ratio it should be smaller than the quantization noise. The quantization noise power is (see chapter 1)

$$v_q^2 = \frac{\Delta^2}{12} = \left(\frac{FS}{2^{n_{tot}}}\right)^2 \cdot \frac{1}{12} = \frac{FS^2}{2^{2n_{tot}+2}} \cdot \frac{1}{3} \tag{9-46}$$

By making the thermal noise smaller than the quantization noise we have

$$v_{n,1}^2 < \frac{FS^2}{2^{2n_{tot}+2}} \cdot \frac{1}{3} \Rightarrow C_1 \geq \frac{3 \cdot 2^{2 \cdot n_{tot}+2} \cdot kT}{2^{n_1} \cdot FS^2} \tag{9-47}$$

The total sampling capacitor in the first stage is thus independent of the stage resolution. We have for the i-th stage

$$v_{n,i}^2 < \frac{FS^2}{2^{2r_i+2}} \cdot \frac{1}{3 \cdot 2} \Rightarrow C_i \geq \frac{3 \cdot 2^{2 \cdot r_i+3} \cdot kT}{FS^2 \cdot 2^{n_i}} \tag{9-48}$$

where r_i is the resolution of the stages from i to the last stage. Hence the total sampling capacitor can be smaller for large stage resolutions (except for $i = 1$).

When deriving the above limits we considered only one stage at a time. All the thermal noise sources will contribute to the total noise in the converter. The rms value of the total thermal noise referred to the input can be expressed as [18]

$$v_{n,tot}^2 = \sqrt{\frac{v_{n,1}^2}{1} + \ldots + \frac{v_{n,i}^2}{(G_1 \ldots G_{i-1})^2} + \ldots + \frac{v_{n,m}^2}{(G_1 \ldots G_{m-1})^2}} =$$

$$= \sqrt{\frac{v_{n,1}^2}{1} + \ldots + \frac{v_{n,i}^2}{2^{2(n-1)(i-1)}} + \ldots + \frac{v_{n,m}^2}{2^{2(n-1)(m-1)}}} \tag{9-49}$$

where $v_{n,i}^2$ is the noise power contribution from the MDAC in stage i, G_i is the residue gain factor in stage i and m is the number of stages. If the gain factors, G_i are large the thermal noise in the later stages is attenuated when referred to the input. Hence the capacitor sizes according to (9-48) will be very small and must be larger for other reasons such as clock feedthrough. Therefore, if the stage resolution is large the noise in the first stage will dominate. For small stage resolutions several stages contributes to the total noise and the capacitor sizes must be larger than the size given by (9-48). It should also be noted that when the thermal noise power is equal to the quantization noise power the SNR decreases by 3 dB. The capacitors should therefore in the actual implementation be larger to give some margin for other errors sources.

From the above discussion we conclude that a large stage resolution implies that we can use small capacitors in all but the first stage. This is advantageous from a power consumption point of view.

9.6.3   Limited Bandwidth

The relative settling error assuming a single pole system is given by

9.6 Impact of Non-Ideal SC Circuits

$$\varepsilon_r = e^{-t_s \cdot \omega_{-3dB}} \qquad (9\text{-}50)$$

where t_s is the available settling time and ω_{-3dB} is the -3dB bandwidth of the circuit.

The first stage is most critical since it needs the highest accuracy. The closed loop -3 dB bandwidth of the MDAC in the first stage, assuming a single-stage opamp is

$$\omega_{-3dB} = \omega_u \cdot \beta = \frac{g_m}{C_{load}} \cdot \beta = \frac{g_m}{C_{load}} \cdot \frac{2}{(2^{n_1}+1)+C_p/C_1} \qquad (9\text{-}51)$$

where C_{load} is the total load capacitance at the opamp output, n_1 is the number of bits in the first stage, C_1 is the unit-capacitor in the first stage MDAC and C_p the parasitic capacitance at the input of the opamp. The size of the load capacitor C_{load} has a large impact on the speed of the circuit. It consists of parasitic capacitors at the opamp output, the load due to the capacitors in the MDAC and the sampling capacitor of the following stage. The load capacitance is given by

$$C_{load} = 2^{n_2} \cdot C_2 + \frac{2(C_1(2^{n_1}-1)+C_p)}{(2^{n_1}+1)+C_p/C_1} + C_{p,out1} \qquad (9\text{-}52)$$

where C_2 is the unit-capacitor in stage 2, n_2 the number of bits in stage 2, and $C_{p,out1}$ the parasitic capacitor at the output of the opamp in the first stage. For simplicity we ignore the parasitic capacitors. The bandwidth can now be approximated as

$$\omega_{-3dB} \approx \frac{2g_m}{2^{n_2} \cdot C_2 \cdot (2^{n_1}+1)+2C_1(2^{n_1}-1)} \qquad (9\text{-}53)$$

The smallest possible size on C_2 is determined by the thermal noise requirement (9-49) and we have

$$C_2 > \frac{2^{n_1} \cdot C_1}{2^{n_2} \cdot G_1} = \frac{2^{n_1} \cdot C_1}{2^{n_1-1} \cdot 2^{n_2}} = \frac{C_1}{2^{n_2-1}} \qquad (9\text{-}54)$$

Sometimes when the power consumption is not critical all the stages in the pipeline can be identical to save design time. The sampling capacitor is now much larger than required by the thermal noise requirement. The sampling capacitor in stage 2 should never be larger than in stage 1 and we therefore have an upper limit given by

$$C_2 < 2^{n_1-n_2} \cdot C_1 \qquad (9\text{-}55)$$

1. *The sampling capacitors are as small as required by the thermal noise requirement (see (9-54))*

By using (9-50), (9-53), (9-54) and requiring an error smaller than LSB/2 referred to the remaining resolution in the pipeline, we have

$$\omega_{-3dB} > \ln(2^{n_{tot}-n_1+1})2f_s \Rightarrow$$

$$\ln(2^{n_{tot}-n_1+1})2f_s < \frac{g_m}{C_1 \cdot 2^{n_1+1}} = \frac{g_m \cdot FS^2}{3 \cdot 2^{2n_{tot}+3} \cdot kT} \Rightarrow \quad (9\text{-}56)$$

$$f_s < \frac{g_m \cdot FS^2}{3 \cdot 2^{2n_{tot}+4} \cdot kT \cdot \ln(2^{n_{tot}-n_1+1})}$$

where f_s is the sampling frequency.

2. *All stages are equal making the sampling capacitor in stage 2 have its largest value.*

By using (9-50), (9-53), (9-55) and requiring an error smaller than LSB/2 referred to the remaining resolution in the pipeline, we have

$$\omega_{-3dB} > \ln(2^{n_{tot}-n_1+1})2f_s \Rightarrow$$

$$\ln(2^{n_{tot}-n_1+1})2f_s < \frac{2g_m}{(2^{2n_1}+3\cdot 2^{n_1}-2)\cdot C_1} =$$

$$= \frac{2g_m \cdot FS^2}{(2^{n_1}+3-2^{1-n_1}) \cdot 3 \cdot 2^{2n_{tot}+2} \cdot kT} \Rightarrow \quad (9\text{-}57)$$

$$f_s < \frac{g_m \cdot FS}{(2^{n_1}+3-2^{1-n_1}) \cdot 2^{2n_{tot}+2} \cdot 3 \cdot kT \cdot \ln(2^{n_{tot}-n_1+1})}$$

We see that if the load capacitance on the MDAC is small the maximum sampling frequency increases as the stage resolution increases, but the increase is slow due to the log function. If the load capacitance is big the maximum sampling frequency decreases as the stage resolution increases. For many practical cases the load capacitance is somewhere between the two extreme values. Therefore there is usually no strong impact on the speed when the stage resolution is changed. It should be noted that the above derivations assumes a first order linear settling model. In many cases this is not an accurate assumption.

9.6.4   Limited Slew Rate

The results in the previous section are valid if the settling is entirely linear. For most practical cases however the settling is slew rate (SR) limited. It can be shown that the output voltage of the MDAC with slew rate limited settling can be calculated as [17]

$$v(v_s) = \begin{cases} v_s(1-e^{-t_s\omega_{-3dB}}), & |v_s| \le v_{lin} \\ v_s - \mathrm{sgn}(v_s)v_{lin}e^{\left(\frac{|v_s|}{v_{lin}}-t_s\omega_{-3dB}-1\right)}, & v_{lin} < |v_s| \le v_{lin}+t_s\cdot SR \\ \mathrm{sgn}(v_s)\cdot t_s \cdot SR, & v_{lin}+t_s\cdot SR < |v_s| \end{cases} \quad (9\text{-}58)$$

9.6 Impact of Non-Ideal SC Circuits

where v_s is the step size of the output voltage as $t \to \infty$, SR is the slew rate, ω_{-3dB} is the -3dB bandwidth, t_s is the available settling time and v_{lin} is the maximum step size that gives linear settling. The maximum step size for linear settling can be calculated as

$$v_{lin} = SR \cdot \omega_{-3dB} = SR \cdot \omega_u \cdot \beta \qquad (9\text{-}59)$$

If the settling is SR limited the -3dB bandwidth must be higher compared to if the settling is linear. The smaller the SR the larger the required bandwidth. If the bandwidth is too high it becomes very difficult to achieve a large phase margin for the opamp. It is easier to get linear settling if the feedback factor, β, of the MDAC is small (see chapter 7). Therefore the settling is more linear for higher stage resolutions due to the smaller feedback factor as seen in (9-58).

9.6.5 Matching

The capacitor matching in the MDAC ultimately limits the linearity of the converter. A conclusion from Sec. 9.5 was that the DAC linearity in the first stage must be as good as the total resolution of the entire ADC. It may therefore seem unlikely that anything would be gained by changing the stage resolution. However the effect of the component matching on the performance depends on the implementation. This is best illustrated by an example.

We assume that the capacitor matching errors are normally distributed and statistically independent (in reality there is usually some correlation between the errors). We compare the performance of pipelined converters with different resolutions in the first stage and we assume that all stages but the first are error free. In all cases the total resolution of the converter is 15 bits. Fig. 9-19 show the result of MatLab simulations where the resolution in the first stage was changed from 3 to 5 bits and the relative matching was changed from 0.5 - 2%. In the figure the fail percentage, i.e. the percentage of the samples that do not meet a certain ENOB requirement, as function of the ENOB is plotted. The results in the figure indicate that the achievable resolution increases with 0.5 ENOBs as the stage resolution is increased by one bit while the resolution is increased by 1 ENOB if the capacitor matching is improved by a factor 2.

The achievable capacitor matching is of course dependent on the layout and the process, but for a fixed relative capacitor matching we conclude that a converter with a large resolution in the first stage is better. In the example we intentionally made the matching errors large to avoid that the quantization errors influence the result. A relative capacitor matching of 0.1% or better can normally be achieved in an analog CMOS processes.

9.6.6 How to Choose Stage Resolution

The optimum stage resolution is a trade-off between many different factors and it is not easy to find the ADC with the smallest area and power for a certain specification. An attempt to optimize the performance is presented in [18]. In this paper a 5 MS/s 15 bit converter is used as a design example. Choosing the resolution in the two first stages as 5, 4 bits or 4, 4 bits while the following stages have a smaller resolutions opti-

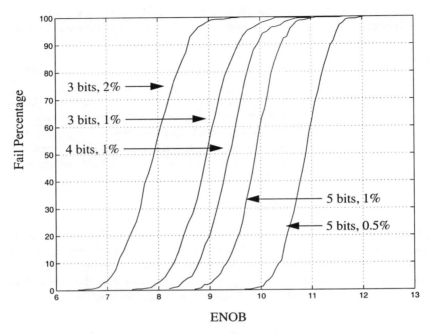

Figure 9-20 Effect of capacitor mismatch in SC MDAC of a pipelined converter.

mizes the power. The area is optimized for a converter where all stages have a resolution of 3 bits. In general a good rule of thumb is to use a low stage resolution of <2-3 bits for converters up to about 10 bits. For higher resolutions the first stages should have a higher resolution of 4-5 bits.

9.7 SUMMARY

In this chapter we have reviewed the principle of pipelined converters, digital correction and calibration. With digital correction, the influence of errors in the comparators is significantly reduced. Hence the dominating error sources are DAC errors and errors in the residue amplifier. For very high resolutions it may be necessary to use calibration. The principle of self-calibration algorithms in pipelined converters was explained in this chapter. It is normally difficult to calibrate dynamic errors such as settling errors.

The effect of the errors in the LSB stages of a pipelined ADC is reduced by the residue gain factors, making the errors in the last stages almost negligible. The most critical stage is thus the first stage. For errors in the first stage MDAC we concluded that

- Errors appearing at the output of the subtractor in the first stage must be within the resolution of the remainder of the pipeline
- Errors at the input of the subtractor must be within the total resolution of the con-

verter.

Whether an error should be modeled as appearing at the input or the output of the subtractor depends on the circuit solution.

We have also presented expressions showing how the stage resolution affects the influence of circuit limitations such as finite opamp gain and limited bandwidth in an SC MDAC. These expressions can be used as guide-lines when designing the circuits. The optimum stage resolution is a trade-off between many different factors and it is not easy to find the ADC with the smallest area and power for a certain specification. The general rule for high-speed ADC however seems to be that a larger stage resolutions should be used in the first stages while the resolution should be reduced in the LSB stages of the pipelined ADC.

REFERENCES

[1] S. H. Lewis, "Optimizing the Stage Resolution in Pipelined, Multistage, Analog-to-Digital Converters for Video-Rate Applications", *IEEE Trans. on Circuits and Systems-II*, vol. 39, no. 8, pp. 516-23, Aug. 1992.

[2] S. H. Lewis and P. R. Gray, "A Pipelined 5-Msample/s 9-bit Analog-to-Digital Converter", *IEEE J. of Solid-State Circuits*, vol. SC-22, no. 6, pp. 954-61, Dec. 1987.

[3] S. H. Lewis, R. Ramachandran and W. M. Snelgrove, "Indirect Testing of Digital-Correction Circuits in Analog-to-Digital Converters with Redundancy", *IEEE Trans. on Circuits and Systems-II*, vol. 42, no. 7, pp. 437-45, July 1995.

[4] B. Ginetti, P. G. A. Jespers and A. Vandemeulebroecke, "A CMOS 13-b Cyclic RSD A/D Converter", *IEEE J. of Solid-State Circuits*, vol. 27, no. 7, p. 957-64, July 1992.

[5] H. S. Lee, D. A. Hodges and P. R. Gray, "A Self-Calibrating 15 Bit CMOS A/D Converter", *IEEE J. of Solid-State Circuits*, vol. SC-19, no. 6, pp. 813-19, Dec. 1984.

[6] S. H. Lee and B. S. Song, "Digital-Domain Calibration of Multistep Analog-to-Digital Converters", *IEEE J. of Solid-State Circuits*, vol. 27, no. 12, pp. 1679-88, Dec. 1992.

[7] E. G. Soenen and R. L. Geiger, "An Architecture and an Algorithm for Fully Digital Correction of Monolithic Pipelined ADC's", *IEEE Trans. on Circuits and Systems-II*, vol. 42, no. 3, pp. 143-53, March 1995.

[8] S. H Lee and B. S. Song, "Interstage Gain Proration Technique for Digital-Domain Muti-Step ADC Calibration", *IEEE Trans. on Circuits and Systems-II*, vol. 41, no. 1, pp. 12-18, Jan. 1994.

[9] P. Rombouts and L. Weyten, "Comments on "Interstage Gain Proration Technique for Digital-Domain Muti-Step ADC Calibration"", *IEEE Trans. on Circuits and Systems-II*, vol. 46, no. 8, pp. 1114-1116, Aug. 1999.

[10] H. S. Lee "A 12-b 600 ks/s Digitally Self-Calibrated Pipelined Algorithmic ADC", *IEEE J. of Solid-State Circuits*, vol. 29, no. 4, pp. 509-515, April 1994.

[11] A. N. Karanicolas, H. S. Lee and K. L. Bacrania, "A 15-b 1 Msample/s Digitally Self-Calibrated Pipeline ADC", *IEEE J. of Solid-State Circuits*, vol. 28, no. 12, pp. 1207-15, Dec. 1993.

[12] U. K. Moon and B. S. Song, "Background Digital Calibration Techniques for Pipelined ADCs", *IEEE Trans. on Circuits and Systems-II*, vol. 44, no. 2, pp. 102-9, Feb. 1997.

[13] T. H. Shu, B. S. Song and B. Bacrania "A 13-b 10-Msample/s ADC Digitally Calibrated with Oversampling Delta-Sigma Converter", *IEEE J. of Solid-State Circuits*, vol. 30, no. 4, pp. 443-52, April 1995.

[14] J. M. Igino and B. A. Wooley "A Continuously Calibrated 12-b, 10/MS/s, 3.3-V A/D Converter", *IEEE J. of Solid-State Circuits*, vol. 33, no. 12, pp. 1929-52, Dec. 1931.

[15]　B. S. Song, M. F. Tompsett and K. R. Lakshmikumar, "A 12-bit 1-Msample/s Capacitor Error-Averaging Pipelined A/D Converter", *IEEE J. of Solid-State Circuits*, vol. 23, no. 6, pp. 1324-33, Dec. 1988.

[16]　P. C. Yu and H. S. Lee, "A 2.5-V, 12-b, 5-Msample/s Pipelined CMOS ADC", *IEEE J. of Solid-State Circuits*, vol. 31, no. 12, p. 1854-61, Dec. 1996.

[17]　L. A. Williams, III and B. A. Wooley "A Third-Order Sigma-Delta Modulator with Extended Dynamic Range", *IEEE J. of Solid-State Circuits*, vol. 29, no. 3, pp. 193-202, Mar. 1994.

[18]　J. Goes, J. C. Vital and J. E. Franca, "Systematic Design for Optimazation of High-Speed Self-calibrated Pipelined A/D Converters", *IEEE Trans. on Circuits and Systems-II*, vol. 45, no. 12, pp. 1513-26, Dec. 1998.

10 TIME-INTERLEAVED A/D CONVERTERS

10.1 INTRODUCTION

By using time-interleaving techniques very high sampling rates can be achieved. Unfortunately any mismatch between the time-interleaved channels will give rise to distortion. At high signal frequencies the most difficult mismatch error to handle is the delay skews between the clock signals to the S/H circuits of the different channels. In Sec. 10.2 and Sec. 10.3 we discuss the basic principle and review the effect of the error sources in time-interleaved ADCs. Solutions to reduce these errors are considered in Sec. 10.4. In Sec. 10.5 and Sec. 10.6 we discuss a passive sampling technique in more detail. This technique can reduce the effect of delay skew errors between the channel. Tedious derivations are covered in Sec. 10.7.

10.2 PRINCIPLE OF TIME-INTERLEAVED ADCS

An attractive way to increase the conversion rate of ADCs is to use time-interleaving techniques where several ADCs, using different clock phases, are operated in parallel [1]. This enables higher total conversion speed but mismatch between the channels introduces distortion.

The concept of time-interleaving is illustrated in Fig. 10-1. The converters can be of any type and are operated at f_s/M where f_s is the total sampling frequency of the time-interleaved ADC and M is the number of channels. The speed requirements on each converter are relaxed by a factor M but the number of converters is at the same time increased by the same factor. This may result in a large chip area and a high power consumption. The increase in power and area is not necessarily a linear function of M compared to using only one converter, since smaller bias currents can be used in the opamps (if SC circuits is used) due to the lower speed. It is also sometimes possible to share opamps between the channels [2]. The performance of the time-interleaved converter is of course limited by the accuracy of the channel ADCs but there are additional errors caused by mismatch between the channels. There are three main sources of error in time interleaved systems, phase skew errors, gain errors and offset errors [1], [3], [4].

In the analysis below the SNDR is used as a measure on how the errors effect the performance of the parallel converter. When characterizing ADCs it is common to use

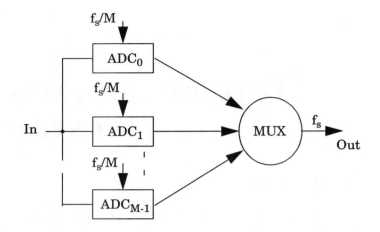

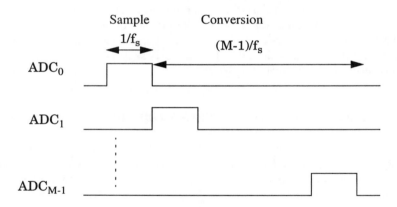

Figure 10-1   Parallel ADCs and corresponding sub-ADC timing.

histogram methods. It should be noted that errors in time-interleaved converters will not be detected in such a test. Therefore FFT spectrum testing methods are better suited [5].

The samples in channel $m = 0, \ldots, M-1$ is ideally sampled at time instants (see Fig. 10-2)

$$t_m, t_{m+M}, t_{m+2M}, \ldots \tag{10-1}$$

where $t_m = mT$ and T is the sampling period of the time-interleaved ADC. Due to imperfections in the sampling circuits the actual sampling instant will deviate from the ideal sampling instant. We assume that the delay skew error in channel m is constant,

10.2 Principle of Time-Interleaved ADCs

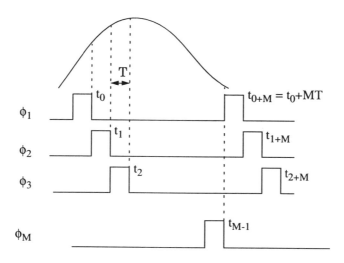

Figure 10-2 Sampling of the input signal by a time-interleaved ADC.

i.e. it does not change with time. The actual sampling instant in channel m can be expressed as

$$t_m = mT - r_m T \tag{10-2}$$

where r_m is the relative error in the sampling instant with respect to the sampling period of the time-interleaved ADC.

Assuming the analog input signal to have the spectrum $G_a(\omega)$, the digital spectrum of the time-interleaved system can be expressed as [3]

$$G(\omega) = \frac{1}{MT} \sum_{m=0}^{M-1} \left(\sum_{k=-\infty}^{\infty} G_a\left(\omega - k\frac{2\pi}{MT}\right) e^{-j\left(\omega - k\frac{2\pi}{MT}\right)t_m} \right) \cdot e^{-jm\omega T} \tag{10-3}$$

which can be rewritten as

$$G(\omega) = \frac{1}{T} \sum_{k=-\infty}^{\infty} \left(\frac{1}{M} \sum_{m=0}^{M-1} e^{-j\left(\omega - k\frac{2\pi}{MT}\right)r_m T} e^{-jmk\frac{2\pi}{M}} \right) G_a\left(\omega - k\frac{2\pi}{MT}\right) \tag{10-4}$$

If there are no errors in the time-interleaved ADC we have [3]

$$G(\omega) = \frac{1}{T} \sum_{k=-\infty}^{\infty} G_a(\omega - 2\pi k) \tag{10-5}$$

which corresponds to a signal sampled at sampling rate T.

10.3 ERRORS IN TIME-INTERLEAVED ADCS

Based on the equations in the previous section we can derive the effects of mismatch in the channels.

10.3.1 Offset Errors

If we assume that there are no delay skew or gain errors in the channels the digital spectrum is expressed by

$$G(\omega) = \frac{1}{T} \sum_{k=-\infty}^{\infty} \left(\frac{1}{M} \sum_{m=0}^{M-1} G_a\left(\omega - k\frac{2\pi}{M}\right) \cdot e^{-jmk\frac{2\pi}{M}} \right) \quad (10\text{-}6)$$

With a sine wave input signal, $A \cdot \sin(\omega_0 \cdot t)$, and channel offsets the Fourier transform of the analog signal, sampled by channel m is given by

$$G_a(\omega) = \pi j A(\delta(\omega + \omega_0) - \delta(\omega - \omega_0)) + 2\pi o_m \delta(\omega) \quad (10\text{-}7)$$

where o_m is the offset in channel m and A is the amplitude of the sine wave. By using (10-6) and (10-7) the digital spectrum can be expressed as

$$\begin{aligned} G(\omega) &= G_s(\omega) + \frac{1}{T} \sum_{k=-\infty}^{\infty} A(k) \cdot 2\pi\delta\left(\omega - k\left(\frac{2\pi}{MT}\right)\right) \\ &= \frac{1}{T} \sum_{k=-\infty}^{\infty} \pi j A(\delta(\omega + \omega_0 - 2\pi k) - \delta(\omega - \omega_0 - 2\pi k)) + \\ &\quad + \frac{1}{T} \sum_{k=-\infty}^{\infty} A(k) \cdot 2\pi\delta\left(\omega - k\left(\frac{2\pi}{MT}\right)\right) \end{aligned} \quad (10\text{-}8)$$

where $G_s(\omega)$ is the digital spectrum of the sinusoidal sampled at the sampling rate $f_s = 1/T$ and

$$A(k) = \sum_{m=0}^{M-1} \left[\frac{1}{M} \cdot o_m\right] \cdot e^{-j2\pi\frac{km}{M}} \quad (10\text{-}9)$$

The first term of (10-8) corresponds to the input signal while the second term correspond to the distortion caused by channel offsets. The distortion is not signal dependent and appears at

$$f_s/M \cdot m, \; m = 0, 1, ..., M-1 \quad (10\text{-}10)$$

This is illustrated by an example shown in Fig. 10-3. The input signal is a 10 MHz sine

10.3 Errors in Time-Interleaved ADCs

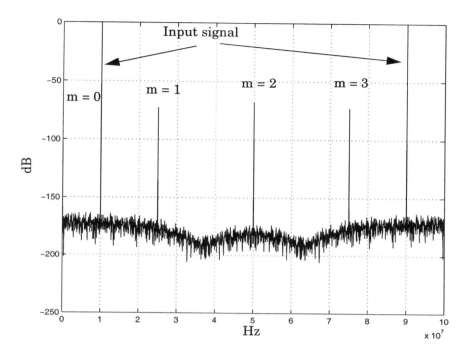

Figure 10-3   Output spectrum with offset errors.

wave sampled at 100 MS/s and the time-interleaved ADC has 4 channels. The simulation was done in MATLAB and the channel offsets were one set of samples of random variables with zero mean and a standard deviation of 0.001. The distortion tones appear at 0, 25, 50, 75 MHz while the input signal shows as two tones at 10 MHz and 90 MHz.

The signal power of the input signal is

$$P_s = A^2/2 \qquad (10\text{-}11)$$

We see in (10-8) that the distortion consists of a sum of impulses. Each impulse corresponds to a complex exponential signal in the time domain, i.e. $e^{j\omega}$. The power of the exponential signal is 1. The factors $A(k)$ determines the amplitude of the signal and we can therefore calculate the total distortion power as

$$P_d = \sum_{k=0}^{M-1} |A(k)|^2 \qquad (10\text{-}12)$$

The factors $A(k)$ in (10-9) can be interpreted as the Discrete Fourier Transform (DFT) of the sequence o_m/M, $m = 0, ..., M-1$. From Parseval's relation we have

$$\sum_{m=0}^{M-1}(o_m/M)^2 = \frac{1}{M}\sum_{m=0}^{M-1}|A(m)|^2 \Rightarrow \sum_{m=0}^{M-1}|A(m)|^2 = \frac{1}{M}\sum_{m=0}^{M-1}o_m^2 \qquad (10\text{-}13)$$

Assuming that the channel offsets are Gaussian random variables with zero mean and a variance of σ_o^2 the expected value of the distortion power is

$$P_d = E\left[\sum_{k=0}^{M-1}|A(k)|^2\right] = E\left[\frac{1}{M}\sum_{m=0}^{M-1}o_m^2\right] = \sigma_o^2 \qquad (10\text{-}14)$$

Hence the distortion power is independent of the signal amplitude. The SNDR for a sinusoidal input signals is limited by

$$SNDR < 10\cdot\log\left(\frac{P_s}{P_d}\right) = 10\cdot\log\left(\frac{A^2}{2\sigma_o^2}\right) \qquad (10\text{-}15)$$

The ENOB as function of σ_o is shown in Fig. 10-4. In the plot we assumed that

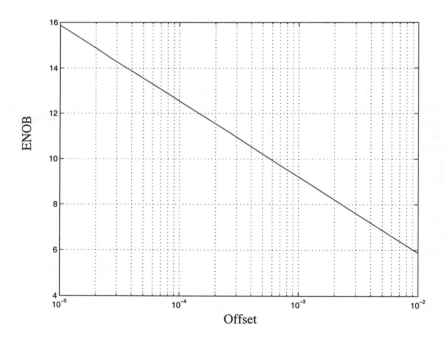

Figure 10-4   ENOB as function of offset errors.

$A = 1$.

10.3.2 Gain Errors

If we assume that there are no delay skew or offset errors between the channels the digital spectrum is expressed by

$$G(\omega) = \frac{1}{T} \sum_{k=-\infty}^{\infty} \left(\frac{1}{M} \sum_{m=0}^{M-1} a_m G_a\left(\omega - k\frac{2\pi}{M}\right) \cdot e^{-jmk\frac{2\pi}{M}} \right) \quad (10\text{-}16)$$

where a_m is the gain in channel m. The Fourier transform of a sinusoidal signal with an amplitude of 1 is given by

$$G_a(\omega) = \pi j(\delta(\omega + \omega_0) - \delta(\omega - \omega_0)) \quad (10\text{-}17)$$

The digital spectrum can now be rewritten as

$$G(\omega) = \frac{1}{T} \sum_{k=-\infty}^{\infty} A(k) \cdot j\pi \left(\delta\left(\omega + \omega_0 - \frac{2\pi k}{MT}\right) - \delta\left(\omega - \omega_0 - \frac{2\pi k}{MT}\right) \right) \quad (10\text{-}18)$$

where $A(k)$ is determined by

$$A(k) = \sum_{m=0}^{M-1} \left[\frac{1}{M} \cdot a_m \right] \cdot e^{-j2\pi\frac{km}{M}} \quad (10\text{-}19)$$

If there are no gain errors $A(k)$ is

$$A(k) = \begin{cases} a & \text{if } k = 0, M, 2M, \ldots \\ 0 & \text{for all other } k \end{cases} \quad (10\text{-}20)$$

where a is the nominal gain in the channels. This means that there is no distortion in the output signal since all tones appear at frequencies corresponding to the input signal. However, with gain errors $A(k) \neq 0$ for all k and according to (10-18) there are tones at

$$f_{in} + k \cdot f_s/M \text{ and } (f_s - (f_{in} + k \cdot f_s/M)), k = 1, \ldots, M-1 \quad (10\text{-}21)$$

This is illustrated by an example in Fig. 10-5, showing an FFT of a time-interleaved ADC. The input signal is a 10 MHz sinusoid sampled at 100 MS/s and the time-interleaved ADC has 4 channels. Random gain errors with a mean value of 1 and a standard deviation of 0.001 were introduced. The fundamental obviously corresponds to $k = 0$ while $k = 1, \ldots, M-1$ corresponds to the distortion. All distortion tones are folded into the Nyquist band 0 to $f_s/2$. The signal amplitude is determined by $A(0)$ while the distortion amplitudes are determined by $A(m)$, $m = 1, \ldots, M-1$.

The factors $A(k)$ can be interpreted as the Discrete Fourier Transform (DFT) of the sequence a_m/M, $m = 0, \ldots, M-1$. From Parseval's relation we have the total out-

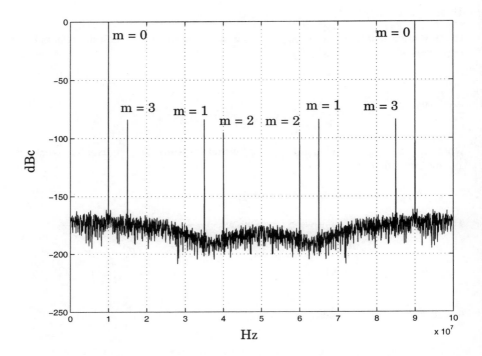

Figure 10-5 Output spectrum with gain errors.

put signal power is given by

$$P_{tot} = \sum_{m=0}^{M-1} \frac{|A(m)|^2}{2} = \frac{1}{2M} \sum_{m=0}^{M-1} a_m^2 \qquad (10\text{-}22)$$

The fundamental corresponds to $k = 0$ and the signal power is given by

$$P_s = \frac{|A(0)|^2}{2} = \frac{1}{2} \sum_{m=0}^{M-1} \left[\frac{1}{M} \cdot a_m \right] \qquad (10\text{-}23)$$

Assuming the gain errors to be random variables with a mean value of a and a standard deviation of σ_a the expected value of the total power is

$$P_{tot} = E\left[\frac{1}{2M} \sum_{m=0}^{M-1} a_m^2 \right] = \frac{1}{2M} \sum_{m=0}^{M-1} E[a_m^2] = \frac{E[a_0^2]}{2} = \frac{\sigma_a^2 + a^2}{2} \qquad (10\text{-}24)$$

and the expectation of the signal power is

10.3 Errors in Time-Interleaved ADCs

$$P_s = \frac{E[A(0) \cdot A^*(0)]}{2} = \frac{1}{2M^2} \sum_{m=0}^{M-1} \sum_{n=0}^{M-1} E[a_m \cdot a_n] =$$

$$= \frac{1}{2M} E[a_m^2] + \frac{1}{2M^2}(M^2 - M)(E[a_m])^2 \qquad (10\text{-}25)$$

$$= \frac{1}{2M}(\sigma_a^2 + a^2) + \frac{1}{2M^2}(M^2 - M)a^2 = \frac{1}{2M}(\sigma_a^2 + M \cdot a^2)$$

The distortion power is the total power minus the signal power and the SNDR is limited by

$$SNDR = 10 \cdot \log\left(\frac{P_s}{P_{tot} - P_s}\right) = 10 \cdot \log\left(\frac{\frac{1}{2M}(\sigma_a^2 + M \cdot a^2)}{\frac{(\sigma_a^2 + a^2)}{2} - \frac{1}{2M}(\sigma_a^2 + M \cdot a^2)}\right)$$

$$= 10 \cdot \log\left(\frac{\frac{1}{M}(\sigma_a^2 + M \cdot a^2)}{\left(1 - \frac{1}{M}\right) \cdot \sigma_a^2}\right) \qquad (10\text{-}26)$$

Since the standard deviation of the gain error usually is much smaller than the nominal gain, (10-26) can be simplified to

$$SNDR \approx 20 \cdot \log(a/\sigma_a) - 10 \cdot \log\left(1 - \frac{1}{M}\right) \qquad (10\text{-}27)$$

The last term in (10-27) depends on the number of channels, but will only change about 3 dB as M goes from 2 to ∞. The ENOB as function of gain error is shown in Fig. 10-6. The figure shows the result for $M = 2, 4, 8$ and $a = 1$ in all cases. The approximation in (10-27) is based on that linear gain errors are allowed which is common in many applications. If gain errors are not allowed the SNDR can be approximated as the ratio of the input signal power and the difference in power between the input and output signal. The expected value of the SNDR is then given by

$$SNDR \approx 10 \cdot \log\left(\frac{a^2}{|a^2 - (\sigma_a^2 + a^2)|}\right) = 20 \cdot \log\left(\frac{a}{\sigma_a}\right) \qquad (10\text{-}28)$$

which corresponds to the result in [4].

10.3.3 Phase Skew Errors

If we assume that there are no gain or offset errors between the channels the digital spectrum is expressed by (10-4) repeated here for convenience

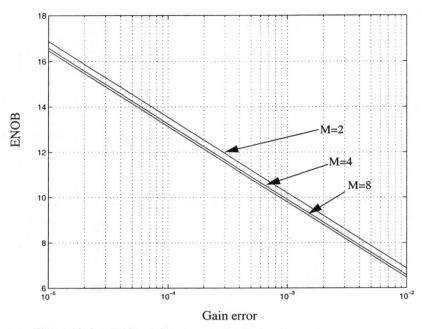

Figure 10-6 ENOB as function of gain errors.

$$G(\omega) = \frac{1}{T} \sum_{k=-\infty}^{\infty} \left(\frac{1}{M} \sum_{m=0}^{M-1} e^{-j\left(\omega - k\frac{2\pi}{MT}\right)r_m T} e^{-jmk\frac{2\pi}{M}} \right) G_a\left(\omega - k\frac{2\pi}{M}\right) \quad (10\text{-}29)$$

The Fourier transform of a sinusoidal signal with an amplitude of 1 is given by

$$G_{a,m}(\omega) = \pi j(\delta(\omega + \omega_0) - \delta(\omega - \omega_0)) \quad (10\text{-}30)$$

The digital spectrum can now be rewritten as [3]

$$G(\omega) = \frac{1}{T} \sum_{k=-\infty}^{\infty} A(k) \cdot j\pi\left(\delta\left(\omega + \omega_0 - k\frac{2\pi}{MT}\right) - \delta\left(\omega - \omega_0 - k\frac{2\pi}{MT}\right) \right) \quad (10\text{-}31)$$

where $A(k)$ is determined by

$$A(k) = \sum_{m=0}^{M-1} \left[\frac{1}{M} \cdot e^{-j\omega_0 r_m T} \right] \cdot e^{-j2\pi\frac{km}{M}} \quad (10\text{-}32)$$

This is a similar expression as for gain errors and the distortion tones for phase skew errors will appear at signal frequencies

10.4 Error Reduction Techniques

$$f_{in} + m \cdot f_s/M \text{ and } (f_s - (f_{in} + m \cdot f_s/M)), \, m = 1, \ldots, M-1 \quad (10\text{-}33)$$

If the phase skew errors are treated as Gaussian random variables with zero mean and a variance of σ_t^2 the same principle as for gain errors can be adopted to derive the SNDR. The SNDR can be approximated as [3]

$$SNDR = 20 \cdot \log\left(\frac{1}{\sigma_t 2\pi f_{in}}\right) - 10 \cdot \log\left(1 - \frac{1}{M}\right) \quad (10\text{-}34)$$

Just as for gain errors the last term in (10-34) will disappear if the SNDR is calculated as the ratio of the input signal power and the total error power. Fig. 10-7 shows how

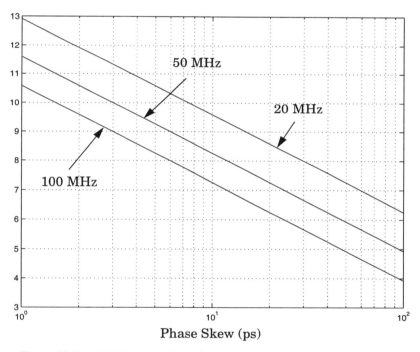

Figure 10-7 ENOB versus phase skew in a four channel ADC.

the ENOB is affected by phase skew for $M = 4$ at some different input signal frequencies. At 20 MHz and a resolution of 10 bits σ_t should be less than 8 ps. This is difficult to achieve in a CMOS process.

10.4 ERROR REDUCTION TECHNIQUES

There are several techniques to reduce the effect of mismatch in the channels. Offset and gain errors can be removed by calibration. Phase skew errors are usually more difficult to handle since it is difficult to measure the actual skews.

10.4.1 Two-Rank Sample-and-Hold

The phase skew can almost be eliminated by using a S/H circuit at the input as shown in Fig. 10-8. The output of this S/H circuit is a discrete-tome analog signal and it is not

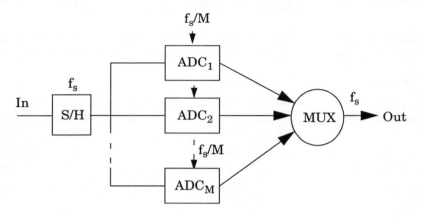

Figure 10-8 Two-Rank S/H circuit.

continuously changing. The phase skew errors have no effect on the result as long as the input S/H settles to its final value. The obvious disadvantage of this method is that the S/H circuit must be operated at the full speed and it is difficult to design high-speed, high-accuracy S/H circuits in CMOS. This technique was used in [6] and [7] to get 10 bits at 100 MHz in CMOS and 6 bits at 1 GHz in GaAs. The two-rank sampling will only improve delay skew errors and have no effect on gain and offset errors.

10.4.2 Averaging

If the input signal is oversampled, averaging can be used to reduce the channel mismatch [8]. The concept of averaging for $M = 2$ is illustrated in Fig. 10-9. The transfer function of the averaging block is

$$H(z) = \frac{1}{2}(1 + z^{-1}) \qquad (10\text{-}35)$$

and the Fourier transform

$$|H(j\omega)|^2 = \frac{1}{2}(1 + \cos(\omega T_s)) \qquad (10\text{-}36)$$

There is a zero at $f_s/2$ and the channel mismatch errors, located in the frequency band $[f_s/4, f_s/2]$ will be low-pass filtered. The distortion term at $f_s/2$ caused by offset errors is completely removed while gain and phase skew errors are reduced by a factor

10.4 Error Reduction Techniques

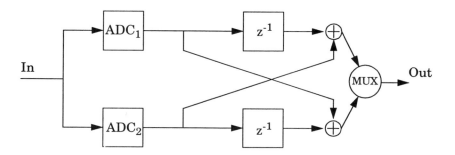

Figure 10-9 The principle of averaging.

$4/(\omega_{in}T_s)$. The concept of using digital processing after the time interleaved ADC can be generalized. With an oversampling ratio of M and M time-interleaved channels no distortion terms appear in the signal band. The distortion can thus be removed by digital filters. The drawback of this method is that even though we have increased the sampling rate by parallelism the input bandwidth is not increased due to the oversampling.

10.4.3 Randomization

By randomly choosing the channel to convert the next sample, the distortion caused by gain and offset mismatch will be reduced [9] since the correlation between the channel errors and the input signal is reduced. However the noise floor will increase since the distortion is randomized to noise. This method can not reduce the effect of phase skew errors and a serious drawback is that the individual converters must use a clock signal which is at least as fast as the total sampling speed of the time-interleaved ADC. Another problem is that when having many time-interleaved channels it may be difficult to do the layout.

10.4.4 Calibration

There are several calibration techniques to improve the linearity of pipelined ADCs, e.g. [10], [11]. A drawback with these techniques is that the conversion must be interrupted when the ADC is calibrated. It is sometimes preferable to use a calibration algorithm that works in the background [12], [13]. When using these calibration techniques the linearity of the converter can be very good but in time-interleaved converters the matching between the channels must also be considered. Both analog and digital calibration techniques for time-interleaved ADCs have been reported which can reduce the effect of gain and offset errors [14]. The phase skew errors are more difficult to calibrate, but in [15], a method of estimating the phase skew errors is proposed. However this method can not easily be implemented as a background calibration technique.

10.4.5 Filter Banks

By introducing signal processing before and after the individual ADCs the performance can be improved. In [16] a quadrature mirror filter (QMF) was used to reduce the effect of gain and offset mismatch as shown in Fig. 10-10. The discrete-time filters

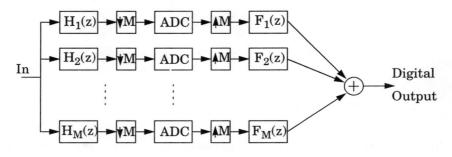

Figure 10-10 Time-interleaved ADCs with a QMF bank.

$H_i(z)$ are SC filters that are operated at the full speed of the time-interleaved ADC. This is a serious drawback since it makes the SC filters design too difficult. However by using polyphase structures it is possible to run the filters at the lower channel speed which makes a practical implementation feasible. Channel gain and offset errors appearing after the filters $H_i(z)$ will be suppressed and have a small effect on the performance, but unfortunately the error suppression in the transition band of the filters is very small. Another limitation is that delay skew errors will not be improved by this type of filter bank. The discrete-time filters at the input can be replaced by continuous-time filters [17]. These hybrid filter banks have the advantage that also delay skew errors will be suppressed (except in the transition bands). When using passive filters very high speeds can be achieved. However for monolithic implementations it is very difficult to make high performance passive components. Similar methods for sigma-delta ADCs where filter banks and modulation have been used to improve performance with several parallel channels have been proposed in [18] and [19].

10.5 PASSIVE SAMPLING TECHNIQUE

One limitation of time-interleaved ADCs is the sampling phase skew distortion. One way to overcome this problem is to place a track-and-hold amplifier at the input. This is not desirable since it requires a high-gain operational amplifier driving a large capacitive load at a very high frequency. In this section, we discuss a passive sampling technique for SC circuits to reduce the influence of the sampling phase skew. Since it does not require operational amplifiers, it is suitable for high-speed applications and yet simulations indicate that it can reduce the sampling-phase-skew-related distortion by 10~20 dB in a high-speed time-interleaved SC ADCs.

10.5.1 Sampling in SC Time-Interleaved ADCs

The time-interleaved ADC shown in Fig. 10-11 consists of M identical sub ADCs

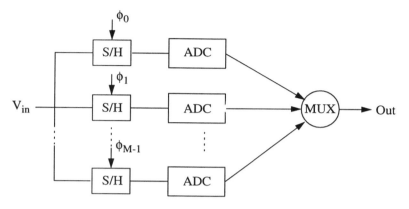

Figure 10-11 Time-interleaved ADC, the block diagram.

preceded by a passive sampling circuit. The passive sampling circuit and its corresponding clock phases are shown in Fig. 10-12. The first sub-ADC samples the input

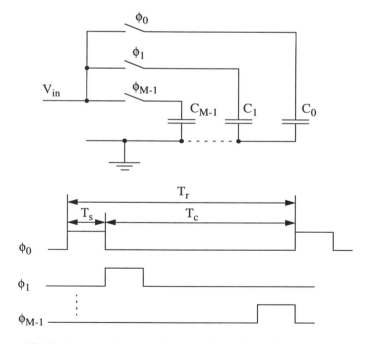

Figure 10-12 Input passive sampling circuit, and clock phases.

voltage V_{in} on clock phase ϕ_1, the second sub-ADC samples the input voltage V_{in} on clock phase ϕ_2, and the M-th sub-ADC samples the input voltage V_{in} on clock phase ϕ_M.

Assume that the sampling duration for every phase is T_s, and the repetition period for every phase is T_r. The time for every sub-ADCs to convert the sampled analog value is T_c, given by

$$T_c = T_r - T_s = (M-1) \cdot T_s \tag{10-37}$$

The only high-speed part is the passive sampling circuit, consisting of capacitors and switches, which needs to track and sample the analog input during the time interval T_s. This passive sampling circuit can be designed to be very fast and can handle a very large input bandwidth. The problem with phase skew errors arises when the M different clock phases shown in Fig. 10-12 is generated. Very small differences in delay between the clock phases generates large distortion for high signal frequencies.

10.5.2 Improved Passive Sampling Technique

The use of two-rank sampling solves the problem with delay skews distortion. But it naturally requires an opamp. This opamp must run at the full speed of the ADC and it is therefore difficult to design and power consuming. If we can have a global sampling technique without an opamp we would improve performance without much penalty. This leads to the improved passive sampling technique where a single clock phase is used to define the sampling instant. This is illustrated in Fig. 10-13.

The improved sampling circuit is controlled by a global clock phase ϕ and it defines the sampling instant. When the clock phase ϕ is high and ϕ_m is high, the input is sampled by the i-th sub-ADC. When the clock phase ϕ goes low, the analog value is sampled by the sampling capacitor since one plate of the sampling capacitor is floating. The clock phases ϕ_i always goes low after the clock phase ϕ goes low. Even if there are large phase skew between successive clock phases ϕ_i, they do not have any influence on the sampling instant and therefore the effect of the phase skew is eliminated. However, due to the parasitic capacitance, the charge stored on the sampling capacitor still changes with the analog input even when the clock phase ϕ is low.

10.5.3 Effect of Parasitic Capacitors

When an extra switch is introduced in the sampling, the parasitic capacitance has influence on the sampling instant even for a single channel. We show the parasitic capacitors associated with one channel in Fig. 10-14. All the switches are assumed to be NMOS transistors for simplicity. C_{p1} represents the parasitic capacitance at the right hand side of the sampling capacitor, and C_{p2} represents parasitic capacitance between switch transistors M_2 and M_3. Fig. 10-15 shows the circuit after the sampling switch (M_3) is opened while the other switches (M_1 and M_2) controlled by ϕ_m are still closed. The switch-on resistance of the switches is neglected for simplicity. The time instant when the sampling switch (M_3) is opened is denoted t and the instant when switch M_2 is opened is denoted $t + \tau$. The charge stored on C_{p2} will cause a signal dependent error on the output signal. At time t the total charge on the right hand side

10.5 Passive Sampling Technique

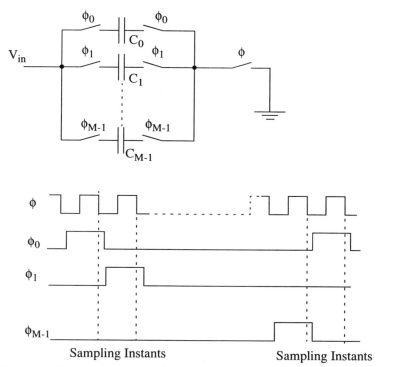

Figure 10-13 Proposed passive sampling technique for time-interleaved ADCs.

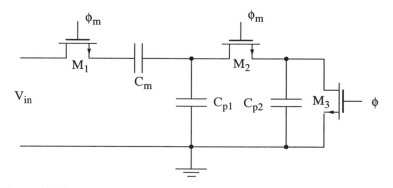

Figure 10-14 Parasitic capacitors in one channel.

node of C_m is

$$q(t) = q_{C_m}(t) + q_{C_{p1}}(t) + q_{C_{p2}}(t) = \\ = -C_m V_{in}(t) + 0 + 0 = -C_m V_{in}(t) \tag{10-38}$$

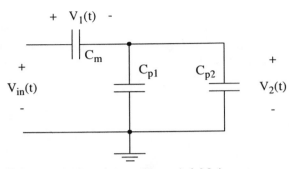

Figure 10-15 The circuit when the sampling switch M_3 is open.

At $t + \tau$, when switch M_2 is opened, the total charge on the right hand side of the sampling capacitor C_m is given by

$$q(t+\tau) = q_{C_m}(t+\tau) + q_{C_{p1}}(t+\tau) + q_{C_{p2}}(t+\tau)$$
$$= (V_2(t+\tau) - V_{in}(t+\tau)) \cdot C_m + V_2(t+\tau) \cdot (C_{p_1} + C_{p_2}) \quad (10\text{-}39)$$

Due to charge conversation, the charge in equations (10-38) and (10-39) should be equal. Therefore, we have the voltage across the parasitic capacitors at $t + \tau$ given by

$$V_2(t+\tau) = (V_{in}(t+\tau) - V_{in}(t)) \cdot \frac{C_m}{C_{p1} + C_{p2} + C_m} \quad (10\text{-}40)$$

The charge stored on C_{p2} is given by

$$q_{C_{p2}}(t+\tau) = C_{p2} \cdot V_2(t+\tau) =$$
$$= (V_{in}(t+\tau) - V_{in}(t)) \cdot \frac{C_i \cdot C_{p2}}{C_{p1} + C_{p2} + C_m} \quad (10\text{-}41)$$

After the switch M_2 opens, the charge stored on C_{p2} will be lost while all the charge stored on the sampling capacitor C_m and the parasitic C_{p1} will be transferred during the hold phase when an opamp is used. Assuming an ideal opamp, all the charge stored on the sampling capacitor C_m and the parasitic C_{p1} will be completely transferred. The only error source is due to the lost charge stored on C_{p2} at $t + \tau$. Therefore the analog output voltage after the sampling is given by

$$V_{out} = -\frac{q(t+\tau) - q_{C_{p2}}(t+\tau)}{C_m} = V_{in}(t) \cdot (1-\alpha) + \alpha \cdot V_{in}(t+\tau) \quad (10\text{-}42)$$

where

10.5 Passive Sampling Technique

$$\alpha = \frac{C_{p2}}{C_{p1} + C_{p2} + C_m} \tag{10-43}$$

It is seen that using this sampling technique for one channel, an error is introduced due to the parasitic capacitance. When we apply the passive sampling technique to a time-interleaved ADC, mismatch in the parameters α and τ between the channels will introduce distortion. Referring to Fig. 10-13 and Fig. 10-14, we assume that there are M time-interleaved channels and that the switch M_3 controlled by clock phase ϕ opens at time instants

$$T_s \cdot n, \, n = 0, \ldots, \infty \tag{10-44}$$

and that the switch controlled by clock phase ϕ_m in channel m ($m = 0, \ldots, M-1$) opens at

$$m \cdot T_s + n \cdot M \cdot T_s + \tau + t_{skew, m}, \, n = 0, \ldots, \infty \tag{10-45}$$

where T_s is the sampling period, τ is the average delay between the turn off of switch M_3 and switches M_2 in the m-th channel and $t_{skew, m}$ is the relative clock skew of clock phase ϕ_m. If the parasitic capacitors and sampling capacitors for all the channels are assumed to be equal, i.e. the factor α is equal for all the channels, and the time skews are assumed to be independent random variables with normal distribution and variance σ_t^2 the SNDR can be approximated as (see Sec. 10.7)

$$SNDR = 20 \cdot \log\left(\frac{1}{\sigma_t \cdot 2\pi f_{in}}\right) - 10 \cdot \log\left(1 - \frac{1}{M}\right) - 20 \cdot \log(\alpha) \tag{10-46}$$

for small α and small values on $f_{in}\tau$, where f_{in} is the input signal frequency. Eq. (10-46) shows that with parasitic capacitors the effect of phase skew errors is not completely removed but it is reduced by the factor $1/\alpha$ compared to time-interleaved ADCs using the ordinary sampling techniques. To investigate the effect of mismatch in the capacitors it is assumed that there are no phase skew errors and that α_m is the capacitor ratio factor for channel m. If the factors α_m are independent random variables with normal distribution, mean value α and variance σ_α^2 the SNDR due to capacitor mismatch can be approximated as (see Sec. 10.7)

$$SNDR = 20 \cdot \log\left(\frac{1}{\sigma_\alpha}\right) - 10 \cdot \log\left(1 - \frac{1}{M}\right) - 10 \cdot \log(|e^{j2\pi\tau f_{in}} - 1|^2) \tag{10-47}$$

for small σ_α and small $f_{in}\tau$. In order to maintain a large SNDR, we need a small τ.

A conventional sampling technique for two time-interleaved channels is shown in Fig. 10-16. Two non-overlapping clock phases are used. We also use bottom plate sampling, i.e. the rightmost switch is opened slightly before the leftmost switch, since this will minimize the signal dependent clock feedthrough. In the clock phase when the capacitors are disconnected from the input they are usually connected to an opamp before they are converted by a sub ADC. Here the capacitors are simply connected to AC-ground for simplicity. The AC-ground voltage is chosen to 1 V while. All the

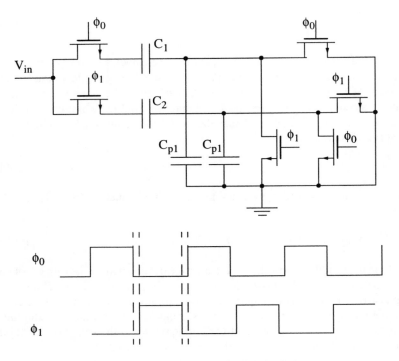

Figure 10-16 The circuit when the sampling switch M_3 is open.

bulks of the transistors are connected to ground. The time gap between the non-overlapping clock phases is 2 ns and the clock voltage swing is 0 to 5 V. The clock phase for the right most switches in Fig. 10-16 is delayed 1 ns to reduce clock feedthrough errors. The rise and fall times for all clock phases are 200 ps. The sampling capacitors are 1 pF and all the transistor sizes are 50 µm / 0.6 µm. The parasitic capacitors C_{p1} are approximately equal to C_{gs} of the switches, here approximately 50 fF. The passive sampling circuit was simulated using a sampling rate of 50 MS/s, i.e. each channel samples at 25 MS/s, and a 18 MHz input signal. A 50 ps delay skew was introduced in one of the channels. From [3] we have that for a two channel S/H with delay skew errors the SNDR is limited by

$$SNDR = 10 \cdot \log\left(\frac{\cos^2(\pi \cdot t_{skew} \cdot f_{in})}{\sin^2(\pi \cdot t_{skew} \cdot f_{in})}\right)$$
$$= 10 \cdot \log\left(\frac{\cos^2(\pi \cdot 50p \cdot 18M)}{\sin^2(\pi \cdot 50p \cdot 18M)}\right) = 51 \text{dB}$$
(10-48)

A 500-point FFT plot from the simulation both with and without delay skew is shown in Fig. 10-17 and Fig. 10-18. The distortion caused by the delay skew appears at 7 MHz and is approximately 51 dB. The other distortion tones in the FFT is harmonic distortion from clock feedthrough errors and nonlinearities in the transistors.

10.5 Passive Sampling Technique

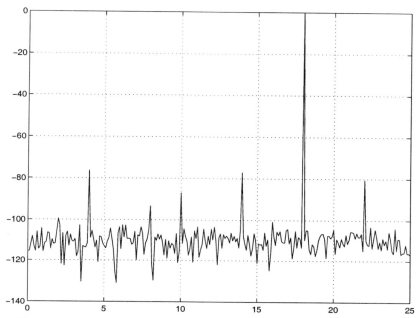

Figure 10-17 Simulated spectrum without phase skew. The conventional sampling technique is used.

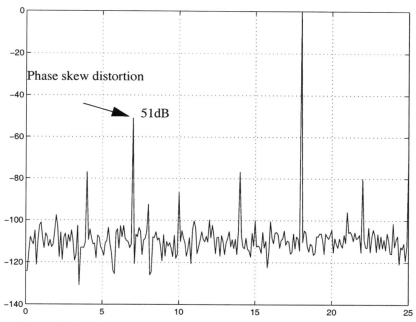

Figure 10-18 Simulated response with 50ps phase skew. The conventional sampling technique is used.

The circuit used for simulating the improved switching technique and the corresponding clock phases are shown in Fig. 10-19. The clock phases for the conventional sam-

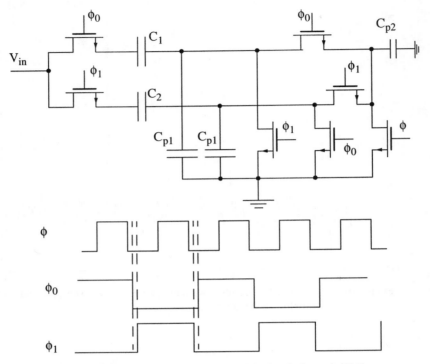

Figure 10-19 Improved sampling technique applied to 2-channel ADC.

pling technique are used. The only difference is the clock phase used for the additional sampling transistor. The additional sampling switch is turned off 1 ns before the bottom plate switch in the channels. The parasitic capacitors are approximately $C_{p1} = 50$ fF and $C_{p2} = 150$ fF where C_{p2} is the gate-source capacitance of a switch transistor and an additional 100 fF capacitor to account for parasitic elements in the layout. According to (10-46) we can expect the distortion to be reduced by a factor α. It is given by

$$\alpha = \frac{C_{p2}}{C_{p1} + C_{p2} + C_m} = \frac{150}{50 + 150 + 1000} = 0.125 \tag{10-49}$$

which corresponds to an 18 dB improvement. Thus we expect the phase skew distortion to be approximately 69 dB.

A 500-point FFT both with and without a 50 ps phase skew is shown in Fig. 10-20 and Fig. 10-21. The phase skew distortion is approximately 68.5 dB which is close to the predicted value. Fig. 10-20 shows that the harmonic distortion is larger than for the conventional sampling technique. This is mainly caused by the additional 100 fF

10.5 Passive Sampling Technique

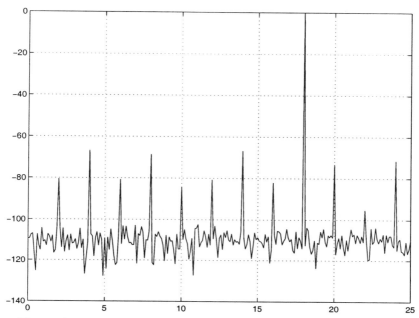

Figure 10-20 Simulated response and without phase skew. The improved sampling technique is used.

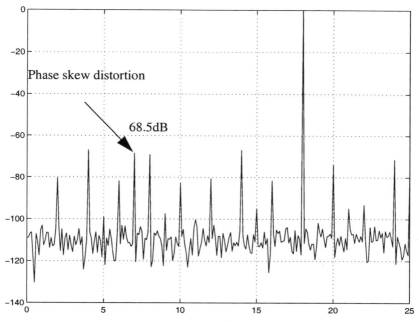

Figure 10-21 Simulated response and with 50ps phase skew. The improved sampling technique is used.

capacitor (C_{p2}). A simulation without the additional capacitor showed a harmonic distortion similar to the conventional sampling technique and a phase skew distortion of less than 80 dB. Table 10-1 summarized the results from the simulations.

Table 10-1 Phase skew distortion in simulation.

	Improved	Conventional
Phase skew = 50 ps, C_{p2} = 50fF	80 dB	51 dB
Phase skew = 50 ps, C_{p2} = 150fF	68.5 dB	51 dB

10.6 REALIZATION OF THE IMPROVED SAMPLING TECHNIQUE

In the previous section only the passive part of the S/H circuit was considered. To process the sampled analog value in the sub-ADCs, the sampled value must be held constant and for high accuracy one opamp is needed for every sub-ADC. Fig. 10-22.

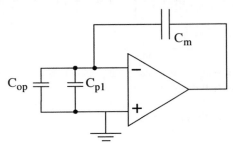

Figure 10-22 The opamp in the hold phase.

shows how the sampling capacitor is connected in the hold phase. The bottom plate parasitic capacitor C_{p1} is connected between the input of the opamp and ground as well as the input parasitic capacitance of the opamp C_{op}. Assuming a finite DC gain A of the opamp and using (10-42), the output voltage in the hold phase is given by

$$V_{out} = (V_{in}(t) \cdot (1 - \alpha) + \alpha \cdot V_{in}(t + \tau)) \cdot b \qquad (10\text{-}50)$$

where

$$a = \frac{C_{p2}}{C_{p1} + C_{p2} + C_m} \text{ and } b = \frac{1}{1 - \left(1 + \frac{(C_{p1} + C_{op})}{C_m}\right)\frac{1}{A}} \qquad (10\text{-}51)$$

Hence the finite gain in the opamp cause gain errors in the channels. Notice that the settling time for the operational amplifier corresponds to the slower channel sampling rate and therefore the speed requirement of the opamp is not high.

10.6.1 M-Channel S/H Circuit

Fig. 10-23. shows how the S/H circuit is implemented for M channels. The clock

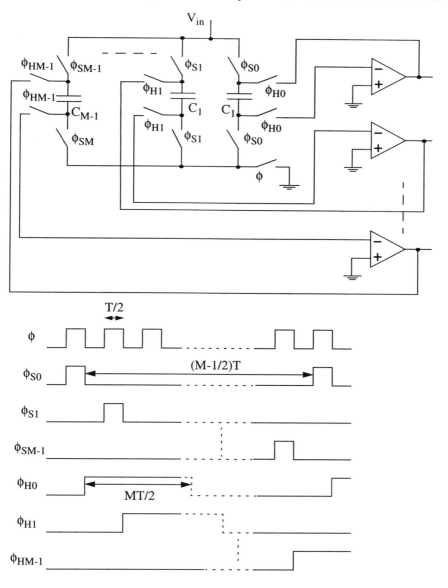

Figure 10-23 The improved sampling technique for an M-channel ADC.

phase defining the sampling instant ϕ runs at the full speed of the time-interleaved ADC, i.e. the clock period is T. The sampling phases ϕ_{Sm}, $m = 0, ..., M-1$ have a clock period of $T \cdot M$. The sampling phases are shown in Fig. 10-23 where the on-time of a sampling phase is only $T/2$. The hold clock phases ϕ_{Hm}, $m = 0, ..., M-1$ have a clock period of $M \cdot T$ but their on-time is $TM/2$. This means that the settling time for the opamps are $T \cdot M/2$. This is typically the case when the circuit is used with pipelined ADCs. If other types of converters are used it may be possible increase the hold-time to $T(M-1)$.

10.6.2 Opamp Sharing Technique

The opamps are power consuming parts of the ADC. The number of opamps can be reduced by using opamp sharing techniques. For a 2-channel ADC, one opamp can be shared by both channels [2], [20] as show in Fig. 10-24. The technique can in principle

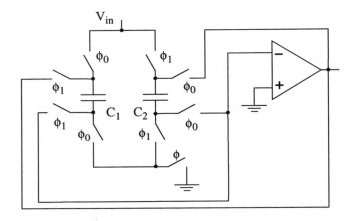

Figure 10-24 Opamp sharing for a 2-channel ADC.

be extended to a M-channel time-interleaved ADC. It can reduce the number of opamps by a factor 2.

10.7 DERIVATIONS

In this section we derive the effect of parasitic capacitors in the improved sampling technique discussed previously in this chapter.

10.7.1 Output Spectrum

According to Sec. 10.5.3 the sampled signal in channel m is

$$v_s = v(t) \cdot (1 - \alpha_m) + \alpha_m \cdot v(t + \tau_m) \qquad (10\text{-}52)$$

where τ_m is the delay between the sampling switch turn-off time and the channel switch turn-off time (see Sec. 10.5.3) and

10.7 Derivations

$$\alpha_m = \frac{C_{p2}}{C_{p1} + C_{p2} + C_m} \tag{10-53}$$

Due to circuit imperfections the gain factors and delays are different in the channels. The delay in channel m can be expressed as $\tau_m = t_d + r_m/T$ where t_d is the nominal delay, r_m the relative error in the delay and T the sampling period of the time-interleaved ADC. By using the results in [3], the digital spectrum for v_s can be expressed as

$$G(\omega) = G_1(\omega) - G_2(\omega) + G_3(\omega) \tag{10-54}$$

where

$$G_1(\omega) = \frac{1}{T} \sum_{k=-\infty}^{\infty} G_a\left(\omega - \frac{2\pi k}{T}\right), \tag{10-55}$$

$$G_2(\omega) = \frac{1}{T} \sum_{k=-\infty}^{\infty} \left(\frac{1}{M} \sum_{m=0}^{M-1} \alpha_m e^{-jkm2\pi/M} \right) G_a\left(\omega - \frac{2\pi k}{T}\right) \tag{10-56}$$

and

$$G_3(\omega) = \frac{1}{T} \sum_{k=-\infty}^{\infty} \left(\frac{1}{M} \sum_{m=0}^{M-1} \alpha_m \cdot e^{-j\left(\omega - \frac{2\pi k}{MT}\right)(r_m T - t_d) - jmk\frac{2\pi}{M}} \right)$$
$$\cdot G_a\left(\omega - k\frac{2\pi}{M}\right) \tag{10-57}$$

$G_a(\omega)$ is the analog spectrum of the input signal $v(t)$ and $G_1(\omega)$ is the spectrum of the sampled input signal. $G_1(\omega)$ corresponds to the spectrum of a signal sampled with channel gain mismatch, i.e. $\alpha_i \cdot v(t)$. $G_3(\omega)$ is the spectrum of a signal with both gain and phase skew errors and corresponds to the last term in (10-52), $\alpha_i \cdot v(t + \tau_i)$. For a complex sinusoidal input signal $g(t) = e^{j\omega_0 t}$ the expressions can be simplified to

$$G_1(\omega) = \frac{1}{T} \sum_{k=-\infty}^{\infty} 2\pi\delta\left(\omega - \omega_0 - \frac{2\pi k}{T}\right) \tag{10-58}$$

$$G_2(\omega) = \frac{1}{T} \sum_{k=-\infty}^{\infty} A_2(k) \cdot 2\pi\delta\left(\omega - \omega_0 - \frac{2\pi k}{MT}\right) \tag{10-59}$$

$$G_3(\omega) = \frac{1}{T} \sum_{k=-\infty}^{\infty} A_3(k) \cdot 2\pi\delta\left(\omega - \omega_0 - \frac{2\pi k}{MT}\right) \tag{10-60}$$

where

$$A_2(k) = \frac{1}{M} \sum_{m=0}^{M-1} \alpha_m e^{-jkm2\pi/M} \tag{10-61}$$

and

$$A_3(k) = e^{j2\pi f_0 t_d} \sum_{m=0}^{M-1} \left[\frac{1}{M} \cdot \alpha_m \cdot e^{-j2\pi r_m \frac{f_0}{f_s}}\right] \cdot e^{-j2\pi \frac{km}{M}} \tag{10-62}$$

10.7.2 Phase Skew Errors

All the factors α_i are assumed to be equal, i.e. $\alpha_i = \alpha$ which gives

$$G_1(\omega) = \frac{1}{T} \sum_{k=-\infty}^{\infty} 2\pi\delta\left(\omega - \frac{2\pi k}{T}\right) \tag{10-63}$$

$$G_2(\omega) = \frac{1}{T} \sum_{k=-\infty}^{\infty} \alpha \cdot 2\pi\delta\left(\omega - \frac{2\pi}{T}\right) \tag{10-64}$$

$$G_3(\omega) = \frac{1}{T} \sum_{k=-\infty}^{\infty} A_3(k) \cdot 2\pi\delta\left(\omega - \frac{2\pi k}{MT}\right) \tag{10-65}$$

where

$$A_3(k) = \alpha \cdot e^{j2\pi f_0 t_d} \sum_{m=0}^{M-1} \left[\frac{1}{M} \cdot e^{-j2\pi r_m \frac{f_0}{f_s}}\right] \cdot e^{-j2\pi \frac{km}{M}} \tag{10-66}$$

The contributions from $G_1(\omega)$ and $G_2(\omega)$ to the output spectrum corresponds to a sinusoid with amplitude $1 - \alpha$. $G_3(\omega)$ is the spectrum of a non-uniformly sampled sinusoid due to the delay skews. The amplitude of the fundamental in $G_3(\omega)$ is determined by $|A_3(0)|$. Hence the total fundamental power is

$$|1 - \alpha + A_3(0)|^2 \tag{10-67}$$

The distortion power is the difference between the total power in $G_3(\omega)$ and the power of the fundamental of $G_3(\omega)$, i.e.

10.7 Derivations

$$\alpha^2 - |A_3(0)|^2 \qquad (10\text{-}68)$$

Hence the SNDR is limited by

$$SNDR = 10 \cdot \log\left(\frac{|1 - \alpha + A_3(0)|^2}{\alpha^2 - |A_3(0)|^2}\right) \qquad (10\text{-}69)$$

The numerator can be written as

$$\begin{aligned}|1 - \alpha + A_3(0)|^2 &= (1 - \alpha + A_3(0))(1 - \alpha + A_3^*(0)) = \\ &= (1 - \alpha)^2 + (1 - \alpha)(A_3(0) + A_3^*(0)) + |A_3(0)|^2\end{aligned} \qquad (10\text{-}70)$$

If the delay skew errors are Gaussian random variables with zero mean and a standard deviation of σ_t the following approximations can be used [3]

$$E[A_3(0)] = \alpha \cdot e^{j2\pi f_0 t_d} \cdot e^{-2\sigma_t^2 \pi^2 f_0^2} \approx \alpha \cdot e^{j2\pi f_0 t_d} \cdot (1 - 2\sigma_t^2 \pi^2 f_0^2) \qquad (10\text{-}71)$$

and

$$\begin{aligned}E[|A_3(0)|^2] &= |E[A_3(0)]|^2 + \frac{1}{M} \cdot (\alpha^2 - |E[A_3(0)]|^2) \\ &\approx \alpha^2 - \alpha^2 \cdot 4\sigma_t^2 \pi^2 f_0^2 \cdot \frac{M-1}{M}\end{aligned} \qquad (10\text{-}72)$$

The approximations are based on the Taylor series expansion. If we assume that $\sigma_t f_0$ is small the signal power can be approximated as

$$\begin{aligned}P_s &\approx (1-\alpha)^2 + (1-\alpha) \cdot \alpha (e^{j2\pi f_0 t_d} + e^{j2\pi f_0 t_d}) + \alpha^2 \\ &= (1-\alpha)^2 + (1-\alpha) \cdot \alpha \cdot 2\sin(2\pi f_0 t_d) + \alpha^2\end{aligned} \qquad (10\text{-}73)$$

The expression can be further simplified by assuming $f_0 t_d$ to be small which yields

$$P_s \approx (1-\alpha)^2 + 2(1-\alpha) \cdot \alpha + \alpha^2 = 1 \qquad (10\text{-}74)$$

We can now approximate the SNDR as

$$SNDR \approx 10 \cdot \log\left(\frac{1}{\alpha^2 - |A_3(0)|^2}\right)$$

$$= 10 \cdot \log\left(\frac{1}{\alpha^2 - \left(\alpha^2 - \alpha^2 \cdot 4\sigma_t^2\pi^2 f_0^2 \cdot \frac{M-1}{M}\right)}\right) \qquad (10\text{-}75)$$

$$= 10 \cdot \log\left(\frac{1}{\alpha^2 \cdot 4\sigma_t^2\pi^2 f_0^2 \cdot \frac{M-1}{M}}\right)$$

$$= 20 \cdot \log\left(\frac{1}{\alpha \cdot 2\pi f_0 \sigma_t}\right) - 10 \cdot \log\left(\frac{M-1}{M}\right)$$

10.7.3 Gain Errors

We assume that there are no phase skew errors and that we have a complex sinusoidal input signal $g(t) = e^{j\omega_0 t}$. The digital spectra from section Sec. 10.7.1 can now be written as

$$G_1(\omega) = \frac{1}{T}\sum_{k=-\infty}^{\infty} 2\pi\delta\left(\omega - \omega_0 - \frac{2\pi k}{T}\right) \qquad (10\text{-}76)$$

$$G_2(\omega) = \frac{1}{T}\sum_{k=-\infty}^{\infty} A_2(k) \cdot 2\pi\delta\left(\omega - \omega_0 - \frac{2\pi k}{MT}\right) \qquad (10\text{-}77)$$

$$G_3(\omega) = \frac{1}{T}\sum_{k=-\infty}^{\infty} A_3(k) \cdot 2\pi\delta\left(\omega - \omega_0 - \frac{2\pi k}{MT}\right) \qquad (10\text{-}78)$$

where

$$A_2(k) = \frac{1}{M}\sum_{m=0}^{M-1} \alpha_m e^{-jkm2\pi/M} \qquad (10\text{-}79)$$

and

$$A_3(k) = \frac{e^{j2\pi f_0 t_d}}{M}\sum_{m=0}^{M-1} \alpha_m \cdot e^{-j2\pi\frac{km}{M}} \qquad (10\text{-}80)$$

The output spectrum can now be written as

10.7 Derivations

$$G(\omega) = G_1(\omega) + \frac{1}{T} \sum_{k=-\infty}^{\infty} A(k) \cdot 2\pi\delta\left(\omega - \omega_0 - \frac{2\pi k}{MT}\right) \quad (10\text{-}81)$$

where

$$A(k) = \frac{e^{j2\pi f_0 t_d} - 1}{M} \sum_{m=0}^{M-1} \alpha_m \cdot e^{-j2\pi\frac{km}{M}} \quad (10\text{-}82)$$

The total signal power is given by

$$P_s = |1 + A(0)|^2 = (1 + A(0))(1 + A^*(0)) =$$
$$= 1 + A(0) + A^*(0) + |A(0)|^2 \quad (10\text{-}83)$$

If the gain factors are assumed to be independent Gaussian random variables with a mean of α and a standard deviation of σ_α, we have the following expectations

$$E[A(0)] = \frac{e^{j2\pi f_0 t_d} - 1}{M} \cdot E\left[\sum_{m=0}^{M-1} \alpha_m\right] = (e^{j2\pi f_0 t_d} - 1) \cdot \alpha \quad (10\text{-}84)$$

and

$$E[|A(0)|^2] = E[A(0) \cdot A^*(0)] = \left(\frac{|e^{j2\pi f_0 t_d} - 1|}{M}\right)^2 \cdot \sum_{m=0}^{M-1}\sum_{n=0}^{M-1} E(\alpha_m \cdot \alpha_n)$$
$$= \left(\frac{1}{M} \cdot E[\alpha_m^2] + \left(1 - \frac{1}{M}\right) \cdot E[\alpha_n] \cdot E[\alpha_m]\right) \cdot |e^{j2\pi f_0 t_d} - 1|^2 \quad (10\text{-}85)$$
$$= \left(\frac{1}{M} \cdot (\sigma_\alpha^2 + \alpha^2) + \left(1 - \frac{1}{M}\right) \cdot \alpha^2\right) \cdot |e^{j2\pi f_0 t_d} - 1|^2$$
$$= \left(\frac{1}{M} \cdot \sigma_\alpha^2 + \alpha^2\right) \cdot |e^{j2\pi f_0 t_d} - 1|^2$$

where we have used that $E[\alpha_m^2] = \sigma_\alpha^2 + (E[\alpha_m])^2 = \sigma_\alpha^2 + \alpha^2$. Thus the signal power is

$$P_s = 1 + (e^{j2\pi f_0 t_d} - 1) \cdot a + (e^{-j2\pi f_0 t_d} - 1) \cdot a + \left(\frac{1}{M} \cdot \sigma_a^2 + a^2\right) \cdot |e^{j2\pi f_0 t_d} - 1|^2$$
$$= 1 + \alpha(\sin(2\pi f_0 t_d) - 1) + \left(\frac{1}{M} \cdot \sigma_\alpha^2 + \alpha^2\right) \cdot |e^{j2\pi f_0 t_d} - 1|^2 \approx 1 \quad (10\text{-}86)$$

where the last approximation is valid if $f_0 t_d$ and σ_α are small. The distortion power is given by

$$P_d = \sum_{m=0}^{M-1} |A(m)|^2 - |A(0)|^2 \tag{10-87}$$

The following expectation can be calculated (we use the fact that $A(m)$ can be interpreted as a DFT)

$$E\left(\sum_{m=0}^{M-1} |A(m)|^2\right) = E\left(\frac{|e^{j2\pi t_d f_o} - 1|^2}{M} \sum_{m=0}^{M-1} \alpha_m^2\right) =$$

$$= |e^{j2\pi t_d f_o} - 1|^2 \cdot (\sigma_\alpha^2 + \alpha^2) \tag{10-88}$$

The SNDR can now be calculated as

$$SNDR \approx 10 \cdot \log\left(\frac{1}{\left((\sigma_\alpha^2 + \alpha^2) - \left(\frac{1}{M} \cdot \sigma_\alpha^2 + \alpha^2\right)\right) \cdot |e^{j2\pi t_d f_o} - 1|^2}\right)$$

$$= 10 \cdot \log\left(\frac{1}{\left(\sigma_\alpha^2 - \frac{1}{M} \cdot \sigma_\alpha^2\right) \cdot |e^{j2\pi t_d f_o} - 1|^2}\right) \tag{10-89}$$

$$= 20 \cdot \log\left(\frac{1}{\sigma_\alpha}\right) - 10 \cdot \log\left(\frac{M-1}{M}\right) - 10 \cdot \log(|e^{j2\pi t_d f_o} - 1|^2)$$

10.8   SUMMARY

In this chapter we have discussed the time-interleaved ADC. The performance is limited by mismatches between the channels. Gain and offset errors can be calibrated. A more severe problem is the phase skew errors which increases at high signal frequencies. The most effective way to reduce these errors is to use an input S/H circuit which usually needs an opamp that runs at the full speed of the ADC. The opamp is therefore difficult to design and power consuming. An improved global passive sampling technique that does not need an opamp was introduced. The limitations of the technique was also discussed. The phase skew distortion can be reduced by 10 to 20 dB compared to not using a global sampling technique.

REFERENCES

[1]   W. C. Black and D. A. Hodghes, "Time Interleaved Converter Arrays", *IEEE J. of Solid-State Circuits*, vol. SC-15, no. 6, pp. 1022-29, Dec. 1980.

[2] K. Nagaraj, J. Fetterman, J. Anidjar, S. Lewis and R. G. Renninger, "A 250-mW, 8-b, 52-Msamples/s Parallel-Pipelined A/D Converter with Reduced Number of Amplifiers", *IEEE J. of Solid-State Circuits*, vol. 32, no. 3, p.312-20, March 1997.

[3] Y. C. Jenq, "Digital Spectra of Nonuniformly Sampled Signals: Fundamentals and High-Speed Waveform Digitizers", *IEEE Trans. on Instrumentation and Measurement*, vol. 37, no. 2, pp. 245-51, June 1988.

[4] A. Petraglia and S. K. Mitra, "Analysis of Mismatch Effects Among A/D Converters in a Time-Interleaved Waveform Digitizer", *IEEE Trans. on Instrumentation and Measurement*, vol. 40, no. 5, pp. 831-5, Oct. 1991.

[5] J. B. Simoes, J. Landeck and C. M. B. A. Correia, "Nonlinearity of a Data-Acquisition System with Interleaving/Multiplexing", *IEEE Trans. Instrumentation and Measurement*, vol. 46, no. 6, pp. 1274-79, Dec. 1997.

[6] K. Y. Kim, N. Kusayanagi and A. A. Abidi, "A 10-b, 100-MS/s CMOS A/D Converter", *IEEE J. of Solid-State Circuits*, vol. 32, no. 6, pp. 302-11, Dec. 1997.

[7] K. Poulton., J. J. Corcoran and T. Hornak, "A 1-GHz 6-bit ADC System", *IEEE J. of Solid-State Circuits*, vol. SC-22, no. 6, pp. 962-70, Dec. 1987.

[8] K. Nakamura, M. Hotta, L. R. Carley and D. J. Allstot, "An 85 mW, 10 b, 40 Msample/s CMOS Parallel-Pipelined ADC", *IEEE J. of Solid-State Circuits*, vol. 30, no. 6, pp. 173-83, Dec. 1995.

[9] H. Jin, E. Lee and M. Hassoun, "Time-Interleaved A/D Converter with Channel Randomization", *IEEE Intern. Symp. on Circuits and Systems, ISCAS-97*, vol. 1, pp. 425-8, 1997.

[10] A. .N. Karanicolas, H. S. Lee and K. L. Bacrania, "A 15-b 1 Msample/s Digitally Self-Calibrated Pipeline ADC", *IEEE J. of Solid-State Circuits*, vol. 28, no. 12, pp. 1207-15, Dec. 1993.

[11] E. G. Soenen and R. L. Geiger, "An Architecture and an Algorithm for Fully Digital Correction of Monolithic Pipelined ADC's", *IEEE Trans. on Circuits and Systems-II*, vol. 42, no. 3, pp. 143-53, March 1995.

[12] T. H. Shu, B. S. Song and K. Bacrania, "A 13-b 10-Msample/s ADC Digitally Calibrated with Oversampling Delta-Sigma Converter", *IEEE J. of Solid-State Circuits*, vol. 30, no. 4, p. 443-52, April 1995.

[13] U. K. Moon, B. S. Song, "Background Digital Calibration Techniques for Pipelined ADCs", *IEEE Trans. on Circuits and Systems-II*, Vol. 44, No. 2, pp. 102-9, February 1997.

[14] K. Dyer, D. Fu, P. Hurst and S. Lewis, "A Comparison of Monolithic Background Calibration in Two Time-Interleaved Analog-to-Digital Converters" *IEEE 1998 Intern. Symp. on Circuits and Systems, ISCAS-98*, vol. 1, pp. 13-16, 1998.

[15] Y. C. Jenq, "Digital Spectra of Nonuniformly Sampled Signals: A Robust Sampling Time Offset Estimation Algorith for Ultra High-Speed Waveform Digitizers Using Interleaving", *IEEE Trans. on Instrumentation and Measurement*, vol. 39, no. 1, pp. 71-75, Feb. 1990.

[16] A. Petraglia and S. K. Mitra, "High-Speed A/D Conversion Incorporating a QMF Bank", *IEEE Trans. on Instrumentation and Measurement*, vol. 41, no. 3, pp. 427-31, June 1992.

[17] S. R. Velazques, *Hybrid Filter Banks for Analig/Digital Conversion*, Ph.D. Thesis at Massachusetts Institute of Technology, June 1997.

[18] R. Khoini-Poorfard, L. B. Lim and D. A. Johns, "Time-Interleaved Oversampling A/D Converters: Theory and Practice", *IEEE Trans. on Circuits and Systems-II*, vol. 44, no. 8, pp. 634-45, Aug. 1997.

[19] S. K. Kong and W. H. Ku, "Effects of Non-Ideal Hadamard Modulators on the Performance of PDS ADC", *Electronics Letters*, vol. 33, no. 2, p. 109-10, 16th Jan. 1997.

[20] T. C. Choi and R. W. Brodersen, "Considerations for High-Frequency Switched-Capacitor Ladder Filters", *IEEE Trans. on Circuits and Systems.*, vol. cas-27, no. 6, pp. 545-52, June 1980.

11 OVERSAMPLING A/D CONVERTERS

11.1 INTRODUCTION

Oversampling ADCs (OSADC) are commonly used in many telecommunications applications such as ADSL. Due to the use of DMT signals in ADSL applications, the requirement on the analog front end (AFE) is generally increased compared with single tone applications. It has also been shown that the linearity is more important than quantization noise for ADCs in DMT applications and OSADCs usually have higher linearity than Nyquist ADCs. An additional advantage of using OSADCs is the very low quantization noise power at low frequencies, which is beneficial for DMT applications. This is because more bits can be modulated on the lower frequency carriers where the signal is less attenuated by the telephone line. The major drawback of OSADCs is that a relatively high clock frequency must be used if the signal bandwidth is large, making the implementation difficult. Therefore techniques to reduce the oversampling ratio (OSR) of the converter are of interest. In Sec. 11.2 the basic properties of the first and second order sigma-delta converter are reviewed while sigma-delta converters for high signal bandwidths are treated in Sec. 11.3. SC implementation of sigma-delta converters are considered in Sec. 11.4 while the impact of circuit imperfections are reviewed in Sec. 11.5. Finally a design example is given in Sec. 11.6.

11.2 BASICS OF OVERSAMPLING SIGMA-DELTA CONVERTERS

The basic principle behind oversampling sigma-delta modulators is to trade signal bandwidth for resolution. The quantization noise of a low resolution ADC is high-pass filtered to yield a low quantization noise at low frequencies [1]. The noise at high frequencies is removed by a digital filter before the signal is decimated to generate the final output of the converter.

A first order sigma-delta modulator is shown in Fig. 11-1. It contains an ADC, an integrator, a DAC and a subtractor. The resolution of the ADC is low, usually only 1 bit. The digital output $y(nT)$ is fed back through the DAC and subtracted from the input signal. The output of the subtractor is accumulated in the integrator and quantized by the ADC. The integrator can be either continuous-time or discrete-time, but a discrete-time integrator is more common and will only be considered here. The performance of the modulator can be investigated by using a linear model for the modulator. We assume that the quantization error $e(nT)$ in the ADC is white noise uncorrelated to

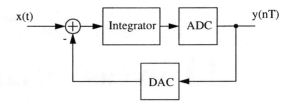

Figure 11-1 First order sigma-delta modulator.

the input signal. A linear model of the modulator is shown in Fig. 11-2. If a 1 bit quan-

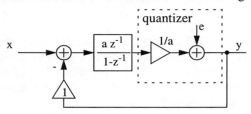

Figure 11-2 Linear model for first order sigma-delta modulator.

tizer is used, the gain in the integrator, a, will only affect the voltage swing at the output of the integrator not the output signal. Therefore this gain factor must be compensated for in the linear model of the quantizer. The z-domain output signal is given by

$$Y(z) = z^{-1} \cdot X(z) + (1 - z^{-1}) \cdot E(z) \qquad (11\text{-}1)$$

Hence the input signal is only delayed while the quantization noise is high-pass filtered. If the signal bandwidth is small compared to the sampling frequency only a small portion of the quantization noise appears in the signal band and a high resolution results. The ratio between the signal bandwidth and the Nyquist frequency is called the oversampling ratio (OSR), i.e.

$$OSR = \frac{f_s/2}{BW} \qquad (11\text{-}2)$$

where f_s is the sampling frequency and BW the signal bandwidth.

For the first order modulator the resolution increases by 1.5 bits for every doubling of the sampling frequency. To get a high resolution a very high OSR is needed. The required OSR can be reduced by using a second order modulator. The linear model of a second order modulator is shown in Fig. 11-3. There are now two integrators in the loop and the output signal is given by

$$Y(z) = z^{-2} \cdot X(z) + (1 - z^{-1})^2 \cdot E(z) \qquad (11\text{-}3)$$

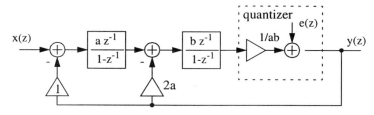

Figure 11-3   Linear model of second order sigma-delta modulator.

Hence the order of the noise shaping function is two. We now gain 2.5 bits for every doubling of the sampling frequency. An additional advantage of the second order modulator is that the assumption of white uncorrelated quantization noise is more accurate than for the first order modulator.

The order can be further increased and for an L-order modulator we gain $L + 1/2$ bits for every doubling of the sampling frequency [1]. There are however problems with increasing the order of the modulator due to stability.

11.3   OVERSAMPLED SIGMA-DELTA CONVERTERS FOR HIGH SIGNAL BANDWIDTHS

Achieving a high resolution over a large signal bandwidth is almost impossible using a second order modulator since the sampling rate would be very high. The OSR can be reduced in different ways.

One way to reduce the OSR is to increase the order of the modulator. This can be done by adding more integrators in the loop. However the modulator may be unstable unless the integrator gains and the feedback signals are scaled properly. The resulting noise shaping function is of the same order as the number of integrators but it is multiplied by a large constant [2]. This increases the noise power.

The performance can be somewhat improved by using a loop filter with feedforward and feedback of internal signals and thus introducing zeros in the transfer function. However the resulting SC implementation will have a large spread in capacitor values which makes the design difficult [2]. The resulting noise shaping function is also in this case multiplied by a large constant. Hence, we should consider other ways of decreasing the OSR.

The resolution of the modulator can also be increased by increasing the resolution of the quantizer. A problem with this solution is that nonlinearity errors in the DAC will limit the performance of the modulator unless calibration or error averaging is used [3].

Another option, avoiding the need for calibration, is to use a cascaded sigma-delta modulator [1]. This type of converter is sometimes referred to as a MASH structure. In this type of converters several low order modulators are cascaded to achieve a high order noise shaping function. The low order quantization noise is cancelled by a digital cancellation logic. A drawback of the cascaded architecture is that mismatch between the analog and digital circuits limits the performance. The matching need not be within the full resolution of the converter and cascaded sigma-delta converters with resolutions of 15-16 bits and a signal bandwidth in the range of 1 MHz have been reported [2], [4]. Therefore this type of converters is interesting for ADSL applications. In the following we focus on architectures based on cascading low order modulators.

11.3.1 2-2 Fourth Order Cascaded Modulator

A fourth order cascaded modulator consisting of two second order modulators [5] is shown in Fig. 11-4. The analog output signal of the second integrator in the first mod-

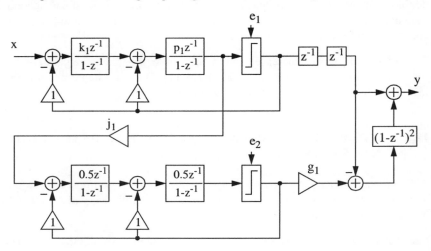

Figure 11-4 4th order cascaded sigma-delta modulator.

ulator is fed to the input of the second modulator. The digital outputs of the modulators are fed to the correction logic which generates the output y. The digital gain factor g_1 must be chosen such that the low order quantization noise is cancelled. This is achieved when the product of the gain in the first integrator ($k_1 = 0.5$), the gain in the second integrator ($p_1 = 0.5$), the gain for the signal between the two stages ($j_1 = 1$) and g_1 equals 1. Thus we have for the modulator in Fig. 11-4

$$k_1 \cdot p_1 \cdot j_1 \cdot g_1 = 1 \Rightarrow g_1 = 4 \qquad (11\text{-}4)$$

The output signal is given by [5]

$$\begin{aligned} Y(z) &= z^{-4} \cdot X(z) + g_1 \cdot E_2(z) \cdot (1 - z^{-1})^4 \\ &= z^{-4} \cdot X(z) + 4 \cdot E_2(z) \cdot (1 - z^{-1})^4 \end{aligned} \qquad (11\text{-}5)$$

which corresponds to a 4th-order noise shaping function. We see that g_1 should be as

11.3 Oversampled Sigma-Delta Converters for High Signal Bandwidths

small as possible to minimize the quantization noise. This is achieved by increasing the integrator gains and the gain between the two stages. However this can usually not be done for two reasons. Firstly, if the input signal to the second stage is too large the second modulator saturates and the SNR is reduced. Secondly in the actual implementation we must consider the signal swings in the circuits. The swing at the output of an opamp should not be too close to V_{dd} or V_{ss} to avoid distortion. The relation between the feedback signal swing and the swing at the output of an integrator is determined by the integrator gain. To minimize the noise it is usually desirable to have approximately the same signal swing in all the integrator outputs. The integrator gains can therefore not be chosen arbitrarily.

We must also consider the swing of the input signal. For OSADCs the maximum input signal amplitude without overloading the modulator is below the full-scale (FS) of the converter. The FS is determined by the reference levels in the DAC. For a second order modulator the maximum signal swing at the input is approximately -3 dBFS. If desirable a gain factor can be added at the input to compensate for the smaller swing of the input signal. For an SC implementation this is easily done by changing a capacitor ratio.

The signal swings in the modulator can be determined by simulating the modulator using a behavioral model. The modulator in Fig. 11-4 was simulated using MatLab. The reference levels of the DAC are +1 and -1. A sinusoidal input signal with an amplitude of FS/2 and a frequency of 118.56 kHz is used. The modulator sampling rate was 30 MS/s. A 64k-point FFT of the output signal is shown in Fig. 11-5. The

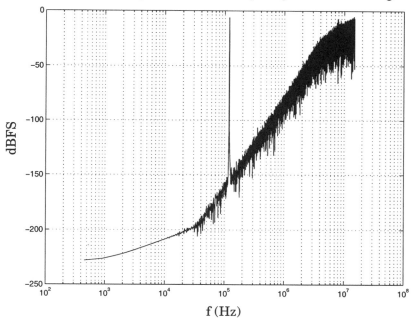

Figure 11-5 64k-point FFT of integrator output.

quantization noise increases by 80 dB/decade as expected for a 4th order noise shaping function.

The signal-to-noise ratio as a function of the input signal amplitude for different OSRs is shown in Fig. 11-6. In the figure we see that the SNR improves by slightly less than

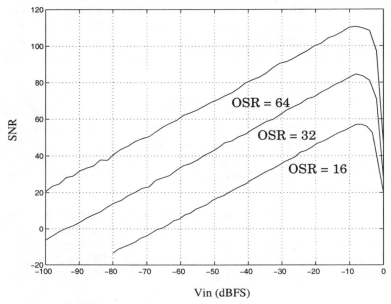

Figure 11-6   SNR vs. input amplitude.

30 dB when the OSR is doubled, which corresponds to approximately 4.5 bits. The peak SNR appears at approximately -6 dBFS which corresponds to half the full-scale amplitude.

The signal swings at the integrator outputs for the cascaded modulator in Fig. 11-4 were estimated by calculating the histogram from the MatLab simulation. The result is shown in Fig. 11-7. The swing at the output of the first integrator is approximately 1.5 times the feedback signal amplitude. The gain in the first integrator can therefor be decreased by approximately 1/3 to make the signal swing close to 1. We also see that the swing of the signal fed to the second modulator is approximately 1. This is slightly too large and the second modulator is overloaded. This explains why the peak SNR occurred at a relatively low input signal amplitude in Fig. 11-6.

From the above discussion we conclude that to improve the signal swings in the modulator it is necessary to decrease some of the gain factors. But smaller gain factors means that g_1 becomes larger, reducing the SNR.

11.3.2   Improved 2-2 Cascaded Modulator

It is possible to improve the performance of the cascaded modulator in Fig. 11-4 by modifying the architecture slightly. The architecture in Fig. 11-8 feeds only the quan-

11.3 Oversampled Sigma-Delta Converters for High Signal Bandwidths

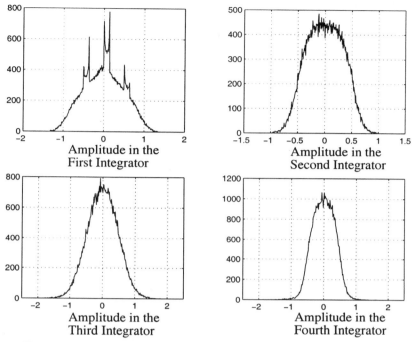

Figure 11-7 Histogram plots for integrator outputs.

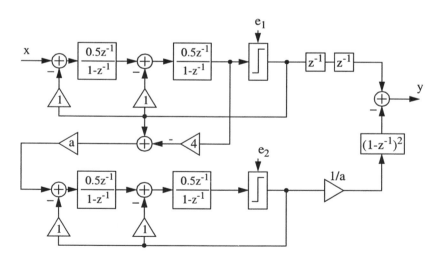

Figure 11-8 4th order cascaded sigma-delta modulator feeding only the quantization noise to the second stage.

tization noise of the first modulator to the second stage. The output signal is given by

$$Y(z) = X(z) \cdot z^{-4} + \frac{(1-z^{-1})^4}{a} \cdot E_2(z) \qquad (11\text{-}6)$$

The gain factor a should be approximately 1/4 to avoid saturation in the second stage. In this case the gain factor in the digital domain is only determined by the gain between the two stages not the integrator gains.

11.3.3 2-1-1 Cascaded Modulator

It is not necessary to use only second order modulators in a cascaded modulator. There may actually be advantages in using first order modulators instead. The modulator in the first stage is normally a second order modulator since the quantization noise is less correlated to the input signal than the noise in a first order modulator. Since the input signal to the later stages in the cascaded modulator are the quantization noise in the previous stages, the quantization error in all the stages show a small correlation to the input signal even if the later stages are of the first order. A 4th order 2-1-1 cascaded modulator [4] is shown in Fig. 11-9. The transfer function is given by [4]

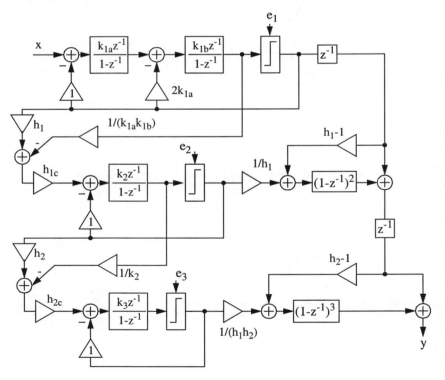

Figure 11-9   4th order 2-1-1 cascaded sigma-delta modulator.

$$Y(z) = z^{-4} \cdot X(z) + \frac{(1-z^{-1})^4 \cdot E_3(z)}{h_{1c} \cdot h_{2c}} \quad (11\text{-}7)$$

The coefficient h_{2c} is normally chose as 1 ([4]) which means that the fourth order noise shaping function is multiplied by the gain factor at the input of stage two. This is exactly the same as for the 2-2 modulator in Sec. 11.3.2. The maximum input signal swing without causing overloading of a first order modulator is larger than for a second order modulator. The gain factor h_{1c} can therefore be larger when the second stage is a first order modulator and the quantization noise is therefore smaller. The coefficient h_{1c} is normally chosen as 3 ([4]).

11.3.4 Cascaded Modulator with Multi-Bit Quantizer

If the required dynamic range is very high and a small OSR is required due to a high signal bandwidth, a 4th order cascaded modulator may not be sufficient to meet the requirements even if the improved architectures in section Sec. 11.3.2 and Sec. 11.3.3 are used. The quantization noise can be reduced by adding more stages in cascade. The order of noise shaping function will then be increased which may be a disadvantage when considering the decimation filter that follows the modulator. If the order is high the noise increases rapidly at high frequencies making the required attenuation in this filter high which is undesirable.

Another solution is to use a multi-bit quantizer. In a single stage modulator a multi-bit quantizer is problematic since the linearity of the multi-bit DAC in the feedback path directly affects the linearity of the modulator. This is not the case in a cascaded modulator if the multi-bit quantizer is used in the final stage. The input signal to the final stage in a cascaded modulator shows a very weak correlation to the input signal which means that the errors caused by a non-ideal DAC will appear as a increased noise floor in the output signal, not harmonic distortion [1]. In addition to this the error from the DAC will be noise shaped. If the DAC error is modelled as white uncorrelated noise, e_d, the output is affected by an additional noise term $H_d(z) \cdot E_d(z)$, where $H_d(z)$ is the noise shaping function for the DAC error. A multi-bit quantizer in the second stage of a 2-2 modulator would give an output signal

$$Y(z) = z^{-4} \cdot X(z) + k_1 \cdot (1-z^{-1})^4 \cdot E_2(z) + k_2 \cdot (1-z^{-1})^2 \cdot E_d(z) \quad (11\text{-}8)$$

where k_1 and k_2 are constants determined by the signal scaling in the modulator. Hence the DAC errors are shaped by a second order function. The required linearity of the DAC can therefore be significantly smaller than the total resolution of the OSADC. It should also be noted the $E_2(z)$ is smaller if the resolution of the quantizer is higher.

11.4 SC IMPLEMENTATION

Since sigma-delta converters are normally used in high accuracy applications they are almost always implemented using fully differential circuits. An SC fully differential second order modulator is shown in Fig. 11-10. There are four different clock signals

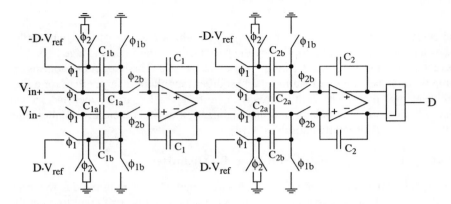

Figure 11-10 Fully differential SC implementation of second order modulator.

controlling the switches in the circuit. ϕ_1 and ϕ_2 are non-overlapping clock phases and are used to switch the capacitors to the input signals of the integrator. The clock phases ϕ_{1b} and ϕ_{2b} used for the switches at the opamp input are turned off slightly before ϕ_1 and ϕ_2 respectively to reduce the signal dependent clock feedthrough. The comparator output, D, is +1 or -1 depending on the sign of the second integrator output. It determines the sign of the reference voltage, V_{ref}, in the feedback loops. The capacitor ratios in the circuit determine the integrator gains. It may be more convenient to slightly modify the linear model of the modulator as shown in Fig. 11-11 to

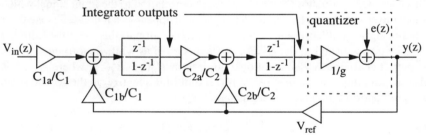

Figure 11-11 Linear model of second order Sigma-Delta Modulator.

better reflect the operation of the circuit in Fig. 11-10. The gain factor in the comparator is given by

$$g = \frac{C_{1b} \cdot C_{2a} \cdot V_{ref}}{C_1 \cdot C_2} \tag{11-9}$$

The voltage swing at the integrator outputs can now be scaled to their desired levels by changing the proper capacitance values. The swing at the first integrator can for

11.4 SC Implementation

instance be doubled by doubling the size of C_{1a} and C_{1b}. The only restriction is that the capacitor C_{2b} must be chosen according to the relation

$$C_{2b} = 2 \cdot \frac{C_{1b} \cdot C_{2a}}{C_1} \tag{11-10}$$

The easiest way to check for proper signal swing in the circuit is to simulate the linear model in Fig. 11-10 by using MatLab.

11.4.1 Integrator Modifications

There are some modifications that can be applied to the integrator to improve the performance. If for instance the capacitors C_{1a} and C_{1b} are equal the integrator can be implemented as shown in Fig. 11-12. The same capacitor is now used to sample both

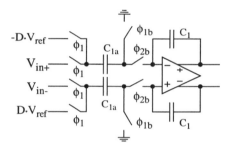

Figure 11-12 Simplified Integrator with only one.

the input signal and the reference signal [1]. In this way we can save chip area and increase the speed since the feedback factor is larger and the total load is smaller. In addition to this the noise is smaller since the noise contribution from one capacitor is removed. A potential problem with this solution is that the charge taken from the reference is signal dependent which makes the design of the reference buffers more difficult (the same problem exists for the integrator in Fig. 11-10).

There are other ways to reduce the size of the capacitors. The circuit in Fig. 11-13 uses a differencing input sampling to reduce the size of the capacitors by a factor 2 [6]. The input voltage is sampled on both clock phases but with different sign. In this way the effective voltage swing is increased by a factor of 2. Therefore it is possible to decrease the capacitor while still transferring the same amount of charge. Again both the speed and noise are improved. However, the signal dependent load on the reference buffers remains.

The signal dependent loading can be avoided by changing the switching of C_{1b} as shown in Fig. 11-14. This technique is also used in circuits with a single reference voltage [7]. The left hand side of the capacitor C_{1b} is now always switched between $+V_{ref}$ and $-V_{ref}$ making the load on the reference buffers signal independent. The sign of the reference signal is changed by connecting the right hand side of C_{1b} to the positive or negative input of the opamp. The digital signal B controlling the switches

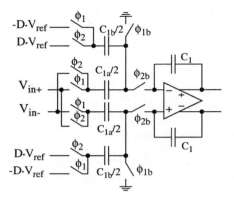

Figure 11-13 Integrator with differencing input sampling.

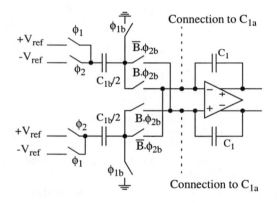

Figure 11-14 Integrator with signal independent load on reference buffers.

is

$$B = \begin{cases} 0, & D = -1 \\ 1, & D = 1 \end{cases} \qquad (11\text{-}11)$$

A slightly different approach to improve the settling performance is to use double sampling at the input of the integrator [8]. This is achieved by using two sampling capacitors, one sampling the input on ϕ_1 and the other on ϕ_2. Since it is now possible to integrate on both clock phases the whole sampling period is now available for settling, reducing the speed requirements on the opamp. A problem with this solution is that any mismatch between the double sampling capacitors cause distortion. For the input signal the distortion appears out of the signal band and is usually not a problem. However the signal coming from the quantizer must also be double sampled. Mismatch in this signal path may cause distortion in the signal band and therefore some kind of calibration or error averaging is needed [9]. For fully differential SC circuits the problem can be avoided by using only one capacitor in the feedback path alternately connected between the two inputs of the opamp [10]. The circuit now act as a bi-linear integrator which changes the transfer function of the modulator. This change is acceptable in some applications.

11.5 NON-IDEAL EFFECTS

There are several effects that deteriorate the performance of the sigma-delta modulator, e.g. finite opamp gain, finite bandwidth, nonlinear settling and capacitor mismatch. These non-ideal effects are considered in the following.

11.5.1 Finite Opamp Gain

In chapter 6 the transfer function of an integrator with a finite DC gain and a parasitic capacitor at the opamp input was derived. The integrator is shown in Fig. 11-15. The

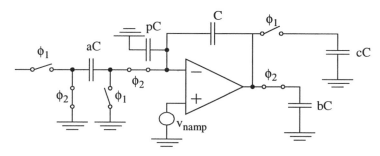

Figure 11-15 SC Integrator.

parasitic capacitor is represented by capacitor pC. According to chapter 6 the resulting transfer function is given by [4]

$$H(z) = \frac{r_2 \cdot z^{-1} \cdot a}{1 - \frac{r_2}{r_1} \cdot z^{-1}} \tag{11-12}$$

where

$$r_1 = \frac{\beta_1 \cdot A_0}{1+\beta_1 \cdot A_0} = \frac{1}{\left(\frac{p+1}{A_0}+1\right)}$$

$$r_2 = \frac{\beta_2 \cdot A_0}{1+\beta_2 \cdot A_0} = \frac{1}{\left(\frac{p+1+a}{A_0}+1\right)} \qquad (11\text{-}13)$$

The first integrator in a sigma-delta converter is the most critical since errors in the following integrators are noise shaped. For the first integrator in the SC modulator in Fig. 11-10 the factors r_1 and r_2 are given by

$$r_1 = \frac{1}{\left(\frac{C_p/C_1+1}{A_0}+1\right)}$$

$$r_2 = \frac{1}{\left(\frac{C_p/C_1+1+(C_{1a}+C_{1b})/C_1}{A_0}+1\right)} \qquad (11\text{-}14)$$

where C_p is the parasitic capacitor at the input of the opamp. By using that $1/(1+x) \approx 1-x$ for small x and $A_0^2 \gg A_0$ we get the following approximations

$$r_2 \approx 1 - \frac{C_p/C_1+1+(C_{1a}+C_{1b})/C_1}{A_0}$$

$$\frac{r_2}{r_1} \approx 1 - \frac{(C_{1a}+C_{1b})/C_1}{A_0} \qquad (11\text{-}15)$$

Thus the relative gain error for the integrator is approximately

$$\varepsilon_{gain} \approx \frac{C_p/C_1+1+(C_{1a}+C_{1b})/C_1}{A_0} \qquad (11\text{-}16)$$

while the relative error in the denominator, causing leakage in integrator, is approximated as

$$\varepsilon_{leak} \approx \frac{(C_{1a}+C_{1b})/C_1}{A_0} \qquad (11\text{-}17)$$

Usually quite large errors can be tolerated but the acceptable size of these errors depends on the architecture and the desired resolution. In for instance [4] a 4th order cascaded sigma-delta converter with $OSR = 24$ and a total resolution of 15 bits is designed. In this case a opamp gain of only 60 dB is sufficient which shows that the modulator is relatively insensitive to this kind of errors. However to provide some margin for other errors in the circuit the opamp gain should be chosen significantly larger than the lower limit (in [4] a gain of 80 dB is used). The easiest way to find the

11.5 Non-Ideal Effects

limit on the DC gain is normally to simulate the behavioral level model of the modulator using for instance MatLab.

11.5.2 Finite Bandwidth

From chapter 6 we know that a finite opamp bandwidth introduces a gain error in the integrator. We have that

$$H(z) = \frac{(1-\varepsilon_{r,2}) \cdot a \cdot z^{-1}}{1-z^{-1}} \tag{11-18}$$

where $\varepsilon_{r,2}$ is the relative settling error in the integration phase. According to chapter 6 we have

$$\varepsilon_{r,2} = e^{-\omega_{-3dB,2} \cdot t_s} = e^{-\frac{g_m}{(b \cdot (a+p+1)+(a+p)) \cdot C} \cdot t_s} \tag{11-19}$$

where t_s is the available settling time and g_m the transconductance of the input device in the opamp. For the first integrator of the modulator in Fig. 11-10 we have no load (bC) in the integration phase and hence

$$\varepsilon_{r,2} = e^{-\frac{g_m}{(C_{1a}+C_{1b}+C_p)} \cdot t_s} \tag{11-20}$$

where C_p is the parasitic capacitor at the input of the opamp. It should be noted that in an actual implementation the physical bottom plate of the integrating capacitor C_1 will add a parasitic capacitor at the output of the opamp. Since C_1 may be quite large this parasitic capacitor may add a significant load to the opamp output. In a more accurate analysis this capacitor should be taken into account. In [4] it is shown that if both finite opamp gain and bandwidth are present the finite bandwidth will also affect the integrator leakage. In a simplified analysis we have [4]

$$H(z) = \frac{r_2 \cdot (1-\varepsilon_{r,2}) \cdot z^{-1}}{1 - \frac{r_2}{r_1} \cdot \left(1-\varepsilon_{r,2} \cdot \left(1-\frac{r_1}{r_2}\right)\right) \cdot z^{-1}} \tag{11-21}$$

However the term $\varepsilon_{r,2} \cdot (1-r_1/r_2)$ corresponds to the product of the relative settling error and the error caused by finite opamp gain. Therefore it should be quite small and the transfer function with both limited gain and bandwidth can be approximated as

$$H(z) = \frac{r_2 \cdot (1-\varepsilon_{r,2}) \cdot z^{-1}}{1 - \frac{r_2}{r_1} \cdot z^{-1}} \tag{11-22}$$

As long as the settling is linear, relatively large settling errors are acceptable and the required settling accuracy is usually significantly smaller than the accuracy of the OSADC.

11.5.3 Non-linear Settling

For most practical cases the settling will not be entirely linear. For large input signals the first part of the settling is slew rate limited while the last part is linear. This means that the relative settling error is signal dependent and will therefore generate distortion. The required settling accuracy must usually be increased when the settling is non-linear. Assuming a single stage opamp the slew rate for the first integrator in Fig. 11-10 is determined by

$$SR = \frac{I_b}{C_{load}} = \frac{I_b}{\frac{(C_{1a}+C_{1b}) \cdot C_1}{C_{1a}+C_{1b}+C_1}} \qquad (11\text{-}23)$$

where I_b is the bias current in the differential input stage of the opamp. For a more accurate analysis the parasitic load capacitance at the output of the opamp should be taken into account. The -3 dB bandwidth of the integrator is given by

$$\omega_{-3dB} = \frac{g_m}{C_{load}} \cdot \beta = \frac{g_m}{C_{1a}+C_{1b}} \qquad (11\text{-}24)$$

where g_m is the transconductance of the input device of the opamp and β is the feedback factor. It can be shown that the output signal of the integrator at time $n \cdot T$ can be calculated as

$$V_{out}(n \cdot T) = V_{out}((n-1) \cdot T) + \\ + g(a_1 \cdot V_{in}((n-1) \cdot T) - D((n-1) \cdot T) \cdot a_2) \qquad (11\text{-}25)$$

where a_1 is the gain from the input of the integrator, V_{in}, to the output, a_2 the gain of the signal from the quantizer output to the output of the integrator and D is the quantizer output (± 1). $g(x)$ is a function describing the non-linear settling of the opamp. Assuming slew rate limited single pole settling it is given by [11]

$$g(x) = \begin{cases} x(1 - e^{-\omega_{-3dB}t_s}) & |x| \leq \frac{SR}{\omega_{-3dB}} \\ x - \frac{\text{sgn}(x)SR}{\omega_{-3dB} \cdot e} e^{\left(\frac{|x|}{SR} - t_s\right) \cdot \omega_{-3dB}} & \frac{SR}{\omega_{-3dB}} < |x| \leq \left(\frac{1}{\omega_{-3dB}} + t_s\right) \cdot SR \\ \text{sgn}(x) \cdot t_s \cdot SR & \left(\frac{1}{\omega_{-3dB}} + t_s\right) \cdot SR < |x| \end{cases} \qquad (11\text{-}26)$$

Based on (11-25) and (11-26) a behavioral model which include the effect of non-linear settling can be simulated in for instance MatLab. In this way the required settling accuracy can be determined. This was done in [11], where the required slew rate as function of the number of settling time constants were plotted for a third order cascaded modulator. The results show that there are two regions where a high resolution can be achieved without needing a very large slew rate. This is illustrated in Fig. 11-

11.5 Non-Ideal Effects

16. In [11] these regions are called the slow regime and the fast regime. The slow

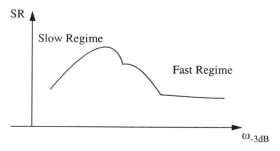

Figure 11-16 Slew rate requirement vs. bandwidth of integrator.

regime corresponds to linear settling, i.e. no slew limiting occurs. In this regime a low bandwidth is needed and the number of settling time constants need only be in the order of 2-3 to achieve more than 100dB SNDR in the converter. The number of settling time constants is defined as

$$n_\tau = t_s \cdot \omega_{-3dB} \qquad (11\text{-}27)$$

where t_s is the available settling time and ω_{-3dB} is the -3 dB bandwidth of the integrator. It is interesting to note that if the bandwidth of the integrator is increased without increasing the slew rate the resolution of the converters is reduced. This clearly illustrates that the required settling accuracy is increased when the settling is slew rate limited. The worst case slew rate requirement may be several times larger than what is required with linear settling. If the bandwidth is increased even further the required slew rate will decrease and finally reach a value that may be as low or even lower than in the case of linear settling. This corresponds to the fast regime. Depending on the resolution of the converter the fast regime is usually entered at about 8 - 12 settling time constants.

It is usually not possible to directly use the result in Fig. 11-16 to design the opamp in the integrator. The reason is that the bandwidth and slew rate of the integrator can not be chosen independently of each other. They are connected through the effective gate voltage of the input devices of the opamp. Therefore it may not be possible to use the slow regime since the effective gate voltage becomes too large. If we design the integrator in the fast regime it may not be possible to use the small slew rate indicated in Fig. 11-16, since the effective gate voltage would be too small.

The above results are based on a model with only one pole in the integrator. In all practical cases there are always several poles. The parasitic poles should be at a high frequency compared to the bandwidth of the integrator to avoid that they influence the settling. Therefore the bandwidth of the integrator should not be larger than necessary. The effects of parasitic poles and switch resistance can with some effort be included in a behavioral model to get a more accurate result [12].

11.5.4 Capacitor Mismatch

The gains from the input of the modulator and from the comparator feedback signal to the output of the integrator are given by C_{1a}/C_1 and C_{1b}/C_1 respectively. Mismatch in these capacitor ratios will change the corresponding gain factors. The errors can be modelled as a gain error in the integrator and a gain error in the input signal. The gain error at the input does normally not cause any problem at all while the integrator gain error has the same effect as a finite opamp bandwidth (se previous section). A capacitor matching in the order of 0.1% or better is achievable in a modern CMOS process. This is normally sufficient to reach resolutions in the order of 15 bit in cascaded modulators ([4]).

11.5.5 Gain Errors in Cascaded Modulators

Due to capacitor mismatch the actual gains in the integrator deviate from their ideal values. This effect will degrade the performance of the cascaded modulator due to the mismatch between the analog and digital parts. For the 2-2 cascaded modulator in Fig. 11-4 the digital gain factor g_1 must meet the relation [5]

$$g_1 \cdot k_1 \cdot p_1 \cdot j_1 = 1 \qquad (11\text{-}28)$$

where k_1 is the gain in the first integrator in the first modulator, p_1 the gain in the second integrator of the first modulator and j_1 the gain at the input of the second modulator. If there are gain errors in the coefficients the output signal becomes [5]

$$Y(z) = z^{-4} \cdot X(z) + g_1 \cdot (1-z^{-1})^4 \cdot E_2(z) + \\ + \varepsilon \cdot (1-z^{-1})^2 \cdot E_1(z) \qquad (11\text{-}29)$$

where $\varepsilon = 1 - g_1 \cdot k_1 \cdot p_1 \cdot j_1$ is the gain error. With gain errors the quantization noise from the first modulator appears in the output signal. However, the noise is spectrally shaped by a second order noise-shaping function and therefore relatively large mismatch errors are acceptable unless a very high resolution is required.

11.5.6 Thermal Noise

There are two types of noise in CMOS circuits, thermal noise and 1/f noise. If the bandwidth is fairly high the thermal noise will dominate while the 1/f noise dominates at low frequencies. If the 1/f noise is large it can be reduced by circuit techniques such as correlated-double-sampling [3]. Therefore we focus on the thermal noise in this section. The input referred thermal noise of all integrators except the first in a cascaded modulator is noise shaped. The capacitor sizes in all but the first integrator are therefore usually determined by other reasons than thermal noise. From chapter 6 we have that the input referred thermal noise of an integrator can be approximated as

$$v_n^2 = \frac{4}{3} \cdot \frac{kT}{C_{load}} \cdot \beta \cdot (1 + n_t) + \frac{kT}{C_s} \qquad (11\text{-}30)$$

where k is the Boltzmann's constant, T the absolute temperature, C_{load} the total

11.5 Non-Ideal Effects

effective load capacitor at the opamp output in the integration phase, β the feedback factor in the integration phase, C_s the total sampling capacitor and n_t a noise contribution factor that depends on the opamp architecture. The first term in (11-30) correspond to the noise from the opamp while the second term corresponds to the noise from the sampling capacitor. It is worth noting that the noise only depends on the capacitor sizes in the circuit. For the integrator in Fig. 11-10 we have two input signals. It is convenient to refer all noise sources (one from opamp, one from C_{1a} and one from C_{1b}) to the input signal applied to capacitor C_{1a}. It can be shown that

$$v_{n,in}^2 = \frac{4}{3} \cdot \frac{kT}{C_{load}} \cdot \beta \cdot (1 + n_t) \cdot \frac{(C_{1a} + C_{1b})^2}{C_{1a}^2} + \frac{kT}{C_{1a}} + \frac{kT}{C_{1b}} \cdot \frac{C_{1b}^2}{C_{1a}^2} =$$
$$= \frac{kT}{C_{1a}} \cdot (1 + K_1) \tag{11-31}$$

where K_1 is a factor accounting for the noise in the opamp and the sampling capacitor C_{1b}. The factor K_1 is given by

$$K_1 = \frac{4}{3} \cdot \frac{\beta}{C_{load}} \cdot (1 + n_t) \cdot \frac{(C_{1a} + C_{1b})^2}{C_{1a}} + \frac{C_{1b}}{C_{1a}}$$
$$= \frac{\frac{4}{3} \cdot (1 + n_t)}{\frac{C_{1a}C_L(C_{1a} + C_{1b} + C_1)}{(C_{1a} + C_{1b})^2 C_1} + \frac{C_{1a}}{(C_{1a} + C_{1b})}} + \frac{C_{1b}}{C_{1a}} \tag{11-32}$$

where C_L is the load capacitance at the opamp output when excluding the load from the feedback network. If the improved switching in Fig. 11-12 is used C_{1b} is zero and the factor K_1 is reduced to

$$K_1 = \frac{4}{3} \cdot \frac{\beta}{C_{load}} \cdot (1 + n_t) \cdot C_{1a} = \frac{\frac{4}{3} \cdot (1 + n_t)}{\frac{C_L(C_{1a} + C_1)}{C_{1a}C_1} + 1} \tag{11-33}$$

In (11-33) we see that if the noise contribution factor n_t is small and the load capacitor, C_L is large, the factor K_1 is small and the noise is dominated by the kT/C noise from the sampling capacitor.

The noise requirement is relaxed due to the oversampling since only a small part of the noise falls in the signal band. To make the thermal noise power smaller than the quantization noise we have

$$v_{n,in}^2 < \frac{FS^2}{2^{2n_{tot}+2} \cdot 3} \cdot OSR \tag{11-34}$$

where n_{tot} is the desired number of bits in the modulator and FS is the full-scale swing of the input signal.

Taking the thermal noise into account the signal-to-noise ratio of the converter can be calculated as

$$SNR = 10 \cdot \log\left(\frac{P_{signal}}{P_q + P_n}\right) \qquad (11\text{-}35)$$

where P_{signal} is the signal power, P_q the quantization noise power and P_n the thermal noise power. We see that if the thermal noise power is equal to the quantization noise power the SNR will decrease by 3 dB. Therefore it is necessary to make the thermal noise significantly smaller than the quantization noise to avoid that the SNR is decreased. The thermal noise is reduced by increasing the capacitors in the circuit.

11.5.7 Comparator Errors

When using a single bit quantizer the modulator is relatively insensitive to comparator offset and hysteresis. From [4] we have that an offset in the order of 100mV and a hysteresis of 40 mV is acceptable for a total resolution of 15 bits in a 2-1-1 cascaded modulator.

11.5.8 Non-linear Effects

In addition to the gain errors discussed in the previous sections a number of non-linear phenomena may degrade the performance. As is the case for most errors sources in a sigma-delta modulator the errors in the first integrator are most critical and require the most accurate components.

In the actual implementation the capacitors are voltage dependent and this will cause distortion. This non-linearity generate a large second harmonic distortion term while higher order harmonics are smaller. For high resolutions it is therefore important to use accurately balanced differential circuits that suppress the large second harmonic.

Clock feedthrough from the switches is one more source of distortion. By using "bottom plate" sampling where the switch at the opamp input is turned off slightly before the switch at the input of the integrator a large portion of the signal dependent clock feedthrough is removed and the remaining error is mainly an offset. There is however still distortion and again it is important to use accurately balanced differential circuits to suppress even order harmonics.

A third source of distortion is the non-linear resistance of the switches. The input signal to the OSADC is normally a continuous-time signal. If the signal frequency is high and the bandwidth of the passive sampling circuit is too low the variation in bandwidth caused by signal dependent switch resistance will give large distortion. The distortion increases with signal frequency. It is therefore important to make on-resistance of the switches small enough to avoid this problem. The signal dependent on-resistance is mainly a problem for sampling circuits with continuous-time input signals. For all other switches the resistance must only be small enough to ensure that the final value

11.6 Design Example

at the end of the sampling phase is within the required settling accuracy which is usually much less than the resolution of the OSADC. The resistance of the switches may also affect the settling behavior of the opamp. This must be verified by simulation.

The influence of the non-linear error sources discussed in this section can be investigated by a circuit simulators such as SPICE or SPECTRE. However, the results should not be trusted blindly since especially clock feedthrough errors are not always very accurately modelled.

11.6 DESIGN EXAMPLE

In this section we give a design example of a cascaded sigma-delta modulator. The aim is to design a 14-bit converter with a signal bandwidth of 0.5 MHz which corresponds to the requirements for G.lite. We use the 2-1-1 architecture since it gives a low quantization noise. The 2-1-1 modulator architecture is repeated in Fig. 11-17 for

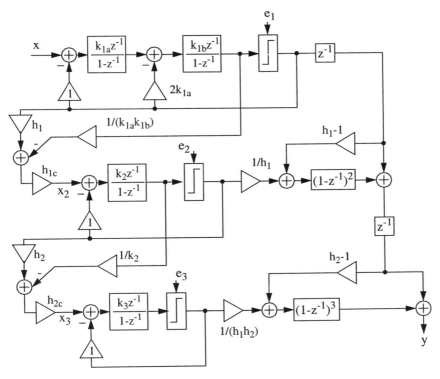

Figure 11-17 4th order 2-1-1 cascaded sigma-delta modulator.

convenience. We start by choosing the coefficients in the modulator. The transfer function of the modulator is given by [2]

$$Y(z) = z^{-4} \cdot X(z) + \frac{(1-z^{-1})^4 \cdot E_3(z)}{h_{1c} \cdot h_{1c}} \qquad (11\text{-}36)$$

where $E_3(z)$ is the quantization noise in the third stage. To make the quantization noise small, h_{1c} and h_{2c} should be as large as possible. These two coefficients also determines the gain factors in the digital cancellation logic. To simplify the implementation the factors should be powers of 2. All other gain factors in the modulator are in an SC implementation determined by capacitor ratios. The capacitor ratios should be chosen such that the circuit is easy to layout using unit-capacitors.

To make the modulator less sensitive to circuit noise the voltage swings in all the nodes of the circuit should be approximately the same. The maximum signal swing at the input of last stage can be determined by [2]

$$x_3 = h_{2c} \cdot \left(x_2 + \frac{\Delta}{2} - h_2 \frac{\Delta}{2} \right) \qquad (11\text{-}37)$$

where Δ is the step size in the D/A converters and x_2 is the maximum swing at the input of the second stage. To maximize h_{2c}, h_2 should be 1. To make the swings equal we also have $h_{1c} = 1$. Coefficients h_1 and h_{1c} influence the performance of the cascaded modulator and the best choice is $h_1 = 3$ and $h_{1c} = 0.5$ ([2]). The integrator gains are chosen to give signal swings at the opamp outputs that are equal to or slightly less than the swing of the feedback signals of the DAC. Suitable values on all the coefficients are now summarized in Table 11-1 ([2]).

Table 11-1 Gain factors in the modulator.

Coeff.	Value
k_{1a}	1/4
k_{1b}	1/2
h_1	3
h_{1c}	0.5
k_2	1/4
h_2	1
h_{2c}	1
k_3	1/4

Behavioral simulations of the modulator were used to plot the SNR as function of signal amplitude for different OSR. The result is shown in Fig. 11-18. We see that an oversampling ratio of 24 is sufficient for 14 bits resolution. A closer look at the plot for OSR = 24 reveals that the peak SNR appears for an input signal of approximately -3 dBFS. The peak SNR is approximately 86.5 dB which corresponds to a resolution of slightly more than 14 bits. Therefore we decide to use OSR = 24. The desired bandwidth is 0.5 MHz which makes the oversampling frequency, $f_s = 2 \cdot 0.5 \cdot 24 = 24$

11.6 Design Example

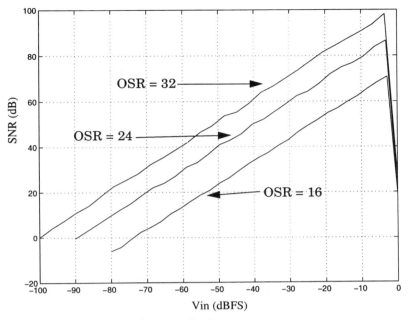

Figure 11-18 SNR vs. input amplitude.

MHz. We assume that the reference voltage levels in the DAC are $\pm 1.25\,V$ and we use a fully differential SC circuits for the implementation.

We can calculate the capacitor ratios in the implementation by using the gain factors in Table 11-1. To get reasonable absolute values on the capacitors we must consider the thermal noise and capacitance matching. The thermal noise in the first integrator dominates. It can be estimated as (see Sec. 11.5.6)

$$v_{n,\,in}^2 = \frac{kT}{C_{1a}} \cdot (1 + K_1) \tag{11-38}$$

where K_1 is a factor accounting for the thermal noise in the opamp and the noise stored on the capacitor sampling the reference voltage. In this case it is possible to use the improved integrator in Fig. 11-12 where the same capacitor is used to sample both the input signal and the reference signal. The factor K_1 is given by

$$K_1 = \frac{4}{3} \cdot \frac{\beta}{C_{load}} \cdot (1 + n_t) \cdot C_{1a} = \frac{\frac{4}{3} \cdot (1 + n_t)}{\frac{C_L(C_{1a} + C_1)}{C_{1a}C_1} + 1} \tag{11-39}$$

The thermal noise must be smaller than the quantization noise yielding

$$v_{n,in}^2 < \frac{FS^2}{2^{2n_{tot}+2} \cdot 3} \cdot OSR \Rightarrow C_{1b} > \frac{2^{2n_{tot}+2} \cdot 3 \cdot kT \cdot (1+K_1)}{FS^2 \cdot OSR} =$$

$$= \frac{2^{2 \cdot 14 + 2} \cdot 3 \cdot 1.38 \times 10^{-23} \cdot 300 \cdot (1+K_1)}{2.5^2 \cdot 24} \approx K_1 \cdot 90 fF$$

(11-40)

The capacitor size in (11-40) is calculated for a single ended implementation. The SNR in a fully differential implementation is at least 3dB better. To account for that the factor K_1 is larger than 1 and to give a design margin we choose a sampling capacitor in the first stage of 1 pF. For the integrators in the following stages the thermal noise is not so critical and the capacitor sizes can be scaled down to save power. Here we decrease the integration capacitors in the second integrator by a factor four, the capacitors in both the third and fourth integrators by a factor eight. The fully differential SC-implementation is shown in Fig. 11-19 ([2]). The capacitor sizes can now be determined as shown in Table 11-2. The signal swings in the integrators were simu-

Table 11-2 Capacitor values in the modulator.

Capacitor	Value (pF)
C_1	4
C_{1a}	1
C_2	1
C_{2a}	0.25
C_{2b}	0.25
C_3	0.5
C_{3a}	0.1875
C_{3b}	0.1875
C_{3c}	0.125
C_4	0.5
C_{4a}	0.25
C_{4b}	0.125
C_{4c}	0.125

lated using a behavioral model and assuming that the reference levels are ±1.25 V. The input signal is a 0.875V sinusoid (~ -3 dBFS). The result is shown in Fig. 11-20. A closer look at this figure reveals that the peak voltage in integrator 1 is 0.95V while the peak voltage in all other integrators is approximately 0.85V. The peak SNR is 86.7 dB which corresponds to 14.1 ENOB.

The influence of capacitor mismatch was simulated. The relative capacitor errors were assumed to be random Gaussian variables. Mote Carlo analysis was used to determine the yield. The standard deviation of the errors for a 90% yield is shown in Table 11-2.

11.6 Design Example

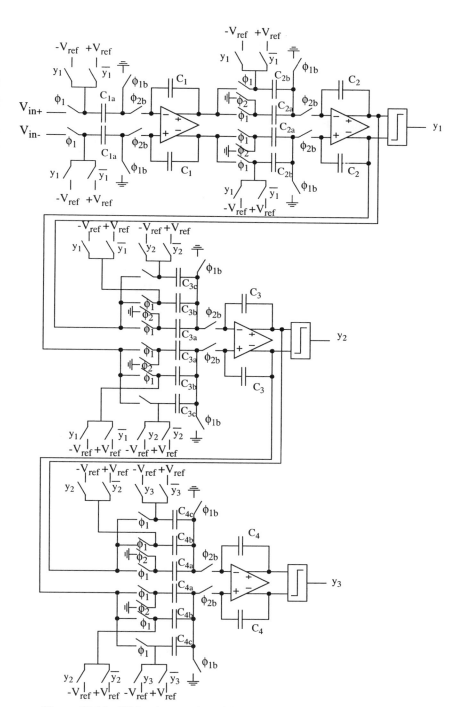

Figure 11-19 SC-implementation of 2-1-1 modulator.

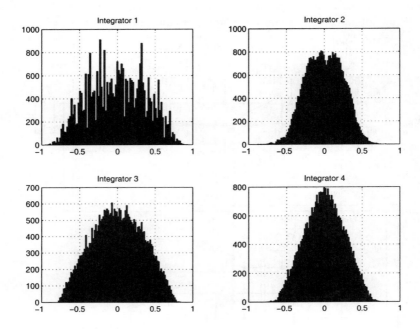

Figure 11-20 Signal swings in 2-1-1 modulator.

Table 11-3 Required relative matching error.

Integrator	Matching
1	0.25%
2	0.25%
3	0.5%
4	2%

The SNR is allowed to drop 1 dB in the yield simulation, i.e. the SNR can drop to 85.7 dB.

To determine the required opamp gain and speed of the SC circuits we must determine the capacitive loads and feedback factors of all the integrators. We must also take the parasitic capacitors into account. We assume that the parasitic capacitor at the output of a integrator is 30% of the integration capacitor The parasitic capacitor at the opamp input in the first integrator is assumed to be 0.5pF. All the other integrators are assumed to have a parasitic capacitor of 0.25 pF at the opamp input. The feedback factor during the integration phase of integrator i can be calculated as

$$\beta_i = \frac{C_i}{C_i + C_{ia} + C_{ib} + C_{ic} + C_{ip}} \qquad (11\text{-}41)$$

11.6 Design Example

where C_{ip} is the parasitic capacitor at the input of the opamp. The capacitive load during the integration phase can for integrator i be calculated as

$$C_{il} = \beta_i \cdot (C_{ia} + C_{ib} + C_{ic} + C_{ip}) + C_i \cdot 0.3 \tag{11-42}$$

The result of the calculation is shown in Table 11-4.

Table 11-4 Capacitors and feedback factors.

Capacitor	Value (pF)
β_1	0.72
C_{11}	2.3 pF
C_{1p}	0.5 pF
β_2	0.57
C_{21}	0.73 pF
C_{2p}	0.25 pF
β_3	0.4
C_{31}	0.45 pF
C_{3p}	0.25 pF
β_4	0.4
C_{41}	0.45 pF
C_{4p}	0.25 pF

According to the previous discussions in this chapter finite opamp gain will change the transfer function of the integrator to

$$H(z) = \frac{r_2 \cdot z^{-1} \cdot a}{1 - \frac{r_2}{r_1} \cdot z^{-1}} \tag{11-43}$$

where

$$r_1 = \frac{\beta_s \cdot A_0}{1 + \beta_s \cdot A_0} \qquad r_2 = \frac{\beta_i \cdot A_0}{1 + \beta_i \cdot A_0} \tag{11-44}$$

and β_s is the feedback factor in sampling phase, β_i is the feedback factor in the integration phase and A_0 the opamp DC gain. The feedback factor for integrator i during the sampling phase can be calculated as

$$\beta_{si} = \frac{C_i}{C_i + C_{ip}} \tag{11-45}$$

The calculated values on the feedback factors are shown in Table 11-5.

Table 11-5   Feedback factors in sampling phase

Capacitor	Value (pF)
β_{s1}	0.89
β_{s2}	0.8
β_{s3}	0.67
β_{s4}	0.67

The effect of finite opamp gain was simulated using a behavioral model. The result is shown in Fig. 11-21. We see that the required opamp gain is approximately 60 dB in

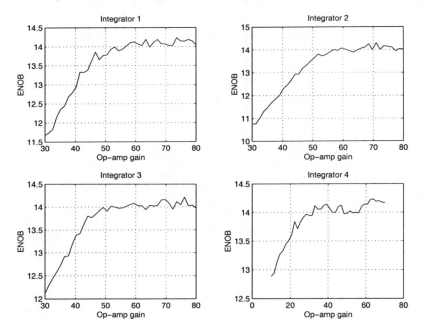

Figure 11-21   Signal swings in 2-1-1 modulator.

the two first integrators, 50 dB in the third and only 30 dB in the fourth. In the design, at least 10-20 dB should be added to the required opamp gains to provide some margin for other errors and to account for the poor modelling of the output conductance in SPICE simulations.

The finite bandwidth due to the opamp, introduces gain errors in the integrators. The equations in Sec. 11.5.2 can be used to find a lower limit on the bandwidth. However, the equations are only valid if the settling is linear. In most practical applications the settling is slew rate limited. The requirements on the bandwidth is increased when the settling is slew rate limited. Therefore the linear settling model is not so useful. It should be noted that when designing the opamp the slew rate and the bandwidth can not be chosen independently. Assuming a single pole opamp the -3 dB bandwidth is

$$\omega_{-3dB} = \frac{g_m}{C_{load}} \cdot \beta \qquad (11\text{-}46)$$

where g_m is the transconductance of the input device of the opamp, C_{load} the total load at the output of the opamp and β the feedback factor. The transconductance can be calculated as

$$g_m = \frac{I_d}{V_{eff}} \qquad (11\text{-}47)$$

where I_d is the bias current in the opamp and $V_{eff} = V_{gs} - V_T$ the effective gate voltage of the input device in the opamp. The slew rate is determined by

$$SR = \frac{I_d}{C_{load}} \qquad (11\text{-}48)$$

Hence the ratio between the slew rate and the bandwidth is

$$\frac{SR}{\omega_{-3dB}} = \frac{V_{eff}}{\beta} \qquad (11\text{-}49)$$

We see that this ratio is determined by V_{eff}. This voltage can often not change very much. If it is too small the input devices will be very large and if it is too big the voltage swing is reduced. Assuming a effective gate voltage of 0.3-0.4 V and simulating a behavioral model including slew rate (see Sec. 11.5.3) shows that the current in the first integrator must be approximately 0.3-0.4 mA. In the simulation we used typical process parameters from a 0.8u CMOS process. To provide a design margin the current must be larger than the simulated value. This current serves as a starting point for the design of the opamp. When taking all the parasitics into account the current must be increased.

11.7 SUMMARY

In this chapter we have discussed oversampling sigma-delta converters. They are suitable for high resolutions but the high oversampling ratio (OSR) limits the signal bandwidth. To increase the signal bandwidth cascaded modulators have been discussed. Gain errors in the integrators cause leakage of low order quantization noise at the output which limits the resolution of the converter. The integrator gain errors need not be as accurate as the full resolution and converters with up to 15 - 16 bit have been successfully implemented without calibration. To further reduce the quantization noise in the modulator a multi-bit quantizer can be used in the last stage of the converter. For a single stage modulator a multi-bit quantizer would require calibration since the DAC linearity directly limits linearity of the OSADC. For cascaded modulator however, the signal in the last stage has a weak correlation to the input signal and linearity errors in the multi-bit DAC will cause noise, not distortion. In addition to this the DAC linearity error is noise shaped, which significantly reduces the required linearity of the DAC.

We have also discussed SC implementations of oversampled sigma-delta converters and investigated some of the most important limitations and how they affect the performance of the modulator. Based on the discussions and design example was presented.

REFERENCES

[1]   S. R. Norsworthy, R. Schreier and G. C. Temes, *Delta-Sigma Data Converters: Theory, Design and Simulation*, IEEE Press, 1997.

[2]   G. Yin and W. Sansen, "A High-Frequency and High-Resolution Fourth-Order $\Sigma\Delta$ A/D Converter in BiCMOS Technology", *IEEE J. of Solid-State Circuits, vol. 29*, no. 8, pp. 857-865, Aug. 1994.

[3]   M. Sarhang-Nejad and G. C. Temes, "A High-Resolution Multibit $\Sigma\Delta$ ADC with Digital Correction and Relaxed Amplifier Requirements", *IEEE J. of Solid-State Circuits*, vol. 28, no. 6, pp. 648-660, June 1993.

[4]   Y. Geerts, A. M. Merques, M. Steyaert and W. Sansen "A 3.3-V, 15-bit, Delta-Sigma ADC with a Signal bandwidth of 1.1 MHz for ADSL Applications", *IEEE J. of Solid-State Circuits*, vol. 34, no. 7, pp. 927-36, July 1999.

[5]   H. Baher and E. Afifi, "Novel Fourth-Order Sigma-Delta Convertor", *Electronics Letters*, vol. 28, no. 15, pp. 1437-38, 16th July 1992.

[6]   R. Unbehauen and A. Cichocki, *MOS Switched-Capacitor and Continuous-Time Integrated Circuits and Systems*, Springer-Verlag Berlin Heidelberg, 1989.

[7]   G. Yin, F. Stubbe and W. Sansen, "A 16-b 320-kHz CMOS A/D Converter Using Two-Stage Third-Order $\Sigma\Delta$ Noise Shaping", *IEEE J. of Solid-State Circuits*, vol. 28, no. 6, pp. 640-7, June 1993.

[8]   P. J. Hurst and W. J. McIntyre, "Double Sampling in Switched-Capacitor Delta-Sigma A/D Converters", *Intl. Symp. on Circuits and Systems*, ISCAS'90, pp. 902-905, 1990.

[9]   L. Yu and M. Snelgrove, "Mismatch Cancellation for Double-sampling Sigma-Delta Modulators", *Intl. Symp. on Circuits and Systems*, ISCAS'98, pp. 356-9, 1998.

[10] D. Senderowicz, G. Nicollini, S. Pernici, A. Nagari, P. Confalonieri and C. Dallavalle, "Low-Voltage Double-Sampled $\Sigma\Delta$ Converters", *IEEE J. of Solid-State Circuits*, vol. 32, no. 12, pp. 1907-1919, Dec. 1997.

[11] L. A. Williams, III, and B. Wooley, "A Third-Order Sigma-Delta Modulator with Extended Dynamic Range", *IEEE J. of Solid-State Circuits*, vol. 29, no. 3, pp. 193-202, Match 1994.

[12] R. Naiknaware and T. Fiez, "Power Optimization of $\Sigma\Delta$ Analog-to-Digital Converters Based on Slewing and Partial Settling Considerations", *Intl. Symp. on Circuits and Systems*, ISCAS'98,

12 MODELING OF NYQUIST D/A CONVERTERS

12.1 INTRODUCTION

High-speed DACs are important building blocks in communications systems. They are key components in e.g. xDSL applications. For these and similar applications, traditional static performance measures such as offset error, gain error, integral non-linearity (INL), differential non-linearity (DNL), are not sufficient. For communications applications, the dynamic performance determines the quality of DACs [1, 2]. The most important frequency-domain performance measures are spurious-free dynamic range (SFDR), inter-modulation distortion (IMD), signal-to-noise ratio (SNR), and signal-to-noise-and-distortion ratio (SNDR, or SINAD).

We present an overview of behavioral-level modeling of error sources in Nyquist-rate DACs. Prior modeling [3, 4, 5, 6] of DAC errors have more been focused on INL and DNL requirements, and only a few of them have discussed the impact of DAC errors on the frequency-domain measures. We show how specific circuit errors as for example matching errors, finite output impedance, and settling errors, affect the performance of the converter. The impact of DAC errors on DNL and INL is also briefly discussed in this chapter.

Since the current-steering DAC is a suitable candidate for high-speed applications, the modeling in this chapter is focused on this type of DAC. Process variations and other parasitics will influence the matching between current sources and will introduce noise and distortion. The output impedance of the current-steering DAC is depedendent on the number of current sources connected to the output and is therefore signal-dependent.

The DAC and its major errors sources are outlined in Sec. 12.2. In Sec. 12.3 we show how a finite output resistance influences the performance and in Sec. 12.4 we discuss matching errors. The circuit noise, discussed in Sec. 12.5, and especially thermal noise, limits the resolution. In Sec. 12.6 we also discuss and model the influence of errors in the time-domain and dynamic errors such as nonlinear slewing, bit skew, and glitches.

In Chapter 13 we present the design of a current-steering wideband CMOS DAC for telecommunications applications where the models can be applied to the design strategy.

12.2 ERRORS IN CURRENT-STEERING DACS

The current-steering DACs are widely used. For convenience, the circuit structure of an N-bit DAC is shown in Fig. 12-1 and we will briefly recapture some of its proper-

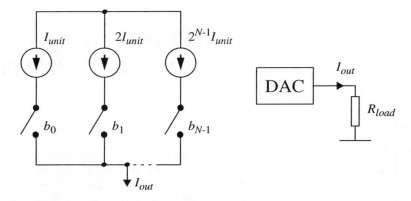

Figure 12-1 a) An N-bit binary weighted current-steering DAC. b) The output is terminated over a 50-ohm load.

ties. To improve matching, unit current sources are used to form the binary weighting. Hence for the i-th bit, we use 2^i unit current sources in parallel instead of a transistor 2^i times wider than the LSB transistor. If the matching errors are uncorrelated, this will improve the element matching and the relative error will improve by a factor of $\sqrt{2^i}$ for the i-th current source. The switches in Fig. 12-1, are controlled by the digital bits, b_i, of the code k

$$k = \sum_{l=0}^{N-1} 2^l \cdot b_l \qquad (12\text{-}1)$$

where b_{N-1} is the MSB and b_0 is the LSB. The minimum value of k is 0 and the maximum value is $k = 2^N - 1$. The current associated with the i-th LSB position is given by

$$I_i = 2^i \cdot I_{unit} \qquad (12\text{-}2)$$

and the total output current is given by

$$I_{out}(k(nT)) = I_{unit} \cdot k(nT) \qquad (12\text{-}3)$$

where T is the sampling period and the output is held during the period T. To increase readability, we will use the notations

$$I_{out}(k) = I_{unit} \cdot k \qquad (12\text{-}4)$$

12.2.1 Major Error Sources in Current-Steering DACs

Error sources that limit the current-steering DAC performance are

- *Finite output resistance.* The finite output resistance strongly effects the linearity of the converter. This is primarily due to the fact that the output resistance of the converters is signal-dependent.

- *Matching errors.* Since variations in the process cause the oxide thickness and threshold voltage, t_{ox} and V_T, to vary, the unit currents are unequal, which also affects the linearity. The matching errors are of both stochastic and deterministic nature [8].

- *Circuit noise.* A fundamental limit on the resolution is given by the signal-to-noise ratio (SNR) and we have to guarantee that the circuit noise is below the quantization noise floor.

- *Slewing and settling errors.* For DACs, the output is analog and therefore a limited settling time will introduce a settling error which may be signal dependent and hence introduce distortion. The settling errors arise due to the limited RC constant at the output of the converter.

- *Glitches.* Due to the non-ideal switches and different capacitive load on different bits, time skew between the bit switches will for a short period of time create a transition code at the output. This introduces a current or voltage step, referred to as a glitch.

- *Clock feedthrough (CFT).* Due to capacitive coupling in the current switches between the switching signals (ϕ_i) and the current output, current spikes are added to the signal.

We will address some of these errors in this chapter. In Chapter 13, where we discuss the design of wideband current-steering DACs, we use the results. Most of the models and formulas presented can be applied and generalized to most binary weighted DACs and not only to the current-steering DAC. The models may also easily be extended to cover thermometer coded converters without any large increase of complexity.

12.3 OUTPUT RESISTANCE VARIATIONS

The output impedance and the parasitic impedance of interconnections and switches in the converter will strongly influence the performance [4]. Any non-ideal current source has a finite output resistance and can be modeled as shown in Fig. 12-2. In the figure the current source is terminated over a resistive load at the output.

When the different current sources are switched to the output, the total output resistance is changed and when only static values are considered, the AC current through the load, I_{load}, is

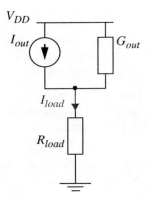

Figure 12-2 Generalized view of a current source with a non-zero output conductance.

$$I_{load} = \frac{I_{out}}{1 + R_{load} \cdot G_{out}} \qquad (12\text{-}5)$$

where I_{out} is the nominal output current from the DAC given by (12-4), $G_{out} = 1/R_{out}$ is the output conductance, and R_{load} is the load resistance. From (12-5) it is seen, that if the output conductance of the DAC is constant, there is only a gain error, which does not degrade linearity. If the output conductance depends on the input, it will give rise to distortion. Using (12-5), and assuming a signal-dependent output conductance of the DAC, $G_{out}(k)$, the current delivered to the load is

$$I_{load}(k) = \frac{I_{out}(k)}{1 + G_{out}(k) \cdot R_{load}} \qquad (12\text{-}6)$$

where k is the DAC's digital input given by (12-1).

The output conductance, $G_{out}(k)$, is determined by the number of parallel unit current sources that are switched to the output. The output conductance of one unit current source is assumed to be $G_{unit} = 1/R_{unit}$ and with the i-th LSB we have the corresponding conductance $G_i = 2^{i-1} \cdot G_{unit}$. The total output conductance, $G_{out}(k)$, of the converter is given by

$$G_{out}(k) = G_{unit} \cdot \sum_{l=0}^{N-1} 2^l \cdot b_l = G_{unit} \cdot k \qquad (12\text{-}7)$$

We refer to the product between the unit output conductance and the load resistance as the conductance ratio, i.e.,

$$\rho = G_{unit} \cdot R_{load} \qquad (12\text{-}8)$$

We also use the resistance ratio, R_{ratio}, given by

12.3 Output Resistance Variations

$$R_{ratio} = \frac{1}{\rho} = \frac{R_{unit}}{R_{load}} \tag{12-9}$$

Combining (12-1), (12-4), (12-6), and (12-8), we have that the load current can be written as

$$I_{load}(k) = I_{unit} \cdot \frac{k}{1 + \rho \cdot k} \tag{12-10}$$

Rewriting (12-10), we also end up at

$$I_{load}(k) = \frac{I_{unit}}{\rho} \cdot \left(1 - \frac{1}{1 + \rho \cdot k}\right) \tag{12-11}$$

The effect of the finite impedance ratio is illustrated in Fig. 12-3. The figure shows the

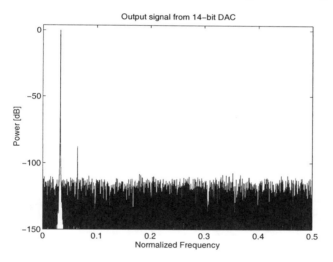

Figure 12-3 Effect of limited resistance ratio (10^8) when a full-scale sinusoid is applied to the DAC.

Matlab simulation result when applying a full-scale sinusoid to a 14-bit DAC. The maximum output impedance of one unit current source is 5 GΩ and the load resistance is 50 Ω. We see in the figure that the SFDR is approximately 88 dBc.

12.3.1 DNL and INL vs. output resistance

We will now show how the nonlinearity affects the differential and integral nonlinearities. The DNL and INL definitions were given in Chapter 1 and we have

$$DNL_k = I(k) - I(k-1) - I_{unit} \tag{12-12}$$

and

$$INL_k = \sum_{i=0}^{k} DNL_i = I(k) - I(0) - k \cdot I_{unit} \qquad (12\text{-}13)$$

k is the input code in an offset binary representation. Note that no best-fit line compensation has been done. By inserting (12-11) in (12-12) we get

$$\begin{aligned} DNL_k &= I_{unit} \cdot \left(\frac{k}{1 + \rho \cdot k} - \frac{k-1}{1 + \rho \cdot (k-1)} \right) - I_{unit} \\ &= I_{unit} \cdot \frac{k \cdot (1 + \rho \cdot (k-1)) - (k-1) \cdot (1 + \rho \cdot k)}{(1 + \rho \cdot k)(1 + \rho \cdot (k-1))} - I_{unit} \qquad (12\text{-}14) \\ &= \frac{I_{unit}}{(1 + \rho \cdot k)(1 + \rho \cdot (k-1))} - I_{unit} \approx \frac{I_{unit}}{(1 + \rho \cdot k)^2} - I_{unit} \end{aligned}$$

The *DNL* is illustrated in Fig. 12-4a) for a 14-bit DAC with resistance ratio

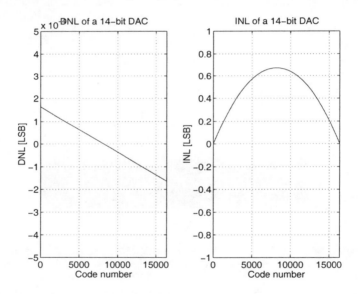

Figure 12-4 Simulated DNL and INL as a function of input code for a resistance ratio of 10^8.

$R_{ratio} = 10^8$ and 50-Ω load resistance. The *DNL* and *INL* values in the figures have been calculated with respect to a best-fit line. The best-fit offset and gradient values were approximately $0.447 \cdot I_{unit}$ and $0.9998 \cdot I_{unit}$, respectively.

We see that the *DNL* in (12-14) is a decreasing monotonic function, since the term

$$\frac{I_{unit}}{(1 + \rho \cdot k)^2} \qquad (12\text{-}15)$$

12.3 Output Resistance Variations

is positive and decreasing towards zero for increasing k. This implies that *DNL* is always larger than -1 LSB. As long as the resistance ratio is reasonably large, monotonicity is guaranteed, and we can see in Fig. 12-4a) that the *DNL* is still very good, but from Fig. 12-3 we found that the *SFDR* is low and may limit the resolution. The *INL* for the same DAC configuration is shown in Fig. 12-4b) where the *INL* is given by (12-13). Also, especially in communications, these measures cannot directly be used to analyze the true impact of the resistance error.

12.3.2 SNDR vs. Output Resistance

The error current, $\Delta I(k)$, is given by

$$\Delta I(k) = I_{load}(k) - I_{out}(k) = I_{unit} \cdot k - \frac{I_{unit}}{1 + \rho \cdot k} \cdot k = \frac{\rho \cdot k^2}{1 + \rho \cdot k} \cdot I_{unit} \quad (12\text{-}16)$$

where $I_{load}(k)$ is the actual current through the load resistance and $I_{out}(k)$ is the desired output current (the sum of the currents from the current sources). The expression in (12-16) can be compared to the ideal continuous-time current, i.e., a ramp. The error power introduced is given by the quantization noise power and the time averaged power from (12-16).

$$P_n = P_{qn} + P_\varepsilon = \frac{I_{unit}^2}{12} + \overline{[\Delta I(k)]^2} \quad (12\text{-}17)$$

where

$$P_{qn} = \frac{I_{unit}^2}{12} \text{ and } P_\varepsilon = \overline{[\Delta I(k)]^2} \quad (12\text{-}18)$$

Assume that the input signal is a sinusoidal signal

$$k = K_{dc} + K_{ac} \cdot \sin\theta + \nu \quad (12\text{-}19)$$

where K_{dc} is the DC level of the signal, K_{ac} is the amplitude of the sinusoid, θ is the normalized signal frequency times the sequence index, and ν corresponds to the quantization error, which is AWGN for converters with a larger number of bits. The AC power of the sinusoid at the output is given by

$$P_s = \frac{K_{ac}^2}{2} \cdot I_{unit}^2 \quad (12\text{-}20)$$

The average error power P_ε is found by approximating (12-16) by using the first order Taylor expansion and the average value of $k = K_{dc}$ from (12-19), i.e.,

$$P_\varepsilon \approx \{(\rho \cdot K_{dc}^2) \cdot I_{unit}\}^2 \quad (12\text{-}21)$$

Now we can find the signal-to-noise-and-distortion ratio (*SNDR*) as a function of the output impedance, hence

$$SNDR = \frac{P_s}{P_n} = \frac{P_s/P_{qn}}{1 + P_\varepsilon/P_{qn}} \approx \frac{6K_{ac}^2}{1 + \{12 \cdot \rho \cdot K_{dc}^2\}^2} \qquad (12\text{-}22)$$

In Fig. 12-5 we show the simulated SNDRs of 10-bit, 12-bit, and 14-bit DACs when

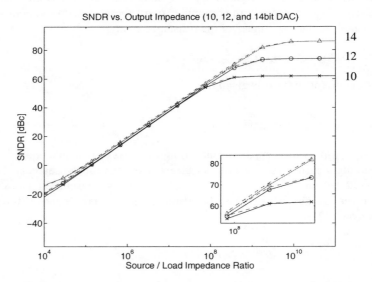

Figure 12-5 Simulated and calculated SNDR vs. resistance ratio for 10-bit, 12-bit, and 14-bit DACs. To illustrate the similarity with the calcualted, a portion of the curves is magnified.

varying the resistance ratio, R_{ratio}. The input signal is a full-scale sinusoid. In the simulation we use $R_{load} = 50\ \Omega$ and $I_{unit} = 1.25\ \mu A$. The *SNDR* results is compared with the calculated results from (12-22) and the formula is verified. In the figure we see that the curves are saturated at both lower and higher resistance ratios. The lower saturation arises in the simulator (Matlab) since it becomes difficult to separate signal power from noise power at such low resistance ratios. The higher saturation arises due to the fact that the quantization noise is dominating at higher ratios and the *SNDR* is given by $SNDR \approx 6.02 \cdot N + 1.76$ dB.

12.3.3 SFDR vs. Output Resistance

The size of the distortion tone in Fig. 12-3 can be found by considering the case of a single sinusoid input as given in (12-19). Then the output current as given by (12-11) becomes

$$I_{load}(k) = \frac{I_{unit}}{\rho} \cdot \left[1 - \frac{1}{1 + \rho \cdot (K_{dc} + K_{ac} \cdot \sin\theta)}\right] \qquad (12\text{-}23)$$

where the noise term ν has been neglected. The equation is rewritten as

12.3 Output Resistance Variations

$$I_{load}(k) = \frac{I_{unit}}{\rho} \cdot \left(1 - \frac{1}{1 + \rho \cdot K_{dc} + \rho \cdot K_{ac} \cdot \sin\theta}\right)$$

$$= \frac{I_{unit}}{\rho} \cdot \left(1 - \frac{1}{1 + \rho \cdot K_{dc}} \cdot \frac{1}{1 + \frac{\rho \cdot K_{ac}}{1 + \rho \cdot K_{dc}} \cdot \sin\theta}\right) \quad (12\text{-}24)$$

Examining (12-24) we find that only the second term within paranthesis contains AC frequency information, and we have

$$I_{load}(k) = \frac{I_{unit}}{\rho} \cdot \left(1 - \frac{1}{1 + \rho \cdot K_{dc}} \cdot I_{ac}(k)\right) \quad (12\text{-}25)$$

The gain factors inside and outside the parantheses can be neglected since we are considering only power *ratios* when determining the *SFDR*, and the AC signal is

$$I_{ac}(X) = \frac{1}{1 + A \cdot \sin\theta} \quad (12\text{-}26)$$

where

$$A = \frac{\rho \cdot K_{ac}}{1 + \rho \cdot K_{dc}} \quad (12\text{-}27)$$

Comparing (12-24) with (12-25) it is clear that $I_{load}(k)$ and $I_{ac}(k)$ will have the same distortion, since they only differ in offset and constant gain.

To avoid signal clipping we have that $K_{ac} < K_{dc}$ for a binary offset code, and we have

$$0 \leq A < 1 \quad (12\text{-}28)$$

and obviously that

$$|A \cdot \sin\theta| < 1 \quad (12\text{-}29)$$

This implies that we may find a converging Taylor series expansion of $I_{ac}(k)$ from (12-26)

$$I_{ac}(k) = \sum_{n=0}^{\infty} (-A \cdot \sin\theta)^n$$

$$= 1 - A\sin\theta + \sum_{n=1}^{\infty} (A\sin\theta)^{2n} - \sum_{n=1}^{\infty} (A\sin\theta)^{2n+1} \quad (12\text{-}30)$$

The DC level and the gain can once again be neglected. By using trigonometric formulas, we find that (12-30) may be written as

$$I_{ac}(k) = f(k) - \sin\theta \cdot A \cdot \left[1 + \sum_{n=1}^{\infty} \left(\frac{A}{2}\right)^{2n}\binom{2n+1}{n}\right] - \ldots$$

$$\ldots - \cos 2\theta \cdot 2 \sum_{n=1}^{\infty} \left(\frac{A}{2}\right)^{2n}\binom{2n}{n-1} \tag{12-31}$$

where $f(k)$ contains the DC component and higher order harmonics that do not influence the *SFDR*. The *SFDR* is now found in (12-31) as the power ratio between the fundamental and the 2nd harmonic as

$$SFDR = \frac{A^2}{4} \cdot \frac{\left[1 + \sum_{n=1}^{\infty}\left(\frac{A}{2}\right)^{2n}\binom{2n+1}{n}\right]^2}{\left[\sum_{n=1}^{\infty}\left(\frac{A}{2}\right)^{2n}\binom{2n}{n-1}\right]} = \left(\frac{1 + \sqrt{1-A^2}}{A}\right)^2 \tag{12-32}$$

By substituting back A from (12-27) in (12-32) this becomes

$$SFDR = \left[\frac{1 + \rho \cdot K_{dc}}{\rho \cdot K_{ac}} + \sqrt{\left(\frac{1 + \rho \cdot K_{dc}}{\rho \cdot K_{ac}}\right)^2 - 1}\right]^2 \tag{12-33}$$

which may also be written as

$$SFDR = \left[\frac{R_{ratio} + K_{dc}}{K_{ac}} + \sqrt{\left(\frac{R_{ratio} + K_{dc}}{K_{ac}}\right)^2 - 1}\right]^2 \tag{12-34}$$

where $R_{ratio} = 1/\rho$. When R_{ratio} is very large, we have that (12-34) can be approximated

$$SFDR \approx \left[2 \cdot \frac{R_{ratio} + K_{dc}}{K_{ac}}\right]^2 \tag{12-35}$$

The result in (12-35) is important. If $R_{ratio} \gg K_{dc}$ we have that the *SFDR* is not strongly dependent on the DC level. On the other hand we see that the *SFDR* is strongly dependent on the AC amplitude. For the case with equal amplitude and DC level, i.e.,

$$K_{ac} = K_{dc} \tag{12-36}$$

we have that

$$SFDR = \left[1 + \frac{R_{ratio}}{K_{ac}} \cdot \left(1 + \sqrt{1 + \frac{2 \cdot K_{ac}}{R_{ratio}}}\right)\right]^2 \tag{12-37}$$

When R_{ratio} is large (12-37) becomes approximately

12.3 Output Resistance Variations

$$SFDR \approx 20\log R_{ratio} - 20\log \frac{K_{ac}}{2} \text{ dBc} \qquad (12\text{-}38)$$

We see that by decreasing the AC amplitude, X_{ac}, of the signal we get an improvement of the linearity of the circuit. In the special case with a full-scale sinusoid, i.e.,

$$K_{ac} = K_{dc} = \frac{2^N - 1}{2} \approx 2^{N-1} \qquad (12\text{-}39)$$

we get

$$SFDR = \left(1 + \frac{R_{ratio}}{2^{N-1}} \cdot \left[1 + \sqrt{1 + \frac{2^N}{R_{ratio}}}\right]\right)^2 \approx \left(2 \cdot \frac{R_{ratio} + 2^{N-1}}{2^{N-1}}\right)^2 \qquad (12\text{-}40)$$

The result found in (12-40) is based on that the second harmonic is dominating the distortion [9, 10, 11]. To illustrate this, we show in Fig. 12-6 the simulated output spec-

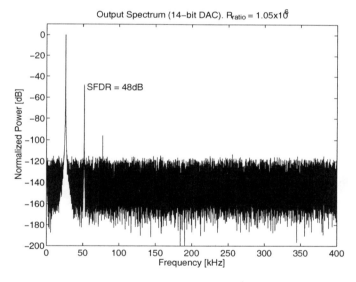

Figure 12-6 Output spectrum when applying a full-scale sinusoid on a 14-bit DAC. The resistance ratio is $R_{ratio} = 1.05 \cdot 10^6$.

trum from a 14-bit DAC when applying a full-scale sinusoid and having a resistance ratio of $R_{ratio} = 1.05 \times 10^6$. If $R_{ratio} \gg 2^N$, (12-40) can be approximated

$$SFDR \approx 20\log R_{ratio} - 6(N-2) \text{ dBc} \qquad (12\text{-}41)$$

From (12-41) we realize that with a doubling of the load resistance, the *SFDR* is decreased by 6 dB. With a maintained resistance ratio, the linearity will also deteriorate with an increased nominal number of bits.

In Fig. 12-7 we show the simulated and calculated *SFDR* vs. resistance ratio for a 10-

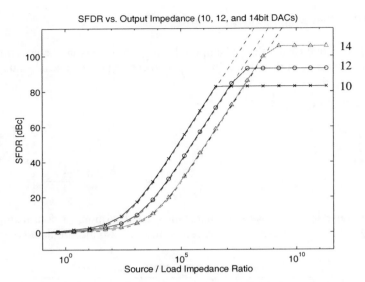

Figure 12-7 Simulated (solid) and calculated (dashed) SFDR vs. output over load resistance ratio for 10-bit (x), 12-bit (o), and 14-bit (triangle) DACs.

12-, and 14-bit DAC. At high ratios the simulated values are saturated since the spuriouses are hidden in the noise floor and at low ratios the distortion becomes very large (compare Fig. 12-5). It is seen that the simulated *SFDR* follows the mathematical result well.

When using differential signals, harmonics of even order will be cancelled and the 3rd harmonic dominates. It can be found [9, 10] that the distortion with respect to the 3rd harmonic is

$$SFDR = \left[1 - 2 \cdot \frac{K_{dc} + R_{ratio}}{K_{ac}} \cdot \left(\frac{K_{dc} + R_{ratio}}{K_{ac}} + \sqrt{\left(\frac{K_{dc} + R_{ratio}}{K_{ac}}\right)^2 - 1}\right)\right]^2 \quad (12\text{-}42)$$

For high ratios, this is approximately

$$SFDR \approx \left(2 \cdot \frac{K_{dc} + R_{ratio}}{K_{ac}}\right)^4 \quad (12\text{-}43)$$

For $K_{dc} = K_{ac} \approx 2^{N-1}$ and $R_{ratio} \gg K_{ac}$ we get

$$SFDR \approx 40 \log R_{ratio} - 12(N-2) \text{ dBc} \quad (12\text{-}44)$$

From (12-44) we see that the *SFDR* with respect to the third harmonic is increasing faster with respect to increase of a nominal number of bits, N, than for the *SFDR* with respect to the second harmonic (12-41). However, it is obvious that if differential signals are used, the requirement on the output impedance can be relaxed.

12.3.4 Influence of Parasitic Resistance

The models in the previous sections only considered the load resistance. The resistance of internal wires and switches has not been considered. However, these may be incorporated as well, see the modified current source in Fig. 12-8 a). For the i-th LSB

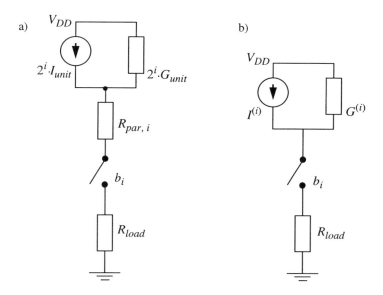

Figure 12-8 Model of the i-th LSB current source at with (a) parasitic resistance, $R_{par,i}$, from switches and internal wires and (b) modified model.

current source, we also include the parasitic resistance, $R_{par,i}$, which is associated with the interconnection wires from the current source to the output and the switch-on resistance of the switch for the bit position. We see that the switch-on resistance of the switch will introduce larger errors for the MSB current sources, since the output impedance is much smaller than for the LSB current sources.

Using Norton's theorem, we can transform the circuit to be similar to the current source as illustrated in Fig. 12-8b). Using that schematic view, we have a modified current source with the value

$$I^{(i)} = \frac{2^i \cdot I_{unit}}{1 + 2^i \cdot G_{unit} \cdot R_{par,i}} \tag{12-45}$$

and a modified output conductance

$$G^{(i)} = \frac{2^i \cdot G_{unit}}{1 + 2^i \cdot G_{unit} \cdot R_{par,i}} \tag{12-46}$$

Using superposition with the current sources of other bit positions and using the motivation leading to (12-6), we have that the total output current is

$$I_{load}(k) = \frac{\sum_i I^{(i)} \cdot b_i}{1 + R_{load} \cdot \sum_i G^{(i)} \cdot b_i} = \frac{\sum_i \dfrac{2^i \cdot I_{unit} \cdot b_i}{1 + 2^i \cdot G_{unit} \cdot R_{par,i}}}{1 + \sum_i \dfrac{2^i \cdot G_{unit} \cdot b_i \cdot R_{load}}{1 + 2^i \cdot G_{unit} \cdot R_{par,i}}} \qquad (12\text{-}47)$$

Comparing (12-47) with (12-6) we see that for each bit b_i, a bit weight of

$$w_i = \frac{2^i}{1 + 2^i \cdot G_{unit} \cdot R_{par,i}} \qquad (12\text{-}48)$$

is associated instead of the desired 2^i. For large i, hence for the MSBs, the weight is decreasing, hence a non-linear transfer function which introduce distortion. Assume that we are able to design the current switches to have a parasitic resistance which is exponentially lower for the MSB and higher for the LSB, as

$$R_{par,i} = \frac{R_{par,LSB}}{2^i} = \frac{R_{par}}{2^i} \qquad (12\text{-}49)$$

where $R_{par} = R_{par,LSB}$ is the parasitic resistance associated with the LSB current source. The bit weight, according to (12-48), turns into

$$w_i = \frac{2^i}{1 + G_{unit} \cdot R_{par}} \qquad (12\text{-}50)$$

which only gives a gain error, since using these weights, (12-47) turns into

$$I_{load}(k) = \frac{\sum_i \dfrac{2^i \cdot I_{unit} \cdot b_i}{1 + G_{unit} \cdot R_{par}}}{1 + \sum_i \dfrac{2^i \cdot G_{unit} \cdot b_i \cdot R_{load}}{1 + G_{unit} \cdot R_{par}}} =$$

$$= \frac{1}{1 + G_{unit} \cdot R_{par}} \cdot \frac{I_{unit} \cdot k}{1 + \dfrac{G_{unit} \cdot R_{load}}{1 + G_{unit} \cdot R_{par}} \cdot k} \qquad (12\text{-}51)$$

Comparing with (12-10) we see that (12-51) contains an additional gain factor and a *parasitic conductance ratio* as

$$\rho_p = \frac{G_{unit} \cdot R_{load}}{1 + G_{unit} \cdot R_{par}} \qquad (12\text{-}52)$$

We also define the parasitic resistance ratio as

$$R_{p,ratio} = \frac{1}{\rho_p} = \frac{1 + G_{unit} \cdot R_{par}}{G_{unit} \cdot R_{load}} = \frac{R_{unit} + R_{par}}{R_{load}} \qquad (12\text{-}53)$$

From (12-51) we see that for a high parasitic resistance, there is less current directed to the output, and this gain factor itself will not influence the SFDR and SNDR, but at the same time the parasitic conductance ratio becomes smaller and the linearity is improved.

The modified ratios, (12-52) and (12-53), can be used to find the approximate SNDR and SFDR in the formulas (12-22), (12-34), and (12-44).

12.4 CURRENT SOURCE MISMATCH

A current source with a mismatch error can be modeled as an additional current source in parallel with the nominal current source, as shown in Fig. 12-9 for the current

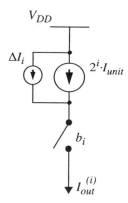

Figure 12-9 Modeling of current source with error current source, ΔI_i.

source corresponding to bit i. The actual output is then given by

$$I_{out}^{(i)} = b_i \cdot (2^i \cdot I_{unit} + \Delta I_i) \qquad (12\text{-}54)$$

where an infinite output impedance is assumed. All individual error sources can be summed and modeled as one total error current source connected to the output. The output current delivered to the load, $I_{load}(k)$, can be written as the sum of the nominal output current, $I_{out}(k)$, and the error current, $\Delta I(k)$

$$I_{load}(k) = I_{out}(k) + \Delta I(k) \qquad (12\text{-}55)$$

where k is the digital number given by (12-1). Considering the individual bit positions, we can rewrite (12-55) as

$$I_{load}(k) = I_{out}(k) + \sum_{i=0}^{N-1} \Delta I_i(b_i) \qquad (12\text{-}56)$$

where $\Delta I_i(b_i)$ is given by

$$\Delta I_i(b_i) = 2^i \cdot b_i \cdot \delta_i(b_i) \cdot I_{unit} \tag{12-57}$$

where $\delta_i(b_i)$ is the relative error current in the i-th bit. In the static case, we assume that $\delta_i(b_i) = \delta_i$ for all different inputs, k. This gives the total output current as

$$I_{load}(k) = I_{out}(k) + I_{unit} \cdot \sum_{i=0}^{N-1} 2^i \cdot \delta_i \cdot b_i$$
$$= I_{unit} \cdot \left[k + \sum_{i=0}^{N-1} 2^i \cdot \delta_i \cdot b_i \right] = I_{unit} \cdot \sum_{i=0}^{N-1} 2^i \cdot (1 + \delta_i) \cdot b_i \tag{12-58}$$

In reality, mismatch errors of transistors due to process variations include both graded linear and stochastic errors [5, 8, 12]. Gradients in oxide thickness [5] and along wires or voltage drops over supply lines [4, 13] create linear matching errors which are strongly dependent on the layout of the current sources. These type of errors will not be discussed here since it can be avoided by proper layout. Correlated matching errors are briefly discussed in Sec. 12.4.3, but interested readers can find more work in [14] on linearly distributed matching errors in DACs. In this text, we focus on stochastic errors. For one specific DAC the matching error associated with one current source is fixed, but comparing a large number of chips (of same design) the matching error will be gaussian distributed and more or less uncorrelated with other the matching error in other chips.

For a CMOS current source, the β- and V_T-mismatch can be characterized by their distribution, transistor sizes, and physical mutual distance [8, 12]. This is further addressed in Chapter 13.

$E_\delta\{A\}$ is used to denote the expectation value of A with respect to the stochastic variable δ, and $\overline{E_\delta\{k\}}$ denotes the average value with respect to the input signal k, as

$$\overline{E_\delta\{k\}} = \lim_{P \to \infty} \frac{1}{P} \sum_{p=0}^{P-1} I(k(pT)) \tag{12-59}$$

First, we assume that the matching errors in the different unit current sources are uncorrelated. This is a very coarse assumption, and in reality, transistors close to each other will have highly correlated errors. With the unit current source $I_{unit}^{(m)}$, $m = 0, \dots, 2^N - 1$, we associate a relative error

$$\delta_{unit}^{(m)} = \delta_{unit} \tag{12-60}$$

Its mean value and standard deviation are

$$E_\delta\{\delta_{unit}^{(m)}\} = \mu_{unit}^{(m)} = \mu_{unit} = 0 \text{ and } E_\delta\{[\delta_{unit}^{(m)}]^2\} = \sigma_{unit}^{(m)} = \sigma_{unit} \tag{12-61}$$

A non-zero mean value will only give rise to a DC error in the output and can be neglected in these calculations. The expectation value, with respect to the matching error, of the output current from the i-th bit, which consists of 2^i current sources in

12.4 Current Source Mismatch

parallel, follows from (12-61)

$$E_\delta\{I_{out}^{(i)}\} = E_\delta\left\{\sum_{m=1}^{2^i} I_{unit} \cdot (1 + \delta_{unit}^{(m)})\right\} = \ldots$$

$$\ldots = 2^i \cdot I_{unit} + I_{unit} \cdot \sum_{m=1}^{2^i} E_\delta\{\delta_{unit}^{(m)}\} = 2^i \cdot I_{unit} \quad (12\text{-}62)$$

The variance of the same current is given by

$$E_\delta\{[I_{out}^{(i)} - E_\delta\{I_{out}^{(i)}\}]^2\} = E_\delta\left\{\left[\sum_{m=1}^{2^i} I_{unit} \cdot \delta_{unit}^{(m)}\right]^2\right\} = \ldots$$

$$\ldots = I_{unit}^2 \cdot \sum_{m=1}^{2^i} E_\delta\{[\delta_{unit}^{(m)}]^2\} = I_{unit}^2 \cdot 2^i \cdot \sigma_{unit}^2 \quad (12\text{-}63)$$

From (12-62) and (12-63) we find the mean value and standard deviation for the normalized error δ_i in the current source corresponding to the i-th bit as

$$\mu_i = 0 \text{ and } \sigma_i = \frac{\sqrt{I_{unit}^2 \cdot 2^i \cdot \sigma_{unit}^2}}{I_{unit} \cdot 2^i} = \frac{\sigma_{unit}}{\sqrt{2^i}} \quad (12\text{-}64)$$

12.4.1 SNDR vs. Mismatch

Since the mismatch errors are assumed to be uncorrelated, the expectation value of the power is given by

$$E_\delta\{I_{load}^2(k)\} = E_\delta\{I_{out}^2(k)\} + E_\delta\left\{\left(I_{unit} \cdot \sum_{i=0}^{N-1} b_i \cdot 2^i \cdot \delta_i\right)^2\right\} =$$

$$= I_{out}^2(k) + I_{unit}^2 \cdot \sum_{i=0}^{N-1} b_i^2 \cdot 2^{2i} \cdot E_\delta\{\delta_i^2\} \quad (12\text{-}65)$$

For the mismatch error we have from (12-64)

$$E_\delta\{\delta_i^2\} = \sigma_i^2 = \frac{\sigma_{unit}^2}{2^i} \quad (12\text{-}66)$$

Since $b_i \in \{0, 1\}$, we have that $b_i = b_i^2$. Using (12-66) in (12-65) we get

$$E_\delta\{I_{load}^2(k)\} = I_{out}^2(k) + I_{unit}^2 \cdot \sum_{i=0}^{N-1} b_i \cdot 2^i \cdot \sigma_{unit}^2 =$$

$$= I_{unit}^2 \cdot k^2 + I_{unit}^2 \cdot \sigma_{unit}^2 \cdot k \quad (12\text{-}67)$$

which actually can be derived directly from the result in (12-63). We can now find the code-averaged power value for the expression in (12-67) (also compare with Fig. 12-5) and we have that

$$\overline{E_\delta\{I_{load}^2(k)\}} = I_{unit}^2 \cdot \overline{k^2} + I_{unit}^2 \cdot \sigma_{unit}^2 \cdot K_{dc} \qquad (12\text{-}68)$$

where $K_{dc} = \overline{k}$ is the average value of the input code, i.e., the DC value. We have the error power given by

$$P_\varepsilon = I_{unit}^2 \cdot \sigma_{unit}^2 \cdot K_{dc} \qquad (12\text{-}69)$$

and we see that by reducing the DC value of the signal, we also reduce the error power, however the DC value is moslty fixed to $K_{dc} \approx 2^{N-1}$. By using (12-18), (12-20), (12-22), and (12-69), we can calculate the SNDR as

$$SNDR = \frac{P_s}{P_{qn} + P_\varepsilon} = \frac{P_s/P_{qn}}{1 + P_\varepsilon/P_{qn}} = \frac{6 \cdot K_{ac}^2}{1 + 12\sigma_{unit}^2 \cdot K_{dc}} \qquad (12\text{-}70)$$

With a full-scale sinusoid we have from (12-39) that the *SNDR* (12-70) becomes

$$SNDR = \frac{(3/2) \cdot 2^{2N}}{1 + \sigma_{unit}^2 \cdot 2^N \cdot 6} \qquad (12\text{-}71)$$

In dB we have that

$$SNDR \approx 6N + 1.76 - 10\log(1 + 6 \cdot \sigma_{unit}^2 \cdot 2^N) \qquad (12\text{-}72)$$

In Fig. 12-10 we show the simulated and calculated *SNDR* vs. mismatch for a 10-, 12, and 14-bit DAC. The results are found by taking the average value of 1024 simulations for each mismatch value. We find that the simulated values match the calculated ones well.

12.4.2 SFDR vs. Mismatch

To find the *SFDR*, we have to investigate how the mismatch error power is distributed in the frequency domain. Consider the Fourier series coefficients, $C_{p,i}$, for bit b_i

$$C_{p,i} = \frac{1}{M} \sum_{m=0}^{M-1} b_i(m) \cdot e^{-2\pi j \cdot \frac{m}{M} \cdot (p+1)} \qquad (12\text{-}73)$$

where m is the sequence index and M is the period in number of samples. The coefficients for the total error current, I_{Cp}, are given by using (12-58) and (12-73)

$$I_{Cp} = I_{unit} \cdot \sum_{i=0}^{N-1} 2^i \cdot \delta_i \cdot C_{p,i} \qquad (12\text{-}74)$$

The power of each tone from (12-74) is given by

12.4 Current Source Mismatch

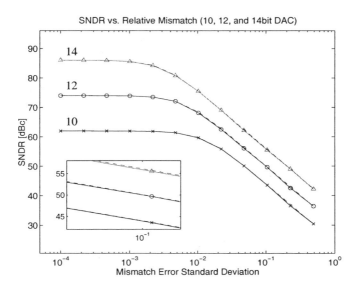

Figure 12-10 Calculated and simulated *SNDR* vs. mismatch error standard deviation for a 10-, 12-, and 14-bit DAC.

$$P_p = 2 \cdot |I_{Cp}|^2 \tag{12-75}$$

Since all mismatch errors are uncorrelated, we have that the expectation value of the power of the p-th tone is

$$E_\delta\{P_p\} = E_\delta\{2 \cdot |I_{Cp}|^2\} = 2I_{unit}^2 \cdot \sigma_{unit}^2 \cdot \sum_{i=0}^{N-1} 2^i \cdot |C_{p,i}|^2 \tag{12-76}$$

For a full-scale single-tone sinusoid the AC power is

$$P_s = \frac{1}{8} \cdot I_{unit}^2 \cdot 2^{2N} = 2^{2N-3} \cdot I_{unit}^2 \tag{12-77}$$

The harmonic distortion with respect to the p-th harmonic is approximately

$$HD_p = \frac{P_s}{P_p} = \frac{2^{2N-3} \cdot I_{unit}^2}{2I_{unit}^2 \cdot \sigma_{unit}^2 \cdot \sum_{i=0}^{N-1} 2^i \cdot |C_{p,i}|^2} \approx \frac{2^{2N}}{16\sigma_{unit}^2} \cdot \left[\sum_{i=0}^{N-1} 2^i \cdot |C_{p,i}|^2\right]^{-1} \tag{12-78}$$

For each bit we have the error given by a pulse-shaped waveform. Now, we assume that the matching error dominates in the MSBs (compare to (12-63)). If the input signal is a full-scale sinusoid, the MSB will have a pulse width of $M/2$ and a period of M and its Fourier series coefficients are given by

$$|C_{p,N-1}|^2 = \begin{cases} 0 & p \text{ is even} \\ \dfrac{M^2}{\sin^2(p\pi/M)} \approx \dfrac{1}{(p\pi)^2} & p \text{ is odd} \end{cases} \quad (12\text{-}79)$$

where M is the period. Since in the static case M is larger than $p\pi$ the approximation holds. For the lower significant bits, the frequency spectrum is more noise-like. To find the harmonic distortion we use the approximation in (12-79) to rewrite (12-78) as

$$HD_p \approx \frac{2^{2N}}{16 \cdot \sigma_{unit}^2} \cdot \frac{(p\pi)^2}{2^N - 1} \approx \frac{2^N}{16 \cdot \sigma_{unit}^2} \cdot (p\pi)^2 \quad (12\text{-}80)$$

In dBc we find that

$$HD_p \approx 20\log\frac{p\pi}{4} + 3N - 10\log\sigma_{unit}^2 \quad (12\text{-}81)$$

Thereby we also find the *SFDR* for the full-scale signal to be given by the minimum value of (12-81) ($p \geq 3$), i.e.,

$$SFDR \approx 20\log\frac{3\pi}{4} + 3N - 10\log\sigma_{unit}^2 \quad (12\text{-}82)$$

In Fig. 12-11 we show the average simulated and calculated *SFDR* for a 10-, 12-, and

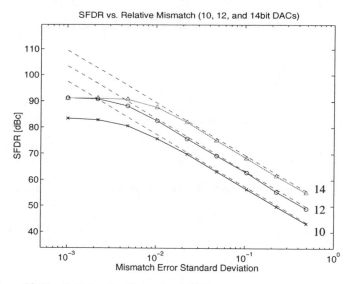

Figure 12-11 Calculated and simulated *SFDR* vs. mismatch for a 10-, 12-, and 14-bit DAC.

14-bit DAC. As we see there is a small error between the simulated and calculated values. This depends on the error in (12-80) where only the MSB has been considered. Since we are considering uncorrelated errors, the error power from all bits should be

12.4 Current Source Mismatch

added to the specific tone. This implies a slightly lower *SFDR*. The curves are saturated at low mismatch, since in the simulations the harmonics are hidden in the noise floor.

In Fig. 12-12 we show the typical output spectrum of a 14-bit DAC when applying

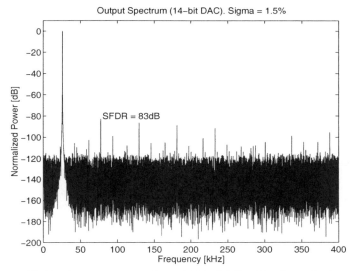

Figure 12-12 Output spectrum for a 14-bit DAC with mismatch size approximately 1.5%

matching errors with standard deviation $\sigma_{unit} \approx 1.5\%$. The SFDR is approximately 83 dBc for a full-scale sinusoid input.

12.4.3 Impact of Correlated and Graded Matching Errors

The matching errors are not independent Gaussian distributed variables. In reality, the matching errors in two adjacent current sources may be correlated [5, 8]. We may also have a linear graded distribution of the matching errors over the array of unit current sources [5, 8, 14].

Previously, we found that if all the matching errors for the unit sources were uncorrelated, we would for the i-th bit have a relative matching error with mean and standard deviation as in (12-64), hence

$$\mu_i = 0 \text{ and } \sigma_i = \frac{\sqrt{I_{unit}^2 \cdot 2^i \cdot \sigma_{unit}^2}}{I_{unit} \cdot 2^i} = \frac{\sigma_{unit}}{\sqrt{2^i}} \tag{12-83}$$

Let the mismatch error for the m-th unit source associated with the i-th LSB be denoted $\delta_i^{(m)}$. If the errors are correlated with each other, the expectation value of the output current for bit i is still $2^i \cdot I_{unit}$. The expectation value of the squared output error current becomes

$$E_\delta\{\Delta I_i^2\} = E_\delta\left\{\left[I_{unit} \cdot \sum_{m=1}^{2^i} \delta_i^{(m)}\right]^2\right\} = I_{unit}^2 \cdot \sum_{n=1}^{2^i} \sum_{m=1}^{2^i} E_\delta\{\delta_i^{(m)} \cdot \delta_i^{(n)}\} \tag{12-84}$$

For the i-th bit we denote the covariance or correlation between two unit current sources as

$$r_p = E_\delta\{\delta_i^{(m)} \cdot \delta_i^{(m+p)}\} = E_\delta\{\delta_i^{(m)} \cdot \delta_i^{(m-p)}\}, \text{ and } r_{-p} = r_p \tag{12-85}$$

Using (12-85) in (12-84) we get

$$\begin{aligned} E_\delta\{\Delta I_i^2\} &= I_{unit}^2 \cdot \sum_{n=1}^{2^i} \sum_{m=1}^{2^i} r_{m-n} \\ &= I_{unit}^2 \cdot 2^i \cdot \sigma_{unit}^2 + 2I_{unit}^2 \cdot \sum_{m=1}^{2^i-1} (2^i - m) \cdot r_m \\ &= I_{unit}^2 \cdot 2^i \cdot \sigma_{unit}^2 \cdot \left[1 + \sum_{p=1}^{2^i-1} \frac{p}{2^i} \cdot \frac{2r_p}{\sigma_{unit}^2}\right] \end{aligned} \tag{12-86}$$

since $r_0 = \sigma_{unit}^2$. The squared value of the nominal current through the i-th binary weighted current sources is $I_i^2 = I_{unit}^2 \cdot 2^{2i}$ and we have that

$$I_i^2 + E_\delta\{\Delta I_i^2\} = I_{unit}^2 \cdot 2^{2i} \cdot \left[1 + \frac{\sigma_{unit}^2}{2^i} \cdot \left(1 + \sum_{p=1}^{2^i-1} \frac{p}{2^i} \cdot \frac{2r_p}{\sigma_{unit}^2}\right)\right] \tag{12-87}$$

From (12-87) we identify the standard deviation for relative matching error in the ith bit

$$\sigma_i^2 = \frac{\sigma_{unit}^2}{2^i} \cdot \left(1 + \frac{2}{2^i \cdot \sigma_{unit}^2} \cdot \sum_{p=1}^{2^i-1} p \cdot r_p\right) \tag{12-88}$$

If the correlation factors are all zero, we have the same result as in (12-64). If the correlation factors, r_m, are positive, the standard deviation σ_i is larger than the one used in the previous model. However, we need to know the corellation to be able to find an approximation for the result in (12-88). Naturally, there is a correlation of the matching errors between different bits as well and the correlation is strongly dependent on the layout style. More work on modeling and characterization of the influence of linearly graded matching errors on DAC performance is found in [5, 14, 16].

12.5 INFLUENCE OF CIRCUIT NOISE

A fundamental limit on resolution and performance is the noise found in the output signal [17]. The resolution of the converter is guaranteed as long as the circuit noise

12.5 Influence of Circuit Noise

is much less than the average quantization noise. For an ideal N-bit uniform DAC, the *SNR* (determined by the quantization noise) over the Nyquist band is

$$SNR \approx 6.02 \cdot N + 1.76 \text{ dB} \tag{12-89}$$

We model the noise similar to the case of a matching error as in Fig. 12-9, hence as an error current source in parallel with the unit current source. In this case, opposite to the matching error, the noise error is varying as function of time.

With the unit current source m an AWGN noise current, $i_{unit}^{(m)}$, is associated. Within a certain bandwidth, BW, its mean value and variance are given by

$$E\{i_{unit}^{(m)}\} = 0 \text{ and } E\{[i_{unit}^{(m)}]^2\} = \overline{i_{unit}^2} = \sigma_n^2 \tag{12-90}$$

The expectation of the noise power in the i-th LSB is $2^i \cdot \sigma_n^2$ if the noise in the individual noise sources is uncorrelated. The expectation of the normalized total output noise power, P_n, is given by the sum of all noise current sources

$$P_n = E\{(\sum i_{unit}^{(m)})^2\} = \sigma_n^2 \cdot \bar{k} \tag{12-91}$$

where $K_{dc} = \bar{k}$ is the mean value of the digital input, hence the number of sources that are in average connected to the output. The normalized signal power for a sinusoid is given by (12-20) and is

$$P_s \approx I_{unit}^2 \cdot \frac{K_{ac}^2}{2} \tag{12-92}$$

The SNR is found by comparing (12-91) and (12-92)

$$SNR = \frac{P_s}{P_n} = \frac{1}{2} \cdot \frac{I_{unit}^2 \cdot K_{ac}^2}{\sigma_n^2 \cdot \bar{k}} \tag{12-93}$$

Assuming that the input signal is full-scale sinusoid and that the mean value is $K_{dc} = \bar{k} = 2^{N-1}$, (12-91) turns to approximately

$$P_n = 2^{N-1} \cdot \sigma_n^2 \tag{12-94}$$

and with (12-77) we have that (12-93) becomes

$$SNR \approx \frac{1}{2} \cdot \frac{(2^N-1)^2}{2^{N-1}} \cdot \frac{I_{unit}^2}{\sigma_n^2} = 2^{N-2} \cdot \frac{I_{unit}^2}{\sigma_n^2} \tag{12-95}$$

The result in (12-95) holds if the circuit noise is dominating. We can also add the quantization noise to the formula and we have

$$SNR \approx \frac{P_s}{P_q + P_n} = \frac{I_{unit}^2 \cdot K_{ac}^2/2}{\frac{I_{unit}^2}{12} + 2^{N-1} \cdot \sigma_n^2} = \frac{K_{ac}^2}{\frac{1}{6} + 2^N \cdot (\sigma_n^2/I_{unit}^2)} \tag{12-96}$$

where we have assumed that the noise bandwidth is equal to the Nyquist range.

For a full-scale sinusoid and dominating circuit noise, the SNR in dB is

$$SNR \approx 3(N-2) + 20\log I_{unit} - 10\log \sigma_n^2 \quad \text{dB} \tag{12-97}$$

If the number of bits is increased by one the signal power is increased by a factor of 6 dB, if σ_n is fixed, but at the same time, the noise is increased by a factor 3 dB. This is the essence of (12-97).

For wideband CMOS DACs it is mostly the thermal noise from the MOS transistor that dominates [17]. First we will show an example on the typical noise for a CMOS transistor and current source. The corresponding transistor thermal noise power spectral density (PSD) is

$$S_i(f) \approx \frac{8}{3}kT \cdot g_m \tag{12-98}$$

where k is the Boltzmann constant, T is the absolute temperature, and g_m is the small-signal transconductance of the current source transistor. Since the load resistance most likely is much smaller than the output impedance of the current source, (12-98) holds for both single transistor and cascoded current sources The transconductance, g_m, is approximately [17]

$$g_m \approx \sqrt{\mu_0 C_{ox} \cdot S \cdot I_{unit}} \tag{12-99}$$

where μ_0 is the charge mobility, C_{ox} is the gate capacitance per area, and S is the transistor size aspect ratio. Using the noise and transconductance definitions in (12-98) and (12-99) and assuming that the thermal noise in the transistors dominates, the noise spectral density in the unit current source is

$$S_i(f) \approx \frac{8}{3}kT \cdot \sqrt{\mu_0 C_{ox} \cdot S \cdot I_{unit}} \tag{12-100}$$

If we assume a certain noise bandwidth of BW, the normalized total output noise power, $\overline{i_{unit}^2}$, from the unit current source is

$$\overline{i_{unit}^2} = S_i(f) \cdot BW = \frac{8}{3}kT \cdot BW \cdot \sqrt{\mu_0 C_{ox} \cdot S \cdot I_{unit}} \tag{12-101}$$

Substituting the values from (12-101) into (12-97) gives the approximate expression

$$SNR \approx 15\log I_{unit} + 3N - 5\log S - 10\log BW - 10\log\left(\frac{8}{3}kT \cdot \sqrt{\mu_0 C_{ox}}\right) \tag{12-102}$$

In Fig. 12-13 the simulated and calculated (from (12-97)) SNR for a 14-bit DAC vs. the current through the unit current source, I_{unit}, is shown. Data from a standard 0.6μm CMOS process was used. It can be seen from Fig. 12-13 that for high unit currents, the thermal noise is lower than the quantization noise. If the quantization noise is added to the calculated value, as in (12-96), the curves will fit. At lower currents the

12.6 Influence of Dynamic Behavior

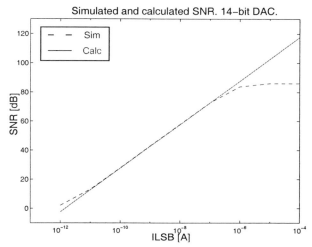

Figure 12-13 Simulated and calculated SNR vs. the I_{unit} current for a 14-bit DAC. Only thermal noise is simulated.

simulated curve saturates since the signal, due to the low *SNR*, cannot be separated from noise.

12.6 INFLUENCE OF DYNAMIC BEHAVIOR

In the previous sections, we have discussed how the output current is dependent on errors in static values. This implies that the derived formulas apply for low frequencies. At lower oversampling ratios – closer to the Nyquist range – the dynamic properties as for example glitching, and capacitive parts of the current sources, will influence performance. For high-speed DAC the performance is not only determined by the final values of each sample, but also determined by how each sample settles.

We will briefly discuss and model the influence of settling errors and the influence of bit skew and glitches.

12.6.1 Settling Error

Previously, we considered only the output resistance of the current source. In general, the output *impedance* of the current-steering DAC varies with the signal. Consider the modified model of the i-th LSB current source in Fig. 12-14. Parasitic capacitances have been added at the switch, to the output node of the current source, and at the load. The capacitive elements added to the current sources is further discussed in Chapter 13.

Analysing the dynamic behavior of the DAC, we can no longer only consider the settled values at the output. In this analysis we will consider the *linear* settling error. *Non-linear* settling errors, due to dominant signal-dependent capacitances, will introduce

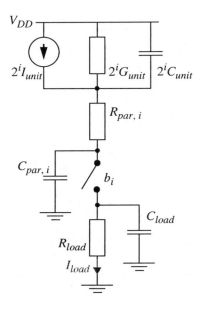

Figure 12-14 Current source in the ith LSB position with output conductance and capacitance as well as parasitics and load components.

distortion. This is not the case for linear settling errors, but this analysis gives a guidance for studies on how parasitic capacitances may influence performance.

We use a first order approximation, i.e., one dominant pole at the output. Let the start and end value errors be given by settling errors only. We can for example let the actual (or true) output current at the time instant $T_n = nT$ be denoted

$$\tilde{I}_{load}(T_n) \tag{12-103}$$

and the wanted output current through the load resistance at the same time instant is given by $I_{load}(T_n) = I_{unit} \cdot k(T_n)$. The absolute settling error found at the switching instant is given by

$$\varepsilon(T_n) = I_{load}(T_n) - \tilde{I}_{load}(T_n) \tag{12-104}$$

Let the DAC be approximated by a single-pole system with the time constant given by $\tau = RC$. At the transition from T_n to the next sampling instant, $T_n + T = T_{n+1}$, the output signal will settle towards a new end value. We have that

$$\tilde{I}_{load}(t) = \tilde{I}_{load}(T_n) + [I_{load}(T_{n+1}) - \tilde{I}_{load}(T_n)] \cdot \left(1 - e^{-\frac{t-T_n}{\tau}}\right) \tag{12-105}$$

for $T_n \le t \le T_{n+1}$. From the discussion above, we have that the output current at the sampling instant T_{n+1}

12.6 Influence of Dynamic Behavior

$$\tilde{I}_{load}(T_{n+1}) = \tilde{I}_{load}(T_n) + [I_{load}(T_{n+1}) - \tilde{I}_{load}(T_n)] \cdot (1 - e^{-T/\tau})$$
$$= \tilde{I}_{load}(T_n) \cdot e^{-T/\tau} + I_{load}(T_{n+1}) \cdot (1 - e^{-T/\tau})$$
(12-106)

Further using (12-106) in (12-104) we have the settling error given by

$$\varepsilon(T_{n+1}) = I_{load}(T_{n+1}) - [\tilde{I}_{load}(T_n) \cdot e^{-T/\tau} + I_{load}(T_{n+1}) \cdot (1 - e^{-T/\tau})]$$
$$= -[\tilde{I}_{load}(T_n) \cdot e^{-T/\tau} + I_{load}(T_{n+1}) \cdot (-e^{-T/\tau})]$$
(12-107)
$$= \varepsilon(T_n) \cdot e^{-T/\tau} + [I_{load}(T_{n+1}) - I_{load}(T_n)] \cdot e^{-T/\tau}$$

We can use the z-transform on the expression above, and we have that

$$\varepsilon(z) = z^{-1} \cdot \varepsilon(z) \cdot e^{-T/\tau} + [1 - z^{-1}] \cdot I_{load}(z) \cdot e^{-T/\tau}$$
(12-108)

This gives

$$\varepsilon(z) = \frac{1 - z^{-1}}{1 - z^{-1} \cdot e^{-T/\tau}} \cdot e^{-T/\tau} \cdot I_{load}(z)$$
(12-109)

If we use the transfer function, $H(z)$, describing the relation between the settling error and the load current, we have

$$\varepsilon(z) = H(z) \cdot I_{load}(z)$$
(12-110)

where

$$H(z) = \frac{1 - z^{-1}}{1 - z^{-1} \cdot e^{-T/\tau}} \cdot e^{-T/\tau}$$
(12-111)

If a sinusoid at the normalized frequency $\omega_0 T$ and with an amplitude $K_{ac} \cdot I_{unit}$ is fed into the DAC, the settling error signal will also be given by a sinusoid with the same frequency as

$$\varepsilon(nT) = |H(e^{j\omega_0 T})| \cdot I_{unit} \cdot K_{ac} \cdot \sin(\omega_0 T \cdot n + \arg H(e^{j\omega_0 T}))$$
(12-112)

where n is the sequence index and $z = e^{j\omega T}$ on the unity circle. The requirement is that the maximum settling error should be less than half an LSB, hence

$$\max|\varepsilon(nT)| = |H(e^{j\omega_0 T})| \cdot I_{unit} \cdot K_{ac} < \frac{1}{2} I_{unit}$$
(12-113)

This equation can be rewritten as

$$[|H(e^{j\omega_0 T})| \cdot K_{ac}]^2 < \frac{1}{4}$$
(12-114)

giving

$$|H(e^{j\omega_0 T})|^2 < \frac{1}{4 \cdot K_{ac}^2} \tag{12-115}$$

The largest settling error is found when a full-scale sinusoid is applied at the input, hence in the worst case we have

$$|H(e^{j\omega_0 T})|^2 < \frac{4}{4 \cdot (2^N - 1)^2} \approx 2^{-2N} \tag{12-116}$$

By using $z = e^{j\omega T}$ on the unit circle in (12-111) we have

$$|H(e^{j\omega T})|^2 = \left|\frac{e^{j\omega T} - 1}{e^{j\omega T} - e^{-T/\tau}}\right|^2 \cdot e^{-\frac{2T}{\tau}} = \left|\frac{e^{\frac{j\omega T}{2}} - e^{-\frac{j\omega T}{2}}}{e^{\frac{j\omega T}{2} + \frac{T}{2\tau}} - e^{-\frac{j\omega T}{2} - \frac{T}{2\tau}}}\right|^2 \cdot e^{-T/\tau}$$

$$= \frac{4 \cdot \sin^2\frac{\omega T}{2} \cdot e^{-T/\tau}}{4 \cdot \left|\sinh\left(\frac{j\omega T}{2} + \frac{T}{2\tau}\right)\right|^2} = \frac{\sin^2\frac{\omega T}{2} \cdot e^{-T/\tau}}{\sinh^2\frac{T}{2\tau}\cos^2\frac{\omega T}{2} + \cosh^2\frac{T}{2\tau}\sin^2\frac{\omega T}{2}} \tag{12-117}$$

In the equation we can identify two important parameters, the T/τ and the $\omega_0 T$ factors. The first factor expresses the time constant related to the update time and the latter factor expresses how the signal frequency is related to the update time. Setting (12-117) in (12-116) and reordering, we have that

$$\frac{1}{|H(e^{j\omega_0 T})|^2} = \cosh^2\frac{T}{2\tau} \cdot \left[1 + \frac{\tanh^2\frac{T}{2\tau}}{\tan^2\frac{\omega_0 T}{2}}\right] \cdot e^{T/\tau} > 2^{2N} \tag{12-118}$$

For a fixed value of T/τ the maximum value of the left-hand expression in the equation is found when $\omega_0 T = \pi$, hence the signal is placed at the Nyquist frequency. Therefore, we can instead investigate the worst-case as

$$\cosh^2\frac{T}{2\tau} \cdot e^{T/\tau} = \left[\frac{e^{T/2\tau} + e^{-T/2\tau}}{2}\right]^2 \cdot e^{T/\tau} \approx \frac{e^{2T/\tau}}{4} > 2^{2N} \tag{12-119}$$

where we assume (or require) that $T/2\tau > 1$. From (12-119) we get

$$\tau < \frac{T}{(N+1)\ln 2} \approx \frac{T}{0.69 \cdot (N+1)} \tag{12-120}$$

Note that we in the frequency domain also have the sinc weighting from sample&holds at the DAC output. This will add a factor $\pi/2$ to the left-hand expression in (12-119), however, if N is large, this will not strongly influence the final result in (12-120).

Some other complexities add to these result. First, if we consider the current source in Fig. 12-14 we find that the time constant of the DAC is likely to be signal dependent,

12.6 Influence of Dynamic Behavior 349

hence $\tau = \tau(k)$, which will introduce distortion. The RC time constant of the current source itself (withouth switches and interconnection) is constant, since when connecting a number of unit current sources in parallel, the output resistance decrease with the number of sources, but the capacitance increases with the number of sources. We have that

$$\tau_{unit} = \frac{C_{unit}}{G_{unit}} = \frac{2^i \cdot C_{unit}}{2^i \cdot G_{unit}} = \tau_i \qquad (12\text{-}121)$$

where τ_i is the time constant for the i-th LSB current source and τ_{unit} is the time constant associated with each unit current source.

12.6.2 Bit Skew and Glitches

As indicated in the previous section the settling time is crucial and a finite settling time introduces errors to the output signal. From (12-120) we have a limit on the possible settling time to achieve a full resolution bandwidth up to the Nyquist frequency. To meet this requirement we must also guarantee that the switching instant T_n for all bits is accurate. Errors will add voltage/current spikes to the output signal (i.e., glitches as discussed in Chapter 1) and the settling behavior will be affected in a non-linear way. For instance at the major code transition

011...11 → 100...00

we may have the following transition code for a short period of time

111...11

This generates a glitch of amplitude $2^{N-1} \cdot I_{unit}$. Naturally, in a good design the glitch energy – or glitch impulse – is kept as low as possible [7], by using segmentation and proper switching schemes [18]. The induction of a high current or charge also affects the settling since the start value at T_n (12-103) changes dramatically.

To model and investigate the behavior of glitches, we introduce a switching time uncertainty for the i-th LSB

$$\tau_i(k(T_n)) \qquad (12\text{-}122)$$

where $T_n = nT$ and T is the sampling period. Notice that this error is dependent on both the bit position i and the input code, k. In Fig. 12-15 we show how this skew can be characterized for the i-th LSB. The ideal (dashed) voltage pulse applied to the switch is compared with a linearized actual pulse (solid) also indicated with an offset. The switching activity for the i-th LSB is denoted

$$\Delta b_i(T_n) = b_i(T_n) - b_i(T_{n-1}) \qquad (12\text{-}123)$$

which expresses the number of switching bits at the time instant T_n. If the i-th LSB is switching, we will have the change in amplitude given by

$$2^i \cdot \Delta b_i(T_n) \qquad (12\text{-}124)$$

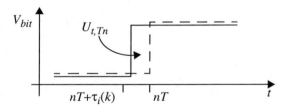

Figure 12-15 Modeling of the timing uncertainty. The ideal switching signal (- -) is compared with the actual signal (—).

and for all bits, we have

$$\sum_{i=0}^{N-1} 2^i \cdot \Delta b_i(T_n) = k(T_n) - k(T_{n-1}) = \Delta k(T_n) \quad (12\text{-}125)$$

As illustrated in Fig. 12-15 we let the glitch impulse be expressed by a squared-wave pulse with the duration $\tau_i(k)$. Dependent on the sign of τ_i, the pulse is given by

$$U_{\tau i, Tn}(t) = \begin{cases} 1 & T_n \leq t \leq T_n + \tau_i(k(T_n)) & \tau_i > 0 \\ 1 & T_n + \tau_i(k(T_n)) \leq t \leq T_n & \tau_i < 0 \\ 0 & \text{for } t \text{ outside these intervals} \end{cases} \quad (12\text{-}126)$$

An expression for the total glitch amplitude, $G(t)$, at the output is given by

$$G(t) = I_{unit} \cdot \sum_{n \geq 0} \sum_{i=0}^{N-1} U_{\tau i, Tn}(t) \cdot 2^i \cdot \Delta b_i(T_n) \quad (12\text{-}127)$$

This is a pulse-train, where the widths and amplitudes at the time instants nT depend on the unit current I_{unit}, the switching activity, and the time uncertainty τ. An approximation of (12-127) would be to assume that the pulses $U_{\tau i, Tn}$ are in-dependent on the bit position. Hence, we let the glitch at time instant nT be given by the amplitude $\Delta k(T_n) \cdot I_{unit}$ and an average width $\tau(T_n)$ as

$$G(t) = I_{unit} \cdot \sum_{n \geq 0} U_{\tau, Tn}(t) \cdot \Delta k(T_n) \quad (12\text{-}128)$$

which is an rough approximation. The timing errors may also be modeled as damped sine waves instead of pulse shaped errors [6]. The timing errors are typically dependent on the voltages across the switches and mismatch of the switch sizes.

12.7 SUMMARY

We have given an overview of different models for current-steering D/A converters.

12.7 Summary

The modeling should be used as a guidance in determining circuit structure, component sizes, etc., in current steering DACs. From the models, we can also calculate the effect of matching errors.

Using the results in this chapter, we are able to simulate and predict the performance of the DAC on a high abstraction level, with for example Matlab, which saves time. over circuit-level simulations. Mostly, static requirements have been addressed but in the same results can be modified to cover AC as well.

Extending the modeling, we can also use measured result from a DAC, and extract information on the process and the parasitic resistance and capacitance in the circuit.

REFERENCES

[1] P. Hendriks, "Specifying communication DACs," IEEE Spectrum, Vol. 34, No. 7, pp. 58-69, July 1997

[2] P. Hendriks, "Tips for Using High-Speed DACs in Communications Design", IEEE Electronic Design, no. 2, pp. 112-8, Jan. 1998

[3] E. Liu and A. Sangiovanni-Vincentelli, "Verification of Nyquist Data Converters Using Behavioral Simulation," IEEE Transactions on Computer-Aided Design of Integrated Circuits and Systems, Vol. 14, No. 4, April 1995

[4] T. Miki, Y. Nakamura, M. Nakaya, S. Asai, Y. Akasaka, Y. Horiba, "An 80 MHz 8-bit CMOS D/A Converter," IEEE Journal of Solid-State Circuits, Vol. 21, pp. 983-988, Dec. 1986.

[5] H.J. Schouwenaars, D. W. J. Groeneveld, and H. A. H. Termeer, "A Low-Power Stereo 16-bit CMOS D/A Converter for Digital Audio," IEEE Journal of Solid-State Circuits, Vol. 23, pp. 1290-1297, Dec. 1988.

[6] J. VandenBussche, G. Van der Plas, G. Gielen, M. Steyaert, W. Sansen, ",," in Proc. of IEEE 1998 Custom Integrated Circuits Conference, CICC'98, pp.473-6

[7] R.J. van de Plassche, Integrated Analog-to-Digital and Digital-to-Analog Converters, Boston: Kluwer Academic Publishers, 1994.

[8] M.J.M Pelgrom, A. C. J. Duinmaijer, and A. P. G. Welbers, "Matching Properties of MOS Transistors," IEEE Journal of Solid-State Circuits, vol. 24, no. 5, pp. 1433-9, Oct. 1989

[9] J.J. Wikner and N. Tan, "Influence of Circuit Imperfections on the Performance of Current-Steering DACs," Analog Integrated Circuits and Signal Processing, pp. 7-20, Jan. 1999

[10] J.J. Wikner, "Measurement and Simulations of a CMOS DAC Chipset," Internal Report, LiTH-ISY-R-2086, Linköping University, Sweden, Dec. 1998.

[11] J. J. Wikner, "Measurement and Simulations of a CMOS DAC Chipset," Internal Report, LiTH-ISY-R-2704, Linköping University, Dec. 1998

[12] J. Bastos, M. Steyaert, A. Pergoot, and W. Sansen, "Mismatch Characterization of Submicron MOS Transistors," Analog Integrated Circuits and Signal Processing, Vol. 12, pp. 95-106, 1997

[13] B.E. Jonsson, "Design of Power Supply Lines in High-Performance SI and Current-Mode Circuits," in Proc. of the 15th Norchip Conf., NORCHIP'97, pp. 245-50, Tallinn, Estonia, Nov. 10-11, 1997

[14] K.O. Andersson and J.J. Wikner, "Modeling of the Influence of Graded Element Matching Errors in CMOS Current-Steering DACs," in Proc. of the 17th NorChip Conf., NORCHIP'99, Oslo, Norway, Nov. 8-9, 1999.

[15] C.A.A. Bastiaansen, D. W. J. Groeneveld, H. J. Schouwenaars, and H. A. H. Termeer, "A 10-b 40-MHz 0.8-mm CMOS Current-Output D/A Converter," IEEE Journal of Solid-State Circuits, Vol. 26, No. 7, pp 917-921, July 1991

[16] J.J. Wikner and N. Tan, "Modeling of CMOS Digital-to-Analog Converters for Telecommunication," in *Proc. 1998 IEEE International Symposium on Circuits and Systems*, ISCAS'98, Monterey, California, USA, May, June 1998

[17] D.A. Johns and K. Martin, *Analog Integrated Circuit Design*, John Wiley & Sons, New York, NY, USA, 1997, ISBN 0-471-14448-7

[18] D. Mercer, "A 16-b D/A Converter with Increased Spurious Free Dynamic Range," *IEEE Journal of Solid-State Circuits*, vol. 29, pp. 1180-1185, Dec. 1994.

13 IMPLEMENTATION OF CMOS CURRENT-STEERING D/A CONVERTERS

13.1 INTRODUCTION

One of the most suitable candidates for high-speed and high-resolution is the current-steering DAC and in this chapter an overview of different common current-steering DAC topologies is given in Sec. 13.2. We highlight the advantages and disadvantages with the different structures. For some of the topologies there are inhereted good properties for high performance.

In the previous chapter we found that the DAC performance is strongly dependent on for example the finite output resistance of the current sources, R_{out}. The output resistance and circuit noise are dependent on the DC output current of the current source, I_{out}, and the DC current is dependent on the transistor size, etc. Matching is practically determined by the transistor size, aspect ratio W/L or the gate area WL. When increasing the output current with the ambition to improve the SNR, the output resistance decreases and thereby the non-linearity increase, etc. Therefore, we have a delicate relation between the three major error sources; noise, linearity and matching. The noise degrades the SNR and the non-linearity and mismatch degrade the SFDR. Usually trade-offs have to be done in order to meet the frequency-domain specifications.

Further the influence of clock feedthrough (CFT) in switches as well as their switch-on resistance are crucial for the dynamic performance. Also the influence of the interconnecting wires must not be neglected, etc. In Sec. 13.3 we discuss practical design issues for determining proper unit current sources, current switches, digital circuits, etc.

To further illustrate the design methodology, we present the design of a 1.5-V to 5-V, 10-bit to 14-bit wideband CMOS DAC chipset for xDSL and wideband radio applications in Sec. 13.4. We show design trade-offs and examples on how to implement the different circuit building blocks.

Measured and simulated results from the DAC chipset are presented and discussed in Sec. 13.5. We also compare measurement results from two similar DACs of the chipset. They have small variations in the design and layout of the current sources and interconnection wires. With a small variation in layout, an SFDR improvement of over 12 dB is achieved.

13.2 CURRENT-STEERING DAC TOPOLOGIES

The structure and operation of the binary-weighted current-steering DAC was outlined in Chapter 4 and Chapter 12 and it is recaptured in Fig. 13-1. There are N binary

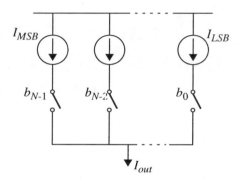

Figure 13-1 An N-bit current-steering DAC.

weighted current sources and N switches. The switches are dependent on the input code, $X_d = \{b_i\}_{i=0}^{N-1}$, and they determine which current source that should be directed to the output. The total current is terminated by an off-chip resistance or I/V converted (and filtered) using an output buffer.

As was derived in the previous chapter, the matching between current sources is crucial to the performance. Instead of using a current source with the nominal value I_{src} one should use M unit current sources in parallel, i.e., $I_{src} = M \cdot I_{unit}$. This improves the relative matching of the current sources by a factor of $\sqrt{M}$ if the matching errors are uncorrelated [1].

One should be aware of the influence of stochastic as well as linear graded matching errors (also referred to as short-distance and long-distance variations). Linear errors can often be approximately modeled by a plane [2]. Hence the size of the error is given by an expression as

$$\varepsilon(x, y) = g_x \cdot x + g_y \cdot y \qquad (13\text{-}1)$$

where x and y are the coordinates on the chip surface. g_x and g_y are the gradients in the respective direction and typically, they are in the same order of magnitude.

By choosing a special layout technique for the current sources, e.g., interdigitized or common-centroid, the converter can be made more or less sensitive to linear graded errors [2 - 6]. The stochastic error can be minimized by considering the results from Pelgrom [1] and the choice of the transistor sizes in the current sources is important.

To achieve high performance, we have to decide how to design and lay out the unit current sources and below we outline some different popular current-steering DAC structures.

13.2.1 Array Structure

The first, and perhaps simplest, approach is to follow Fig. 13-1 and use a "flat" array structure. The MSB current source is formed by placing 2^{N-1} parallel unit current sources in a row. The next MSB is formed in a similar way by placing 2^{N-2} sources in the next row, etc. Each row is interconnected and fed to a current switch as is illustrated in Fig. 13-2 for a single-ended structure.

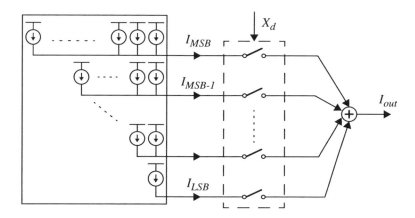

Figure 13-2 "Flat" array based layout of unit current sources.

It is obvious that the array becomes very wide for a large number of bits. The natural and obvious way to circumvent this is for the MSBs to use more than one row of current source and to use the same row for several LSBs. This also makes the converter less sensitive to linear graded mismatch errors [3].

The size of the current switch should be determined by the current flowing through it. For a large current we need a low switch-on resistance and vice versa. To make the array become more narrow, we can split the MSBs rows into several rows with less current in each row, and for each row we use one switch. Flat array structures are not suitable for resolutions over 6-8 bits, due to the large glitching and sensitivity to current source matching errors.

13.2.2 Segmented Structures

To reduce the glitching it is preferred to use a thermometer code representation. However, it is very difficult to use a full thermometer code for a high-resolution converter, since the number of switches and the size of the interconnection wires, etc., will be very large for resolutions above 7-8 bits.

The better choice is a DAC structure where the M most significant bits are thermometer coded and the $N-M$ least significant bits are binary weighted. This is referred to as a segmented structure and it is illustrated in Fig. 13-3. With the thermometer code we can use a number of equally large current sources, and hence the layout can be

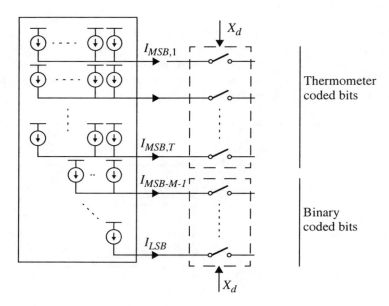

Figure 13-3 Segmented current source array. The M MSBs are thermometer coded which gives $T = 2^M - 1$.

more regular. The current switches are equally large for the thermometer coded bits and the current that flows through one switch will be smaller than for the unsegmented MSB.

The segmentation guarantees monotonicity for the MSBs and improves DNL since for an increase of the input, additional current sources are connected to the output. Matching of the thermometer coded current sources can further be improved by using dynamic randomization, averaging, or calibration techniques, as discussed in Chapter 4.

An extension to the segmented structure is to use multi-segmentation. For example, the M MSBs are thermometer coded in one cluster, the K LSBs are kept binary coded, and the $N - M - K$ intermediate bits are also thermometer coded in one separate cluster.

The key design issue in a segmented converter is to determine how many of the bits that should be thermometer coded. Roughly, the glitch energy is dependent on the number of bits that are switching in the input signal. By using a cost function, e.g., the power of the glitches or the weighted number of switching bits, we can examine how the power of the glitches vary with the number of thermometer coded bits M. Simulation results when applying a number of different signal types to a 14-bit DAC are shown in Fig. 13-4. It is found that about 4 to 6 thermometer coded bits give a large improvement in glitching performance. For more thermometer coded bits, the improvement is not that high, and more digital circuitry is also introduced, hence higher complexity, power consumption from digital circuits, and switching noise. It

13.2 Current-Steering DAC Topologies

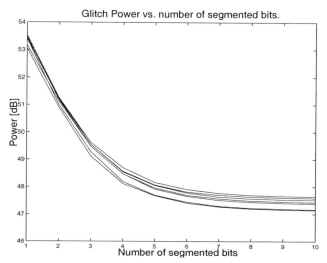

Figure 13-4 Power of glitches as a function of the number of segmented bits in a 14-bit DAC.

should be noted that in this simulation, the influence of the complexity of designing current sources with a high output resistance and switches with a low enough resistance have not been added. In the literature we find several 14-bit high-performance DACs with 6 or 7 thermometer coded bits or multi-segmented structures with similar resolutions [5, 6, 7].

13.2.3 Current Cell Matrix Structures

Another popular structure is the current cell matrix structure [8, 9, 10]. In this structure, as examplified in Fig. 13-5, the unit current sources (or sources of a higher significance if segmentation is used) are selected from a matrix using row and column decoding logic. The select signals determine if a specific current source should be turned on or off. In for example [8] each current cell (Fig. 13-5) uses three control signals, one column and two row signals.

To reduce the effect of linear graded matching errors, such as oxide thickness variations and voltage drops over supply wires [4, 11] the cells can be laid out in such way that these errors are averaged and minimized. With modifications of the row and column decoders, also dynamic randomization techniques can be used on the entire matrix to reduce distortion (see Sec. 4.10.1).

There are some drawbacks with the current cell matrix structure. Within the unit current cell a certain amount of digital logic is needed. This requires digital supply distribution within the cell. At the same time we need the three digital control signals, analog supply, analog bias voltages, and differential output currents, i.e., two more wires. This is a total of 9-10 wires dependening on the choice of current source. To decrease the amount of noise, we want to shield the analog wires from the digital as well as possible.

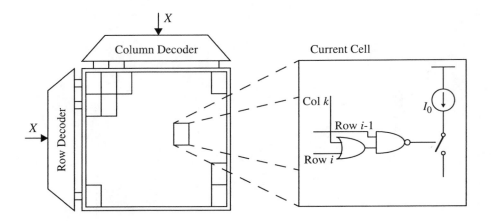

Figure 13-5 Unit current source matrix with decoding circuits.

For DAC structures with resolutions higher than 6-8 bits this technique is somewhat limited due to chip area and complexity. Hybrid DACs use this matrix approach for the MSBs and a binary weighted or thermometer coded structure for the LSBs [8, 9, 11].

13.3 PRACTICAL DESIGN CONSIDERATIONS

We present some practical design considerations concerning the design and layout of circuit elements, such as current source, switch, and digital circuits, in the current-steering DAC.

13.3.1 Implementation of Current Sources

Naturally, the unit current source to be used in the converter can be constructed in several different ways. In Fig. 13-6 we show three versions where a single PMOS transistor is used (a) together with one (b) or two cascode (c) PMOS transistors. It is the source-gate voltage applied on transistor M1 that practically sets the current through the current source. Since the ideal current source should have an infinite output impedance, the cascode transistors are used to increase the output impedance. The bulk connections of all transistors are connected to the positive supply. This decreases the gain of the cascodes, but the capacitive load at the internal nodes of the current source is reduced.

The output current, which typically is set by the DAC specification, is approximately given by the drain current of transistor M1 in the saturation region [12]

$$I_D \approx \frac{\beta}{2} \cdot (V_{SG} - V_T)^2 \cdot (1 + \lambda \cdot V_{SD}) \qquad (13\text{-}2)$$

13.3 Practical Design Considerations

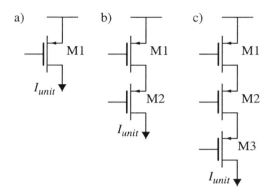

Figure 13-6 Schematic view of PMOS current sources using a) single transistor and b) single cascode and c) double cascode.

where β is the transconductance parameter, $V_{SG} = V_{DD} - V_{bias}$ is the source-gate voltage, V_T is the threshold voltage, λ is the channel length modulation factor, and V_{SD} is the source-drain voltage.

The output resistance of the current sources shown in Fig. 13-6 are given by R_a, R_b, and R_c, respectively, as

$$R_a \approx \frac{1}{\lambda_1 \cdot I_D}, \quad R_b \approx R_a \cdot \frac{(1+\eta_2)\sqrt{2\beta_2}}{\lambda_2 \cdot \sqrt{I_D}}, \quad \text{and} \quad R_c \approx R_b \cdot \frac{(1+\eta_3)\sqrt{2\beta_3}}{\lambda_3 \cdot \sqrt{I_D}} \qquad (13\text{-}3)$$

where I_D is the drain current (the output current of the source), λ is the channel length modulation, β_i is the transconductance parameter of transistor Mi, and η_i is a parameter given by the bulk-source transconductance of the transistors. The transconductance parameter is also given by

$$\beta = \mu_0 C_{ox} \cdot \frac{W}{L} \qquad (13\text{-}4)$$

where μ_0 is the charge mobility, C_{ox} is the gate capacitance per unit area, and W/L is the transistor aspect size ratio. The channel length modulation for transistor is approximately inversely proportional to the transistor channel length, hence $\lambda \sim 1/L$. This implies that, if the current is fixed by the specification, the output resistance is proportional to $\sqrt{W_2 L_2}$ and $\sqrt{W_3 L_3}$ and proportional to the length L_1 of transistor M1. Hence, with larger transistors we increase the output impedance. However, the capacitance which must be kept low, is proportional to the gate areas and the perimeters of the transistors. To illustrate this, we show in Fig. 13-7, the simulated output impedance of the different current source structures from Fig. 13-6. The supply voltage is 5 V and the DC output voltage is 0.5 V. Process parameters from a standard digital CMOS process have been used. The transistor sizes are S1 = 2u/8u, S2 = 2u/4u, S3 = 2u/1.2u. We find that the simulated output resistance (at DC) is approximately R_a = 10 MΩ, R_b = 1 GΩ, and R_c = 50 GΩ, for the respective source, with a current of 1.22 µA.

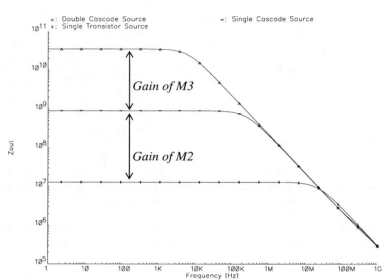

Figure 13-7   Simulated output impedance of the unit current source.

It can be seen from the figure above that the output resistance is increased with more cascodes, but the dominating pole is also lowered and at higher frequencies they basically have the same behavior. This will influence the DAC linearity at higher frequencies and is further addressed in Sec. 13.5.1. Using the results from the previous chapter, we have that for a specified SFDR the output resistance of the unit source is

$$R_{unit} \geq R_L \cdot 10^{\frac{6(N-2)+SFDR}{20}} \tag{13-5}$$

where R_L is the load resistance. With a 50-Ω load, the bound on the output resistance of a unit current source for an 80-dBc SFDR is

$$R_{unit} \approx 2 \text{ G}\Omega \tag{13-6}$$

This is a lower bound, since the influence of parasitic resistance in switches and interconnection wires should be added as well. Dependent on the process parameters, we may find the proper current source structure by using the result from (13-3) and (13-6) as well as the knowledge about the output pole as illustrated in Fig. 13-7.

Note that the bandwidth of the signal is most likely set by the output load due to its dominant capacitive part. The distortion is set by the ratio between the current sources' output impedance and the load impedance.

There are several sources for transistor matching errors, i.e., size errors, threshold voltage variations, supply and bias voltages variations, oxide thickness variations, out-

13.3 Practical Design Considerations

put voltage variations, etc. Differentiating (13-2) with respect to four of its parameters, we have that

$$\frac{\Delta I_D}{I_D} = \frac{\Delta \beta}{\beta} + 2 \cdot \frac{\Delta V_{SG} - \Delta V_T}{V_{SG} - V_T} + \frac{\Delta \lambda \cdot V_{SD} + \lambda \cdot \Delta V_{SD}}{1 + \lambda V_{SD}} \tag{13-7}$$

where ΔP expresses the variation of the parameter P. Assuming that the deviations of the source-gate and source-drain voltages, as well as the channel length modulation, are very small, we identify the so called β- and V_T-matching errors from (13-7)

$$\frac{\Delta \beta}{\beta} \text{ and } \frac{2 \Delta V_T}{V_{SG} - V_T}, \text{ respectively} \tag{13-8}$$

From studies in the literature [1, 13, 14], we know that the β- and V_T-matching can be characterized by their distribution. We have that

$$\sigma^2(\frac{\Delta I}{I}) \approx \sigma^2(\frac{\Delta \beta}{\beta}) + \frac{4 \cdot \sigma^2(\Delta V_T)}{(V_{SG} - V_T)^2} \tag{13-9}$$

where

$$\sigma^2(\frac{\Delta \beta}{\beta}) = \frac{A_\beta}{WL} + S_\beta^2 \cdot D_x \text{ and } \sigma^2(\Delta V_T) = \frac{A_{V_T}}{WL} + S_{V_T}^2 \cdot D_x \tag{13-10}$$

where A_β, A_{V_T}, S_β, and S_{V_T} are process dependent constants, WL is the gate area of the transistor, and D_x is the distance between the objects that should be matched. Hence, for good matching of two (or more) objects, we want them to be as close to eachother as possible and they should be as large as possible.

From a geometrical point of view the terms in (13-10) are strongly dependent on the layout style of the transistors. For large transistors, special layout styles, e.g., common-centroid or interdigitized layout, are used to reduce the influence of these kinds of mismatch. In an array of current sources as illustrated in Fig. 13-2, the matching errors between two neighboring unit current sources may still be strongly correlated and therefore the binary weighted current sources should be constructed by using unit current sources from different positions in the array [15]. This will however increase the wire resistance and capacitance since the interconnection wires becomes longer and the routing becomes more complex.

The output resistance is dependent on the bias voltage controlling the gate voltage of the current source (13-3). These voltages may be generated using current mirrors and the reference current through the primary side of the mirror is determined by a termination resistance. But, we have to guarantee proper stable and accurate bias and supply voltages to all the unit current sources [4, 11, 16]. In Fig. 13-8 we have illustrated the effects of voltage loss over the supply distribution wire, which has been modeled as a series of resistances. This will give rise to a voltage drop along the wire, and we have $V_{DD} > V_1 > ... > V_N$, which further influence the V_{SG}-matching of the transistors. At the nodes V_i the current sources are connected. If we assume that the supply wire

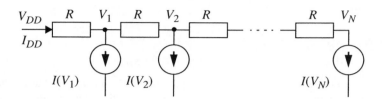

Figure 13-8 Model of the voltage supply wire connected to a number of DAC current sources.

between two adjacent current sources is equally long, the wire resistance, R, is also the same. The currents, $I(V_i)$, generated from a number of unit current sources, are determined by the square of the source-gate voltage according to (13-2), hence dependent on the squared voltages V_i. Therefore we will have currents that are deviating from their nominal values and we have to design the supply wire width and length to minimize this voltage drop.

13.3.2 Current Switches

The current switches influence the DAC performance. We have to guarantee that the switches are fast enough and that they do not introduce distortion.

Since we are switching a current source, we must ensure that the current switch at no time-instant turns the current source completely off. Otherwise, when it is turned off, the potential at the output of the current source will move towards the supply voltage. When the current source is turned on again, the potential difference between the current source output and the DAC output is large and a glitch occurs. To avoid this, we use differential switches, so that the current source always can deliver current. In Fig. 13-9a) we show a differential switch and in Fig. 13-9b) a possible MOS implementa-

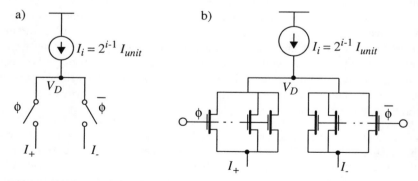

Figure 13-9 Differential current switch as a) circuit model and b) transistor implementation.

tion of the switch.

As outlined in the previous chapter, in a flat array structure (Sec. 13.2.1) the switch-on resistance is important. Especially, for larger currents through the switches, i.e., for

13.3 Practical Design Considerations

MSB current sources, the switch-on resistance must be low, since the output impedance of the current source is low. For an MOS implementation, this implies that the size aspect ratio of the transistors must be large. However, large switches also increase clock feedthrough (CFT) [17] due to the gate capacitances. A trade-off has to be done. The lower the switch-on resistance, the more transistors are connected in parallel, as illustrated in Fig. 13-9b).

In the linear region, the switch-on resistance of an MOS transistor is

$$R_{sw} \approx \frac{1}{\beta \cdot (V_\phi - V_T - V_D)} \quad (13\text{-}11)$$

where β is the transconductance parameter, V_ϕ is the gate or switch voltage, V_T is the threshold voltage, and V_D is the drain voltage. The switching voltages, $\phi, \overline{\phi}$, are usually given by the supply voltages, but for low supply voltages the switching voltage can be increased by using a charge-pump technique [18]. The charge-pump technique is also described in Sec. 8.4.

13.3.3 Digital Circuits

It is important to never turn the current source completely off and therefore the switching signals have to be properly matched to improve the glitch performance. A proper switching scheme for an NMOS implementation of the differential current switch is shown in Fig. 13-10a) [8, 17]. The switching signals have to be slightly overlapping

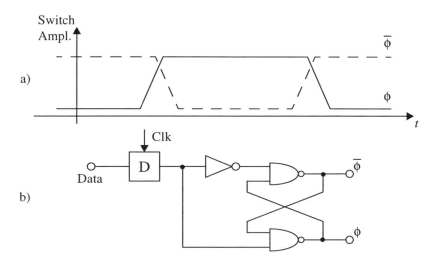

Figure 13-10 a) Wanted switch signals for the differential current switch and b) a possible circuit implementation for generation of switch signals.

and they can for example be generated with a latch as shown in Fig. 13-10b).

The input data, i.e., input bits, to the switch signal generator is aligned globally with a latch to reduce the timing skew. Therefore, a good clock distribution is needed. In Fig. 13-11 we show two different clock distribution approaches, which are suitable for

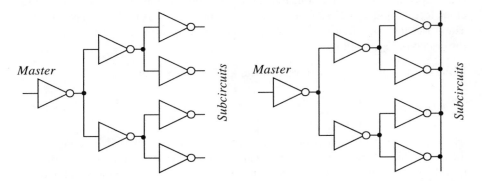

Figure 13-11 Clock tree structures to improve buffering and equal clock delay to the circuit.

high-speed and high-accuracy. Both clock trees improve the skew compared to a single-line clock distribution. The configuration in Fig. 13-11a) improves speed since it has a low capacitive load. The tree in Fig. 13-11b) has a higher capacitive load due to the additional wire capacitance, but the skew can be minimized if there are mismatch between the different inverters. The trees can be optimized for maximum speed or optimum clock edge behavior, and the tapering factor is determined by the speed requirements and process parameters.

As we have seen in the previous chapters, to improve monotonicity, to reduce the glitching, and relax the design requirements, segmented DAC structures are preferred. The most significant bits are thermometer coded, hence a binary to thermometer encoder is needed. A binary code consisting of M bits is converted into a thermometer code of $2^M - 1$ bits.

This encoding can be realized by expanding the boolean expressions and using logic. This approach becomes tedious for a higher number of input bits since the expressions and the number of outputs are increasing exponentionally. Another method is similar: We use an iterative approach, since an M-to-$(2^M - 1)$ binary-to-thermometer encoder can be constructed using an $(M - 1)$-to-$(2^{M-1} - 1)$ encoder as illustrated in Fig. 13-12. The logic depth of this encoder is equal to $M - 1$ and the propagation delay may limit the maximum achievable speed. There are 2^{M-1} AND and 2^{M-1} OR gates. With pipelining we ensure that a high sampling frequency can be reached. Due to the propagation delay through the encoder, the binary coded LSBs have to be delayed.

In mixed analog/digital designs, such as segmented DACs that have a larger digital portion, low power consumption for digital circuits and low switching noise are attractive features. Disturbances from the digital to the analog part spread along supply lines and the substrate. The substrate coupling may be strong depending on the substrate doping. It is therefore necessary to do careful designs with proper shielding, which is

13.3 Practical Design Considerations

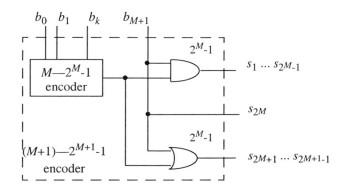

Figure 13-12 Iterative implementation of a binary-to-thermometer encoder. Note that there are $2^k - 1$ AND and OR gates.

done in several different ways [19], i.e., guard rings, grounding, etc.

It is also necessary to shield analog signals and separate analog and digital supplies [20]. The analog pins should be placed from the digital as far as possible. Grounding pins as shields should be used between analog and digital pins, etc.

13.3.4 Current Source Calibration

Some special techniques, such as calibration or trimming techniques, can be used to minimize the influence of mismatch in current-steering DACs [21, 22, 23].

Consider PMOS transistors as current sources in the DAC (Fig. 13-6). One technique is to not use the same bias voltage for all current sources. Instead we use the principle of holding the source-gate voltage, V_{SG}, during a period of time by using the charge stored on the source-gate capacitance, C_{sg}, as illustrated in Fig. 13-13. This implies that we have to charge the capacitance with a proper voltage. This is done during a calibration phase when the capacitance is charged by driving a reference current, I_{ref}, through the current source. The source-gate voltage will be set by the reference current according to the current formula (13-2) and transistor size errors of the transistor will not affect the output current, since V_{SG} is dependent on the reference current only. During the operation phase, as shown in the figure, the output current is determined by the held source-gate voltage. Due to leakage currents through the transistors and CFT, the true value cannot be guaranteed for a longer period of time and the source-gate voltage has to be refreshed on a regular basis.

This technique can be used for the thermometer coded MSBs in a segmented DAC or for the current sources in an R-2R ladder structure [21], where the current sources all are equally large. The advantage of this technique is that the influence of matching errors can be significantly reduced [17, 22]. The disadvantages are that additional switches are needed, i.e., hence a larger load and layout complexity, and for high res-

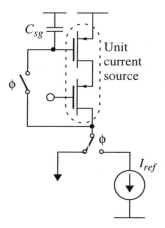

Figure 13-13 Example on circuit solution to calibrate the unit current sources during specific calibration phases.

olutions a large number of current sources has to be calibrated. This also implies that the calibration phase may need a long time, which may not be possible for certain high-speed applications. This requirement can be relaxed by using additional current sources which can be calibrated when other current sources are in operation.

13.4 A CMOS CURRENT-STEERING DAC CHIPSET

In this section we discuss the design and implementation of a CMOS DAC chipset for wideband applications. The chip's supply voltages range from 1.5 V to 5 V. The nominal resolutions are 10 bits up to 14 bits and the update frequency is specified from 10 MHz up to 100 MHz. The bandwidth should be in the range of 1 to 20 MHz. All DACs are segmented current-steering structures with the four MSBs thermometer coded. The output currents are terminated over a 50-Ω load.

As a comparison, we briefly describe the implementation of three different converters in the chipset and to simplify the reference to the different converters throughout the text, we use the notations DAC A, B, and C as summarized in Table 13-1.

Measurement results and conclusions of the chip set are presented in Sec. 13.5 and we highlight the results from a closer comparison between two of the designs (A and B), where the current sources and interconnection slightly differ.

The knowledge about the fundamental limits given by the behavioral-level models in Chapter 12, was used when designing the DACs. Knowing how the output current affects transistor sizes, output impedance, and noise, lets us choose a proper value, etc. However, no process information on matching of transistors as in (13-10) was available.

The chip floor plan of the converters [24, 25, 26] is shown in Fig. 13-14 and from the floorplan we identify three different parts; the digital part consisting of clock, flip-

13.4 A CMOS Current-Steering DAC Chipset

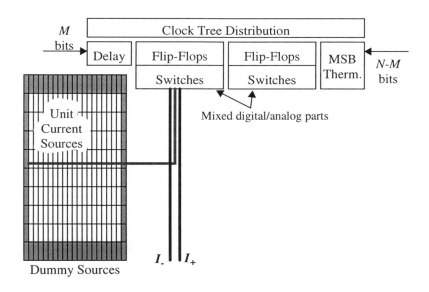

Figure 13-14 Chosen DAC segmented structure.

Table 13-1 Brief information on the DACs of the chipset.

DAC	Description
A	Segmented structure. 3.3 - 5 V supply. 14-bit resolution. 0.6um CMOS. 20 mA peak current outputs. Double and Single cascode PMOS transistors as current sources.
B	Segmented structure 3.3 - 5 V supply. 14-bit resolution. 0.6um CMOS process. 20 mA peak current outputs. Double cascode PMOS transistors as current sources.
C	Segmented structure 1.5V supply. 10-bit resolution. 0.6um CMOS process. 10 mA peak current outputs. Single cascoded PMOS transistors as current sources.

flops latching the digital input, a 4-bit binary-to-thermometer encoder for the MSBs, and a delay block for the LSBs to equalize the delay in the encoder. The analog part consists of an array of unit current sources and bias current sources, and finally the mixed analog/digital parts consisting of the current switches.

For the i-th LSB, 2^{i-1} unit current sources from the array are connected together and their output is fed to the current switch controlled by the i-th bit. The outputs are two currents and if the corresponding bit is a "1" or "0", the current is routed to the positive, I_+, or the negative output, I_-.

When using a thermometer code for the 4 MSB current sources in a 14-bit DAC, they are encoded into 15 equally large current sources, each one containing 1024 unit current sources. When encoding 5 MSBs we would have 31 equally large current sources each with 512 unit current sources. The segmentation introduces a delay for the MSBs through the encoding tree and to avoid skew between all bits, a digital delay line is used for the LSBs. A clock tree as shown in Fig. 13-11a) is used to get proper clock delay and guaranteeing an alignment of the data to all flip-flops controlling the current switches.

13.4.1 Unit Current Sources and Source Array

In the chipset, both double- and single-cascode PMOS transistors were used as unit current sources. For a 5-V supply double-cascodes were used and for the 1.5-V supply, single-cascode transistors were used. The layout view of a double-cascode current source is shown in Fig. 13-15. To simplify the layout and routing, all transistors have

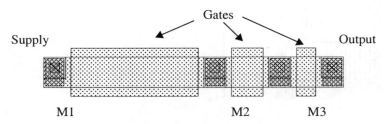

Figure 13-15 Layout view of the PMOS current source with a single cascode.

the same width. For DAC A and B the transistor sizes were given by the equal width $W = 2$ µm and the lengths $L_1 = 8$ µm, $L_2 = 2$ µm, $L_3 = 1.2$ µm for the transistors M1, M2, and M3, respectively. For the single-cascode source, only M1 and M2 were used. Double-cascode transistors were used to guarantee a very high output impedance needed for the 5-V 14-bit converters, and the simulated output resistance (at DC) of the double-cascode unit current source is approximately $R_{unit} \approx 30$ GΩ. The single-cascode unit current source has an approximate output resistance of $R_{unit} \approx 1$ GΩ.

The SFDR dependent on the output resistance will be determined by the odd harmonics and we can use equations (12-42) - (12-44) to find the expected SFDR performance. For a single-ended output, equations (12-38) - (12-41) apply.

In the DACs, the binary weighted current sources were formed by using unit current sources from the array by starting in one corner and use one unit current source for the LSB, the next two adjacent unit sources for the second LSB, etc. In the 14-bit DAC the array consists of 258x66 unit current sources. All edge sources are dummies and hece only used to increase the edge matching of the internal unit current sources. This layout technique was used to optimize the use of the chip area, although the matching properties are not the best, since the influence of linear graded matching errors increases.

The voltage drop across the supply line for all current sources of the array has to be avoided [16] and therefore, a power supply plane is covering the whole current source array. In all converters of the chipset, a higher-level metal layer was used. To reduce

13.4 A CMOS Current-Steering DAC Chipset

the influence of switching noise from the digital circuits shieldings are used between analog and digital parts.

In the 14-bit DACs, the bias circuit is a cascoded current mirror formed by 1024 unit current sources. The reference current is determined by an off-chip resistor. The unit current sources used in the bias circuit are chosen from four adjacent rows of the current source array.

13.4.2 Current Switches

The layout of the current switch for a current source is shown in Fig. 13-16. The same

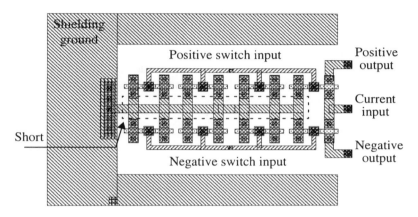

Figure 13-16 Layout view of a differential current switch.

switch is used for all weighted current sources to achieve equal capacitive load. For the LSB (as shown in the figure) only one of the NMOS transistors are used for each channel. The other transistors are shorted. For the MSBs, i.e., thermometer coded bits, all transistors are used in order to reduce the on-resistance of the switch. A shielding metal ground plane, nwell field, and substrate contacts around all switches are used to reduce noise coupling.

The simulated switch-on resistance for different switches and different supply voltages are summarized in Table 13-2 for a standard digital 0.6 um CMOS process. In the simulations we have used the specified DC operating points such as the current flowing through the switches and the voltage outputs, which are 0.5 V for the high-voltage DACs A and B and 0.25 V for the low-voltage DAC C. The DC current through the switches are different for different switch sizes, i.e., LSB current for the smallest switch, etc. The switch on-resistance increases with a lower supply voltage since the gate-source voltage is reduced. In the 10-bit 1.5-V DAC C, the switching voltage was increased by using the charge pump circuit as discussed in Sec. 8.4.1.

13.4.3 Digital Circuits

The digital circuits are constructed in a straight forward way. The 4-to-15 binary-to-

Table 13-2 Swich-on resistance in Ω compared to the voltage supply and number of transistors.

Supply voltage [V]	Number of active NMOS transistors			
	1	2	4	8
1.5	5.25k	2.62k	1.30k	650
2.7	2.42k	1.21k	610	636
3.0	2.09k	1.05k	520	393
3.3	1.87k	930	470	314
5	1.36k	680	340	170

thermometer encoder was constructed by expanding the boolean function describing its 15 output values as functions of its inputs. The gates are realized with unclocked CMOS circuits. The total delay through the encoder is given by four gates with inverter buffers at the output. The difference in the delay is small enough to guarantee over 100 MHz operation. The LSBs are delayed with four cascaded inverters and aligned. All the data are latched before driving the bit switches. A tree-formed clock distribution (similar to the structure in Fig. 13-11a, but with a tapering factor of 3) is used to guarantee as small clock skew as possible to the current switches.

13.4.4 Chip Implementations

The chip layouts of the converters are similar as illustrated in Fig. 13-14. In Fig. 13-17 we show a die photograph of DAC A. The core size is approximately 2x2 mm^2 and including pads, the chip area is 3.0x3.4 mm^2. From the die photo in Fig. 13-17 we also see that by rotating the current source array 90° we can reduce the wire lengths and the core area significantly. In Table 13-3 in Sec. 13.5.1 the chip data is summarized.

13.5 CHIPSET MEASUREMENTS

There are several possible ways of measuring the DACs. However, since we are considering high-speed, high-resolution converters, the best way to measure is to use a low-jitter input data generator, a spectrum analyzer, and a high-bandwidth oscilloscope. The same principle holds for computer simulation of the converters. To perform static measurements, such as DNL or INL measurements of a 14-bit DAC with $2^{14} = 16384$ amplitude levels a lower-speed higher-resolution digitizer is used. The digital data is captured and analyzed by a computer, further reducing time and complexity [27]. This is not practical for communications and instead we apply single-tone or multi-tone inputs and examine the output spectrum. For full-scale single-tone measurements, an update frequency that is relatively prime with the signal frequency is used. Otherwise, all information about the converter's performance cannot be fully extracted. Using a prime frequency ratio also ensures that the distortion terms are not folded back onto the fundamental and interfering with other frequency components and causing inaccuracy in the measurement. For dual-tone measurements the signal

13.5 Chipset Measurements

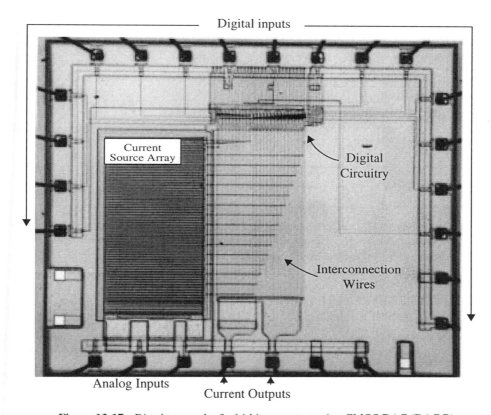

Figure 13-17 Die photograph of a 14-bit current-steering CMOS DAC (DAC B).

frequencies are also chosen to be relatively prime, and they are relatively prime with respect to the update frequency as well. For multi-tone measurements we apply a number of tones with frequencies at multiples of a fundamental frequency according to an orthogonal frequency-division multiplexing (OFDM) or discrete multi-tone (DMT) modulation.

13.5.1 Measurement Results

We present different measurement results from the converters of the chipset. No analog filtering has been applied to the output signal. The sample frequencies vary from 1 MHz up to 100 MHz.

- *Measurement of the 14-bit single- and double-cascode DAC A*

In Fig. 13-18 we show the measured output spectrum for the 14-bit single-cascode DAC A when applying a dual-tone input signal at a supply voltage of 5 V. The update frequency is 20 MHz and the signal frequencies are 3.43 MHz (f_1) and 3.51 MHz (f_2). The SFDR is 49 dBc and the IMD given by the $2f_1 - f_2$ frequency component is -54 dBc. The output is measured over a transformer, transforming the differential signal into a single-ended signal, i.e., even order harmonics are reduced and it is found that the third harmonic limits the SFDR to 48 dBc.

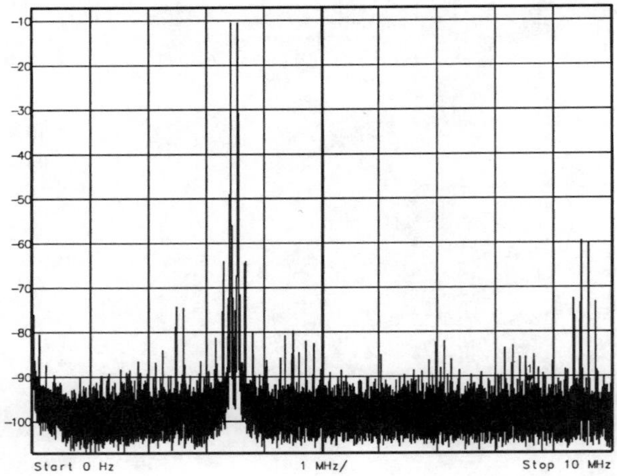

Figure 13-18 Dual tone measurement of a 14-bit double cascode DAC with a 5 V supply and an update frequency of 20 MHz.

From further measurements [31], we find that for the single-cascode 14-bit DAC A, the SFDR decreases from 49 to approximately 42 dBc when the update frequency changes from 5 to 50 MHz. The sample frequency is approximately 6 times higher than the signal frequency. The performance is poor and to find the limitations on performance we study the measurement results from DAC B and C and compare the layouts with the one of DAC A.

- *Measurement of the 14-bit double-cascode DAC B*

In Fig. 13-19 we show the summarized measurement results from DAC B. Single-tone signals at -3 dBFS are applied. The SFDR as a function of update frequency and signal frequency is shown. We find that the SFDR varies from more than 75 dBc at low frequencies down to 45 dBc for higher frequencies. In the figure we have also added measurement results from DAC A and the two lower curves (dashed-star for DAC A and solid-triangle for DAC B) are measured under the same conditions. We find that although the layouts are very similar, the improvement in SFDR is approximately 13 dB to 6 dB with increasing sample frequency and a signal frequency that is approximately 6 times lower than the sample frequency. However, in DAC B the interconnection wires have been made wider and shorter and the parasitic impedance was decreased. There were also changes to the layout of the unit current source. Internal interconnections that increase the capacitive part of the current source output impedance were removed. In Fig. 13-20 we illustrate this by displaying a partial layout view of the unit current source arrays in DAC A and B. The reduction of the internal capacitance increases the output pole frequency of the current source (compare with Fig. 13-7) and at higher frequencies we will have a higher $|Z_{out}/Z_{load}|$ ratio and therefore a higher linearity. Further, the short-distance matching is improved by removing vias and metal wires (supply connections) placed on top of the transistor gates.

13.5 Chipset Measurements

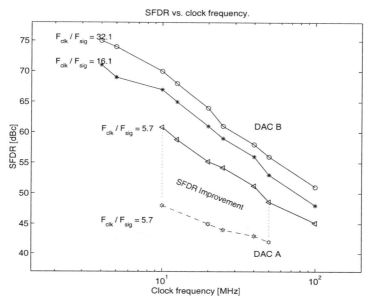

Figure 13-19 Summary of the measured result from DAC B and comparison of measured results from DAC A and B.

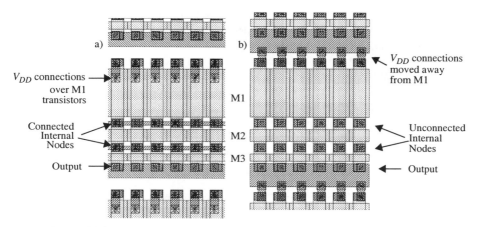

Figure 13-20 Part of current source array for DAC A (a) and DAC (b).

In Fig. 13-21 the measured output spectrum when applying an 18-tone signal to DAC B is shown. The tones have equal amplitude (-18 dBFS) and the phase of each carrier is randomized to reduce the PAR. The update frequency is 20 MHz, the supply voltage is 3.3 V, and the tone frequencies are centered around 2.2 MHz. We find the SFDR to be 57 dBc which in this case is approximately equal to the measured IMD.

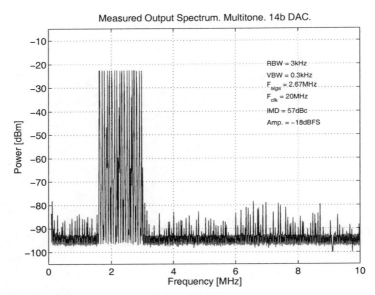

Figure 13-21 Multi-tone measurement of DAC B. 18 tones at -18dBFS are applied. The update frequency is 20 MHz.

- *Measurements of the low-voltage 10-bit single-cascode DAC C*

From the low voltage DAC C, we know that we can reach a rather high linearity. In Fig. 13-22 we show the measured single-ended output when applying a dual-tone input signal. The update frequency is 10 MHz and the signal frequencies are 1.775 MHz and 1.792 MHz. The measured IMD is approximately 52 dBc. The supply voltage is 1.5 V.

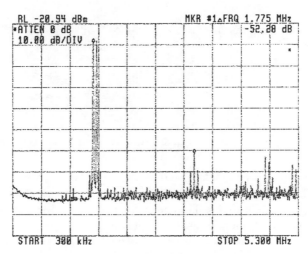

Figure 13-22 Measured dual-tone output from the 1.5 V DAC. Update frequency is 10 MHz and signal frequencies around 1.775 MHz.

13.5 Chipset Measurements

In Fig. 13-23 we show the summarized single-ended measurement results from the 1.5 V 10-bit DAC [18].

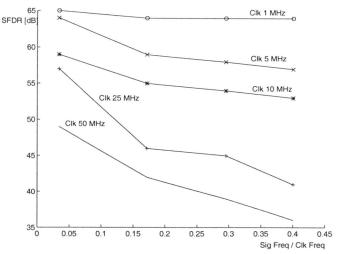

Figure 13-23 Summary of measured results from the 1.5 V DAC C.

We find that the SFDR is over 65 dBc at low frequencies and for high clock frequencies and a low clock to signal frequency ratio, the SFDR is decreasing rapidly. The improved performance over DAC A originates from the fact that the requirements on the output impedance and element matching are greatly reduced over the 14-bit versions of the converters.

13.5.2 Conclusions and Comparison

From the measurement results [31] we conclude:

- In DAC A and B there is no larger change in performance when changing the supply voltage from 5 V to 3.3 V.

- There are no larger variations between the single and double cascode implementations for higher frequencies. In Fig. 13-7 we see that the output impedance at higher frequencies is approximately the same for both current source versions. Hence we have to have a design for a higher pole. The bandwidth is rather low and also indicating the high capacitive part at the output of the current sources. When increasing the pole frequency we have to lower the output resistance.

- The performance can be improved with a more careful layout with respect to linear graded matching errors.

- Time-domain simulation results show low glitching due to good clock distribution and segmentation of the MSBs.

In Table 13-3 we summarize the results from the design of the chipset DACs.

Table 13-3 Chip data for the different DACs of the chipset.

DAC	Supply [V]	Nominal resolution [bit]	Chip area [mm^2]	Measured results
A	3.3 - 5	14	2x2 (core)	SFDR = 49 dBc at 10 MHz update frequency and 1.7 MHz full-scale sinusoid. 14-bit DAC.
B	3.3 - 5	14	2x2 (core)	SFDR = 75 dBc at 5 MHz update frequency and 156 kHz full-scale sinusoid. 14-bit DAC. SFDR = 61 dBc at 10 MHz update frequency and 1.7 MHz full-scale sinusoid. 14-bit DAC. IMD = 57 dBc at 20 MHz update frequency and 18-tone -18 dBFS at frequencies around 2.2 MHz.
C	1.5	10	2x2 w. pads	SFDR = 64 dBc at 1 MHz update frequency and 400 kHz full-scale sinusoid. SFDR = 57 dBc at 10 MHz update frequency and 1.7 MHz full-scale sinusoid. IMD = 52 dBc at 10 MHz update frequency and dual-tone frequencies around 1.7 MHz.

13.6 SUMMARY

In this chapter, we have discussed the design and implementation of current-steering DACs for wideband applications. Different structures have been outlined and for high-speed and high-resolution applications we have found the segmented DAC structure to be most suitable. A key design issue is to find the proper number of bits encode into a thermometer code or if multi-segmented structures are needed. We have shown how the performance of the converter depends on errors in the layout by comparing two similar DACs and highlighted the importance of high frequency poles in the output impedance of the unit current sources. We have shown that it is possible to reach an SFDR of over 75 dBc in a standard 3.3 V digital CMOS process and an SFDR of 65 dBc with a single 1.5-V supply voltage.

REFERENCES

[1] M.J.M Pelgrom, A.C.J. Duinmaijer, and A.P.G. Welbers, "Matching Properties of MOS Transistors," *IEEE J. of Solid-State Circuits*, vol. 24, no. 5, pp. 1433-9, Oct. 1989

[2] H.J. Schouwenaars, D.W.J. Groeneveld, and H.A.H. Termeer, "A Low-Power Stereo 16-bit CMOS D/A Converter for Digital Audio," *IEEE Journal of Solid-State Circuits*, Vol. 23, pp. 1290-1297, Dec. 1988.

[3] K.O. Andersson and J.J. Wikner, "Modeling of the Influence of Graded Element Matching Errors in CMOS Current-Steering DACs," in *Proc. of the 17th NorChip Conference*, NORCHIP'99, Oslo, Norway, Nov. 7-8, 1999

[4] T. Miki, Y. Nakamura, M. Nakaya, S. Asai, Y. Akasaka, Y. Horiba, "An 80 MHz 8-bit CMOS D/A Converter," *IEEE Journal of Solid-State Circuits*, Vol. 21, pp. 983-988, Dec. 1986.

[5] C.H. Lin and K. Bult, "A 10 b 250 M sample/s CMOS DAC in 1 mm^2", in *Proc. of the 1998 International Solid-State Circuits Conference, ISSCC'98*, 1998.

[6] A. Van den Bosch, M. Borremans, J. Vandenbussche, G. Van der Plas, A. Marques, J. Bastos. M. Steyaert, G. Gielen, and W. Sansen, "A 12 bit 200 MHz low glitch CMOS D/A converter," in Proc. of the 1998 International Custom Integrated Circuits Conference, CICC'98, 1998.

[7] Analog Devices, AD9764, TxDAC series, 1999.

[8] H. Kohno, Y. Nakamura, et al., "A 350-MS/s 3.3-V 8-bit CMOS D/A Converter Using a Delayed Driving Scheme," in Proc. of the IEEE 1995 Custom Integrated Circuits Conference, pp. 211-4, 1995.

[9] Ji Hyun Kim and Kwang Sub Yoon, "A 3.3V-70MHz Low Power 8 bit CMOS Digital to Analog Converter with two-stage current cell matrix structure," In Proc. of the , 1997.

[10] M. Otsuka, S. Ichiki, T. Tskuada, T. Matsuura, and K. Maio, "Low-Power, Small-Area 10bit D/A Converter for Cell-Based IC," In *Proc. 1995 IEEE symposium on Low Power Electronics*, pp. 66-67, 1995

[11] Ki-Hong Ryu, et al., "Design of a 3.3V 12bit CMOS D/A Converter with a high linearity," In Proc. of the , 1999.

[12] D.A. Johns and K. Martin, *Analog Integrated Circuit Design*, John Wiley & Sons, New York, U.S.A., 1997, ISBN 0-471-14448-7

[13] H. Tuinhout, M. J. M. Pelgrom, R. Penning de Vries, and M. Vertregt, "Effects of Metal Coverage on MOSFET Matching," in *Proc. of the Int'l Electron Devices Meeting (IEDM'96)*, pp. 735-8, San Fransisco, CA, USA, Dec. 8-11, 1996

[14] J. Bastos, M. Steyaert, A. Pergoot, and W. Sansen, "Mismatch characterization of submicron MOS transistors,", *Analog Integrated Circuits and Signal Processing*, vol. 12, no. 2, pp. 95-106, Feb. 1997

[15] I.H.H. Jørgensen and S.A. Tunheim, "A 10-bit 100MSamples/s BiCMOS D/A Converter," *Analog Integrated Circuits and Signal Processing*, vol. 12, pp. 15-28, 1997

[16] B.E. Jonsson, "Design of Power Supply Lines in High-Performance SI and Current-Mode Circuits," in *Proc. of the 15th Norchip Sem.*, pp. 245-50, Tallinn, Estonia, Nov. 10-11, 1997

[17] C. Toumazou, J. B. Hughes, and N. C. Battersby, *Switched-Currents: an Analogue Technique for Digital Technology*, Peter Peregrinus, Stevenage, UK, 1993, ISBN 0-86341-294-7

[18] N. Tan, "A 1.5-V 3-mW 10-bit 50 MS/s CMOS DAC with Low Distortion and Low Intermodulation in Standard Digital CMOS Process," in *Proc. of the 1997 IEEE Custom Integrated Circuits Conf. (CICC'97)*, pp. 599 - 602, Santa Clara, CA, USA, May 1997

[19] M. Ismail and T. Fiez, *Analog VLSI: Signal and Information Processing*, McGraw-Hill, New York, NY, USA, 1994, ISBN 0-07-113387-9

[20] P. J. Fish, *Electronic Noise and Low Noise Design*, Macmillan, Basingstoke, UK, 1993, ISBN 0-333-57310-2

[21] A. Biman and D. G. Nairn, "Trimming of Current Mode DACs by Adjusting V_t," in *Proc. of the 1996 IEEE Int'l Symp. on Circuits and Systems (ISCAS'96)*, vol. 1, pp. 33-6, Atlanta, GA, USA, May 12-15, 1996

[22] R. J. van de Plassche, *Integrated Analog-to-Digital and Digital-to-Analog Converters*, Kluwer Academic Publishers, Boston, MA, USA, 1994, ISBN 0-7923-9436-4

[23] C. Toumazou, J.B. Hughes, and N.C. Battersby, *Switched-Currents - an analogue technique for digitial technology*, IEE, Peter Peregrinus, London, UK, 1993

[24] N. Tan, E. Cijvat, and H. Tenhunen, "Design and implementation of High-Performance CMOS D/A Converter," In *Proc. 1997 IEEE International Symposium on Circuits and Systems*, ISCAS'97, Hong Kong, Vol I. pp. 421-424, 1997

[25] J.J. Wikner and N. Tan, "A CMOS Digital-to-Analog Converter Chipset for Telecommunication," *IEEE Magazine on Circuits and Devices*, Vol. 13, No. 5, pp. 11-16, Sep. 1997

[26] N. Tan and J.J. Wikner, "A CMOS Digital-to-Analog Converter Chipset for Telecommunication," *IEEE Magazine on Circuits and Devices*, vol. 13, no. 5, pp. 11-16, Sept. 1998

[27] M.D. Dallet, *Contribution à la Caractérisation des Convertisseurs Analogique-Numériques: Evaluation des Méthodes et Mise en Ouvre de Noveaux Procédés*, These no. 1249, L'universite de Bordeaux I, France, Jan. 1995

[28] J.S. Chow, J.A.C. Bingham, and M.S. Flowers, "Mitigating Clipping Noise in Multi-Carrier Systems," in *Proc. of the 1997 IEEE Int'l Conf. on Communications (ICC'97)*, vol. 2, pp. 715-19, Montreal, Canada, June 8-12, 1997

[29] D.J.G. Mestdagh, P. Spruyt, and B. Biran, "Analysis of Clipping Effect in DMT-Based ADSL Systems," in *Proc. of the 1994 IEEE Int'l Conf. on Communications (ICC'94)*, vol. 1, pp. 293-300, New Orleans, LA, USA, May 1994

[30] C. Tellambura, "Upper Bound on the Peak Factor of N-Multiple Carriers," *Electronics Letters*, vol. 33, no. 19, pp. 1608-9, 1997

[31] J.J. Wikner, "Measurements of a CMOS DAC Chipset," Internal Report, LiTH-ISY-R-1983, Linköping University, Sweden, 1998.

[32] G. Stehr, F. Szidarovsky, and O.A. Paulusinski, and D. Andersson, "Performance Optimization of binary weighted current-steering D/A converters," *Applied Mathematics and Computation*, to be published, 1999.

LURE OF PASSION

Adam's eyes narrowed. "You aren't dreaming of some fairytale existence with my cousin Christian, are you? Because I can assure you it will never come about. He is not a man to be satisfied with one woman for any length of time."

"Why, how—how can you say such a thing to me!" I spluttered. "I have barely met your cousin. Is it your opinion that I am the type of woman who sets her cap for a man moments after meeting him?"

Adam took me by the arms in a grip of iron. "Do you deny that ever since we set out on this ride you have been wishing that it was he accompanying you, rather than I?"

It was the impact of his eyes, just now turned slate-blue, more than his nearness that covered me with confusion. That, and a helpless feeling that I could have no secrets from him. I cast about for words of sharp denial, but could find none.

"Ah," he said grimly, "I have hit the mark." He pulled me closer. "Nevertheless, I will have something of this day for myself."

With that, he bent his head and kissed me—kissed me angrily, punishingly. Its harshness stirred some deep, hidden need in me. I was caught up in it, powerless to resist. To my shame, it was he, not I, who pulled away at last.

His eyes still gleamed with anger, but there was triumph in them, too . . .

THE BEST IN GOTHICS FROM ZEBRA

THE BLOODSTONE INHERITANCE (1560, $2.95)
by Serita Deborah Stevens

The exquisite Parkland pendant, the sole treasure remaining to lovely Elizabeth from her mother's fortune, was missing a matching jewel. Finding it in a ring worn by the handsome, brooding Peter Parkisham, Elizabeth couldn't deny the blaze of emotions he ignited in her. But how could she love the man who had stolen THE BLOODSTONE INHERITANCE!

THE SHRIEKING SHADOWS OF PENPORTH ISLAND (1344, $2.95)
by Serita Deborah Stevens

Seeking her missing sister, Victoria had come to Lord Hawley's manor on Penporth Island, but now the screeching gulls seemed to be warning her to flee. Seeing Julian's dark, brooding eyes watching her every move, and seeing his ghost-like silhouette on her bedroom wall, Victoria knew she would share her sister's fate—knew she would never escape!

THE HOUSE OF SHADOWED ROSES (1447, $2.95)
by Carol Warburton

Penniless and alone, Heather was thrilled when the Ashleys hired her as a companion and brought her to their magnificent Cornwall estate, Rosemerryn. But soon Heather learned that danger lurked amid the beauty there—in ghosts long dead and mysteries unsolved, and even in the arms of Geoffrey Ashley, the enigmatic master of Rosemerryn.

CRYSTAL DESTINY (1394, $2.95)
by Christina Blair

Lydia knew she belonged to the high, hidden valley in the Rockies that her father had claimed, but the infamous Aaron Stone lived there now in the forbidding Stonehurst mansion. Vowing to get what was hers, Lydia would confront the satanic master of Stonehurst—and find herself trapped in a battle for her very life!

Available wherever paperbacks are sold, or order direct from the Publisher. Send cover price plus 50¢ per copy for mailing and handling to Zebra Books, Dept. 2022, 475 Park Avenue South, New York, N.Y. 10016. Residents of New York, New Jersey and Pennsylvania must include sales tax. DO NOT SEND CASH.

RIVERSBEND

By Helen B. Hicks

**ZEBRA BOOKS
KENSINGTON PUBLISHING CORP.**

ZEBRA BOOKS

are published by

Kensington Publishing Corp.
475 Park Avenue South
New York, NY 10016

Copyright © 1987 by Helen B. Hicks

All rights reserved. No part of this book may be reproduced in any form or by any means without the prior written consent of the Publisher, excepting brief quotes used in reviews.

First printing: March 1987

Printed in the United States of America

Chapter One

I had always known that Father would one day send for me.

Even when I was only nine and he had left me in my aunt's care after Mother's death, I knew that at some glorious future date he would summon me to join him in his travels about this sprawling country of ours, and that I would at last be allowed to leave the wholesome but simple and staid existence which I knew with Aunt Blanche, and could again take up the exciting life our little family had led before my mother had been lost to us.

I never dreamed that when the invitation finally came I would not be free to take it.

I remember that it was a crisp autumn day when the postman delivered my long-awaited letter, and bright leaves were swirling in the brisk wind that swept across our small front lawn and down the walk to pile against the front step of the porch. I saw him coming from the big front window of the parlor, and a tremor of anticipation ran through me as it always did at the sight of the small batch of envelopes he carried in one hand. The weather had been doing strange things to me; I was both exhilarated by the cold bite in the air and yet depressed at the thought of the nearness of winter and snow and the enforced inactivity it would bring to us here in this quiet New York suburb, and I had been up from my chair and pacing about our small, too cozy parlor.

Aunt Blanche, who had not been well, was lying on the settee while awaiting a visit from the doctor. She was well covered with a comforter, though the crackling fire in the marble-fronted fireplace was putting out more heat than was comfortable to me. When, at seeing the postman, I gave a little cry of pleasure and started toward the door, she said with the gentle remonstrance that I found so irritating, "Is it the postman, Beth, dear? There is no need to rush out and greet him yourself, you know. Sarah will see to it."

With difficulty I checked my headlong flight.

"Of course, Aunt Blanche. It is just that I think he is carrying a long envelope, such as Father uses, in his hand, and I am anxious to get it. A letter from him is a trifle overdue."

"Then a few more moments can't matter, can they?" she said serenely. "The neighbors might think it unseemly if you were to greet him halfway down the walk as you used to do as a child." Her voice, though it still held its prim, slightly schoolteacherish tone, was somewhat weak from the effects of her illness, and this quelled my natural impatience. I waited with clenched teeth and outward calm while the doorbell jangled, deliberate footsteps went to answer it, the door opened, and then, after what seemed an interminable length of time, closed again. The heavy, deliberate footsteps reached the parlor door at last, and Sarah came in, her maid's cap slightly askew as always, and her red Irish face beaming in anticipation of our pleasure.

"There's one for each of yez, mum," she said to my aunt. "It's easy to tell Mr. Jamison's handwritin', bold and black as it is. I've come to know it after all this time." She handed each of us one of the long envelopes, then added, "Would yez like me to bring the tea tray now, or later on, after the doctor comes?"

Neither Aunt Blanche nor I moved to open our envelopes. "After Dr. Mathews comes, I think," Aunt Blanche answered. "He will be chilled, and a cup of hot tea will be most welcome, I imagine."

Sarah nodded and left, and after we were sure she would not be coming back, Aunt Blanche and I exchanged envelopes, smiling conspiratorially as we did so. Sarah, who could not read, had given us each other's, but not for the world would we have embarrassed her by pointing it out to her.

Father's letter to me, which was long and newsy, as always, was from a California town with the exotic name of San Diego, a place where he had situated some years before. It was the first time he had stayed in one spot for any length of time, and after the first few months of receiving letters with that postmark I had begun to hope that he might settle down and allow me to join him there. But he never had, and in answer to my wistful questions had made it clear that until my schooling was over he considered it best that I remain with my aunt in New York. Now, however, I had, this last June, matriculated from Mrs. Hornsby's Academy for Young Ladies, and my hopes had risen again.

And with reason, I saw to my delight, for in the last paragraph of Father's letter he issued the long-awaited invitation. He wrote:

> I feel that, now, at last, you are grown enough to come to me, and I am in a position to care for you as you should be cared for. I have written of this to your aunt, also, and have urged her to consider coming with you to California, at least for a prolonged visit. She would find a pleasant social life here, and the climate is considered to be the best in the world as far as health reasons are concerned. I know from your letters that Blanche has not been feeling as well as she might lately, and life at the seashore could be just what she needs. Please assure her that I would be most happy to have her make her home with us for as long as she wishes.

This last was very generous of Father, for he knew very well that his wife's sister did not really approve of him, and had initially urged her younger sister, whom she had helped to raise, not to marry him. But she had mellowed over the years, partly, I supposed, because of the regularity of his letters to me and his generous checks to her, and apparently he didn't hold a grudge against her for it.

Joy and excitement shot through me at the thought of having an adventure at last, and Father's consideration for my aunt made it even better, for I had had some qualms at the thought of leaving her alone here in New York, even though she had a wide circle of friends. But now my last worry was gone, and I looked up, smiling eagerly, to see whether she was as delighted as I.

My happiness was instantly replaced by concern, for Aunt Blanche was wiping away a tear that had crept silently down her cheek.

"Why, Aunt Blanche, what is the matter?" I cried in alarm. It wasn't at all like her to be emotional.

"How very sweet of him," she murmured. "How thoughtful to include me in his plans. I must confess to you, Beth, that I may have made a mistake many years ago when I cautioned Constance against marrying Jamison. Although his mode of life has been highly irregular, he has proved to be utterly responsible. I wish he were here that I might make amends."

These words from her alarmed me further. Aunt Blanche had always been one of those people serenely and totally certain that she was right in all instances. To have her say outright that she had been wrong about anything set me at once to feeling that something was terribly amiss. I got up and went to sit in the chair near her settee.

"Why, you can tell him yourself before long," I remarked with an attempt at lightness. "We can be packed and ready within the month, and it takes scarcely any time to get to San Diego now that a train runs to that destination."

She shook her head, and the soft, lined face lifted a little with her attempt to smile. "I am not well enough just yet, dear," she said. Her smooth, dry hand patted mine. "You must go out to him at once, of course, and perhaps I can join you in the spring."

In spite of her casual, almost offhand tone, I felt a giant hand close tightly just beneath my breastbone. "But you are bound to be better soon, Aunt. Dr. Mathews said so just last week."

"That is not quite how he put it, I think," she said gently. "I believe he said—"

Her words were cut off by the doorbell. Its jingle was followed by Sarah's slow, deliberate step.

"I expect that will be Dr. Mathews now. Perhaps you would like to take a breath of fresh air while he examines me?" Aunt Blanche put this in the form of a request, but I knew she meant for me to go. I rose, albeit unwillingly. "Don't forget your cape, dear, if you do go outside. The wind is quite brisk this time of year."

I was so accustomed to her admonitions that I scarcely heard this one, but I murmured dutifully, "Yes, Aunt Blanche," and in the hall I did pluck my cape from its peg near the door. Dr. Mathews was there with Sarah, and she had turned to lead him into the parlor and Aunt Blanche.

"Ah, Elizabeth," he said to me by way of greeting. He glanced at my cape. "Off on an errand, are you?" His kindly face shadowed a trifle. "I had hoped that we might have a short chat today."

The giant hand at my breastbone tightened. "I—Aunt Blanche is waiting for you in the parlor," I answered lamely. "She wouldn't like to be kept waiting." With near rudeness I hastened to the door, only stopping long enough to fling over my shoulder, "I will be back before you leave." With that, I tore open the glassed front door and stumbled out of the house.

As I stood on the porch, tying the strings of my hood beneath my chin, I forced myself to admire the flaming trees that lined the quiet street, and to think only of the graceful look of their topmost branches as the almost gale-force wind set them swaying. How beautiful they looked! It might be the last time I would ever see the foliage in all its fall colors, for Father had written long ago that southern California didn't have the changing seasons as the East Coast did. He had added that oranges and lemons could be picked year round, and that sometimes blossom, bud and green and ripe fruit could all be seen on the same tree at the same time. I simply could not imagine it.

Slowly I went down the step and along the walk leading to the front gate. More of the bright leaves swept across the lawn; they were blown against me and clung tenaciously to my full skirts before reluctantly moving on to pile in drifts of gold and scarlet against the farther fence. I tried to keep my mind on them, on the whipping trees, on the vision of golden orbs hanging, full of sweet juice, of far-off dark green trees, but it was no use. The knot of apprehension that was the cold, clutching hand beneath my heart would not go away.

The wind, sharp with autumn chill, cut through my cape and I shivered suddenly. For all that I had longed to leave my aunt's quiet and unadventurous household, it had never been because I didn't care for her. Though she was so unlike my mother, her younger sister, who had been gaiety and soft laughter and warmth and love, she had done her best to make me happy while I was in her care. I loved her in return perhaps more than I had realized.

I stood for a long time clutching the carved finials atop our gateposts and staring blindly into the deserted street. My dream of joining Father had always been just that—a dream—and I had never really dealt with the fact that in order to be with him I must leave Aunt Blanche behind. Father's invitation to her had solved that problem even before I had fully acknowledged that there was one, but now the shadow of fear had fallen over my happiness. Yet even now I didn't face the substance which lay behind it.

At last I sighed deeply and let go the finials. My hands were stiff with cold and the force with which I had been gripping the pointed posts, and I massaged them against each other as I reluctantly retraced my steps through the dead and dying leaves. Once on the porch again, I stood a moment before twisting the door's carved brass knob in my hand and going in.

The hall was empty and I felt a slight sense of reprieve. From the parlor came the faint clink of cups and saucers and the murmur of conversation. The examination must be over and the doctor having his well-earned cup of tea, then. For a moment I almost fled up the stairs to my room, but checked myself and, with one last straightening of shoulders, entered the room.

Sarah was still serving, and Dr. Mathews and Aunt Blanche were accepting tea and cakes from her and carrying on a light, inconsequential conversation about recent events and mutual friends. As I came in and took a chair nearby, they greeted me casually and continued to talk and sip and make decent inroads into the petit fours Sarah had decorated so prettily, and I thought with welling hope that perhaps I had imagined that something into the doctor's tone which hadn't really been there. I consumed my tea and cake with a better appetite than I had expected.

But when Dr. Mathews rose at last to go he looked toward me with unmistakable significance, and my heart dropped again. And when my aunt, with her usual pleasant insistence, said, "You will see the good doctor out, won't you, Beth, dear?" I knew I had not imagined it after all. My knees went suddenly weak, but silently I did as I was bid.

Dr. Mathews waited until we had nearly reached the

front door before he laid his hand, with its short, strong-looking fingers, on my arm.

"I must talk with you about your aunt's condition," he said in a low, kindly voice. "I had thought it time, anyhow, but she told me of your letter from your father, and that renders it imperative. She tells me that he has asked you to come to California to make your home with him. I am afraid I cannot advise you to leave her at this time."

My tongue seemed to want to cleave to the roof of my mouth with dryness. "But—she—she said she would come to us in the spring," I stammered. "If I were to wait a while before going—" My throat closed over the words and I couldn't go on.

"Your aunt is dying, Beth." Dr. Mathews put it as gently as he could. "Surely you have noticed that she grows weaker every day? It is her heart, which is simply worn out. I give her six months, perhaps a trifle more."

Here it was at last, the thing that I had been dreading to hear. I was almost thankful that I didn't have to try and fend it off any longer. I closed my eyes, and for a moment, couldn't think of anything at all to say.

"Does she—" I took a breath. "She knows she is going to die, doesn't she?"

He nodded. The gray eyes regarded me keenly. "Yes. But we haven't spoken of it. There has been no need."

I felt ashamed, as though I had been lacking in some vital sensitivity. "How could I have been so blind?" I asked bitterly. My tears were rising.

The hand on my arm closed tightly, and he shook me a little. "Now, now!" he said sharply. "None of that! It is perfectly natural that we sometimes fail to see what we desire not to be. This is not the time for the luxury of self-recrimination. Your aunt needs you now. She needs your strength and your dependability, but above all she needs your humor and your laughter—and your love. She has been there for you, Beth, these past ten years. I would hope that you will be here for her now."

There was no need for him to remind me of my duties. I would stay here as long as she needed my help. The glorious adventure that I hoped to have with Father simply must wait a little longer.

"Is she—will she be in much pain?" I ventured. My heart ached at the thought of seeing her suffer.

"Not really. What pain she has can be eased by medication. There will be shortness of breath, a steadily increasing weakness—eventually she will not want to leave her bed. She will depend on you—and Sarah—for everything."

From somewhere outside my own misery I took notice of the pain in the doctor's own face. He and Aunt Blanche had been friends for many years. These moments must be as difficult for him as they were for me. I squared my shoulders.

"Don't worry, Doctor. My aunt will have the best care that Sarah and I can give her. And I will be cheerful if it kills me!" This last was spoken over a sob that nearly cut the words off in my throat.

Dr. Mathews nodded approvingly. "Good girl. I knew I could count on you." He went out, leaving me standing in the hallways without the slightest idea of how I was going to carry out my brave words. I didn't feel that I could even go in and face my aunt. What on earth could I say to her?

As though she knew my dilemma, her soft voice reached me from the parlor. "Has the doctor gone yet, Beth?"

I braced myself. "Yes, Aunt, he has." Without any more hesitation I went into the parlor. As I approached, she looked intently at me, and I met her eyes steadily. And in that moment, she knew that I knew, and the look of strain left her face. I never saw it there again.

"Ah, sit down, dear," she said almost gaily. "But first, will you pour me a cup of tea? I find that I fancy another cup, after all."

I came to be grateful for the months that followed, for they helped me to overcome the feelings of guilt that nearly overwhelmed me. I had been most impatient with my aunt all these past ten years, and critical of her quiet, unadventuresome way of life, for it differed so sharply from that which I had known before. Even at the tender age of eight and before, I had often been allowed to accompany my parents into the gay and glittering nightlife offered by New York City in 1885. My mother and father, to my gratification, were of the opinion that a child learns the social graces as much, if not more, by being exposed to them than by deportment lessons and the continual pointing out, however gently, of one's deficiencies. Aunt Blanche, however, meant to be kind despite her obvious disapproval of my upbringing, and I knew it had been her genuine concern that had prompted her to insist that I be left with her when Mother died, instead of my being dragged hither and yon, as she put it, in Father's erratic and sometimes turbulent wake. I had never quite forgiven her, however. I had been so totally immersed in my own dreams and longings that I hadn't realized that she might have once had dreams of her own, but had of necessity set them aside when her young niece needed her. I had reacted with the unconscious cruelty of childhood, but at last I was able to make amends.

She stayed well enough until after Christmas. We had a quiet but festive day on the holiday, and she seemed to enjoy the familiar rituals of opening gifts and admiring the tree which I had stayed up late to decorate. One gift from Father gave her special delight, for it was a large, carefully packed carton of fragrant oranges and lemons and limes, and when I opened it their pungent scents perfumed the little room.

"Oh!" Aunt Blanche exclaimed. "They smell so good! I should like so much to have one of them. Do you think I could, Beth? Right now?" Her pleading, though said merrily, was almost like a child's.

"Of course, Aunt. Father would be happy to see you so pleased with them." I rose at once and prepared one of the golden orbs for her. As she ate it daintily, I reflected on Father's response when I had told him I couldn't come to him just yet. Not only had he been understanding and sympathetic, but had said that he felt "most proud" that I had chosen to be so thoughtful and unselfish. "As for us," he had added, "we have all the time in the world to be together."

I pulled myself away from my brooding thoughts and went to the kitchen to see if I could help Sarah, for Dr. Mathews was to be our guest at dinner. Sarah waved me away with an indignant gesture.

"Sure, I'll be needin' no help from you or anyone else, child," she said. "'Tis me privilege and honor to be makin' this one last Christmas dinner for the dear soul." Tears trickled down her flushed red cheeks. She impatiently wiped them away. "Now get on with you. You must be with her every minute that you can."

Sarah's elaborate dinner was delicious, and Dr.

Mathews a pleasant and entertaining guest. Aunt Blanche seemed to enjoy herself immensely.

It was the last time she came to the table for a meal.

On a bright May morning Aunt Blanche died peacefully in her sleep.

I could not grieve overmuch, after all, for she had been oh so tired at the last, and I knew she was glad to go. And her death and funeral were not the trauma for me that they might have been, for she had long ago seen to it that everything was in order, and her many friends offered all the assistance that I could have wanted. My duties were simply to dispose of her furniture and personal effects, for the house was a rented one and no more needed to be done than to turn it back to its owner. I did all this, putting into storage those things that I was disposed to keep, then packed my valise and small trunk and started on my long-anticipated journey to meet Father in California.

I had been surprised that he hadn't come east upon hearing of Aunt Blanche's death, but he had written with many apologies that an important business deal made it next to impossible. "I am sorry to have to place such a sorrowful burden on your shoulders," he had written, "but Blanche would have been the first to understand. Her most fervent desire was that I should arrange a secure and abundant future for you, and that is what at this time I am attempting to do." He did not elaborate, but I sensed an urgency in his words that somewhat eased the hurt of his not being with me. Besides, he could not have possibly reached New York in time for her funeral.

My journey by rail to California was said to be a tedious

but not dangerous one. Several trains were to take me all the way across the United States to Los Angeles, where I was to board yet another which would carry me to San Diego and Father. My friends from my school days, gay, giddy girls with no more serious thoughts in their heads than those of attracting equally giddy young men, and of finery, and boating excursions and picnics now that summer was nearly here, alternately commiserated and envied me my coming adventure, and spoke breathlessly of my bravery at risking myself with the Red Indians and Mexicans and rough, unschooled cowboys I would no doubt meet in my travels and at my destination. In vain I tried to assure them that San Diego was a perfectly civilized town and that this was 1895 and the danger of my being scalped or carried off to become the bride of a red-skinned savage was somewhat slight, for their romantic notions afforded them more excitement than the truth would have.

At least I hoped it was the truth. As I settled into my first-class seat after waving farewell to the quite gratifying crowd that had gathered to see me off, I reflected that Father hadn't really written much of what I could expect to encounter on my journey west. It was San Diego, that pleasant and amiable city of perhaps sixteen thousand souls that he had often described in glowing terms. Of what lay between I really had no idea.

As it turned out, I wouldn't have missed the experience for the world. I saw cities and small towns and farms and vast prairies, I passed through long dark tunnels and over towering mountains and across trestle bridges so high and fragile-looking they stopped my heart with fear. And as we journeyed farther west I *did* see Indians and cowboys, both of whom often raced alongside the train to the amusement and excitement of its passengers, though I wasn't at any time in danger from them. And I ate foods in the dining cars that I had never tasted before, such as roast quail and stuffed grouse, and chicory and lobster salad, for Father had made sure that I traveled in comfort and style. But most of all I saw space, miles and miles of empty land, much of it not so much as sporting a tree or a farmhouse or a lone windmill. Though I considered myself to be well read, I had never dreamed that America was so vast, so solitary. There seemed no reason that we should ever deny admittance to anyone, for there was room for all who cared to come.

But for all of the beauty that I saw, as the days passed I found myself wearying of the long slow hours, the dust, and the heat which even in our well-appointed Pullman car was appreciable. I was vastly relieved when we at last pulled into the Los Angeles station and I could transfer to the Santa Fe train that was to take me to San Diego.

This last leg of my journey, which was short compared to the distance I had already come, was also far more comfortable, for much of the time we were within sight and smell of the Pacific Ocean. The breeze coming in from the open windows was cool and fresh and tingled with salt, and as I tasted it on my lips, tingles of excitement shot through me. In no more than an hour or so I would be with Father again!

He must have been as eager as I was, for when I emerged from the train he was waiting alongside it. Surprisingly, I knew him instantly in spite of the years we had been apart.

"Father!" I called excitedly, and the tall, distinguished-looking man at the other end of the platform turned and stared, then started toward me with long, determined strides. I couldn't wait circumspectly for him to reach me, and, forgetting the decorum I had been so carefully taught by Aunt Blanche, I picked up my skirts and ran joyously toward him. We met halfway, and without a moment's hesitation I threw myself into his arms. "Oh, Father!" I cried, tears of happiness welling in my eyes, "I can't believe it is really you at last!"

He hugged me tightly for a long moment, then held me at arm's length and surveyed me. "Can this be my little Beth? You have grown to be such a young woman. And a beautiful one, at that! What a pleasure it will be to present you to my friends!"

"I would like that, too," I said, "if I am allowed to bathe and change clothing first." I stepped back and looked down to ruefully survey my gray traveling suit. "Would you believe, Father, that this suit was white when I began my journey?"

Father laughed heartily and took my arm. "Let us say that I would *almost* believe it. I, too, have traveled long distances on trains, and with less of the amenities than you had on your journey. But you needn't worry about your appearance. First you may spend some time in your room at the hotel, and then we will have a quiet dinner together. I want you all to myself for one evening, at least."

Father had written that he had taken up permanent residence at one of San Diego's finer hotels, though I somehow thought that perhaps this would change now that I had come to join him.

"Is the hotel near the train depot?" I asked as he was seeing to my trunk and valise. Now that the excitement of meeting Father had lessened somewhat, I was eager to reach my room and relax for a time. A bath, I thought longingly, and a change into a crisp, fresh gown—

"Hardly," Father answered absently. He was busy hailing one of the hacks that waited before the depot. "It is on Coronado Island, which we must reach by ferry. But it is well worth the journey. It is truly the most beautiful hotel I have ever had the pleasure of entering."

I was impatient to see it after hearing this, and I could hardly wait until my luggage had been stowed on the rear rack and Father and I were settled in the hired buggy. Once we were on our way, however, I became engrossed in looking at the sparkling bay and the plantings of young palm trees which dotted the avenue, and then, when we had driven onto the ferry, the splendid view of Coronado Island up ahead, though I was shaken a bit when the whistle sounded and our horse started and reared slightly. The driver steadied him with a practiced hand.

"He never gets used to it," the man said in explanation. "Guess it must hurt his ears or something."

We reached the island without further incident, and as I breathed the pure, clear air, I regretted once again that Aunt Blanche could not have lived to accompany me, for surely here she would have mended and become well again.

With some effort I pushed these sad thoughts aside.

"Thank you for asking me to come here, Father," I said impulsively. "I know that I will be truly happy here."

Just then, in spite of the warm afternoon, a little chill ran over me, and I felt as though a shadow had momentarily passed before the sun. But just then I spied the red roofs and the turrets with their banners flying, and in my excitement at discovering that they indeed belonged to the Hotel del Coronado, I forgot all about it.

"It looks like a fairyland," I breathed.

Father smiled indulgently. "It is, in a way. It has every amenity one could ask for."

As the hack slowly rounded the curve of road that ran along the strand, I gazed at the huge building with its towers and gables and hundreds of gleaming windows and thought that I had never seen anything so spectacular. Father was highly gratified by my enthusiasm.

"I expect you *will* be happy here, Beth," he said.

I leaned forward and said to the driver, "Oh, hurry, will you? I simply can't wait to get there!"

How could either of us have known just how very wrong Father really was?

Chapter Two

The first time I saw Christian Rivers was that same evening in the immense dining room of the Hotel del Coronado, and my initial impression was not a favorable one.

He was sitting at a table near to the one Father and I occupied, and he was rudely and quite openly watching me.

I had not noticed him at first, for I was completely bedazzled by the opulent surroundings and the vastness of the dining room itself, which Father had said could seat a thousand at one time, and whose arched ceiling had not a single pillar holding it up. Even though I was from New York, where extravagant architecture is not unusual, the huge hotel with its nearly four hundred bedrooms and elaborate furnishings was overwhelming. Added to that was the fact that I still could not believe I was actually here with Father after all those years of dreaming about it. He sat across from me, even more handsome and imposing than I had remembered, and though I was nearly nineteen I had to remind myself not to giggle like a schoolgirl because my childhood fantasy had come true. All during those tedious and circumspect years of living with Aunt Blanche, of eating wholesome but dull foods such as oatmeal and skim milk, of attending the staid Presbyterian church with its overabundance of genteel spinsters who all smelled faintly of lavender and peppermint drops, of experiencing nothing more exciting than a lecture by an occasional visiting missionary who spent more time deploring the natives' lack of covering than their lack of spiritual knowledge, I had visualized the life I would be leading if only I had been allowed to cast my lot with Father after Mother's death. And my fantasies of being taken to elegant dinners and into the exciting and romantic world of the ballet while wearing gowns of rich velvet and glowing, whispering silks were apparently not so farfetched after all, for I felt as though at least one of them was being fulfilled right now.

I said as much to Father.

"I have dreamed of this moment for so long," I began. "I know Aunt Blanche meant well in insisting I remain with her, but it has taken me all this time to really forgive her for it. I must have made life quite difficult for her at times."

Father arched his eyebrows at me. "If that was your attitude, then I am sure you did. It is too bad. She had already given up so much, so many of the things dear to her, by being good enough to take you in."

I sat for a moment in silent astonishment. I had always secretly thought that at the root of her insistence on keeping me with her was the desire for a companion to assuage her loneliness. Now Father was putting a different light on it altogether. "What things?" I asked rather stupidly. "What exactly do you mean, Father?"

He laid down his fork and regarded me with a quizzical look. "Your aunt Blanche was a spinster, Beth, and much older than your mother. Have you never thought what it cost her to give up her freedom, her staid and settled routine, the peace and welcome silence which she was accustomed to?" At my stricken look he reached over and patted my hand. "Don't misunderstand me, Beth. She was willing to do it, and I know that you were generally a good child, but any child is a burden of sorts, even if a welcome one. There are always such things as colds, childhood illnesses, terrible nightmares that must be soothed away to the detriment of one's sleep—" He picked up his fork again. "Your aunt never had a high regard for me, I am afraid, and possibly with reason, but I always liked and admired her. She was forthright and courageous, and she considered it her duty to warn your mother that my profligate life-style might not be a happy or productive one, yet when Constance persisted in her determination to marry me, Blanche didn't turn her back on her."

His pause to sip from his wineglass gave me the opportunity to say, "Oh, but she had changed her mind about you. She wanted you to know that. She told me so not long before she died."

"I am happy to hear that. For when Constance was so prematurely taken from us, she was there for you." Father took a few bites of food as though the subject were closed, then added softly, "Much as I wanted you to come out to be with me last fall, I was proud and pleased that you chose to stay with your aunt now that she needed you. It reveals character that I am happy to see in my daughter."

My discomfort at the things he had pointed out lessened a little then, and I could go back to my food with more appetite.

"Now that I am here, Father," I ventured later, "what are your immediate plans? Are we to stay here at the hotel indefinitely? Not that I would mind," I added hastily, though it wasn't really the truth, "but I thought that you might be considering taking a house in town. I am quite experienced by now in managing one."

He got an odd little smile and took a sip of wine before answering me. "I am not quite sure as yet, Beth, but I have been thinking along those lines. May I answer your question more fully in the future?"

Something in the secretive way he said this piqued my curiosity, but of course I smiled and let the matter drop.

Our dinner was delicious and richly deserved our full attention, and we were halfway through our serving of succulent pheasant stuffed with wild rice before I first noticed Christian watching me. He was sitting with a pretty young woman in a shockingly low-cut gown who was paying rather more avid attention to her food than was well bred, but he paid none at all to his. Instead, he moodily toyed with a long-stemmed wineglass which he filled occasionally from a bottle that stood in a silver holder. I had glanced at him as I was admiring the elegant diners surrounding us, but now, as my gaze once more crossed his, I realized with a little shock of surprise that he was deliberately staring at me.

I am, I think, not unattractive, and this was not the first time I had been aware of a man's attention in a public place, but this look that he gave me was different. This was a stare of rude male speculation, and I blushed and lowered my eyes quickly to my plate.

Father saw my reaction, and stopping in the middle of a sentence, glanced behind him.

"What is it, Beth?" he asked with concern. "You've turned red as a poppy."

"Why, it's—nothing, really." I felt suddenly unable to go on with the excellent dinner. "It is just that—a young man at a table behind you was looking at me so strangely." I added, as Father craned around again, "The handsome one, with the large, dark eyes."

As he saw Father looking at him, Christian smiled sardonically and raised his glass in a little toast. When Father turned back to me, a disgusted look had replaced his usual pleasant expression.

"Pay no attention," he said. "That young man is a prime example of good stock going to seed. Decent women like yourself have nothing to do with him." Dismissing the subject, Father went back to enjoying his meal.

I supposed Father was right, though I felt a small twinge of rebellion at the slightly dictatorial tone of his remark. It seemed that the handsomest males were invariably those most disapproved of by society, whether or not they actually deserved it.

I could not resist another quick glance at the young man. He was still watching me with that sardonic smile, and I decided that in this case Father might well be right. Determinedly I fastened my gaze on Father's face and went on with our conversation, and when, after a long while, my glance again strayed in that direction, both he and the young woman were gone.

I was happy that Father seemed genuinely pleased to have me here with him. Even though I had been heartbroken when he had left me with Aunt Blanche, I had understood the difficulties in a traveling man attempting to raise a young girl by himself. And, though I didn't really admit it even to myself, I knew that Father's life hadn't always run smoothly. There had been times when the elegant apartments had been replaced by rundown boardinghouses and Mother had had to use her appreciable skills as a seamstress to keep us all looking the fashion plates that Father's business dealings required. I never quite understood just what his business was, but it had to do with the buying and selling of property, and sometimes, I remembered Mother remarking to my aunt, his dream got in the way of his judgment. To which Aunt Blanche had replied with some asperity that Mother was indeed charitable since his dream never included a permanent and secure home for his family.

"*He* is our security, Blanche," Mother had reprimanded her gently. The soft blue eyes had not lost their serenity. "Jamison would never allow us to suffer real want, not even if he had to do manual labor to prevent it. You do understand that, don't you?"

And Father never had, even after she had died and he had crossed the continent to California, for Aunt Blanche had received the agreed-upon monthly stipend for my keep without fail. And in later years, when he had settled permanently in California, she had been pleased that he had not disrupted my schooling by demanding that I join him.

Though her opinion of Father had changed, mine of him hadn't varied at all. To me he was still the glamorous, exciting, beloved figure of my fantasies.

As we sat together in the hotel's great dining room, he seemed in an expansive mood.

"I have long dreamed of having you here with me in

San Diego, Beth," he said, echoing my own thoughts. "California is a marvelous state and San Diego will someday be the jewel in her crown. It has everything anyone could possibly want—a salubrious climate, a good harbor, and, now that the flume has been built, an abundance of pure water. There is no limit to its future growth when word of its delights spreads to the world." He poured more wine, then continued. "And there is nothing to prevent a farseeing man from becoming an integral part of that future."

He continued to expound on the delights and future possibilities of the city while we finished our dinner, then, during dessert, reached into an inner coat pocket and brought out a folded piece of paper.

"You are probably wondering why I've been boring you with all this talk of San Diego and its future," he began. "But I have had a good reason. This paper is the key to *our* future, and it is linked irrevocably to that of the city's." He leaned toward me with a look of controlled eagerness. "I've just completed what I believe to be an incredible investment. If things go properly, it will mean security for our lifetime and even for that of your children."

I made some expression of surprise and delight and reached across to take the paper from him. But Father smiled and returned it to his breast pocket.

"Not just yet, Beth, dear. There are a few more details which need attending to before I can tell you about it. But my meeting tonight should take care of them."

"Tonight?" I repeated, and couldn't keep the dismay from my voice. "Oh, Father, must you go out tonight? I thought that you said we would be spending the evening together."

He patted my hand. "I'm sorry, dear, I really am. I would have taken care of the matter another time, but the young man I am meeting lives in the country and will be in San Diego for only a short while. When I heard that he was in town I called and left a message that I would meet him at nine o'clock. Did you know that this hotel has telephones as well as electric lights? Mr. Edison himself supervised their installation just last year. It is said to be the largest structure outside New York City to be lighted by electricity."

I knew he spoke of this to lighten my mood, but I was sorely disappointed that he planned to leave me so soon. I had so looked forward to being alone with him for a little while, for I had not felt able to talk freely while traveling in the hired hack, nor did one speak of private things in a public dining room. We had so much time to bridge, Father and I, and I wanted to talk with him about the years with Aunt Blanche, and of her death, and of how I had longed to be with him. And I was eager to hear of all the exciting adventures he had had while I was learning to pour tea properly and to deport myself as a lady should. And I most sorely wanted to reminisce with him about Mother in the hope that it would reinforce my fading memories of her. Sometimes in these latter years when I had felt the greatest need of her, I hadn't been able to recall her face.

Father saw that I remained downcast, and as though to put my mind to other things, brought out a small gold pocket watch and consulted it.

"There is some time before I have to leave," he said. "Why don't we go up to our rooms and I will give you the gift I have for you. Then I will hurry into town, take care of my business, and be back before you know it. If you haven't fallen asleep, I will stop by your room and tell you about this." He patted his breast pocket wherein lay the papers. "How does that sound, Little Princess?"

I was still somewhat dejected, but his use of my childhood nickname touched me, and besides, I didn't want him to think that I was going to prove a nuisance. I assumed a bright smile.

"It sounds exciting. And I do love presents. I can't wait to see what you have for me."

"And I can't wait to give it to you." Father's smile was indulgent as he signaled the waiter. He signed our check with his bold, dark signature, then graciously escorted me from the dining room.

As we crossed the crowded lobby to the elevator cage, I again had a sudden chilling sensation that someone was watching me. I glanced around. The dark-eyed young man from the dining room was standing near the main exit, and he was staring at Father and me with that same cynical expression. I felt my cheeks grow warm, but this time, instead of ignoring him, I raised my chin a trifle and gave him a haughty look before preceding Father onto the elevator. However, the uneasy feeling his attention caused in me did not immediately pass.

Once the elevator reached the second floor, we stepped off and proceeded to Father's room, which, as he pointed out, was near to mine, and I watched with not entirely feigned excitement as he unearthed my gift from the great square steamer trunk which I remembered from my childhood. The package which he brought forth was long and slim, and it had been wrapped in tissue paper and clumsily tied with a narrow blue ribbon. Father watched with an eagerness to equal mine as I undid it.

As the object came free from the tissue paper I caught

my breath with delight. It was a parasol, of the type carried almost constantly in the summer by young ladies of fashion who were concerned with preserving their milk-white complexions. But this was no ordinary parasol. The handle was of intricately carved ivory, the shaft an embossed silver rod. The cloth covering the frame was of a fine white silk which had been embroidered with a profusion of multicolored blossoms done in stitches so tiny they blended together like a painting. As I unfurled it, I saw that the design had been worked so cleverly that the pattern on the inside was exactly the same as that of the outside.

"Oh, it's beautiful, Father!" I exclaimed. "Far too beautiful to use!"

"Nonsense," he said sternly, but he was smiling at my pleasure. "It is meant to be used. Our California sun is deceiving, so unless you don't object to having that pert little nose adorned with freckles, I suggest you carry it with you most of the time."

"Oh, I will!" I put it over my shoulder and admired the effect in the ornate mirror which was hung over Father's chest of drawers. "But I shall be so afraid of losing it!"

"Try not to, Princess." He also was admiring my reflection. "I was assured that it is one of a kind. But if such a thing would happen, it would be easy to identify, for I have had your name inscribed on the shaft."

I closed the parasol and examined the slender silver rod more closely. There, near the ivory handle, in letters so gracefully etched that they seemed to be a part of the design, was my name, Elizabeth Darcy Allen, and the year, 1895. Impetuously I threw my arms around Father and kissed him on the cheek.

"It is lovely, Father, and I thank you so much! I will

treasure it forever!"

Father looked inordinately pleased, but made a harrumphing sound and hastily pulled out his pocket watch again. It had, I noticed, an unusual design of ivory inlay on the case. "Um, yes. Well, I am glad it pleases you." He snapped the watch open and gave it a cursory glance. "But I had best take you to your room now if I am to get to town and back before you are asleep."

"All right," I said, albeit reluctantly. Despite my long journey and the emotional turmoil of seeing Father again, I was not in the least tired, and the guests we met in the hallway as he escorted me to my room were so obviously dressed for a gala evening that it was all I could do not to let my envy show. But I managed to smile as, after unlocking the door to my room, he returned my key to me.

"Be sure to slide the bolt after you close the door, Beth," he cautioned me. "Even in a hotel as fine as the Coronado, you should take every safety precaution." After a slight hesitation, he leaned down and kissed me. "I have a good life here, daughter. Your presence will make it perfect." Before I could reply, he ushered me into my room and strode down the hall toward the elevator. I held my door ajar and watched until the grillwork opened and he disappeared inside.

After I had locked and bolted my door as Father had instructed, I was seized by an uncomfortable restlessness. I considered returning to the huge lobby and sitting quietly in a chair so that I could at least have the pleasure of watching the throngs of people come and go, and had even gotten to the point of unlocking the door when common sense prevailed and I relocked it and shot the bolt home again. Father hadn't specifically forbidden me to leave the room while he was away, but I knew that it was because the thought that I would do so hadn't even entered his mind. He would no doubt be surprised as well as displeased to come back and find me there instead of here, where he had left me. I also had the thought, fleeting but still bothersome, that I might again encounter the young man from the dining room, and the prospect of meeting his bold glance while unescorted made my breath quicken with what I assumed to be dismay.

At last, resignedly, I changed into a nightgown and comfortable wrapper, then considered passing some time by taking my belongings from my small, curved-top trunk, but eventually decided against that also, for Father might come home from his meeting and announce that we were to leave the hotel tomorrow. I contented myself with merely unpacking the few articles of clothing I might need in the next day or so and arranging my comb and brush on the dresser.

After I had done this, which took no more than a few moments, I lay on the bed for a time and idly leafed through a magazine which I had bought for my train journey, but it couldn't hold my attention and I soon tossed it aside. After pacing about for a while, admiring the room's unusual features such as the well-proportioned fireplace and the small wall safe, both of which Father had mentioned as being in every bedroom, I got my small lapel watch from the dresser. To my dismay, it was not yet ten o'clock.

Sighing, I laid the watch on my bedside table and opened the heavy draperies. Carriages were still coming and going on the avenue leading to the hotel, but beyond the drive the sea surged restlessly onto a silver stretch of sand. The moon still lay low on the horizon, its pale rays lighting the foamy breakers with an iridescent glow. The deserted beach held a lonely fascination for me, and after turning out the lamp I pushed an easy chair close to the window and curled up in it, covering myself with the warm comforter from my bed.

I didn't feel in the least drowsy, but the movement of the sea must have had a hypnotic effect on me, for I soon fell asleep. I must have dozed for quite some time, for when I was awakened by a slight sound at my door, bright moonlight was flooding the room. Thinking that Father must have returned at last, I threw the comforter aside and eagerly ran to let him in.

I was just reaching for the bolt when I noticed the doorknob. It was being turned, slowly and quietly, by someone in the corridor.

I stared at it, transfixed. It could not be Father out there, for he had told me to bolt the door and knew he would have to knock. Besides, he would never attempt to come into my room without asking my permission first. My heart pounding heavily, I backed away, then froze in fascinated dread as I saw that a small, flat metal object was being inserted between the door and its frame. I choked back a scream. The knob was turned again, and this time the door would have opened if it hadn't been for the bolt which I had securely thrust home.

I have no idea how long I stood quivering with fear, but the would-be intruder must have realized that the bolt was drawn and thus there was no access, for the doorknob did not turn again, and at last I recovered enough to snatch the comforter from the chair and retreat with it to the dubious safety of my bed.

I lay huddled beneath the covers for what seemed an eternity, but finally my terror abated somewhat and I ventured to reach for my watch. Its tiny gold hands pointed to just after midnight. Where on earth could Father be? I could only conclude that either I had slept through his knock, or that he had returned so late that he hadn't wanted to awaken me. I thought longingly of fleeing the short distance to his room, but fear that the intruder might still be lurking in the hall kept me cowering in my bed. And though Father had mentioned telephones, there had been none in evidence during my earlier inspection of this room. With sinking heart I realized that I would simply have to endure my terror until morning.

I didn't expect to sleep again, and indeed it was well toward morning before the chill of terror left me to the point where I was able to at least doze fitfully. But as the first rays of sunlight poured through the open draperies I awoke with a start of fear, and taut nerves set up a silent screaming. I couldn't force myself to stay in bed any longer, and with a murmured prayer of thanks that the long night had actually ended at last, I rose and went to the window.

There below me, bathed in pearly morning light, lay the shining length of sand, and beyond it the great expanse of lace-trimmed, blue-green satin that was the Pacific Ocean. My fear disappeared as though a fresh breeze from the beckoning sea had wafted it away, and I longed to be out on that empty, wind-swept beach where I could smell the sharp, clean odor of salt-laden air and hear the sound of the breakers as they surged and eddied at my feet. Quickly I donned my cotton frock with the fine blue candystripe and fashionable bustle, snatched up my new parasol, and after a quick glance into the empty corridor, hurried to Father's room.

But as I stood before his heavy, paneled door, I was seized by a sudden hesitation. As I saw by my watch, it was still not much past seven o'clock. If he had come back very late last night, would he appreciate being awakened so early? Having been separated from him for so long, I had no idea of his usual morning humor.

At last I knocked once, very gently. When he didn't answer I decided to go on downstairs without him.

Early as it was, there were several people in the spacious lobby, though I ascertained at once that the dark young man was not among them. I nodded to the desk clerk, who eyed me with disdain, and to a young bellboy who tipped his hat to me and smiled in a friendly manner. I considered telling the clerk about what had happened last night, then decided to wait and let Father report it when he came downstairs later on. Instead, I went on out to the wide veranda which stretched across one front section of the hotel.

It was completely deserted. Long rows of wicker chairs, which upon our arrival yesterday had been filled with stylishly dressed people enjoying the invigorating breeze from the ocean, now stood forlorn and empty. Beyond the drive, the wide stretch of beach was devoid of the strolling couples who had thronged it then, and its only occupants were the seagulls who marched about in pigeon-toed pomposity in their ever-engrossing quest for sustenance. It had a solitary, peaceful air, and any last vestiges of my previous night's fear were erased by the serenity of it. I hurried down the drive and ran joyously to the hard-packed strip of sand that lay along the water's edge.

Once there, I stood for several moments breathing in the exhilarating scents of sea and salt as I gazed into the infinite distance stretching before me. The only sounds to break the silence were the rhythmic cadence of the breakers washing against the sloping shore and the mewing of the gulls as they wheeled and dipped low over the placid sea, which reached endlessly toward the distant horizon. I gave a deep sigh of contentment as a sense of peace and freedom swept over me. It was as though this land, this California, was my spiritual home, and I had waited all of my life to discover it.

The early sun had burned away the last vestiges of morning mist, and, mindful of Father's warning, I opened my parasol against its rays. I began to stroll on the hard sand above the water's edge, admiring as I went the treasures that had washed up on the recent tide: an iridescent piece of abalone shell, a length of rubbery seaweed with pear-shaped amber bulbs, a round, flat shell with a perfect starfish etched into its pale gray surface. I exclaimed over each and retrieved the starfish shell to carry back and show Father when he awakened.

I had been so absorbed in all that I saw that I had taken no notice of the passing of time, but the thought of Father reminded me of it, and reluctantly I decided that I must return to the hotel. As I turned to retrace my steps I was startled to see that my walk had taken me much farther from the hotel's environs than I had meant to go.

I also saw that my return path was blocked by the dark young man who had stared at me so rudely last night.

His presence was totally unexpected. I had heard no sound, no swish of sand such as my feet had made. He was simply there, standing wide-shouldered and menacing between me and the safety of the distant hotel. I let out a little scream of shock and fear. This seemed unaccountably to please him, and a half-smile curved the full, sensuous lips.

Seeing this, I felt anger somewhat override the fear that had coursed through me. It sent tingling shock waves through my body clear to my fingertips.

"How dare you startle me like that!" I demanded. I heard my voice, high and unsteady, and I strove to bring it to its normal pitch. "It is most certainly not a gentlemanly thing to do!"

At my words his smile deepened. It seemed to have a hint of danger in it. "That," he said lightly, "is no doubt because I am not a gentleman, and have never professed to be. Didn't Jamison warn you about me last night?"

There was no use pretending that I didn't know what he was talking about. I lifted my chin.

"We did not discuss you," I answered haughtily. It was, after all, nearly the truth. "Apparently he didn't consider you to be of that much importance."

"Oh, ho," he said. "The lady is not above a small lie or two. If, in fact, you are a lady."

I felt the color rise to my cheeks. Though he continued to smile, his unaccountable hostility was apparent, as was his unwarranted innuendo.

"Very well," I returned icily, "if you must have the truth. He told me only that no decent woman would have anything to do with you." I closed my parasol with a snap and started to go on past him toward the hotel.

Immediately he moved directly in my path. Willing myself not to flinch, I tilted my head to stare steadily into his eyes, which were no more than a few inches from mine. In spite of my anger I could not help noticing how clear they were, how soft a brown were the unusually large iris. Their direct stare caused an odd sensation to stir beneath my breastbone.

"What else did he say?" the young man asked softly.

I recovered myself. "Not one word," I answered firmly. I should have left it at that, but his skeptical expression prompted me to add, "But I will say something to your credit. You aren't a liar, at any rate. For as you pointed out, you are certainly no gentleman."

I almost instantly regretted it, for at my words something flickered across his face, and I realized that I had touched a nerve. I thought for a moment that he was going to apologize, but he merely bowed slightly and stepped out of my path.

With as much dignity as I could muster I started for the distant hotel. I could feel his eyes following me, but I didn't look back. Neither did it seem that I breathed again until the lobby doors closed behind me.

Chapter Three

My stroll had taken me much longer than I had meant it to, and when I returned the lobby had begun to fill with guests checking in and out and going into the dining room for their morning meal. I glanced about to see if Father had by chance come downstairs to search for me, but when I didn't see his tall figure, I went immediately to my room.

I had carefully locked it when I left, but upon my return I searched it thoroughly before closing the door and bolting it. Only that bolt had stopped the intruder last night, and there had been no way to guard against his breaking in once I had left it. But I saw no sign that anything had been disturbed while I had been away, so I tossed my parasol on the bed and hurried to wash my hands and smooth my windblown hair. I felt that Father must surely be awake by now, and possibly impatiently awaiting my return.

As I rebrushed my hair and otherwise tidied myself before the beveled mirror, I saw that my cheeks were high with color, partly, I decided, because of sun and wind, but even more because of the frightening yet intriguing encounter with the strange young man. What had been his purpose in following me? It seemed a good deal of bother to go to if he only meant to insult me. I thought back over our encounter and then in dismay let the brush fall to my side as I recalled that it had been I, after all, who had said the first angry words. Was it possible that he had been only taking a morning stroll as I had, and was not there because of me at all? Had his anger been only in answer to mine? It would not be an unusual reaction—I had experienced it myself many times. As for my sensing an air of menace about him—perhaps that had been only my natural apprehension in the situation. A lone girl on a deserted beach was highly vulnerable, and in view of my night of terror, I had felt doubly so. Perhaps, if given the chance, he would have apologized for frightening me, and for his bold and ungentlemanly appraisal last night in the dining room. I felt a pang of regret for lashing out at him, especially when I recalled the odd, little-boy look of hurt when I had said that he was not a gentleman.

I began again to brush my hair. When he had confronted me, frightened though I had been I hadn't failed to notice how attractive he was—the smooth, tanned skin and even teeth, the large dark iris of the well-shaped eyes—

I caught myself sharply, and as I got my small crocheted purse, tried to banish the handsome stranger from my thoughts. His attractive appearance scarcely mattered, for I most certainly would not be seeing him socially. Father had made that abundantly clear. As I closed and locked my door again and went toward Father's room, I ignored the tiny spark of rebellion the thought caused. Even though I had controlled my own affairs ever since Aunt Blanche's death, and had managed to cross the Continent unaccompanied, I had not enjoyed the experience and had often wished that Father had been there to take responsibility. Now that he was assuming that role, I should not complain because with that same responsibility automatically went authority.

This time, when I reached Father's room and knocked at his door, I waited confidently for his answer. When none was forthcoming, I rapped again, harder this time, and called as loudly as I dared, "Father? Are you awake?"

Again there was no answer. And as I stood there, confused and dismayed by the silence, I felt the first dark rush of apprehension, the first icy certainty that something was terribly wrong.

My initial impulse was to throw myself against the door and pound on it with my fists while loudly screaming for Father to open it; my next, and saner one, was to rush downstairs and enlist the aid of the desk clerk. This thought I put into effect immediately, bypassing the elevator in favor of the stairs, which I descended so precipitously that I drew surprised and startled glances from those ascending. I rushed to the huge main desk, where the clerk, with maddening slowness, completed sorting his handful of mail before turning to face me.

Ignoring his surprise and obvious disapproval of my breathless state, I spoke without waiting for his greeting.

"Please, can you help me? It is my father—Jamison Allen. He doesn't answer my knock. There's—I am sure that he must be terribly ill."

The clerk glanced at the bank of pigeonholes behind

him, then regarded me with the exaggerated patience one reserves for dullards and slightly backward children. "Mr. Allen's key is here, which means he isn't in the hotel at present. And since I have been at this desk since early morning, I think it is safe to assume that he didn't return from town last night."

I stared at him in stupefaction. I had not even thought of that possibility.

"But—he must have!" I stuttered. "He knew I was waiting up—" I caught the gleam of prurient interest in the clerk's eyes and amended with more dignity, "I just arrived yesterday from the East. My father had an appointment in San Diego, but he planned to return shortly. I am sure he wouldn't spend the night away without sending word that he was, after all, not to return until today."

The clerk shrugged, his interest fading. "Perhaps he missed the last ferry. Or decided to spend the night with friends. Mr. Allen often does that, you know."

I wasn't in the least reassured. "The hotel has telephones," I reminded him. "Surely he would have called and left a message for me."

He paused, as though arrested by the thought. "That's true," he conceded. Then he brightened. "But perhaps the home of his friend doesn't. Don't worry, Miss Allen," he added, his tone more sympathetic, "I am sure your father will be along any moment now. Why don't you have a bite of breakfast in the dining room while you are waiting for him."

"No," I said with an effort, "I think I will go back to my room and wait."

I left the desk, but didn't immediately go toward the elevator. Instead, I found myself standing idly in the middle of the busy lobby. I simply didn't know what to do. I couldn't bear the thought of returning to my room to pace about like a caged tiger, yet the thought of food appalled me. I knew, even if the clerk did not, that Father would have somehow gotten a message to me when he had found he wasn't returning, unless it had been impossible for him to do so. And any of the causes of that impossibility which immediately came to my mind did not bear scrutinizing.

My stomach felt as if I had swallowed a large, cold stone, and there was a strange, melting sensation in my knees. Perhaps the clerk was right, I thought helplessly, and I should at least have a cup of tea before deciding what course to take. Fighting panic, I started across the lobby to the dining room, and nearly collided with a tall, slim man who was hurrying to the desk. He halted and steadied me with a hand beneath my elbow, but the incredibly blue eyes in the set face didn't really see me. With a murmur of apology he continued past and went immediately to the desk.

I stepped behind the shelter of a potted palm and stared after him. I felt shaken by more than the simple matter of a near collision, for, while I was sure I had never seen the man before, something about him was strikingly familiar. As I watched, he spoke at length to the desk clerk, who looked first surprised, then shocked and concerned. After a quick glance around the room, the clerk answered rapidly, and the tall man nodded and went toward the elevator cage. The clerk stared after him, a sheaf of unsorted letters forgotten in his hands.

My own hands grew cold, and unconsciously I clasped them together for the meager warmth. Quickly I returned to the desk.

"Mr. Orley." I had seen the name on the desk plaque. "Have you had news of my father?"

At my words he turned, then went red with sudden discomfiture.

"Miss Allen! I had no idea you were still downstairs. I thought you had gone back to your room!"

Cold fear clutched me. It was a moment before I could manage to say, "Why? Were you looking for me?"

"Ah—" Mr. Orley evaded my eyes and became very busy shuffling his handful of mail. "As a matter of fact, there is a gentleman to see you. I—I didn't know you were still— Ah—I sent him upstairs to your room." He tried to meet my gaze and failed. "Perhaps you should go up and find out what he has to tell you."

Carefully, as though walking a tightrope, I went to the elevator. The car had just arrived, and as the attendant swung open the grilled gate I brushed past the exiting passengers and went in. Someone else got on, though I was scarcely aware of it, then the gate shut and we rose toward the second floor.

No one needed to tell me. Though my mind seemed a black void in which all thought was suspended, I knew what the tall stranger from the lobby had come to tell me. I had known from the moment that the clerk wouldn't meet my eyes that the reunion I had dreamed of for the past ten years was over. It had consisted of one carriage ride, one leisurely dinner, and a few brief moments with Father in his room. Those short hours and some fading childhood memories were going to have to last me a lifetime.

The elevator concluded its interminable journey. Its latticed gate opened with a painful grinding noise, and I stepped out to confront the blue-eyed man I had seen downstairs. I allowed my gaze to touch him briefly, then looked beyond him at the gold-striped wallpaper of the hall.

"You are looking for me?" I asked politely. My voice still had that dry, precise sound to it.

He spoke gently. "You are Elizabeth Allen? Jamison Allen's daughter?" When I didn't answer, he stepped forward and put a hand beneath my arm. Still gently, he drew me away from the elevator door. His look, his voice, his every movement told me what I did not want to hear, and instinctively I pulled away from his touch. He let me go immediately, then said, "Would it be possible for us to talk together in your room?"

When I didn't answer, he again took my elbow, and this time, like a bewildered child I allowed him to lead me to my door. I stood before it, docilely waiting for it to open, until he asked, "May I?" in that same gentle tone, and took my purse from my hand. He found my key and opened the door, then stepped back to allow me to precede him.

I stepped inside, then, after a moment's hesitation, went to the window which looked out on the Pacific. Far down the beach a couple, she with opened parasol, were strolling arm in arm. A sudden sharp envy touched me, shocking me from my somnolent state. I had thought that this morning I would be walking there with Father.

From behind me the stranger cleared his throat and I turned, this time to look directly at him. Reluctance to perform his difficult task was written in his face, and, feeling obscurely sorry for him, I sought to relieve him of his onerous duty. I said, "My father is dead, isn't he?"

He didn't ask how I knew. He only replied, quite simply, "Yes. I am sorry."

I regarded him numbly. It seemed as though there was something else I should say. At last I asked, "How did it happen?"

He hesitated, the lines deepening around his mouth. "He was found in the street not far from my hotel in San Diego. He had been—"

He couldn't seem to finish the sentence, so again I said it for him.

"Murdered?"

"Yes." Then, as though he had to put something into the emptiness that followed, "The motive apparently was robbery. His purse, his watch—everything was gone."

Though he spoke as gently as possible, his words seemed to slam into my stomach, taking the breath out of me. I closed my eyes and saw Father as he had been last night—tall, handsome, dignified. To have ended his life lying in a dirty street without a penny in his pockets. How he must have hated it.

The stranger must have thought me about to faint, for he spoke again from quite near me. "May I get you anything—a brandy, a glass of water?"

I opened my eyes. He was standing so close that I could easily see the fine lines that some earlier sorrow had etched around the compassionate eyes and firm, uncompromising mouth, and even in my distraught state I recognized that his somber expression was not caused entirely by this present disagreeable duty. He didn't at all look to be a happy man.

I struggled to remember his question.

"No," I said at last. "I need nothing." Nothing that it is possible for you to supply, I amended silently. Beneath the numbness, questions were stirring, and at last I summoned the spirit to ask, "You are from the police?"

"No, but they asked me to bring you the unhappy news. Your father was a family friend." He inclined his head in a slight bow as he added, "Forgive me. I should have told you my name at once. I am Adam Rivers. Of Riversbend."

Riversbend. The moment I first heard that name I should have known, should have sensed that once I became entwined with the place and its occupants my life would never be the same. But in truth I felt no premonition, no hint of danger upon hearing it. I only thought that because of his friendship with Father, here was someone to whom I might turn for help.

Adam must have read my expression, for he said at once, "I want to help in any way I can. I am acquainted with your father's lawyer. Perhaps you will allow the two of us to take care of the funeral arrangements for you."

Relief washed over me. Though I had recently gone through the rigors of a funeral, the situation had been entirely different. Not only had I suffered a far greater shock, but here in Coronado, a full continent away from all that was familiar, I was at a loss as to what to do in the face of this unbelievable tragedy. I hadn't, in fact, even given any thought at all to a funeral; hadn't really, as yet, assimilated the fact that Father was actually dead. But he *was* dead; I knew that on the surface of my mind, and I accepted Mr. Rivers's offer with deep gratitude. "It is incredibly kind of you to go to so much trouble for a stranger," I finished faintly. "I don't know how to possibly thank you."

"Don't try," he said. "Just put it out of your mind." As seemingly an afterthought, he added, "It will be necessary, of course, to consult you about your preferences in various matters pertaining to the service."

I grasped eagerly at the thought that I would be seeing him during the three days which propriety required elapse between death and funeral. Those days stretched before me as cold and lonely, and beyond them an abyss—an emptiness containing nothing at all that I could foresee as a future. My mind recoiled from the disturbing vision and I said eagerly, "Oh, please do. I want to help in any way that I can."

Adam hesitated, reluctance once more shadowing his face. "There is the matter of the police. They are anxious to talk to you." I must have looked apprehensive, for he added soothingly, "There is no reason to be frightened. It is just that they will want to know if you can shed any light on the situation."

"I doubt that I can," I said. "I arrived only yesterday—" My throat seemed to close over the words and I couldn't go on.

Just then there was a knock at the door. Adam said, "That is probably the police now. They are eager to begin an investigation." He looked apologetic. "I know it seems cruel to put you through all this just now, but—your father was well liked in this town, Miss Allen. The police, as do the rest of us, want to see his murderer brought to justice."

I tried not to flinch at the word, "murderer." "I understand," I replied. I squared my shoulders. "Please ask them to come in."

It was not as bad as I had expected. Officer Johnson had a fatherly look, and after he and the young policeman with him, whom he addressed familiarly as Ben, had offered their condolences, he asked me quietly about

Father's and my activities of the previous day, and what Father's plans for the evening might have been.

"I had only just arrived," I told them. My voice threatened to quiver and I sternly controlled it. "After Father met me at the train station he hired a hack to bring us here to the hotel. When we came in, Father took only a moment to point out some of the unique features of the hotel, then escorted me to my room so that I might rest and change for dinner. I presume Father did the same."

"Did he *say* that he had spent this time in his room?" Officer Johnson inquired.

"No, but when we met again he had changed into evening clothes." I remembered how splendid he had looked when I had answered his decisive rap at my door, and the way his eyes had lit in admiration at my elaborate toilette. I had felt that he deliberately chose to take the stairs instead of the elevator because he was proud of me and wanted our leisurely and graceful descent to be seen by those of his acquaintance who might be among the elegantly dressed guests in the lobby below. And I was equally proud to be seen on the arm of such a dignified and handsome man, even though I supposed there were those among the throng who knew he was my father. Tears again threatened at the memory, and I hastily continued. "We went directly into dinner in the hotel dining room, after which he escorted me upstairs again before leaving for San Diego."

"Did he say exactly where he was going?" Officer Johnson asked. "That might be of primary importance."

I told them that he had spoken of an appointment in town. "He talked as though the man didn't live in San Diego. I'm sorry, but I don't believe he mentioned a name."

"That appointment was with me," Adam put in. "I got the message that he was coming to see me, but he never arrived. He must have been on his way to my hotel when—" He left the sentence dangling.

"Can you tell us what the meeting was about, Mr. Rivers?" Officer Johnson spoke very respectfully to Adam.

"I have no idea," Adam said. "The message simply stated that he had an important matter which he wished to discuss with me."

I suddenly recalled Father bringing out the paper at dinner last night. "I believe I can help a little," I said. "It was something about an important investment which Father had made. He was carrying a document concerning it." I looked from one to the other of the men. "He no longer had it when he was found?"

The two policemen and Adam exchanged looks. Ben, the younger one, said slowly, "He had nothing on him when we got there, not even identification. Someone among those who found him recognized him, and the chief decided to call Mr. Rivers, since he knew they were friends."

"Did your father perhaps discuss this investment with you?" Officer Johnson asked. "Or mention anyone who might be involved in it with him?"

I thought a moment, trying to recall that interval at the table. The shock of the subsequent happenings had caused it to recede in time, and I felt as though I were looking at the two of us, sitting there together, through the wrong end of a telescope. "Not that I can recall," I answered with some confusion. "He just said that if things went properly, it would mean security for the rest of our lives." I could not help my voice breaking on the last few words, and Officer Johnson said hastily, "Don't trouble yourself about it just now, Miss Allen. I know how hard this must be for you. But we need all the information we can acquire if we are to clear this matter up, for though it seems at the moment to be a simple case of robbery, the department is not willing to accept that without further investigation. So if you do happen to remember anything else, we would appreciate your cooperation. And, speaking of that—" He cleared his throat and shuffled his feet a little. "There is one more thing I must ask you to do."

I forced back my tears. Though I wanted most desperately to help in any way that I could, I felt that I was coming to the end of my tether. The blessed sedative effect of the original shock was wearing off, and from somewhere deep within, the cruel pain of my loss was beginning to burn and to radiate outward into the rest of my body. "I want to do what I can," I said thickly.

The officer looked pained at what he had to say. "We need to take a look at Mr. Allen's room, and we would like for you to be there when we do."

I had not expected that. Must I go to Father's room, where I had felt such happiness just last night when he had given me the lovely parasol? For a moment I didn't feel as though I could possibly do it, and I looked at Adam Rivers in silent supplication. He said reassuringly, "It shouldn't take long. I will come with you if it will help at all."

I felt a vast relief. "I would appreciate it." It was not that the policemen frightened me, for they were most polite, but Adam had such an air of kindness and dependability that I felt I could rely on him. Besides, he had been a friend of Father, and I clung to that thought as to a lifeline.

Officer Johnson nodded. "Fine with me." He stirred restively. "If you're up to it, Miss Allen, I think we'd best get on with it."

I was grateful for the strength of Adam's arm as I stood once more before Father's door. Officer Johnson had thought to get a key from the desk clerk, and as he brought it from his pocket and inserted it in the lock, my breath quickened. Immediately Adam's hand on my arm tightened a little, and my tension eased. But when the door swung open, I gave a gasp of dismay.

The room had obviously been ransacked. There were books, papers, clothing flung indiscriminately about. Father's big steamer trunk, from which he had taken my parasol, stood with gaping lid, its tray on the floor beside it.

I was shaken to the heart by a sense of violation. "But why? Why have they done this?" I cried, my teeth chattering with the shock of it. "What could they have been looking for?"

"I have no idea, miss." Officer Johnson and the other policeman looked helplessly about. Adam, too, stood staring, consternation and puzzlement mingling on his face. "And won't have, most likely. Unless it is you, I don't know of anyone familiar with Mr. Allen's belongings." The older man turned to Adam. "Or possibly you, Mr. Rivers. Since you *were* friends, you may have visited him here at the hotel from time to time." Adam's consternation had faded somewhat, but he was still wearing a puzzled look. He seemed with difficulty to bring his attention to the police officer's words.

"Strangely enough, I have never been in this room,"

he said slowly. "I have joined him at the hotel occasionally for dinner, but we didn't come upstairs." He added, I assume by way of explanation, "He knew that my time in town was limited, and he generally came to see me in San Diego. Just as he had planned to do last night."

As Officer Johnson regarded Adam, there was a trace of speculation beneath his mild exterior. "You are sure that you have no idea what he wished to discuss with you? Some business deal that you had talked about in the past?"

His words seemed harmless enough to me, but Adam stiffened and the lines of grief or pain that I had noticed before deepened around his mouth until they looked as though someone had carved them there. His eyes narrowed with unmistakable anger as he answered, "Jamison was my friend. Do you suppose that I would withhold any information at all that might shed light on the reason for his murder?"

The officer seemed to recognize some danger point in Adam, for he protested hastily, "No, of course not," then turned to me in an obvious move to change the subject. "Do you suppose that you would notice if something of your father's was missing? Some valuable piece of jewelry, perhaps?"

"I?" I protested. "I have no idea what father might have possessed, other than the gold watch I saw him with last night. And I was only in this room for a few minutes." My nerves were fraying badly, and at mention of those few minutes I felt the tears threaten again.

"I think Miss Allen has told you all that she can," Adam said evenly. "I would like to take her back to her room. Then, if you wish, I can go through some of these papers. I do have some knowledge of Jamison's business interests, and I might be able to give you an idea of what should be here. As to any valuables Jamison might have had, I suggest you look in the wall safe for them. Each of these rooms has one, you know, and it doesn't seem to have been opened."

Officer Johnson looked thoughtful. "I suppose I could get the combination from the hotel manager. But I wouldn't like to open it without Miss Allen's presence."

"Then it will just have to wait," Adam said impatiently. I breathed a sigh of relief. "It is plain to see that she is at the breaking point." He came over and put his arm around me to draw me toward the door. "But I think she would be satisfied if both you and the manager open it together and attest as to its contents."

"Oh, yes," I said quickly. "That would be most satisfactory." I didn't care what was in the safe, didn't for the moment care what the burglar had been looking for or if he had found it. I only wanted to be somewhere where the whole nightmare might go away, at least for the moment. I must have sagged a little, for I felt Adam's arm tighten around me and he began walking with me toward the door. I went willingly, grateful for his strength and assurance. If only he had been with me last night—

Last night. I stopped, and looked up at Adam.

"I had forgotten. There was—someone tried to break into my room last night."

"What?" Consternation sounded in the voices of all three men.

"You had better tell us about it, miss," Officer Johnson said. "It's a possibility the two things are connected."

"It was late—sometime around midnight, I think." A chill crept over me as I began to recall it. "I was waiting for Father—he had said he would stop by my room when he returned—and I fell asleep. I was awakened by a sound at the door, and thinking it was he, I went to open it." I described the turning of the doorknob, and the terrifying moment when I saw the metal object slip between the door and its frame. "The door would have opened if I hadn't had the bolt in place, as Father had told me to." I could not repress a shudder. "What does it all mean?" I cried. "If the two things are connected, what is it that they could have wanted? I have nothing of value, and nothing at all that had belonged to Father."

The three men seemed as puzzled as I. Adam said, "I wish I could answer that question." He looked at the two policemen. "I do know that there should be a guard at her door. If the police department can't provide one, then I will."

"No need for that, Mr. Rivers," Officer Johnson said. "I'll call the station and have somebody sent out right away. In the meantime, Ben here will stand guard." He added reassuringly to me, "You go on to your room, Miss Allen. Don't worry about a thing."

If only that were possible, I thought bleakly. Instead, my mind was reeling with questions for which no one had the answers. Were the mysterious intruder and Father's murderer one and the same? And if so, what had he hoped to find in Father's room, or for that matter, in mine? And if he had managed to get my door open, would I have also been a victim? At this I shuddered again and was glad to allow Adam to lead me from the disheveled room.

Once we were in the corridor, with its closed doors and blessed anonymity, Adam said, "I'll take you to your room, but you will allow me to order you something from the dining room? You must eat, you know, even if it is only tea and toast."

I appreciated his thoughtfulness, but for the moment I simply wanted to reach the dubious sanctuary of my room. "Not just yet," I said faintly. Though I had had nothing since dinner with Father last night, I couldn't even tolerate the thought of food. What I wanted, suddenly and most desperately, was to be alone. Adam seemed to sense this, for, after unlocking my door and escorting me into the room, he said, "Then I will leave you for a time, so that you can get some rest. But I am ordering food to be brought up to you in an hour or so, and I will try to be back to see that you eat it. In the meantime, try not to worry. Ben will be just outside."

It was an effort even to nod my head. Without saying more, Adam left, and with wooden movements I followed him to the door and bolted it. Then I turned and faced the empty room.

My parasol was still lying on my bed. I went over and lay down beside it, and after a moment I pulled it to me and caressed its silken folds.

It was only then that I began to weep.

Chapter Four

Adam was true to his word; it could not have been more than an hour before I heard a knock at my door. It was a gentle tap, calculated not to disturb if I actually were asleep, but of course I had not been fortunate enough to escape so easily the nightmare into which I had been plunged. When the knock was repeated, I forced myself to rise and go to the door. I slid back the bolt and opened it cautiously. Adam was standing outside, and with him was a waiter from the hotel dining room. The latter was carrying a covered tray.

"I trust we haven't awakened you," Adam said, though his voice didn't indicate that he thought it possible. "May we come in? I have ordered a light lunch in the hope of tempting your appetite."

"That is very kind of you." I stepped back to allow them to enter, but when the waiter had set the tray on a nearby table and turned to me for further instructions, his eyes traveled over me and I was suddenly acutely aware of my swollen eyes and generally rumpled state. I stammered self-consciously, "But I really must tidy up a bit, Mr. Rivers. I am most disgracefully disheveled."

I thought he started to demur, but instead he answered pleasantly, "Of course. I'll be within call."

The two men went out again, and I turned to look at myself in the mirror. My eyes were strangely bereft of their usual blue. They looked drained and colorless, as a lake sometimes does on a gray and cloudy day. My fair hair had come loose from its moorings, and the pin-striped cotton dress I had donned so gaily only that morning was crumpled and creased. Some vestige of my usual pride returned to me, and I blushed with embarrassment that a total stranger such as Adam should have seen me in that state.

For he was a stranger, even though our being thrown together by these tragic circumstances had made me temporarily forget it. I reminded myself of this as I quickly removed the cotton dress and replaced it with a gown of gray watered silk. I had seldom worn this dress. Though I liked the way the fitted waist continued into a gentle flare at the hip which flattered my slender figure, I had initially thought the subdued color and high, lace-trimmed neck somewhat mature for me. But today, I thought sadly, it fitted my mood and circumstances more than anything else I owned. I swept my hair into a chignon at my nape and secured it firmly with large bone hairpins, then surveyed myself critically in the mirror. I now looked presentable, I decided, though there was nothing I could do about the redness of my eyes beyond bathing them carefully with a little water from the ewer on my nightstand.

As Adam entered in answer to my summons, those intense blue eyes of his swept over me and he seemed to see me for the first time as a woman, an entity beyond the helpless female to whom he was bound by duty and friendship to offer aid. His usually somber expression lightened and he said, "The color of that gown exactly matches your eyes. Strange, I had thought them blue, like mine."

"Sometimes they are blue," I answered. "They seem to change with my mood, or with what I am wearing." Feeling that our conversation was somewhat personal for the occasion, I added, "I should like some of that tea now, if it hasn't gotten cold."

The brooding look returned as though I had reproved him, and he turned his attention to the tray the waiter had brought. Removing its high, domed cover, he revealed a small silver samovar, as well as a cozy-covered teapot. The tray also held cups and saucers of delicate china, a plate of breakfast rolls, and a small, trembling blancmange. "The Hotel del Coronado thinks of everything," he said, and whipping out a small gold matchbox, lit the candle under the samovar. "It will take just a moment for the water to reheat. In the meantime, let's get you comfortably settled and started on that delicious-looking blancmange." Acting as he spoke, Adam pulled two chairs close to the table and seated me in one, and before I could protest he was putting the plate of dessert into my hands.

I shook my head. "Really, Adam, I don't think I can." My stomach was turning a little at the sight of the pale, jiggling mass. I tried to hand it back to him. "If you don't mind, I'd rather just have the tea."

His arched brows, dark below the fair hair, gathered sternly in a way I hadn't seen before. "But I do mind!" he retorted. "You need more nourishment than tea will provide. Now, take at least a few bites of that, or I will have to feed it to you myself. It is bound to give you some strength."

As he stood over me I could see that he was only half joking, and he looked perfectly capable of carrying out his threat. I said faintly, "Of course, if you insist—" and hastily took a bite. It was deliciously sweet and creamy, and slid down my throat with a soothing sensation. I had no trouble at all in disposing of it. "It really is very good," I admitted as I ate.

Adam handed me a steaming cup of tea and another plate, this with one of the small sweet rolls on it. "And you will feel even better when you have finished these. I long ago learned that most women consider tea to be a panacea."

As I obediently sipped from the steaming cup, I had to admit that I did feel better, but I couldn't help reflecting silently that nothing at all could cure the sorrow I was feeling. Adam seemed to sense my thoughts, for he fell silent, then, with an effort, reopened the subject that was in both our minds.

"I spoke with Jamison's lawyer. He planned to make the funeral arrangements immediately. Also, I have asked the hotel to arrange for a small suite of rooms for you. I think that you should move to them before the afternoon."

"But why?" I asked in dismay. My surroundings made no difference at all to me at the moment, and I felt too utterly spent to endure the rigors of transferring even to another part of the hotel. I couldn't imagine his reason for wanting me to do such a thing.

"Your father had many friends and acquaintances, both business and social, Beth. When they learn of his death, and that he has a daughter here at the hotel, they will be calling to offer their condolences." Adam glanced around the small bedroom. "You can hardly receive them here."

I had not thought of that, though now I recalled that between my aunt's death and her funeral I had received dozens of callers in her small, meticulously appointed parlor. I supposed it was because I had not had time to see my father against a background of what must have been a busy business and social life. For the brief moments of our reunion it had been just the two of us, caught together in a lovely, iridescent bubble of time which had been too quickly shattered. Must I allow the world to intrude even on this memory?

"I know none of those people," I retorted more petulantly than was my usual manner. "What do they mean to me? I much prefer to let it be known that I am in seclusion."

The look of grave concern that he had habitually worn since we met changed to disapproval and almost, I thought, to annoyance. He rose abruptly from the chair where he had been sitting opposite me.

"Of course, Miss Allen, the decision is yours," he said, though his tone denied it. "But I think that out of respect for your father's memory you should reconsider. Jamison was a social man and a proud one. Perhaps he would have liked his friends to be permitted one last chance to pay their respects. And perhaps he would have liked you to know just how well thought of he was. He has no other means of showing you now."

His words struck home, and for that reason annoyed me. I got up and went (or to be honest, flounced) to the window. As I stood there, clutching the drapery and gazing unseeingly at the scene below me, futile tears filled my eyes. Adam was right, of course, and I had no idea as to why I should feel this sudden rebellion. I supposed dimly that it had something to do with the fact that I had once again gratefully accepted help offered me without stopping to think that always with assistance came a certain amount of control. Well, I had surrendered the right to resent it. Without turning around, I said, "I will do whatever you think best."

Adam didn't answer, and after a moment I was aware that he had come to stand behind me.

"Miss Allen—Beth—" he began. "Please forgive my insistence. But I have been in like circumstances several times in the past few years, and I know that later on it is a great consolation to know you have done the right thing, even though it matters not a whit to you at the time."

I turned around then. He was very near and I realized how splendidly tall he was. As I looked up at him I saw again the harsh lines of some older grief that marred his otherwise handsome face. A realization of his kindness and my own selfishness overwhelmed me.

"Oh, Adam, forgive me!" I cried. "How could I be cross with you when you have been so good to me! And you are right about Father. He would most certainly have wanted a correct and dignified funeral."

As it turned out, the move to other quarters was totally painless. The hotel took care of everything and all I had to do was walk sedately from my single room to a compact but pleasant suite at the other end of the corridor. It consisted of no more than a mid-sized sitting room, a bedroom, and a small, elegant bath. The hotel manager had thoughtfully sent up a huge basket of fruit, and already had accepted for me several arrangements of flowers, which he had had placed here and there around the room. I was touched by the kindness of strangers and decided that I was glad, after all, that I had acceded to Adam's wishes.

I had scarcely taken in my new surroundings when the telephone rang. Adam answered it for me.

"It seems that you already have callers," he announced. "May I tell them that they may come up?"

I started at him in sudden panic.

"Oh, heavens," I exclaimed. "Already?" I looked down at my watered-silk gown. "I have scarcely had time to— Do I look all right?" I asked anxiously.

His eyes traveled over me. "You look—quite perfect. Unless you wish to change into something black."

I shook my head decisively. On this issue I wouldn't budge. "I have one black dress, which I will wear to the funeral itself. Father didn't approve of black for children and young women; I remember quite distinctly his remarking that when my mother died. Don't you think this gray is subdued enough?"

For a moment I saw amusement in his eyes. "Hardly subdued, on you. With your fair hair, it takes on the silvery hue worn by an angel." Then the somber look returned. "But I believe Jamison was quite right. I know I have already seen enough black gowns and funereal garb to last me a lifetime." In an abrupt change of subject he asked, "Shall I tell your callers, then, that they may come up?"

I merely nodded, my nervousness momentarily arrested by some bitterness beneath his words. It was the second time that he had alluded to having experienced more than once the pain and loss that I was going through. I wondered briefly if he was a widower, then forgot it as I composed myself for the ordeal that lay before me.

It proved not to be such an ordeal as I had expected. Indeed, I was soon caught up by my interest at the varied types of people who made up Father's extensive collection of friends. Most of them wore elegant attire and had the graceful manners belonging to those in the upper stratum of society, but interspersed among these were tradesmen and workingmen and even several examples of the flotsam that seem to exist almost invisibly on the fringes of any society. These last—emaciated men whose trembling hands and hoarse voices spoke of Demon Rum for once denied, men crippled by some unhappy accident or other, a gaunt woman in widow's weeds whose hands when she clasped mine were rough with toil—touched me the most, for their sorrow and distress at my loss were obviously genuine, and more often than not tears filled their eyes as they spoke of Father's kindness and generosity in a time of need. "'Twas none of us poor folk that harmed him, you may be sure of that!" the widow said. "Your dear father never was known to turn away a soul in trouble. A man such as him will be sorely missed!" Her eyes brimming, she gave my hand one more comforting squeeze before taking her leave.

As emotionally draining as it proved to be, I was gratified to discover this aspect of my father. Although he had always been generous with his family, I had had no way of knowing how deep his kindness really went.

Adam spent a great deal of time with me during the interval before the funeral, and often he stepped forward to introduce the various callers to me. I saw that they treated him with respect and deference, though not with the familiarity I would have expected them to show toward someone whose family, I gathered, had been prominent in the area for many years. And his somber, brooding look had returned and was even more noticeable now than it had been when we were alone together.

Occasionally I heard someone ask him, somewhat diffidently, about his family. "How is dear Rosemary?" one buxom dowager of uncertain age inquired on the second day that I had been receiving. "And the poor dear children? Are we to hope that we may see them at the sad gathering tomorrow?"

I glanced at Adam. His face had taken on an odd, remote look. Briefly I wondered if Rosemary were his wife, but his answer soon set me straight. With a tight-lipped smile he answered, "It is hard to say. My sister is, as you know, Mrs. Waring, unpredictable."

Mrs. Waring seemed somewhat nonplussed by his forthright answer, and murmuring something unintelligible, touched my hand lightly with her fingertips and moved majestically on. I wanted to inquire whether the children were his, or Rosemary's, but something in the forbidding set of his mouth stopped me. It was at times like these that I sensed that Adam's kind and thoughtful manner were either a mask over, or perhaps a facet of, a darker or even dangerous side of his nature. It was a chilling thought and I determinedly put it aside and greeted another visitor.

By late afternoon the stream of guests had dwindled to a trickle, and by five o'clock had stopped completely. I dropped into a chair as I realized with some surprise that I was exhausted.

"Thank heaven that is over," I breathed. I assumed that, as with my aunt's funeral, I wouldn't be disturbed tonight.

"I know how hard it must have been for you," Adam said, "but you will be glad you endured it one of these days."

"Oh, I am already glad!" I answered hastily. "It is comforting to know that Father was loved by so many diverse people. He at least made a difference to the world while he lived." I hesitated, then asked casually, "Will your wife and children be attending the funeral tomorrow?"

Adam rose and went to the sideboard where stood ready a decanter of wine. Without consulting me, he filled two of the crystal glasses and brought one over. As I gratefully took it he said, "I have no wife. And the children are not mine, nor Rosemary's. They are my older brother's children. Both he and his wife were killed some time ago."

"Oh, how dreadful!" I exclaimed. A shock had gone through me at his words. No wonder he sometimes looked morose. "Was it a vehicular accident, then?"

The dark brows furrowed and he stared intently at his wine. "No," he said at last. "The deaths happened on separate occasions."

A silence fell. I didn't dare ask more questions in the face of that brooding look, and he seemed to be lost in some dark world of his own. I quietly sipped my wine.

At last Adam seemed to notice the silence. He roused himself and said with a laudable show of interest, "How long has it been since you and Jamison were together? You must have some wonderful memories of him in earlier days."

That was all it took to make my words come pouring out. During the past two afternoons, when others were sharing with me their memories and experiences with Father, I had longed to do the same, but there had not been time. Now, with an interested and sympathetic listener such as Adam, I had my chance, and it was as though an earthen dam had broken. I went on and on, all through the light dinner which he had had sent up, all through the rest of what would have been a long and lonely evening, I talked of my early childhood, of my odd yet fascinating life with my sophisticated parents, of the mundane years with my aunt when Father's letters were the only really interesting occurrences in my life and I adventured vicariously through them. And when I paused, even momentarily, Adam urged me on, as though he knew instinctively that it was the one thing that would allow me eventually to rest.

And he was right. After he had at last excused himself, after I had undressed and tumbled into bed, I fell instantly into a deep and dreamless sleep.

My first thought upon awakening was of Adam's great kindness. The second was that I had only the funeral to get through and I would be free. And then, leadenly, a third followed on its heels. Free—to do what? Even though I assumed that Father had not left me destitute, I had no idea of how I was to proceed with my life. Young ladies of genteel upbringing like myself were not fitted for more than marrying and managing a household, and undoubtedly if I had remained in my old surroundings I would have married one of the several acceptable young men who had often come to call. But here I knew no one except Adam, and I was already aware that he only came occasionally to San Diego for business purposes. Though he didn't say so, I knew that the days he had devoted to me had been at the expense of his primary interest, Riversbend, and that I couldn't depend on him to ease me into the stream of San Diego society. Nor did I want to return to the East. Those gentlemen callers had been as dull as life with Aunt Blanche had been; besides, I knew instinctively that it was no good going back, that the path lying ahead must always be the better choice.

I sighed and reluctantly got out of bed. That path must eventually reveal itself, and until later then I could only proceed one weary step at a time.

Father's funeral had been set for early afternoon, and I dressed and had partaken lightly of the lunch sent up by the time Adam arrived to escort me downstairs. As I greeted him I couldn't help noticing the way his eyes lit up at sight of me.

"I am not sure your father was right after all, about the black," he said. "You look most fetchingly fragile in that gown."

I couldn't help being pleased by the admiration in his voice. He was, after all, a handsome and apparently eligible male, and I had, at times during these past few days, felt emotionally drawn to him. But I feared that his kindness and my great need might well be at the root of that, so I merely murmured some small response and took the arm he offered me.

As we went from the hotel to the closed carriage awaiting us, I stopped momentarily in the courtyard, attracted by the brilliant display of flowers at a stand situated there.

"Oh, I should like some of those for Father!" I exclaimed.

"I have taken the liberty of ordering, in your name, an arrangement to be sent to the church," Adam reminded me.

I felt suddenly resentful that I hadn't been consulted. There are certain things one wishes to decide for one's self. As I hesitated before the booth, the pleasant-faced woman behind the counter said gently, "Perhaps just one blossom will do—to lay on the coffin as a final farewell." The woman's hand hovered over the various flowers. "Perhaps a carnation? They hold up well in the heat."

"Oh, yes!" I answered, still feeling balky toward Adam. "A red one, please. Father always liked their spicy odor."

The woman selected a large one and held it out to me. When Adam tried to pay her, she demurred. "It is for Jamison Allen, isn't it? Please, let it be my gift."

I felt the tears start to my eyes, and could only murmur my thanks. Adam must have realized that I was overcome, for his arm tightened on mine and without further words he led me to the waiting carriage.

It seemed a long, dreary journey from the hotel to the ferry and then on to the large white clapboard church in the city where the funeral was to be held. I assumed it must be Father's regular church, for the minister spoke of him with pleasant familiarity, and at time had trouble speaking for the genuine emotion in his voice.

I was grateful for Adam's presence with me in the family pew, for otherwise I would have been quite alone. As it was, I was overcome by an all-pervading loneliness, and I scarcely noticed the faces of those in the long line of mourners who paraded by Father's open coffin after the eulogy was over. Then, after the church was emptying, it was my turn to approach the bier.

I had not gone to the funeral parlor to say my farewells as was the custom, for I had done so with my aunt and found it distressing to leave her alone in that hushed and chilly atmosphere. I had not felt that I could bear to do so with Father. But I realized fully the wisdom of the custom as I went forward to say my final farewell to him, for as I saw him lying there, handsome and imposing as ever but so unnaturally still, I felt the full impact of his death for the first time. It was as though a huge weight had struck me in the solar plexus. It was true, then. Father, the one person on whom I had been able to depend, on whom I had fastened all of my hopes and dreams, was gone. I was totally, devastatingly alone.

I couldn't hold back my tears, nor could I stop their flow when they came. At last I felt Adam gently draw me away, but even as he led me up the aisle toward the entrance, scalding tears continued to slip down my cheeks. But, once we were in the vestibule of the church, which had fortunately emptied, he turned me toward him.

"That is enough crying for now," he said sternly. "Your obligation to Jamison is to present a facade of dignity to his friends. You may cry all you like once we are in the carriage again, for it is a long way to the cemetery."

At his presumptuousness, anger swept over me. I didn't stop to think that it effectively dried my tears. Imperiously I withdrew my arm from his and swept from the church and past the waiting mourners to my carriage with all the dignity that even Father could have desired. I started to climb into the carriage unaided, but his hand on my arm detained me.

"We must wait with the others until the coffin has been brought out," Adam reminded me.

Again I was furious. "Who has appointed you my mentor?" I demanded in a seething whisper. "Or my personal funeral director? You are taking advantage, sir, of your friendship with my father!" Nevertheless, I stood where I was, unwilling to break the code of society.

His answering whisper was sardonic. "Attending funerals is not my favorite occupation, Miss Allen. It is only *because* I was Jamison's friend that I am here at all."

His words stung. I had thought that perhaps his attraction to me might have had at least a little to do with the continuing concern that he had shown. But at that moment the solemn pallbearers were bringing Father's casket to the waiting hearse and I was aware that many pairs of eyes were turned to me, and in spite of my anger, or perhaps because of it, I was able to present a dignified, though sorrowful countenance.

No more words were exchanged between Adam and me until we were at last seated in the carriage and the driver had started to follow the slow pace of the hearse. Then Adam said casually, "You may cry all you like now. It is a good fifteen minutes to the cemetery."

I had always thought of myself as having somewhat of an acquiescent nature, but Adam's calm assumption of authority was bringing out a contrary side of me that I hadn't suspected existed. In fact, I was so angry that I had almost forgotten my grief over Father. "It is so good of you to give your permission," I answered, my voice quivering with rage, "but at the moment I have no inclination to do so. I wish only for it to be over so that you may be relieved of your onerous duty."

I half expected an apology, but when none was forthcoming I moved to my corner of the carriage and opened the drawn draperies so that I might breathe the warm summer air in an effort to calm down. For the rest of the interminable journey the only sound was the slow clop, clop of the horses' hooves.

To my surprise, a large number of mourners had followed us to the gravesite. The carriages had kept coming and coming, and it was some time after Adam had taken me to the front row of the folding chairs which had been placed beneath a green and white canvas shelter before they had all gathered and the final rites could begin. I kept my eyes closed during most of it, for tears were threatening again and I was determined that they shouldn't fall. But when at last I went forward to lay my carnation on Father's coffin and heard the accompanying murmur of sympathy I had to acknowledge the great shaft of pain that rose in me. The world blurred and I stumbled and would have fallen except for the strength of Adam's hand beneath my elbow. I couldn't help but be grateful for it.

Chapter Five

Barbaric as they may sometimes seem, there is a purpose to funerals. They are a shutting of the door to the past; no matter how much one may still grieve, one now has the permission, and yes, the obligation to turn about and face the future.

I began a discussion of my future with Adam Rivers the moment we had returned to the hotel.

"Please," I said, once I had removed my gloves and hat and had laid them aside, "tell me what you know of Father's affairs. I have seen that he lived well. I presume that he left some sort of estate and that I am heir to it?"

Instead of answering at once, Adam rose abruptly and went to the wine decanter. He filled one of the crystal glasses, then said, in an odd tone I hadn't heard before, "May I pour a glass for you?"

Without knowing quite why, my heart began a slow pounding. "Perhaps," I began slowly, "you would prefer that I discuss this with someone else? Father's attorney, or possibly his banker?" We had, after all, quarreled earlier, and I had been quite rude. Adam could well feel that his obligation to Father had been fulfilled. At the thought he might be planning to desert me now, I began to regret my display of temper.

It wasn't until he had downed the glass of wine, poured another, and returned to the chair opposite me that I realized with growing alarm that it wasn't annoyance but a reluctance to answer that kept him silent. I clutched the arms of my chair.

"What is it, Adam? Surely Father left something of value. Or is it—" I fell silent, vague thoughts of second wives, mistresses, other offspring tumbling through my mind. I took a deep breath. "Are you hesitating because—because I am not named inheritor in his will?"

"Oh, good heavens, no, Beth. The will is in perfect order and you are named as the sole beneficiary of his estate. The problem is—" he set aside his wine and leaned forward, "we can find almost no evidence that there *is* an estate."

I fought the panic that welled up in me. After several slow, deep breaths I said evenly, "What you are saying, then, is that I am penniless."

Adam shook his head. "It only seems so at the moment. We know that it can't *be* so. Jamison was an astute businessman. He had a natural instinct for real estate investments, and during these last 'boom and bust' years he prospered when most men lost everything. It is well known that he held notes on a great deal of prime property in San Diego." He shifted uncomfortably in his chair before continuing. "It is jut that neither his banker nor his lawyer knows where those notes are, nor did we find them among his papers."

"But how can that be?" I asked in bewilderment. After searching swiftly among my bits and pieces of financial knowledge, I continued tentatively. "Aren't those things usually recorded somewhere?"

"Your father didn't do things the usual way, Beth. He often concluded deals with a handshake, and he considered that or a personal note sufficient." He picked up his glass again and stared into its contents as though seeking an answer there. "Though I hope it is not the case, there may be no formal record anywhere."

The news could not have been more appalling. I had no money beyond the few dollars left from the sum Father had sent for my travel expenses. Moreover, the costs incurred by his funeral service could not be inconsiderable. And then—I felt faint at the thought—there was the cost of my former room and this elegant suite. How on earth was I to pay for them? Terror at the thought emboldened me.

"Is there really nothing at all? They didn't find even a little cash, or some jewelry in his safe?" I bit my lip. "The hotel bills must be staggering, yet I have to have someplace to live until I can think what to do."

Adam's brilliant gaze looked directly at me again.

"Please," he urged, "don't concern yourself about that. The hotel manager has asked me to assure you that there is no hurry about settling the bill. I doubt that it will ever be presented to you." The harsh lines of his face relaxed, as though he were relieved to have brought me at least a little good news. "And Jamison had already set aside an ample sum for his funeral expenses. He had taken care of all that long ago."

A little—though very little—of my terror began to subside. I relaxed my iron grip on the arms of the chair. "That *is* good news," I said, not without a touch of irony. "For now all I have to worry about is my future, which looks terribly uncertain at this moment." In fact, the situation was beyond my worst imaginings. For the last few years Father had supported both my aunt and me, and while he had been generous, Aunt Blanche had not been an astute manager, and the small amount she had managed to save had gone to satisfy a few unexpected funeral expenses. And now I was virtually penniless, and I had no one, either here or in New York, to whom I might turn.

Adam must have seen my panic, for he asked gently, "You have no other relatives? Nor money of your own?"

I shook my head, fearing for a moment to trust my voice. "I am afraid not," I said then. "I must begin to earn my own living, and immediately." I frantically cast about for some latent talent I might develop, but none came to mind. I was reasonably clever with a needle, though not outstanding. I had done well in my schoolwork, but little I had been taught had value in the marketplace. Girls who attended Mrs. Hornsby's Academy for Young Ladies were not expecting to support themselves, and I had been no exception. The only thing I might be suited for, I thought, with only a vague idea as to their duties, was as a governess to young children. I said as much to Adam.

"It is much too soon for anything so drastic," he said. "I am absolutely certain that Jamison had lucrative investments. His conversation about the fabulous future he expected to share with you confirms that. It is simply a matter of locating them. In the meantime, we must decide what is best done for your welfare." He got up and began to pace about the room, his tall figure casting strange elongated shadows against the far wall as he passed back and forth between the lamp that had been lit against the darkness. I watched him idly, my mind empty of any suggestions, my spirit waiting supinely, as it had before, for someone to rescue me from my dilemma. And by so doing, I allowed someone else to once again pick up the reins that guided my destiny, for just then Adam turned abruptly to me and said, "You must come to stay with us, of course, at Riversbend."

I stared at him with widened eyes, for somehow the thought of doing such a thing had never entered my mind. And I wasn't sure that it had entered his to invite me until that minute. There was, after all, very little else that he could do.

I felt the hot tears gather in my eyes and I rose and went to stand at the window, where the path of moonlight on the water blurred before me. "It is so kind of you to offer," I said unsteadily, "but how can I accept? How can I take charity from a stranger no matter how kindly it is meant?"

Adam got up, too, and came to stand behind me. I felt his hands on my shoulders. "Is that what you think of me, Beth?" he asked. His voice was strangely uneven. "What with all the hours we have spent together these past few days, all the grief that we have shared, am I still no more to you than that?"

A tremor ran over me at the odd note in his voice and I turned to look up at him. There was pain in his eyes, yet with it a rising passion. Before I could form an answer he leaned down to me and covered my lips with his. I started to pull away in surprise, but his arms came around me and I found myself a captive of his embrace. Yet a willing captive, for as the kiss deepened I felt myself responding with an ardor that amazed me. It shamed me, too, when at last I had the presence of mind to tear my lips from his.

Trembling in every limb, I shrank back against the velvet draperies of the window.

He must have thought that he had frightened me, for at once he said, "Forgive me, Beth! I had no right to do that. But I have wanted to, almost from the first moment I saw you."

Even though my heart was still palpitating wildly, I decided to press my advantage, for in my naivete I thought his words meant that he hadn't recognized my shamefully passionate response. With as much dignity as I could manage in my agitated state I went past him into the room, then, turning to face him, said, "Is this the treatment I might expect if I avail myself of your hospitality?"

I *was* frightened then, for a dark flush of anger suffused his face. "Is that what you think? That I am inviting you to my home in order to seduce you?" He came toward me in what I thought was almost a menacing manner. "Or are you trying to hide even from yourself the fact that you enjoyed that kiss, too?"

It was so near to the truth that I couldn't answer. Tears once again burned in my eyes, but this time they spilled down my cheeks in a scalding river. At sight of them, Adam's anger seemed to evaporate.

"Dear lord," he said. "What am I thinking of to put you through this, tonight of all nights?" He took a step closer, but at a shrinking motion from me, stopped again. "The invitation was genuine, Beth. I have no ulterior motives, though I do find myself attracted to you more than I will say right now. But I assure you that you will be safe at Riversbend, even from me."

I considered him. Even though he seemed subject to an unpredictable moodiness, I didn't think I was really frightened of him. Besides, what choice did I have? Yet propriety demanded that I make one last protest.

"What about your family? Will my coming there be acceptable to them?" I spoke for manner's sake only, and I wasn't prepared for the abrupt return of his dark and brooding look, nor for his slight hesitation. However, when he answered, his words and tone were reassuring.

"We are so far from town that we have very few visitors. Your presence will be a welcome diversion."

In my desperate need I chose to ignore the vague misgivings that stirred in me. "Then I will accept your kind invitation." I added, glancing around the suite of rooms which bore little evidence of my occupancy, "I can be ready to leave whenever you wish."

He nodded with a return to his formal manner. "Good. I must admit that I am anxious to get back to the Bend. I will come back for you at ten tomorrow morning."

I departed the Hotel del Coronado in considerably less elegance than I had come. I left, in fact, riding on the hard seat of a farm wagon, the back of which was piled high with supplies. They were covered, I supposed against the chance of rain, though I saw no sign of any. I later learned that it was, in fact, dust they were protected from.

Adam apologized for putting me to such an inconvenience. "For whatever reason I come into San Diego," he explained, "I also pick up any needed supplies. Riversbend is such a distance from town that I avoid making unnecessary trips."

"That sounds practical," I returned, though I actually felt uneasy at the thought of being so far from the city life that I knew, "and fortunate for me, since it caused you to have enough room for my baggage." Adam had made no reference by word or look of the happenings of last night, and I was determined not to, also. "Besides, in this lovely weather a wagon ride sounds most pleasant."

But once we had left the city streets, the bumpy wagon ride was decidedly not pleasant, though I tried not to let Adam see that I felt that way. The rolling hills beyond the suburbs had turned brown under the hot June sun and the sparse foliage that dotted them looked dirty and dispirited. Thick clouds of dust rose from beneath our wheels and clung to the folds of my navy wool suit, which I now realized was most inappropriate for the journey, and in spite of my fine parasol I could feel moisture sheening my forehead and upper lip, trapping, I guessed ruefully, more of the fine dust which enveloped us.

"Does it never rain here?" I asked, trying not to sound petulant. "The landscape looks in need of a good wetting."

"If you are still here in the winter you won't ask that," Adam said. "Some years the rain comes down in torrents, and the rivers overflow and wash our crops and livestock out to sea. But it almost never rains between May and September, which, the ladies tell me, helps when planning picnics."

Privately I thought that I would forego the assurance of a dry picnic for a lessening of the dirty brown fog that threatened to choke us, but I maintained a diplomatic silence. In fact, so far our entire journey had been mostly a silent one. I had felt fresh pain at leaving the hotel, for it had seemed a final severing of my ties with Father, and I could not visualize my future. For as long as I could remember, my plans for it had been inextricably bound up with those of Father; in all of our letters to each other we had written of being together again someday, of adventures we might have, of exciting journeys we might one day take. I hadn't thought beyond that time to a day when I might want a life of my own, of the fact that Father might not always be there. And now that he had been taken so suddenly, the way before me seemed empty and barren, and I couldn't yet imbue it with even the shadows of people and events.

And I suppose that is why, though I was relieved and truly thankful for Adam's invitation to stay with his family at Riversbend, the wagon ride seemed to be taking me farther and farther from the kind of life with which I was familiar. I felt lost and frightened and very, very lonely, and the man sitting beside me seemed to be little better than a stranger, a feeling that was deepened by the fact that Adam, beyond making an occasional comment about the weather or the increasingly rugged terrain, seemed to lapse into a preoccupied silence much of the time. Noting this, I couldn't help thinking of the uncertainty I had seen in his eyes when I had asked how his family would feel about my visit, but I didn't bring it up, for I feared to hear that I might not be welcome after all. I simply had noplace else to go.

At last, after one especially long period during which I began searching desperately about for some subject with which to break the silence, he brought the wagon to a halt and said, with a matter-of-factness that I thought concealed a good deal of quiet pride, "There you are, Beth. Riversbend."

We were stopped at the entrance to a wooden bridge spanning a wide, shallow river, beyond which the land rose in verdant rolling swells. Set solidly atop one of the larger hills stood a great square house whose pillars rose a full three stories. It was surrounded by groves of dark green trees, and before it, stretching clear to the riverbank, was a wide expanse of lush lawn. The house itself was of a light, silvery gray, and in the brilliant rays of the afternoon sun it seemed to sparkle much like a diamond set in emeralds.

"Oh, Adam!" I exclaimed with genuine admiration. "It is beautiful, and so cool and inviting! How do you keep the grounds so green without rain?"

He seemed pleased by my admiration.

"It is done with irrigation. I have built a small dam and two paddlewheels upriver. The curve you see here below the house is the source of our name, Riversbend. It is also, of course, a play of words on our family name." He surveyed the surrounding land with the proud air of an earl overlooking his estates. "We are using the water to good purpose instead of just allowing it to run into the sea."

He shook the reins and with a hollow thunk of hooves the team drew our wagon across the bridge. I slanted a glance at Adam. His usual brooding, somber look was completely gone for the moment, and with a tingle of excitement I realized just how handsome he was. Would he really, I wondered, not try to kiss me again without my permission? I felt a little stab of disappointment at the thought.

As we started up the winding gravel road that led to the house my attention was distracted by the small flock of sheep moving about the lawn. As we approached, they stopped grazing and stared at us with benign interest. A few fluffy, half-grown lambs trailed after their mothers.

"Oh, the darlings!" I exclaimed, being a city girl.

Adam smiled indulgently. "Those darlings have a purpose beyond that of being decorative. They keep the lawn short, give us wool, and sometimes an excellent dinner."

Once again the pride was in his voice, and I concealed a smile. Adam, the farmer, came as a surprise, but I could see that he was entirely comfortable in the role. As he caught my look and smiled wryly, the harsh lines of sorrow and bitterness seemed to vanish. His vivid blue eyes had a glint of humor, a touch of laughter that seemed to be directed at himself.

"Do I sound smug?" he asked. "Well, I guess I am, at that. Most people don't think much of the future of agriculture here in the San Diego area, but it is ideal for growing crops if you can bring water to them." He began to rein in the team as we neared the wide porch. "And thanks to my father, who let me have my way in the planning of all this, we are experiencing a certain amount of success."

I thought fleetingly of my own father, who had been so excited about San Diego's future, so sure that glorious things lay before it. Now he would never know whether or not he had been right, nor, I remembered with a pang, would he know what I might do with my life.

"How proud your father must be of you, then!" I spoke with determined cheerfulness.

The eagerness left his face, robbing it of light just as though a cloud had passed over the sun. The deep lines that marred his face instantly returned.

"My father died nearly two years ago," he said briefly.

In spite of the heat I was touched by a sudden chill. I was certainly no stranger to sorrow, but something in his manner hinted at a hidden, unexplained anguish far greater than my own. I dared not even murmur a conventional word of sympathy, but turned instead to admire the great house that stood before us.

It was even larger at close range than I had thought it, and the smooth wooden columns which held up the roof were several feet in circumference. Broad, shallow steps led to the wide veranda which stretched completely across the front of the building. In the exact center were double doors of some heavy, carved wood that had been stained almost oxblood in color, and the heavy brass straps and latches gleamed against their dark tones. Matching knockers of proud lions' heads hung in readiness. The low, multipaned windows had heavy shutters standing guard beside them, and above them marched another row, and yet another. At each side of the doorway, tall green shrubs stood in their plant boxes like soldiers at rigid attention. I was overwhelmed with the enormity of it, but also relieved, for I knew at once that, in the physical sense, at least, I would not be in the way. A house this size could shelter a family of any number, even to its most distant members.

"Why, Adam, Riversbend is a veritable mansion!" I breathed. "Is your family a very large one, then?"

Considering the time we had spent together I knew very little about his family. With the self-centeredness of the recently bereaved I had dominated our personal conversations with remembrances of my father and my past. I knew that he was unmarried, that his sister Rosemary was chatelaine of the estate, and that the orphaned children of a deceased brother apparently made their home there. But beyond that I had only a vague recollection of other names having been mentioned. I had thought it odd that none of his family had attended

Father's funeral, but hadn't mentioned it to Adam for fear that he would take it as a criticism, and I had simply supposed that the distance from San Diego and the tender ages of the children had made it too difficult. Now that I had seen the house, though, the fact troubled me. A house of this size must have many servants. If family members had wanted to come, surely some one of them could have been entrusted with the care of the children.

Adam busied himself with securing the reins. "No," he said finally, in answer to my question, "ours is not a large family. Some of my father's plans didn't come to fruition." He jumped down as if to forestall further questions and came to help me alight. "The third floor is closed off most of the time. Besides, the house isn't as large as it seems. It is built in a square, with the center areas taken up by a courtyard and formal garden."

As Adam lifted me down from the wagon, a slim young Mexican boy came around the side of the house. He broke into a trot as he saw us, and a wide grin split his dusky face.

"Ah, Señor Adam," he exclaimed. "It is good that you are home at last!"

Adam set me down carefully on the bottom step, then replied pleasantly, "It is good to be home, Miguel. This is Miss Allen, who will be our guest for some time. Please bring her luggage into the casa, then take the wagon to the kitchen door and unload the supplies. Tell Jose to water and feed the horses well. The livery stable doesn't provide the excellent care of them that he does."

Miguel responded with an even wider grin and began at once to uncover the wagon's contents, but when he saw Father's trunk, the smile faded. "This trunk, señor!" he exclaimed. I do not think that I can—"

Adam had taken my arm to escort me up the steps, but he paused and glanced back. "Oh, of course. Just bring in the valise. You can find someone to help with the trunks later."

With Adam's hand at my elbow, I climbed the shallow steps to the imposing entrance. As he pressed the latch the heavy door swung open, and I forgot the way my pulse had quickened at his touch, and frankly stared. The large entry hall soared the height of the house, and in its center, hanging from a massive chain, was a huge cut-glass chandelier. Attached to it was a thick hemp rope which slanted downward to one wall and was affixed there. I frowned, then realized that it was by this means that the chandelier was lowered so that it could be cleaned. At each side of the room graceful curved stairways led to a common gallery which crossed the upper width of the hall like that of a baronial mansion.

"It is so beautiful, Adam," I said. "No wonder you are proud."

He looked around as though never tiring of seeing it. "Riversbend is the love of my life," he said. Just then Miguel brought my valise in and set it on the floor beside me, then, after a shy smile at my thanks, went out again. Adam started toward a doorway to our right. "I must find Rosemary, or Maria, our housekeeper. They didn't know when to expect me."

Without saying more he disappeared into the room beyond, leaving me to stare after him with sudden apprehension. Surely he had sent word ahead that he was bringing a guest!

I hadn't long to wonder about it. A slender woman of about thirty, whose severe hairstyle and drab maroon dress ill became her sallow skin, was descending the stairs. Her forbidding expression answered my question immediately.

"Yes?" she inquired, and the one word managed to convey surprise, annoyance, and disapproval. "Who are you, and what is it you want?"

In the face of her obvious hostility, I was at a loss for words, and to my immense relief, Adam reappeared just then with a stocky, pleasant-faced servant woman in tow.

"Oh, there you are, Rosemary," he greeted the woman on the stairs. "I was about to send Maria to fetch you." As Rosemary continued to stand stock-still on the stairs, he added impatiently, "Come down here, will you? I want you to meet our guest." Without waiting for her to comply, he added, "This is Elizabeth Allen, Jamison's daughter." His tone was almost too matter-of-fact for the circumstances, I thought. "I have persuaded her to stay with us for a time at Riversbend."

I braced myself for the annoyance of a woman faced with an unexpected and possibly unwelcome houseguest, but Rosemary's reaction astonished me. Immediately she blanched and clutched the banister for support. The look she gave Adam was not that of anger, but was unquestionably one of complete betrayal.

Adam seemed genuinely bewildered by her reaction. "What is it? Surely you knew—you did receive my message about—the tragedy?"

His sister had pressed her lips together so tightly that they appeared to be bloodless, but after a long moment she said, as though the words were being forced from her, "Of course. How do you do, Miss Allen." With an effort she pulled herself erect and came down the last few steps. "My condolences. The loss of—your father—has no doubt been a terrible blow. I—we shall all miss him." She glanced again at Adam and her tone changed to one of veiled complaint. "I should have liked to have attended his funeral, but I was scarcely told in time. And then, someone had to look after the children."

"Good lord," Adam exclaimed immediately with an irritation which surprised me. "You had only to order the buggy hitched and brought around. And Maria is perfectly capable of caring for Anne and Mark. Are you sure it wasn't just your aversion to going into town that stopped you?"

As though he hadn't spoken she said to me, "I am sure you must want to rest and refresh yourself after your journey. Maria will take you to the east bedroom."

I hardly felt exhausted after our drive, but it seemed a good way to escape the undercurrents of animosity that flashed between brother and sister. And in addition, I was most incredibly dirty. I answered gratefully, "I would appreciate an opportunity to freshen up. I am afraid the dust—"

"Oh, I know all about the dust," she said with an unpleasant toss of her head. "In the summer it is everywhere beyond our own little oasis. And in the winter we have the mud to contend with. That is one of the reasons I seldom venture from Riversbend." Her tone spoke of other, more bitter reasons. "Please follow Maria. She will show you to your room."

The Mexican woman smiled shyly at me and began trudging up the stairs with my valise. As I started to follow her, Adam said, "I think you should try to nap a little, Beth. It will do you good. And dinner isn't until seven."

I must have been more weary than I had thought, for the concern in his voice brought quick tears to my eyes.

"Thank you," I answered. "I will."

As I had the morning after Father's death, I suddenly wanted most desperately to be alone. Something about my uncomfortable situation—this strange house, Rosemary's unspoken hostility, Adam's odd, mercurial behavior—made me feel alienated from them and from life itself. I wished to be out of this house, out of this new and hostile country that had taken all that was dear to me. All at once I longed to return to New York and to my old life there, unexciting as it had been. At least it would have the comfort of familiarity.

But on another level of my mind I knew that that old life was gone forever no matter what my need for it, and I followed Maria's broad back up the stairs and along the gallery with lagging footsteps. My despair was lifted somewhat, though, when through the bank of windows lining the rear wall I caught a glimpse of a courtyard full of flowers and a flash of spray from some central fountain. At least there was beauty here to comfort me.

The housekeeper turned down a corridor leading off the gallery and stopped before one of its many doors. As she opened it, she said in heavily accented English, "This is the east room, señorita." She set the valise on the floor near the large, four-poster bed. "Would you like for me to unpack the suitcase for you?"

"Oh, no, thank you. I can do that," I answered. "But I would like to have a pitcher of water, if I may. It was a long drive and I am thirsty as well as hot and dusty."

"Oh, señorita, forgive me! That you should have to ask!" Maria made the apology sound like a lament. She seized the china pitcher which stood on a cherrywood washstand and went out, closing the door swiftly behind her.

As I laid my pocketbook down I looked about me. I could have been in the bedroom of one of my aunt Blanche's spinster friends. The furniture was of matching cherrywood, the four-poster bed swelled with what was undoubtedly a feather mattress, and on the bedside stand and the tall highboy were small, crocheted doilies that had been starched to rigidity. The ecru counterpane had inserts of the same pattern of crochet, and across the foot of the bed, carefully and symmetrically folded, lay a coverlet which had been painstakingly pieced in blue and white hexagons, then quilted with tiny, careful stitches.

I felt despair settle over me. The room and its furnishings reminded me all too painfully of what I had thought I was eager to escape—the neatness, the sameness, the seeming dullness of my circumspect life with Aunt Blanche. I hadn't realized that they also meant safety. In my eagerness to go adventuring with Father I had thought myself to be bold and full of bravery, but I knew now that it hadn't been bravery at all—I had only been exchanging one form of safety for another.

My bitter thoughts were interrupted by a discreet knock at my door. I opened it to admit Maria, who had returned with my pitcher of water. Following her was a dark-haired young Mexican woman in her early twenties. The silver tray she bore held a cozy-covered teapot, a cup and saucer, and a plate of tiny, delicious-looking cakes.

"This is Emelda, my daughter," Maria said in her soft, liquid tones, "who also wishes to serve you." The girl lifted heavy lashes, and as the dark eyes looked directly into mine, I doubted it. Her mother, unnoticing, set the pitcher on its stand and went on. "She has brought for you a small refreshment."

"That is very thoughtful of you, Emelda," I said.

"Thank you."

The girl set the tray down on a small table before giving me another bold, assessing look.

"There is a bathroom down the hall," she said without preamble. Her voice, unlike her mother's, had only the faintest trace of an accent. "Maybe you would like a bath before dinner."

I longed for a bath, but something about her direct, almost insolent manner made me feel contrary. "Perhaps later." I waved toward the pitcher and basin. "This will do nicely for the moment."

"Whatever you wish." Emelda shrugged indifferently and went out without waiting for her mother, who had begun to turn down the bed. When Maria had finished, she indicated it with a wave of her hand.

"As Señor Adam suggested, you will take a small siesta, no? It will be for you much good." She hesitated, then added, her square brown hands clasping and unclasping before her, "I regret that you have such sorrow. Señor Allen—he was a good man. Often he helped my people when they were in need." Before the tears could spill down her dark cheeks she was gone, leaving me to swallow against the sudden tightness in my throat.

I took off my dusty jacket and hung it on the bedpost, then went to the window. Below me, in a grove of shiny-leaved trees, several white-clad men were picking bright golden fruit and dropping it into sacks which hung over their shoulders. Oranges, I realized, and felt a thrill of excitement such as I had experienced each Christmas upon retrieving one of the fragrant spheres from the toe of my stocking. Its scent had set me dreaming of distant lands and high adventure even before Father had settled in California, and afterward it had seemed to bring him a little closer to me.

I watched the men work for a little while, then, in a lighter mood, began to shed my dusty outer clothing. The furnishings of the room notwithstanding, this was a new and different world, and I must overcome my cowardice and find a way to make a new life in it.

In the meantime, I vowed with new determination, I would enjoy my stay at Riversbend.

Chapter Six

After I had bathed as thoroughly as was possible in the china washbasin, I lay down and tried to follow Adam's suggestion about taking a nap. But try as I might to relax and make my mind a blank, I couldn't drift off to sleep. Images of the past few days—of Father as he walked away from me for the last time, of the strangers who had come to commiserate with me over his death, of the sorrowing faces I had seen filing past his casket at the funeral—one after another presented itself to me. And then other, more frightening images formed behind my closed eyelids—of the moment when that metal blade had slipped so silently through the crack of my hotel door, of the chaos in Father's room and my painful sense of violation at the sight. I had not thought too much about these latter things, taken up as I had been with the funeral, but now that I was safely away from whoever or whatever had presented the danger to me, the nightmarish details of that night and the following day returned to me with painful clarity. I wondered if I would ever forget them completely, or if I would ever know what lay behind them and the mystery of Father's murder. At the thought, a cold stone seemed to settle in the region of my heart.

At last, after changing my position several times on the oversoft mattress, I gave up all hope of sleep. I got up, splashed my face once more with the cool water from the pitcher, dressed in a slightly rumpled cotton print dress which I unearthed from my valise, brushed my hair, and slipped quietly out my door and into the corridor.

I paused there an instant. There was a hushed feel to the house, as though everyone in it was sleeping. I didn't know what it was that I intended to do or where I meant to go, nor was I sure of the propriety of a guest wandering about when her hosts imagined her to be safely sequestered in her room, but I simply could not stay penned up with my thoughts a moment longer.

I made no sound as I went along the hall to the gallery, where the lovely panorama of colorful flowers and bubbling fountain below beckoned to me through the uncurtained bank of windows. I stopped momentarily to admire it.

As I did so, one of the wide wooden gates at the rear of the courtyard opened and a dark man in riding clothes came through. I continued to watch as he slammed the gate carelessly behind him and strode along the pebbled walk toward the house. As he passed the large central fountain he glanced at the upper windows of the gallery as though half expecting someone to be standing there, and instinctively I moved back to the shelter of the brocade draperies which hung at each side of the glass.

But not before I had seen his face.

I fled along the gallery, down the hall, and back to the safety of my room. With shaking hands I closed the door and turned the heavy key in its lock. As I slumped against the door, my breath came quickly and I could feel the rapid pounding of my heart.

My first thought was that he had followed me here, then planned to confront me once more with his strange hostility, but common sense soon returned. Whoever he was, he belonged at Riversbend. The familiar way he had shut the gate, the confidence in his determined stride, even the way he had scanned the upper windows as though expecting to find someone watching him, all spoke of a familiarity with his surroundings. No, this time I was the interloper, and in these circumstances the thought of facing that bold black stare was doubly frightening.

I began pacing back and forth across the floor with quick, agitated steps. Where now was the rationalization of his behavior which I had considered after our confrontation on the strand the morning after Father's death? Just one glimpse of him here, where I had thought to be safe and completely isolated, had shattered it completely, had brought back the unpleasant and somewhat frightening sensations I had experienced then, had even, to my shame, set me to seeking a means of escaping another meeting with him.

After a time I became thoroughly disgusted with my craven attitude and forced myself to a standstill. I was, after all, here at the invitation of Adam Rivers. What did I have to fear from this stranger, forbidding though he might be? I decided, not without trepidation, that I would simply take the initiative and go down to confront him.

Once having made this decision, I knew that I had to

act upon it immediately, or what little courage I had gathered would evaporate. So, pausing only to check my appearance once more in the watery glass above the cherry dresser, I took a deep breath and went again into the hall. I managed to walk with a firm step as I went along the corridor to the gallery, but as I approached the stairs faintness of heart again attacked me, and I found it necessary to hold onto the banister and close my eyes for a moment before proceeding.

When I opened them again a sudden shock ran through my body. Standing before me, his face lifted to gaze thoughtfully at me, was a perfect miniature of the young man in the courtyard. I stood transfixed with amazement until ordinary perception returned and I could see that this was simply a boy of seven or so who bore a remarkable rsemblance to the man. Even the overlarge iris of the soft brown eyes and the little smile just beginning to curve his lips as he looked up at me were the same, though the smile lacked the sardonic amusement of the man's. Standing next to the boy was a slender young girl whom I judged to be about ten years old. She was garbed in an unbecoming frock of the same drab maroon worn by Rosemary, and her fair, straight hair was pulled back in a manner equally severe. Her plain little face wore an expression of concern.

"I'm sorry, did we startle you?" The girl's voice was high and very clear. "You looked as though you might be ill. If you are, I can help you back to your room while Mark runs for help." She said this in a calm, practical manner, as though ladies on the verge of swooning were not unknown to her.

Mark looked as if the idea of streaking off for aid appealed to him, and the muscles of his sturdy young body visibly tensed for action. I said quickly, "Oh, no! Thank you, but there is no need for that. I am not at all ill. I was merely—" I couldn't think of a good way to end that sentence, so I began another. "I would appreciate it, though, if you would guide me to the drawing room. This is such a big house that I fear I might lose my way." My words sounded inane, even to me, but the two children continued to regard me with grave interest.

"We'd be most happy to," the girl said with prim courtesy. "Wouldn't we, Mark?"

"Of course," Mark answered promptly, then added with practicality, "though I don't see how she could get lost. You just have to follow the stairs and turn right."

She gave him a reproving glance, but spoke to me. "I am Anne and this is my brother, Mark. You must be Miss Allen. Aunt Rosemary told us that you had come to stay for a while."

"Yes, I have," I answered. "For a short time, at least." I added casually, "I saw a man come into the courtyard not long ago. A dark-haired man, very handsome. Do you know who it might have been?"

Anne looked puzzled for a moment, then her small face cleared. "Oh, that must have been our cousin Christian!" she said happily. "He is quite handsome, isn't he? I didn't know that he had gotten home."

Both she and Mark seemed quite pleased at the news, but my heart sank. If he was a relative, and by his resemblance to Mark, if not from Anne's words, I knew that he was, then he probably shared the family home. I wondered with something close to anger why Adam hadn't mentioned him to me. I wondered, also, about the relationship. The term "cousin" might mean anything. As we started down the stairs I pressed gently for more information. "You say that he is your cousin? I didn't know that Adam had a brother old enough to have a grown son."

The two children looked puzzled for a moment. Then Anne said slowly, "Actually, he is Uncle Adam's and Father's cousin, which would make him our second cousin, or cousin once removed, or something like that. His father was Grandfather's brother. Aunt Rosemary has explained it to us several times, but even she agrees that it is simpler for us to call him Cousin Christian."

Mark grinned up at me, and once more I saw the hint of Christian's charming smile. "It would be awkward, wouldn't it, to have to say each time, Second-cousin Christian, or Cousin Christian once removed? I would hate it, and I'm sure Cousin Christian would, too."

I agreed gravely that it would be awkward, indeed, but my mind was racing. These, of course, were the dead brother's children, and the boy's likeness to Christian a family one. Adam had the look, too, I realized suddenly, though the difference between his and Christian's coloring made it less apparent. It was no doubt why he had looked familiar to me that first moment in the hotel.

As we reached the bottom of the stairs I smiled at Anne, whose answering beam lit her plain face with a momentary beauty. I could find no family resemblance in her. I did feel more than a slight annoyance with Rosemary for dressing and coiffing the child so unattractively. A frock of some clear color, a loosening of the flaxen hair, perhaps a judicious curl or two at her temples would make such a difference, for, though she might never be a beauty, she had smooth, fair skin and even features. It was as though Rosemary had set out to deliberately make both herself and the child as unattractive as possible.

As the three of us entered the drawing room, which was large and imposing and furnished with more of the heavy cherrywood furniture, I saw that it was unoccupied, but it had an air of frequent usage. There was a grouping of rather stiff upholstered couch and chairs arranged before the massive stone fireplace, which, though now cold, had logs prepared for the touch of a flame, and on the small tables beside each chair were personal touches: a pipe and leather tobacco pouch, a humidor that looked as though it might contain cigars, a bit of partly worked embroidery in a wooden hoop. On a nearby table were several crystal goblets and a decanter of some ruby-red wine. A child's rocker and a small footstool were an intimate part of the grouping, and I was glad to see that the children were included in the family gatherings.

"Won't you sit down, Miss Allen?" Anne asked politely in her clear, precise tones. "Mark and I must tidy up for dinner, but I will have Maria bring you some refreshments, if you like."

"Her date cakes are very good," Mark put in, his face lighting with a healthy child's greed. It fell slightly when I replied, "Thank you, but I will just relax for a while. Someone will no doubt be in directly." I didn't add, even to myself, that it might be Christian, but I found myself strangely reluctant to see the children leave. However, despite her efforts to be a good hostess, I could tell that Anne was anxious to be off. "Please," I said, "do go on upstairs. I will do quite well here by myself."

Though obviously loath to leave me, a guest, to my own devices, Anne nodded with visible relief. She seized Mark's hand and drew him unprotesting from the room, and I settled into one of the high-backed velvet chairs facing the fireplace.

I hadn't had much experience with children, but Mark and Anne seemed singularly bright and pleasant, and Mark, at least, was the fortunate recipient of the handsomeness apparent in the Rivers males. There was about Anne, however, an air of repressed anxiety which I thought hardly natural in so young a child. Remembering Rosemary's pinched and rather severe expression, I decided that she was no doubt an exacting taskmaster, which would probably account for it, but honesty forced me to concede that Adam must be faulted for not seeing this and tempering it in some manner. Then, having naively deduced the Rivers family problems on the basis of a few first impressions, I leaned my head against the chairback and admired the workmanship of the heavy carved mantel over the fireplace. After a while my eyelids grew heavy and the sleep that had eluded me in my bedroom beckoned to me now. I felt myself drifting without resistance toward its sweet surcease.

I didn't hear him come in. The carpeting of the floor must have muffled his footsteps as effectively as had the sand on the beach below the hotel, for I only opened my eyes and realized his presence when he passed my chair on his way to the wine table. I started, then felt the now familar apprehension send a chill down my spine. But I set my jaw. Since I had seen him first, I could compose myself and not let him have the satisfaction of knowing he had frightened me, as he had at our first meeting.

As he stood with his back to me, I had time to observe him closely. I saw that he had changed from the riding clothes he had worn upon arrival to a well-cut dark suit, and I couldn't help admiring the wide shoulders tapering to narrow hips, the surprisingly small feet clad in neat, well-polished boots, the shine of the dark hair which curled against his stiff white collar. I was still staring when he turned, wineglass in hand, and saw me watching him. He started so violently that the ruby liquid jumped and nearly spilled.

"What—!" he exclaimed, and smothered an oath. "What the devil are you doing here?"

I was thankful that I had glimpsed him in the courtyard, for it gave me a slight advantage.

"As it happens, I am a houseguest," I answered in a voice that scarcely trembled. "I am here at the invitation of Adam Rivers."

"Adam!" Disbelief vied with consternation for an instant. Then Christian gave a derisive laugh. "Is my strait-laced cousin sowing some wild oats at last, or are you a cleverer little minx than I had imagined?"

I was at a loss for an answer. As it had before, his attitude, even more than his sharp words, indicated a hostility for which I knew no basis, and while I realized that he must be suffering under some unpleasant misapprehension concerning me, I had no idea of how to dispel it. How can one deny an accusation as yet unspoken?

It was all I could do not to shrink back in my chair when he set his glass down and moved toward me in a manner deliberately menacing. His dark eyes were now black with what I took to be deep anger.

He spoke in a heavy, menacing tone. "I don't know just what your little game is, young woman, but you won't play it here. If you are wise you will develop some condition or other which requires you to return to San Diego immediately."

I have never been sure just why the fear began to leave me; whether it was because I caught a hint of uncertainty beneath the bluster and the bravado, or that his remarkable resemblance to young Mark brought to mind a small boy attempting to assert himself with clenched fists and a show of rage, but I could feel my body relax into the armchair's rigid embrace. I even allowed a small, amused smile to play about my lips.

"Really?" I asked pleasantly. "And if I don't, just what is it that you plan to do?"

For a moment he seemed completely nonplussed by my response, for the fury drained from his face and the scowling black brows arched in surprise. Then, just as quickly it returned, and I feared I had badly misjudged him. He seemed about to utter some threatening retort when there was a sound at the door, and as he glanced toward it, the anger was erased so quickly that I might have imagined it, and was replaced by an expression both bland and noncommittal. Casually he picked up his glass and lifted it in the same sardonic salute he had given Father in the hotel dining room.

"Congratulations, cousin," he said. "Your taste in women is impeccable."

To my intense relief Adam came to stand beside my chair, bringing with him some fresh, pungent scent of sun and air and growing things. But as he glanced at me and then again to Christian, his face darkened.

"Just what the devil is that supposed to mean?" he demanded. "And for God's sake, Christian, where are your manners? Pour Miss Allen a glass of wine."

Christian's ironic smile froze and then faded. "Miss Allen?" he repeated carefully.

"Yes, of course, Miss Allen," Adam repeated impatiently. "Didn't you even introduce yourself to her? Beth, this is my cousin, Christian Rivers. Beth," he added to Christian, "has consented to stay with us until she recovers somewhat from the shock of her father's death. I had expected that you, at least, would attend his funeral, since you were already in San Diego." Then, as Christian continued to look astounded, "Heavens, man! You did hear of the tragedy, didn't you?"

Christian seemed to recover himself with an effort. "I did hear about it, but not until after I was on my way home from the Orestes. I had left San Diego early that same morning, and the news didn't reach the rancho. I have seen no one since I got home." He made me a little bow. "May I offer my condolences, Miss Allen? I know you have sustained a great loss." Abruptly he crossed to the wine table, and taking his time, carefully poured a glass. As he brought it to me, he said with more meaning than the words themselves conveyed, "What I didn't know is that you were Jamison's daughter. Please forgive me."

Adam's dark brows drew down in a frown. "Forgive you? For what?" he asked suspiciously.

Christian glanced at me as though asking for assistance, but I continued to be silent. It was obvious that he didn't want Adam to know of his recent rudeness, nor, no doubt, of his ungentlemanly behavior at the hotel, but I felt no obligation to help him out of his contretemps. Let him extricate himself as best he could.

That he managed to do it with grace and charm did not in the least surprise me.

"It was simply a case of mistaken identity," he said easily. "When I turned around and saw Miss Allen sitting there, I thought for a moment a ghost from my past had caught up with me." He smiled beguilingly, and the dark eyes conveyed apology, amusement, and confidence in my willingness to understand and forgive. "Upon closer inspection, however," he continued, "I realize that my past contains no ghost so beautiful. If it did, I would have contrived to keep her in the present." The charming smile deepened, and again he held the glass out to me, and to my chagrin, I capitulated enough to take it, though I was far from accepting his simplistic explanation of his behavior.

Adam, to judge by his expression, didn't find Christian charming at all, but after a moment's consideration forebore to ask more questions and instead poured a glass of wine for himself, then took the large chair opposite me. Christian went to stand by the fireplace with his foot propped against the fender in a lord of creation pose that seemed to fit him. A strained silence fell.

Adam broke it at last.

"I suppose," he said, directing his words to Christian, "that you have some explanation for your absence this past week." To my surprise, his tone was severe, disapproving, almost as though he were speaking to a wayward son. I felt distinctly uncomfortable.

Christian, however, only raised his eyebrows. "Being several years beyond the age of consent, I wasn't aware that I must account for my whereabouts." His tone was deliberately light. "But since you ask so pleasantly, I will tell you that after leaving San Diego and before stopping at the Orestes, I was at the Harrison rancho. They were delighted by my visit; in fact, they even held a small fiesta in my honor." He winked at me as though presuming me to be an appreciative audience. "You see, Cousin, in some places my presence is welcome."

"It would be welcome here," Adam retorted, "if it ever happened to coincide with work to be done. Or did it slip your mind that a large order of oranges is due to go out next week? We are far behind, and your supervision of the packing would be something of a help, at least."

The look Christian gave him could hardly have been considered remorseful. "I must confess, Cousin, that such earthshaking events don't weigh too heavily on my memory. I will, of course, be happy to help now that I am here."

I saw Adam's look that conveyed his doubts of that, but he merely went on. "Also, it is simply common courtesy to let Rosemary know your plans. She finds it difficult to manage things as it is."

Even though I could not help thinking of his failure to let his sister know he was bringing a guest to stay, I felt a sharp twinge of guilt. Adam was not alluding to my visit, I knew, but not only had my need of him caused him to lose valuable days from his work, but my coming here as I had, unexpected and possibly unwelcome, must have added a good deal to Rosemary's no-doubt-considerable burden. Shame overcame me, and I resolved to be less critical in my thoughts of her.

My resolution lasted only as long as it took Rosemary to join us, which she did as Adam finished speaking. Her coming was heralded by the ostentatious rustling of taffeta. She had changed from the drab maroon into a somewhat rusty-looking black gown of heavy silk which had long sleeves and a high, constricting collar, and whose skirts, fuller than was the current fashion, were bouffant with petticoats. But it wasn't her appearance that caused my resolve to crumble. It was because of the disapproval with which she observed my dress, then said immediately, without greeting or preamble of any kind, "I have had your trunks taken to your room, Miss Allen. There is still time to change before dinner, if you wish." Her tone implied that if I had either good sense or good breeding I would do so at once.

I felt my face grow warm. "I'm sorry," I said immediately, though I was more angry than contrite. "I didn't realize that we were to dress for dinner." I started to rise but Adam stopped me with a gesture.

"You are fine just as you are," he said firmly, and glanced at Rosemary as though he would like to say more. She tightened her lips.

"I merely thought," she said more stridently than seemed necessary, "that in view of Jamison's death she would wish to wear mourning."

"Thank you, but no," I answered. Ignoring the pained look on her face, I went on. "While I did wear black to Father's funeral, I know that he didn't really approve of mourning clothes. And of course I will accede to his wishes."

"Thank God," Christian said fervently. "Why hide youth and beauty beneath sackcloth and ashes?" He eyed Rosemary's gown with sudden distaste. "And by the way, just why are you gotten up in that dismal rig? You haven't worn that since—" He broke off and was abruptly silent.

Rosemary looked as though he had slapped her. Bright spots of color burned along each cheekbone, and she blinked furiously. She turned as though to leave the room, then instead said thickly, "Jamison Allen was a good friend to all of us. Must you ridicule me because I wish to honor him by wearing black?"

Adam set his wineglass down and glared at Christian.

"You know better than to pay any attention to Christian, Rosemary. He is notorious for his bad manners." He got up and, taking Rosemary's elbow, guided her to a chair. "It is good of you to be so considerate." He poured a glass of the ruby wine and handed it to her. "I am sure Beth appreciates it, too."

"Oh, I do!" I agreed hastily. I felt momentarily ashamed of my previous anger. Seemingly mollified, Rosemary took a sip of wine, and then another.

I could not resist glancing at Christian. Instead of looking properly rebuked by Adam's remark, he was watching Rosemary with a gaze that was at once amused and somewhat speculative. In spite of my resolve, I found myself regarding her in much the same way. One generally wore mourning only for a member of the immediate family, and it seemed odd that Rosemary would do so for my father.

Another uncomfortable silence fell. Adam broke it again by saying, "Where are Mark and Anne? They know that it is nearly dinnertime."

"They are probably just outside," Rosemary answered indifferently. "I told them not to join us tonight. Sometimes I feel a desperate need to be free of them for a while."

I bit my lower lip in order to avoid making a protest. Had the woman no feelings? Couldn't she imagine how lonely those children must be since they had lost both father and mother?

To my gratification, Adam immediately got up and strode to the door.

"Anne, Mark," he said peremptorily. "Come in here and greet our guest."

At his bidding the two youngsters entered at once and

approached the group with an air of diffidence. Anne was holding Mark's hand in a touchingly maternal fashion, and as they neared us she cast a sidelong glance at Rosemary, who had leaned back against her chair with a patently martyred air.

I felt a slow, burning anger. I, too, had been deprived of a mother through death, but my aunt had at least made an attempt to take her place, whereas Rosemary seemed to be deliberately withholding that special caring which can be supplied only by a woman. Even Adam, with the best of intentions, could not adequately counteract its loss.

"I have met Anne and Mark," I said. "They were kind enough to accompany me downstairs."

At my words Mark smiled slightly, and Anne cast me a grateful look, as though in answer to the warmth in my voice.

"What do you think of Miss Allen, Mark?" Christian asked. "Don't you think she is pretty?" He glanced at me to see how I would take this sally, but if he expected to discomfit me, he was sadly disappointed. I waited with bland disinterest for Mark's reply.

With a child's forthrightness, Mark said, "Very pretty. Though she has a poor sense of direction. She had to have our help to find the drawing room."

Christian laughed indulgently at this, but Rosemary looked pained and Adam said reprovingly, "That was rude, Mark. You are to apologize to Miss Allen at once."

Mark immediately turned scarlet, and though I tried to protest, murmured a few words of embarrassed apology before going to stand next to Christian, who tousled his hair and gave him a wink of sympathy. Anne, looking as upset as though the rebuke had been directed at her, sank unobtrusively into the child's rocker.

I tried to smooth things over. "I did ask the children to help me, you know," I said lightly. "And it is true that I have a poor sense of direction. Even when I used to ride in the park, I took a friend along so a search party didn't have to be organized to find me." I said all this with a jesting air, but no one smiled, so I added somewhat desperately, "I am sure Adam rides, Miss Rivers, but do you? If so, perhaps you could persuade him to allow us to accompany him on a tour of the ranch."

Rosemary answered briefly, "I am sorry, but I do not ride."

Though I failed to see the humor in it, Christian did laugh then.

"You have yet to learn the idiosyncrasies of the Rivers family, Miss Allen," he said with amused sarcasm, "and far be it from me to apprise you of them in one fell swoop. It is sufficient for the moment to inform you that not only does Rosemary no longer ride, she doesn't venture out of doors, except to the courtyard, and seldom even there. She has decided, somewhat arbitrarily, that nature is bent on destroying the Rivers family."

"Arbitrarily?" Rosemary repeated indignantly. Her voice rising alarmingly, she ignored Adam's warning gesture and added furiously, "You may rationalize the bizarre deaths of three of us in the past two years, but I cannot. Especially Marianne's, which—"

"Not one more word." Adam's stern order cut across her strident voice. I recalled that Marianne had been the name of the children's mother. "I will not have the children upset again." Turning to me, he said in a conversational tone, "I would be happy to show you the ranch, Beth, as soon as the opportunity arises."

He went on to discuss the things that he thought would interest me, but I scarcely heard him. I was too aware of Rosemary's barely controlled agitation and Christian's smug amusement. But what disturbed me most were the expressions on the faces of Anne and Mark. They did, indeed, look upset, and more than a little frightened.

Chapter Seven

I, for one, greeted Maria's announcement of dinner with tremendous relief. I could now see the reason for Adam's hesitation at bringing me to Riversbend. The strong undercurrents of hostility among the family were combining with the shadows cast by the recent and apparently tragic past to render me most uncomfortable, and had I had anywhere else to go I would have already been planning some excuse for cutting my visit short. But as it was, I could only pretend a polite disinterest.

The dinner, a roast of beef with fresh boiled vegetables, was simply but excellently prepared. As we plodded with some difficulty through a conversation centered around Adam's remark that virtually all of the viands and the excellent wine had originated at Riversbend, we were served with quiet efficiency by Maria's daughter, Emelda. It wasn't until I noticed Rosemary eyeing the girl with obvious disapproval that I became particularly aware of her.

Instead of the dark dress she had been wearing when I had met her, Emelda had donned a skirt made from gathered tiers of bright turquoise cloth. At each heading ruffle and around the deep hem was a band of intricate embroidery in threads of bright pink and silver. Her lace-trimmed blouse was gathered so loosely at the neck that it had a tendency to slip tantalizingly off her smooth brown shoulder as she offered the dishes to us, and once, while she was serving Christian, it slipped completely onto her arm, revealing a rich swell of olive breast. I would have dismissed it as an accident had I not seen her sidelong glance at Christian and the provocative smile that accompanied it.

Rosemary saw it, too, for no sooner had Emelda left the room than she exclaimed, "That girl's behavior toward you is disgusting, Christian! Are you sure you do nothing to encourage it?"

"Of course he encourages it," Adam said over Christian's protestations of innocence. "But if it bothers you, you should simply dismiss her."

Immediately, tears started to Rosemary's eyes. "Oh, no," she said fearfully, "I couldn't possibly! If I were to dismiss her, Maria might feel that she had to go, too, and I couldn't manage without her, Adam, you know I couldn't!" She had begun to tremble. "Won't you speak to Emelda? Make her see that it isn't proper to behave like that before the children!"

Anne and Mark, who had noticed nothing untoward until it was pointed out to them, were wide-eyed with curiosity. Adam glanced at them and shook his head in warning, but only said resignedly. "I will speak to her if you like. But it isn't worth getting upset about. Please, go on with your dinner and forget it." However, from the look he gave Christian I felt that Emelda would not be the only one to receive a few stern words from him.

Christian, of course, remained unperturbed, and only grinned satirically as he rose to fetch the decanter from the sideboard. When he had filled Rosemary's glass, she seized it and drank half the wine at a swallow. He winked at me and replenished it, and this time she let it sit undisturbed beside her plate. Christian's implication was that she drank rather more than was good for her, but I saw no evidence of it, and almost defiantly I took a sip from my own glass.

"This is excellent wine," I said with a false enthusiasm born of desperation. My feelings of discomfort were becoming acute. "Do you sell it, or do you make only enough for your own use?"

"For our own use, naturally," Christian answered. "Heaven forbid that we should make a profit on anything."

Adam ignored him. "This is from our experimental vineyard, but recently we have put in more vines. The zinfandel grape, from which this is made, is ideal for the hotter, dryer areas. We haven't yet been able to bring water to all parts of the ranch."

"If the Lord meant this land to produce," Rosemary intoned, "He would send rain." I noticed that her wineglass was once more empty.

"He sent the river, didn't He?" I detected a note of impatience in Adam's voice and felt that I couldn't blame him. "And he gave us the intelligence to put it to use."

Christian, who had cleared his plate, pushed back his chair and settled down as though getting ready to watch a familiar performance. He seemed to wait expectantly for Rosemary's reply.

To my dismay she began, quite visibly, to tremble again. "And He allowed our father to be taken by it!

Don't forget that, Adam! He meant it as a warning!"

Adam had gone white. "It was no warning," he ground out. "It was an accident. No more than that."

"Then it was part of the curse!" she cried, and I felt a stirring at the nape of my neck. "There is a curse against this family, Adam! You know there is! Why must you keep denying it?"

Adam stood up. I had never seen such black anger on anyone's face. He clenched his napkin so tightly that the knuckles of his hand were drained of blood, and the muscle along his jaw stood out like rope. For a long moment it seemed that he didn't dare allow himself to speak.

When at last he did, I was surprised at his quiet tone. "I have asked you not to talk such nonsense before the children," he said. "If you cannot control yourself, perhaps you should consider spending more time in your room."

Rosemary, who had been preparing to sob into her napkin, stared at him openmouthed. Then, slowly, she rose, drawing her dignity around her as though it were a mantle.

"Perhaps you are right," she said evenly. "It has always been abundantly clear that my presence is of no importance to anyone in this family." She bowed her head in my general direction. "Please excuse me, Miss Allen. I am not well. I must retire immediately." Without looking again at any of us, she swept regally from the room.

"Now you've done it," Christian said. "She will sulk for a week. And who will look after the children in the meantime?"

I glanced at them. Mark had gone to stand beside his

sister, who put an arm around him. Both of them looked frightened and uncertain.

"I will, of course," I heard myself say, and added for the children's benefit, "We will have a good time together while Rosemary takes a well-earned rest."

Slowly Adam released the tightly clenched napkin and let it fall to the table. "Thank you," he said. "And please allow me to apologize. Such a display before a guest is unforgivable."

"Oh, come now," Christian drawled with that same detestable amusement, "she was bound to see us in our true light sooner or later. And the fact that Rosemary has gone round the bend is hardly a secret. It is talked of all over San Diego."

For a moment I didn't grasp his meaning. Then, with a gasp which I hoped was inaudible, I got up and said hastily, "If you will excuse us, the children and I will go upstairs. Children, wouldn't you like to show me your nursery?" Without waiting for an answer, I headed for the door, and Anne and Mark scurried after me. Quickly I ushered them up the stairs. I wanted to get them out of earshot of the dining room, for I had seen Adam's face. And, I thought with grim satisfaction, I hope he trounces Christian soundly.

The children's suite, which consisted of two pleasant bedrooms with a spacious nursery between, was not far from my room. The nursery had a well-stocked bookcase, and a low table and chairs were set near the large windows overlooking the courtyard below. A comfortable-looking settee and chair were placed near the brick fireplace at one end of the room, and nearby shelves were stocked with what seemed an ample supply of toys. I sat down on the settee and drew the children to me. Mark was still looking frightened, and Anne's face was pale and set.

"This is a very nice room," I said with an attempt at normality. "It must be pleasant to do your lessons here. Do you have a governess to help you?"

"We did," Anne answered tonelessly. "But she left last month. She said it was too lonely for her."

"But I don't think that was the real reason." Mark pressed nearer to me, as if he felt the need to touch someone. "I heard her tell Emelda that the place gave her the willies. What are the willies, Miss Allen?" His large eyes searched my face. "Could we catch them?"

"Of course not, darling," I answered, wishing I could wring the woman's neck. I drew him to sit beside me on the settee. "It simply means that your governess is a city girl and that the country makes her nervous. I'm from the city, too, you know, but if I see or hear anything that frightens me, I am going to ask you children to explain what it is. I would hate to go running back to town because I was frightened by a noise that turned out to be a cow mooing."

I was relieved when Mark laughed. "I was frightened once when a donkey brayed," he offered. "But of course I was quite little then."

Anne was not to be distracted so easily.

"Do you suppose Cousin Christian is right about Aunt Rosemary, Miss Allen?" she asked anxiously. "She does act rather strangely sometimes. Or—" her small mouth worked with anxiety "—could she be right about the curse? It's true that Mama was killed, and then Grandfather and Papa—" She could not go on, but stood looking at me with her eyes filling with tears and her lower lip caught by small, even teeth.

Again the nape of my neck tingled as though the hair was stirring, but I gave no sign of it. I put my arm around Anne and pulled her close.

"I don't believe there are such things as curses, Anne," I said with what I hoped was honesty. "It is just that when sad things happen, we don't understand it and want to find a reason. Rosemary must be very lonely with your parents and her father gone. Do you suppose it would help if we tried to be especially kind to her?"

"Perhaps," Anne answered slowly, as though she doubted it. "But she seems to like it best when we just stay away from her." She looked at me with the clarity of vision that children sometimes have. "Mark and I know we are a problem to them, but we don't know what to do about it." Her voice broke a little as she added, "We're lonely, too, without Papa."

"I know you are," I said, remembering my own parents. I thought it odd that she hadn't mentioned her mother. "And I know a stranger isn't much comfort. But while I am here, I hope you will let me spend some time with you and help you with your lessons." I supposed that even my somewhat sketchy education would suffice for that. "And I would like to explore this big house one afternoon. Do you think your uncle Adam would allow us to do that?"

Mark snuggled closer. He seemed to fit very nicely into the curve of my arm. "I think he would," he said. "And I might not even be afraid if you were with us."

"And there is the swing, down by the river," Anne exclaimed. "I think she would like that, don't you, Mark? It has very long ropes and you go far out over the river, and very high, as if you were flying." She spoke eagerly now, and her color was returning. "Oh, I hope you stay for a long time!"

I hugged them both. "Let's not worry about tomorrow," I said. "Let's just plan to enjoy whatever time we have." In view of my uncertain future, I thought that I might as well take my own advice. "Now, suppose you get ready for bed. I'll tuck you in and stay until you are asleep, if you like."

Both Anne and Mark seemed enormously relieved at my offer, and prepared for bed with alacrity. When they were ready, Mark in a white lawn nightshirt and Anne in a long-sleeved, high-necked gown, they began to bicker over whose room I was to sit in until they slept, but I solved the problem by moving one of the low chairs to a spot in the nursery where they could see me from their respective rooms.

As I leafed idly through a storybook, I could hear their restless stirrings lessen until at last there was silence. Quietly I laid the book aside and went to Anne's room. She had thrust the cover back and was lying with her knees drawn up and her arms wrapped around her thin chest as though to draw what comfort she could from her own embrace. I thought that even in repose she wore an anxious look.

I drew the quilt over her and went to Mark, who was lying spread-eagled on his back. As I watched, he gave an enormous twitch of his entire body, then returned to his previous position.

Even in the complete relaxation of sleep, he was extraordinarily handsome. Heavy black lashes lay like smudges against the chubby cheeks, and the firmly closed lips were full and well defined. His wide-spaced brows were still spare and light, but hinted at becoming the thick, dark arches of his cousin Christian. The likeness between them was truly striking, especially when one considered that Mark's father and Christian had been only cousins. I could only surmise that the two brothers, Adam's and Christian's fathers, had very much resembled one another.

At the thought of Christian, unease stirred in me. I felt sure that his glib explanation to Adam and me had not even touched on the truth. I was certain that the real story involved my father, and I could not be comfortable in Christian's presence until I learned what it was. I decided, not without trepidation, that tomorrow I must find the opportunity to confront him and demand to know the truth.

Upon returning to my room, I found that the lamp had been lit and my bed had been turned down invitingly. In a shadowy corner of the large room stood two trunks: my small curved-topped one and Father's large steamer trunk. My heart contracted at the sight of the latter. I had last seen it standing half empty in Father's room at the hotel, its violated contents strewn haphazardly around it.

I took the lamp and set it on the bureau near the trunk, and after a long hesitation, opened its lid.

The trunk had been carefully and neatly repacked. The tray had been replaced across its opening, and it held Father's most personal possessions: his military-looking brush and comb set, his stud box, several letters bearing my handwriting, and a large photograph of my mother, which was in an elaborate filigreed frame. She gazed up at me with compassionate eyes above a small, understanding smile. Seeing her brought a rush of bittersweet memories, and I felt the sting of tears.

After a time I closed the trunk and returned the lamp to the table beside my bed. I couldn't bear to investigate its contents tonight. If there was something I should see, some overlooked clue to the mystery of Father's death, it would have to wait for a time when I had my emotions under better control.

With fingers that shook slightly, as though with a chill, I undid the fastenings of my gown and removed it. It took but a moment more to shed my undergarments and slip into a high-necked cambric nightgown, and with a feeling of vast relief I slipped into the billowy feather bed. As I turned down the lamp I supposed I would have trouble getting to sleep, but before I could even shed a lonely tear, blessed oblivion covered me like a warm and comforting blanket.

I awoke to a soft rapping at my bedroom door. When, still bemused by my dreamless sleep, I said, "Come in," Maria entered quietly. Her eyes widened when she saw my drowsy state.

"I am sorry to wake you," she said with her soft musical accent. "I thought you would be waiting for your tea. Or perhaps you would like the chocolate?"

"Tea would be fine," I answered. "But what I would really like first is a bath. Would it be too much trouble to have some hot water brought up for it?"

She smiled. "For that I am prepared. The bath is just down the hall, and in it you will find the hot water already there. It is in the cistern above the bath, and you must only turn the handle." She added proudly, "Señor Adam, he makes many small miracles."

It did, indeed, seem a miracle to have delightfully hot water flow into the tub from what seemed to be a large copper boiler suspended from the ceiling. I had not expected such luxuries so far out in the country. Adam rose even further in my estimation.

I basked in the warm water for some time before scrubbing vigorously with the soap and washcloth which lay nearby. I also lathered and rinsed my hair, which fortunately is naturally curly. After drying off with a heavy towel, I wrapped it around my damp hair and slipped into my nightrobe for the journey back to my room.

As I stepped into the corridor I saw that it was empty. Every door was still tightly closed, even those of the children's quarters. I wondered if they were still sleeping, which seemed unlikely. More probably they had already gone down to breakfast, and even out to play. I felt a sudden eagerness to join them, and hurried to my room.

But as I entered it I got the sudden sense of an alien presence.

I waited just inside the door while I tried to fathom the reason for my unease. As far as I could tell, nothing had been disturbed, nor had Maria returned with my tray. I sniffed questioningly, but detected no definable scent in the air. And yet, some sense beyond the ordinary told me that someone was in this room, or had been only moments before.

I hesitated, then went to the large cherrywood armoire and, with some apprehension, opened it. It stood totally empty, awaiting the contents of my trunk. I glanced about. There was no other possible hiding place, except— I crossed to the bed and, feeling somewhat foolish, dropped to my knees beside it. Gingerly I lifted the dust ruffle.

There was no one there.

I sank back onto my heels, my body going quite limp with a relief which surprised and frightened me. Had I really felt some unfriendly presence, or had I allowed Rosemary's talk of a curse sink into my unconscious, there to mingle with the residual fear left by Father's murder and my terrifying experience with the intruder at the hotel?

I heard a small sound at the door and, still on my knees, whirled to face it.

Maria was standing in the open doorway with my tea tray in her hands. She looked apologetic and somewhat embarrassed.

"I am so sorry to interrupt your prayers. It was only that the door was open—"

Feeling incredibly foolish by this time, I got up quickly. "It is quite all right, Maria," I assured her. "I wasn't praying. I thought I heard something, and was looking— Perhaps it was a mouse," I finished lamely.

She looked aghast. "A mouse! Alas! I will set a trap immediately. Those we do not allow in the casa!" As she brought the tray in and set it on the table, she added, "Here for you is the tea and toast. I will prepare a larger breakfast when you arrive downstairs."

I unwrapped my hair and began slowly to dry it. "Will the others be down soon?"

Amusement flickered across her pleasant face. "They have eaten already. Señor Adam has been out to the fields since early morning, and the children were so impatient to see you that I sent them outside so that they would not disturb your sleep."

I thought of the resolution I had made last night.

'And Christian?"

"Señor Christian has gone for his morning ride, I think. Or else he is with the children."

"I see." I decided that it must be later than I had supposed. "Please don't bother to prepare breakfast for me. The tea and toast will be quite enough. And thank you for bringing them up to me, Maria."

The housekeeper made a deprecating gesture. "Oh, *por nada*," she demurred. Nodding shyly, she went out, closing the door quietly behind her.

I tried to forget my strange unease, but as I continued to dry my hair some small chill persisted. I told myself that it was nonsense, and in an effort to shake the last vestiges of its hold on me, set briskly about completing my morning toilette. I took particular care in selecting a morning gown, for I knew that if I was to confront Christian I would need the fortification of knowing I looked my best, and at last chose a crisp cotton with small blue flowers sprinkled over it, for their color exactly matched my eyes. After catching my hair back with a narrow ribbon of the same clear blue, I surveyed myself in the glass with some satisfaction. The young woman regarding me with a cool, clear gaze looked quite capable of confronting Mr. Christian Rivers.

When I went downstairs, the only person in evidence was Emelda, who was sitting on the steps as she indifferently dusted the banisters. Her response to my greeting was barely civil, and that, coupled with her sullen expression, caused me to assume that Adam had indeed upbraided her about her behavior last night at dinner. I wondered whether she could have possibly been my intruder. It might well have been she; certainly she was in the vicinity, and had I chanced to surprise her coming from my room she would have had a plausible excuse. But if it was she, the girl was a superb actress, for her indifference to me now held not a hint of guilt or self-consciousness.

I considered asking her if she knew Christian's whereabouts, for I was determined to speak to him at once, but decided against it. She would no doubt take it as a sign that I was interested in him, and I didn't wish to arouse needlessly any feelings of jealousy. Instead I went out to the front veranda. It, too, was deserted, and except for the grazing sheep, nothing moved on the broad expanse of emerald lawn, but from the direction of the river I could hear a faint sound of childish laughter. I had started toward it when Christian came around the side of the house.

It was exactly the opportunity I had wished for, but the unexpectedness of our meeting put me off guard, and now that I was suddenly confronted by him my resolve wavered. To my self-disgust I could think of no possible way to open the conversation, so I merely nodded briefly and started to go past him. I had taken only a few steps when he said in a tone oddly conciliatory, "Miss Allen. May I have a word with you?"

I almost didn't answer, for at the sound of his voice my heart had begun pounding furiously. Then, head high, I turned to confront him. His supercilious smile was entirely gone, as was the air of arrogance which generally accompanied it. Upon seeing this my nervousness subsided and instead of the cool refusal which had been on my lips, I found myself answering in reasonable tones, "I suppose so. Unless, of course, you plan to threaten me again."

He had the grace to look embarrassed. "I apologize for that," he said, "and in fact it is part of what I wish to discuss with you. Will you join me here on the veranda for a few moments?"

He indicated a pair of wicker chairs beneath the drawing-room windows, and somewhat reluctantly I sat down. He took the other chair, and then, head bent, thought for a moment as though considering how to begin.

I took the opportunity to consider him. If, indeed, he and Adam had exchanged blows last night, the smooth skin and handsome features showed no evidence of it. And now that the unpleasant arrogance had gone, he appeared to be younger than I had thought, and more vulnerable. His face in repose reminded me touchingly of Mark's sleeping countenance.

"First, I would like to ask your forgiveness for my bad behavior last night," he began. "Adam had every right to be angry. As for my rudeness that evening at the Hotel del Coronado, I do have an explanation, though I can't possibly expect you to excuse it." He paused, as though finding it difficult to go on. "It touches on behavior of mine that I am not proud of, and it isn't a matter that should be brought to the attention of a girl of refinement such as you." He shifted uncomfortably and took on the guilty look of a small boy being brought to task by his elders. "But I don't know of any other way to make you understand, so if you will forgive me—" He hesitated, then went on. "I had found myself in an unfortunate situation regarding a young woman. I desperately needed financial aid, and I turned to your father as an old friend of the family. He reluctantly gave me that help, but—"

Here he paused to smile with rueful embarrassment, "as an old friend of the family he took it upon himself to lecture me at length about my irresponsible behavior, my shocking immorality, and my general lack of the sterling Rivers qualities of industry and diligence." He spread his hands in a self-deprecating manner. "I have had many such lectures these past few years, and while those faults are mind indeed, unfortunately I also have a thin skin and a quick temper, and I confess that I resented his scolding, even though I knew every word to be true. And then, when I saw him with you in the dining room— Forgive me, but I thought it was the old maxim of the pot calling the kettle black. I had no idea that Jamison even had a daughter, let alone that you had joined him in San Diego."

He fell quiet, no doubt allowing time for me to digest his words. And I needed time. I had lived a somewhat sheltered life and had only a vague knowledge of girls who had gotten "into trouble" or of those referred to as soiled doves, and it took a moment for me to realize that he was saying he had been involved with the one and had taken me for the other. I felt a slow wave of warmth reach my face as I stared at him.

"I know it seems impossible that I could have mistaken you for that kind of woman," he went on in answer to my unspoken accusation, "and I apologize deeply for it. I can only excuse myself by saying that my judgment was clouded by the fact that my pride was still smarting from Jamison's criticism. And then, when I saw you accompany him upstairs, my base suspicions were confirmed." He turned from me, as though overcome by remorse. "I only wish I could tell your father how sorry I am."

I had to admit to myself that his story had the ring of

truth to it. And it fitted both Father's comment about him and Christian's own rude remarks on the strand below the hotel. Moreover, I found myself wanting to believe him, for his chastened look was somehow highly appealing.

"Well," I said judiciously, "while I am sure Father would have attempted to thrash you soundly for your remarks to me, he could hardly help being flattered by your mistake." I waited until he looked at me, then allowed myself a small smile. "After all, he *was* old enough to be my father."

Immediately I realized my mistake in injecting even that small touch of humor, for his eyes lit and there was a return to his previous confidence. "Then you will forgive me, after all?" he asked eagerly.

In spite of my inward relenting, I decided perversely that he hadn't yet been punished enough. "Let us say that I will accept your explanation," I amended. "I am afraid you are right in that I still find your behavior inexcusable." I rose. "Good morning, Mr. Rivers." And with that I again started toward the river and the pleasant sound of childish laughter.

Immediately I heard the crunch of his footsteps as he followed me across the gravel drive. "Miss Allen!" he exclaimed. "May I ask where you are going?"

"I plan to join the children," I answered. "I promised to spend some time with them today."

The humbled air was back as he said, "Is it possible that I might accompany you? I would like to show you that I'm not the ogre you assume me to be."

I found his pleading smile hard to resist, but I hardened my resolve. His change of attitude since last night was simply too swift for me to totally believe it. "I think not, thank you," I said firmly.

Disappointment clouded his face. "May I ask why?"

His persistence told me that he wasn't used to having his wishes thwarted. Well, it might do him good to encounter it this once. I raised my chin determinedly.

"Quite honestly, Mr. Rivers, I have seen nothing in your behavior that recommends you to me. Not only your actions, but your attitude toward your family simply reinforces my father's advice concerning you. And while I have accepted your cousin Adam's offer of the hospitality of his home, I don't feel that it obligates me to be more than reasonably civil to you when we happen to meet."

I thought I detected a flash of anger, but it was gone immediately. He merely bowed in graceful submission, then said, with another of his charming smiles, "As you wish. But I give notice that I intend to mend my ways. And if I succeed to your satisfaction, perhaps then—?"

"Perhaps then," I conceded. I couldn't resist allowing him that. This time when I began to cross the lawn he didn't follow, but I knew that if I glanced back I would find him still watching me. Instead of the satisfaction I should have been feeling, I knew a moment of sharp regret that I hadn't allowed him to accompany me, after all.

Chapter Eight

The huge oak stood just at the turn of the river which gave Riversbend its name. Only the topmost branches could be seen from the level of the house, for the ground sloped sharply at the edge of the lawn, but as I reached the lip of the steep incline I could see the tree in its entirety. Its gnarled crown, which rose far above the huge circumference of knotted trunk, looked as though it had cradled birds' nests for a hundred years. From one of its widespread branches hung two lengths of thick manila rope which were fastened at their lower ends to a wide wooden plank. Anne and Mark were seated on the plank, and as I watched they glided forward and up in a long, slow arc which reached its apex mid-river, hesitated as though pinned against the sky, then began a backward swing which lifted almost until the tree itself seemed to embrace them. Again they soared out and over the river, their squeals of delight trailing after them.

I found I was holding my breath. Carefully I expelled it and drew another, but I could not take my eyes from them. They seemed so small, the tree so huge, the river and ground so dangerously far below their fragile bodies. But as I stood transfixed, they continued to swing to and fro without mishap, and at last my heart resumed its normal beat and I continued toward them.

Anne saw me first. "Miss Allen!" she exclaimed with a jerk of excitement which sent the swing lurching crazily. I closed my eyes and swallowed convulsively. "Mark! Stop pumping! Miss Allen must have a turn."

"Oh, that's quite all right," I said faintly. "I don't think that I—" But my protest was lost as the two children became absorbed in slowing the swing by dragging their feet along the ground, a process terribly detrimental to the polish of their black high-buttoned shoes. Before it had completely stopped, they jumped down and ran to me, and seizing my hands, tugged me toward the still-swaying swing.

"You will love it, Miss Allen, I know you will!" Anne exclaimed enthusiastically. "You feel so powerful, as if you were flying!"

"But—isn't it dangerous?" I protested. "If you were to slip off and fall into the river—!"

"I expect that would be better than falling to the ground," Mark said practically. "Besides, we can swim. Uncle Adam taught us in the big reservoir."

"Can *you* swim?" Anne inquired anxiously.

"Why, yes, after a fashion." I looked with distaste at the slow-moving, muddy brown flow. "But the thought is hardly appealing. I would have to bathe immediately after."

They seemed to think that remark enormously funny. After controlling his laughter, Mark said, "You really won't fall, you know. You only need to hold tightly to the ropes."

The swing did look sturdy. The seat was a heavy plank of oak, the ropes new-looking, as though recently replaced.

"Uncle Adam repaired it just last week," Anne said. "I don't think he would let us use it if it was very dangerous."

"No, of course he wouldn't," I agreed immediately. I glanced back at the house. Christian was no longer in sight, so I abandoned my dignity and, grasping the ropes, sat gingerly on the wooden seat. It felt very solid and comfortable.

"Shall I give you a start?" Mark asked eagerly. "I'm quite strong, you know."

"Please do," I answered. The shove he gave me was surprisingly effective, and with the addition of a few unobtrusive pushes of my feet against the ground, I was in motion.

I hadn't been in a swing since my childhood, but the rhythmic pumping movement it required to set it in motion returned to me immediately, and soon I was soaring effortlessly out over the river. After the first few breathless moments I found it not frightening at all. The swing's hypnotic glide combined with sounds of river and children and twittering birds to plunge me into a pleasantly bemused state which was broken only when a subtle change in the children's voices caught my attention. I looked down. Adam, his horse tethered nearby, was standing with them and watching me.

I blushed hotly at having been caught so obviously enjoying a children's pastime and would have stopped immediately, but I realized that there is no graceful way to halt a swing. There is nothing for it but to let it die naturally. It took an interminable time doing it, but when at last it slowed almost completely, Adam came forward and held out his hand. In spite of my embarrassment, there was nothing I could do except take it and step from the swing.

Adam's usually somber expression had softened, and it seemed to me that he held my hand a moment longer than was necessary.

"If I were a poet," he said, "I would have words to dsecribe the way you looked just now. But the closest I can come is to have jumbled thoughts of spring and blossoms drifting gracefully to earth."

It was quite the nicest thing he had ever said to me. My initial embarrassment changed to a warm glow of pleasure. "Thank you," I said. "That is quite enough to flatter any woman. Have you been riding?"

The answer to that was obvious, considering his attire and the waiting horse, but Adam seemed not to notice.

"I have been helping to lay out a continuation of our irrigation system. Perhaps after lunch you would like to ride up there with me and take a look at it. It would give you a chance to see the ranch."

I felt my face light with pleasure, though it was more the thought of going riding on this beautiful summer day than any interest of mine in being with Adam which intrigued me. In truth, ever since we had quarreled the day of Father's funeral I had not been entirely comfortable with him. I felt threatened by the anger which I sensed lay just below his controlled surface, and, in addition, I resented the autocratic way he had imposed his will on me. Besides, the moment I had exclaimed eagerly, "Oh, I would like that!" I saw the children's crestfallen expressions, and knew I couldn't disappoint them. So I added with only partly feigned reluctance,

"But I have already promised Anne and Mark that I would spend the afternoon with them."

Adam considered his young niece and nephew. "Perhaps they would be willing to compromise. How about it, you two? Would you like to come along with us?"

He needed no answer beyond that of their beaming faces.

After Adam had excused himself to complete his morning's work, the children and I continued to enjoy ourselves until it should be time for lunch by taking turns again at the swing, and then, when that activity palled, playing a vigorous game of tag on the sloping lawn. Both Mark and Anne seemed supremely happy, their fears and worries of the evening before quite forgotten and the prospects of a pleasant afternoon lying enticingly before them. They had, at least for the moment, shed the grave and serious manner that I thought totally unnatural in children so young, and I wondered if the fact that Rosemary was still adamantly sequestered in her room had anything to do with it. How could they be lighthearted and gay when her primary emotion toward them apparently was disapproval?

I confess that her absence was a relief to me, at least. I couldn't possibly have run about, pursuing the elusive Mark and Anne in so hoydenish a manner had I thought that she might be watching from one of the upstairs windows, and that I might meet later, at lunch, perhaps, her pointed remarks about the proper decorum for one so recently bereaved. But I felt that I needed desperately to forget for a short while the bitterness of my loss, and I thought with rebelliousness that Father himself would not have disapproved of such innocent pleasure, though I couldn't be sure of Adam's reaction. His concern for appearances at the funeral had been because of the sensibilities of Father's friends, but there was no one here to approve or disapprove of my behavior. Even so, I was glad that his work was to keep him in the farther groves until lunchtime.

As the morning grew warmer our enthusiasm for our game lessened, and by the time Emelda came from the house to fetch us, we were resting in the shade of one of the tall oaks that overhung the lawn. The docile sheep were clustered with us, and Anne and Mark had introduced each of them to me by name.

"Doesn't getting to know them make it difficult when—ah—they have to leave you?" I inquired delicately. It certainly was going to lessen my enjoyment of my next lamb or mutton dinner.

"We try not to think about it," Anne said. "Uncle Adam says that every creature has its place in the world and theirs is on our dinner table." She wrinkled her straight little nose. "But I suppose that is why I don't fancy lamb too much." She sighed. "I wonder why God arranged it so that all of His creatures have to eat one another."

I was at a loss as to an answer for her, and it was with great relief that I saw Emelda crossing the lawn to us.

"Look," I said, "there is Emelda. Luncheon must be nearly ready."

She approached us with that slow, undulating walk that I thought surely must be an affectation developed to attract the admiration of men, but if so, it had become such a habit that she now did it unconsciously, for its charm was lost on the children and me. She reached our small circle and stood, one hand resting casually on a rounded hip, as she gazed down at us.

"It must be nice, eh?" she said impudently, "to have nothing you must do except loll all day in the shade if you choose?" She didn't try to hide the envy in her voice.

Her words might have stirred some guilt in me had I not seen the indolent way she had dusted the stair banisters that morning. "Very nice," I returned, stifling the impulse to retort that I had cared for a large house and a dying aunt for many months with just the help of one servant. "But surely you haven't come all the way out here to ask me that. Please deliver any message so that you may get back to your duties."

I had deliberately used a prim and superior tone, but when I saw the look on her face, immediately regretted it. There shone from her eyes almost a look of hatred, and I knew I had made an implacable enemy. I thought for a moment that she wasn't going to answer me at all.

"Lunch is ready," she said. The words were almost ground out, and she immediately spun on her heel and started back to the house. The undulating sway of her walk was conspicuously absent.

The children were silent and almost motionless, as though they recognized the lightning that had flashed over their heads. "Come on, children," I said lightly, though I couldn't quite meet their eyes, "let's hurry and tidy up for lunch. We wouldn't want to keep the others waiting." I was already ashamed of my behavior. One doesn't allow oneself to be ruffled by the remark of a servant.

By the time we had made ourselves presentable and had returned downstairs, Adam and Christian had arrived in the dining room. In fact, it was apparent to me, at least, that they had had time to engage in a quarrel of some sort, or at least a heated discussion, for Adam wore an angry scowl and Christian had that small superior smile that I was beginning to realize was his means of hiding an ire as great as Adam's.

"What in God's name do you do with your money?" Adam was demanding as we entered the room. "I can't see what there is for you to squander it on, unless it is gambling. And if it is, I'll not be a party to supporting a habit like that!"

Christian cast me a quick glance. "Forget it," he said. "On second thought, I won't be needing it." As he came over and courteously seated me, he continued to Adam. "And you are right about my spending more time here at Riversbend. In the future I plan to do exactly that."

Though Christian was careful not to so much as brush my arm as he pushed in my chair, the warmth and admiration in his voice were obviously meant for me. I felt a tingle of excitement lightly touch my spine, though I was careful to remain expressionless.

Adam seemed to understand the meaning behind Christian's sudden capitulation, also, for his dark frown deepened.

"Try spending it in doing some of the work around here," he said sharply. "Miss Allen hasn't come here to afford you amusement."

Christian's dark eyebrows rose. "Oh? Is that exclusively your prerogative, then? I understand that you have ordered horses to be readied for this afternoon. If there is so much to be done, how is it that you can waste the afternoon riding aimlessly around the ranch?"

I saw Adam's blue eyes darken with fury. "I do the work of three men around here. How dare you question what I do with my time?" he thundered. I gasped as he took an angry step toward Christian. "In fact, there is still picking to be done in the west grove. You can oversee it in my absence."

Though his facial skin turned a dull red, Christian stood his ground. "No thank you, Cousin. I have urgent business to attend to in San Diego. You will excuse me, Miss Allen?" Giving me a slight bow, he left the dining room without another word to anyone.

There was an awkward silence. Adam pulled out his chair and sat down, still looking like a thundercloud. I was angry, too, and I was aware that the bodice of my gown rose and fell with my agitated breathing. I felt that Adam had been unduly harsh with Christian, especially in view of the fact that there had been a guest present, and I further disapproved of such an altercation taking place before the children. As I glanced at their frightened, uneasy faces, I had to press my lips together to restrain myself from voicing a strong opinion.

In spite of my attempts to conceal my emotions, Adam must have seen how I felt, for he said stiffly, "I beg your pardon once again, Beth. We have managed yet again to display an abominable lack of manners."

I couldn't quite be gracious. "I think it's not I, but Anne and Mark to whom you should apologize," I said primly. "It can't be good for their digestions for them to be constantly upset at mealtime."

"They should be used to it," Adam answered curtly. "It has been going on for most of their lives." He glanced at them, and added, as though relenting, "But I do apologize, children. I will make it up to you on our ride this afternoon." And then, as Maria came in with a laden tray, "Now attend to your meal and don't dawdle. There is a lot I plan to show Miss Allen during the next few hours."

Mark and Anne were seemingly reassured by his altered tone and ate their food with normal, hearty appetites.

I had brought two riding habits with me from New York, but it required little thought to choose the dove gray one over the black, for I knew of the heat and dust we would most certainly have to contend with. In fact, both outfits seemed ridiculously formal for riding around the ranch; they had been designed for no more than a sedate canter through a shaded park. And both, I thought gloomily, were most incredibly hot. As I adjusted the requisite white scarf high around my neck, I thought with envy of the low-necked peasant blouses Emelda wore and wondered, not for the first time, just who it was that dictated the bulky, uncomfortable clothing most women generally wore. Probably men, I thought with irritation. They dictated everything else.

I was still feeling grumpy after I had swept my fair hair neatly atop my head and secured the silly little hat carefully atop it, but after surveying myself in the mirror I had to admit that I did look reasonably attractive. Adam would no doubt approve. I pushed away a slight regret that Christian wouldn't also be there when I came downstairs, for I could imagine the way his eyes would have lit with admiration.

Though I didn't link it with this last thought, my irritable mood returned as I went down to join the others, and though I did my best to hide it for the children's sake, I was sure that Adam couldn't help noticing that I assiduously avoided his eyes. As we went outside to mount our horses, which two of the Mexican stablehands had brought around to the veranda, he was unusually silent, and I began to dread the thought of spending an entire afternoon with him. If only Christian were riding with us instead, I thought rebelliously, there would be laughter and teasing and—

I caught myself, appalled. How could I feel this way when Adam had been so kind in my time of need? Did I honestly think that Christian would have set aside his own pursuits in order to aid and comfort a stranger? Hardly. I had the grace to feel ashamed of myself, and resolutely I set about making amends.

"Have you lived at Riversbend always?" I asked Adam, who was riding at a sedate pace beside me, undoubtedly in consideration of my lack of expertise. The children, who rode as though a part of their sturdy little ponies, had trotted ahead. Their excited voices floated back to us.

To my relief, Adam responded willingly to my question.

"No, not always," he answered. "My father didn't buy the land until I was nine or so, and I must have been ten by the time the house was built. But I knew from the moment I first saw it that I wanted to spend my life here. Nothing that has happened since has caused me to change my mind."

I delicately skirted the obvious question as to what exactly *had* happened, as I didn't wish to awaken painful memories. I simply presumed that he meant the deaths of his father and the children's parents and asked instead, "What about the rest of the family? Were they as taken with the place as you?"

"My brother, Morgan, who was the oldest, was

somewhat indifferent. It had already been decided that he was to attend school in the East, and he knew that for some time he would only spend his summers here. Rosemary—I can remember hearing her sob and wail for days after the news was broken. She was in her early teens, and she resented leaving her friends and her school in San Diego." Adam spoke matter-of-factly, and I couldn't tell if he felt any sympathy for her.

"Couldn't she have gone to boarding school?" I asked.

"She begged to be allowed to, but Father wouldn't hear of it. My mother had died the year before and he maintained that Christian and I needed a woman's influence more than Rosemary needed a quite unnecessary education. Poor girl, she has never gotten over it."

I digested this in silence. I was beginning to feel a good deal more sympathy for Rosemary than I had before. No wonder she resented the care of Anne and Mark after having been forced to sacrifice so much for two young boys. Then the full impact of Adam's words sank in.

"Christian? Did he live with you even then?"

Adam's face took on the closed-in look I had noted before. "My uncle had died, and Christian's mother was—not well. So my father took him in. He was accepted without reservation as a member of our family."

Not quite, I thought with a sudden flash of insight. He usurped your position as the youngest. You have never really forgiven him for that.

We had been traveling along a broad wagon road which sliced through the orange groves behind the house. As we emerged from this semishade, Anne and Mark trotted back to us.

"Are we to go toward the river, Uncle Adam?" Mark

cried. "Or are you taking Miss Allen all the way to Lookout Point?"

"Lookout Point, by all means," Adam answered. "She will be able to see most of the ranch from there." His expression remained serious as he added, "Perhaps you and Anne should ride on ahead. To make sure there are no Indians waiting in ambush, you know. But take care. The trail is a bit steep in spots."

"Oh, we will!" Anne seemed as delighted as Mark with this suggestion. "And don't worry. If there are Indians, we will race back and warn you!"

I glanced sharply at Adam, then relaxed in my saddle. He was entering into a fantasy with them, I realized, for he wore a deliberately solemn expression, and the closed-in look had again vanished. How seldom he actually smiles, I reflected, and realized suddenly that I couldn't remember hearing him laugh.

As we started up the slope, Adam took the lead and I followed as closely as I could, clutching my pommel in the steeper spots with what I feared was more desperation than grace. But I forgot to be nervous as we rounded the last curve and emerged on a plateau at the top of the hill.

Spread before us was the whole valley, cleaved at its center by the river and the green ribbon of growth which lined its banks. I could see the flat, red-tiled roof of the house, the huge oak just at the river's bend, and even a few small white specks that were the sheep grazing close to the water's edge. Across the river and beyond the line of foliage, the land was as dry and desolate as was the hill upon which we stood, but below us, on what I presumed to be Rivers land, field after field was planted to crops, and the acres surrounding the house were thick with the glossy leaves of citrus trees. Rosemary had called it an oasis, and from this vantage point it truly looked to be just that. It brought to mind tales of the Arabian desert and of the palm-fringed paradises described therein. All it lacked, I thought with wistful admiration, was a crystal pool where languorous sloe-eyed maidens bathed under the detached gaze of enormous silent eunuchs.

My eye was caught by a sheen of water just below. Set not far from the river's edge, it looked to be a small lake, except that its rectangular shape and sloping sides proclaimed it made by man. A dark line ran from it to the river, where, I noticed for the first time, two paddle-wheels, one small, one large, were lazily churning.

Adam, following my gaze, said, "That is a part of our irrigation system. Water is scooped from the river into the reservoir, to be held until it is needed in the dry season."

Anne and Mark who had dismounted and gone closer to the edge of the plateau than I thought entirely safe, came running back to us. "You see, Miss Allen?" Mark called, pointing to the reservoir. "That's where we learned to swim! It's very deep and very wide, but I can go all the way across!"

Properly impressed, I exclaimed warmly, "That's wonderful, Mark! You must be very brave," but Adam caught the amusement underlying my words and demanded to know what was funny about his reservoir. So I had to explain my fantasy of palm trees and dark-eyed maidens, adding, rather thoughtlessly, "Though I am afraid that is hardly the crystal pool I had in mind."

"It is cleaner than you think," Adam said defensively. "The sediment tends to settle to the bottom. If you are still here in August or September you may find yourself longing to plunge in just as the children and I do."

"You are probably right," I hastened to agree, though privately I thought I would rather swelter than partake of such dubious refreshment.

Somewhat mollified, Adam helped me to dismount, then pointed out the dark lines of canal leading from the river to smaller ditches running throughout the fields and groves. "We are fortunate that the land has almost the exact amount of slope needed for ideal irrigation," he said, and went on to explain the principles of his farming methods. Though I nodded in the right places, I had ceased to really listen and instead had returned to his earlier words. Was it possible that I would still be dependent upon his hospitality in August? I cringed with embarrassment at the thought.

He may have sensed my inattention, for he drew silent. Not wanting him to think I had gone back to my earlier cross mood, I said quickly, "You have thought things out so cleverly. No wonder you are proud of Riversbend. It has everything, even to running water in the bathrooms," then blushed a little for fear I had been indelicate. He seemed not to notice, but answered moodily, "Everything—except a mistress to love it, and a happy family to enjoy it. It is like a tree that has died and rotted away. It may still look strong, but it is hollow at the heart."

"But—Rosemary?" I ventured uncertainly.

"Rosemary!" His tone dismissed her. "Rosemary is too filled with bitterness and resentment to be able to feel love for anything, even—" he glanced toward Mark and Anne and lowered his voice "—for those two orphans. No, Riversbend needs a caring woman to look after it. Someone to whom it means as much as it does to me."

I was looking out over the land as he spoke, and it

struck me that the timbre of his voice had changed. I could feel that he was watching me intently. I said with studied casualness, "It is hard to believe that you have had difficulty in finding someone like that. Or is it that you haven't really tried?"

"I have never found anyone who truly attracted me—until now," he said.

His meaning was unmistakable. I turned to face him. Without my knowing it he had come closer, and his vivid blue eyes were only inches from mine. They moved over my face, lingering longest at my mouth. I knew that he was going to kiss me, and knew also in that moment that if I desired it I could be mistress of Riversbend.

Quickly I spun away from him and turned again as though to admire the expansive view. I wasn't ready to make a decision like that, even though it would mean that I need never fear being alone and destitute again. Besides, though I suspected it was done more often than not, I couldn't accept marrying for convenience rather than for love, and I wasn't sure at all of my feelings toward Adam. His abrupt changes of mood, the uncommunicative, brooding look he often wore hinted at dark and unhappy secrets, and though I had several times felt warmth and even something close to love for him, I wasn't at all sure that it wasn't based more on gratitude than on any romantic emotion.

Again I knew that he had come to stand behind me. After a moment he said, "Does the thought of being my wife repulse you, then?"

I turned to look up at him. His face was set, as though with anger, but I felt that I could detect a deep hurt there, also. I said hastily, "No, of course not, Adam! I am honored that you care for me. But I am—I am just not ready to make such a lifetime commitment, especially so soon after—"

His eyes narrowed. "I see. Well, that is understandable, I suppose, though in view of the fact that it was your father, not a husband, that—" He broke off, then went on, "You are sure the real reason isn't that you have met someone else who takes your fancy?"

I took immediate offense at the accusing tone of his voice. "What do you mean?" I demanded.

"You seemed more amenable until you met my cousin Christian. You aren't dreaming of some fairy-tale existence with him, are you? Because I can assure you it will never come about. He is not a man to be satisfied with one woman for any length of time."

"Why, how—how can you say such a thing to me!" I spluttered. "I have barely met your cousin. Is it your opinion that I am the type of woman who sets her cap for a man moments after meeting him?" The fact that I *had* been attracted to Christian that morning lent perhaps more heat than was warranted to my words.

Adam took me by the arms in a grip of iron. "Do you deny that ever since we set out on this ride you have been wishing that it was he accompanying you, rather than I?"

It was the impact of his eyes, just now turned slate-blue, more than his nearness that covered me with confusion. That, and a helpless feeling that I could have no secrets from him. I cast about for words of sharp denial, but could find none with which to defend myself.

"Ah," he said grimly, "I have hit the mark." He pulled me closer. "Nevertheless, I will have something of this day for myself." With that, he bent his head and kissed me, an angry, punishing kiss that spoke more of hurt pride and anger than of love, and yet that stirred some deep, hidden need in me. I was caught up in it, powerless to resist, and to my later shame it was he, not I, who pulled away at last. His eyes still gleamed with anger, but there was triumph in them, too.

"Try and tell me you were thinking of Christian just then," he said.

Fortunately Mark and Anne came running back with some newfound treasure which they wished to share, and I was spared the necessity of answering him.

Chapter Nine

We were nearly back to the house before Adam and I spoke again. In order to avoid having to even meet his eyes, I had ridden ahead with the children, and he seemed content to let me do so. But as he came within sight of the house, Anne and Mark suddenly surged ahead, and before I could follow them Adam laid one strong tanned hand on my bridle.

"Beth," he said. "Please. I must talk to you."

In a voice quivering with what I hoped he would take for fury, I retorted, "Why? Have you thought of some other rude accusation to hurl at me?"

"I want to apologize. I had no right to say what I did. Or to do what I did. I promised that if you came here you would have nothing to fear, and I meant it. It won't happen again—unless you want it to."

If he had looked properly apologetic I might have forgiven him, but there was still a grim set to his face, and I thought he looked like a man who would do his duty if it killed him. Only his reminder of my position as a guest in his house kept me from hurling some sarcastic answer.

Instead, I bit my tongue until I could answer with proper dignity, "Thank you. I trust you are a man of your word." With that I wrenched my mount about and rode off after Anne and Mark. I had already dismounted and entered the house before he reached the veranda.

Even having been raised as I had in the home of a spinster aunt, I had heard of the antipathy of young boys toward soap and water, so when I got upstairs I was surprised when Mark, who was hot and dirty from our ride, enthusiastically greeted my suggestion that he let Maria prepare the bathroom for his use.

"It is because he seldom gets to use the big bath," Anne explained when I commented about it later. "He usually bathes in a small tub in the nursery." She wrinkled her nose. "Aunt Rosemary doesn't like us to use the water from the roof cistern but makes poor Maria carry well water up from the kitchen pump."

I had helped Anne wash her soft, fair hair and rinse it in lemon water, and now as I gently toweled it dry, I was pleased to see a slight tendency toward a curl. "Oh?" I said absently. "I wonder why that is."

She was silent for a moment, then said more casually than her words warranted, "I suppose because my father drowned in it. The cistern, I mean. I felt rather squeamish about it myself for a while, but Uncle Adam said that it's not good to let things like that rule our lives. Aunt Rosemary has, and she does seem to be getting odder and odder."

I tried to suppress my shock at her revelation. As I glanced hastily into the mirror before which she was sitting, her eyes met mine and I realized that beneath the casual exterior she was anxiously awaiting my reaction.

"He is right, of course," I said as normally as I could. "We must go on with our lives even after we lose someone we love." To my dismay, I heard my voice falter as I added, "I know sometimes it can be very hard."

Her eyes filling with tears, Anne turned on the bench and took my hand in her small ones. "You miss your father very much, don't you?" she asked sympathetically. Unable to speak for a moment, I nodded, and she added, "So does Aunt Rosemary, I think. She cried for a long time after she heard about him."

Surprised and intrigued by this new information, I exclaimed, "She did? Then, were she and my father especially good friends?"

"Oh, yes," she answered innocently. She turned back to the mirror and absently I began brushing her rapidly drying hair. "She always seemed much happier when he was coming to stay. I wish I were prettier," she remarked fretfully in an unexpected change of subject. "Mark is handsome like Papa, so I should be beautiful like Mama was, but I'm not. I really don't think it is fair!" She eyed her reflection with supreme distaste. "I don't look like anyone in the family. Do you suppose I am a changeling, as they have in fairy tales?"

I stopped brushing the silken strands while I studied her image in the mirror. "I don't really think so," I said. "As I recall, they are usually dark. You do, however, look somewhat like I did as a child."

Except for her fair hair this was stretching the truth, but I was rewarded by the smile that lit her face. She gazed entranced at my reflection next to hers in the mirror. "I do? Then I shan't complain anymore, but will just hope that I grow up to be half as lovely as you. Oh, what are you doing?"

"Just coaxing a few curls around your face. See?"

I don't know which of us was the more delighted. The gentle whisper of curl at temple and ears softened her rather chiseled features until she looked actually pretty.

"Oh, thank you, Miss Allen!" she exclaimed. "Now, if only I had something nice to wear! Something with—with—"

"Ruffles?" I supplied. "I have just the thing. It is a collar and cuffs which I hardly wear. If we can find a dress they will fit, it will only take a minute to sew them in."

We did find one, a blue frock a little less severe than her others. Within moments I had set the dainty collar and cuffs into place, and she put it on. The transformation was complete.

Even Mark noticed when he came running in, rather haphazardly buttoned into clean shirt and pants. He stopped and stared at Anne.

"What happened to you?" he asked loudly. He examined her critically. "You look—different." He sniffed the air. "Like—you ought to smell good."

Anne beamed as though it were the ultimate compliment. I said quickly, "And she will, in just a moment," and hurried to my room for my perfume. I snatched it up and ran back into the hall, then stopped just short of colliding with Christian. He was in riding clothes, and he had a large package under one arm. He reached to steady me.

"I'm sorry," I exclaimed, suddenly short of breath. "That was careless of me."

He kept a grave face, though his eyes lit with humor. "The pleasure was mine," he murmured. At just the moment when it would no longer have been circumspect to have held onto me, he removed his hand from my arm. It seemed that I could still feel its warmth through the thin material of my sleeve. "I see you are armed," he added, indicating the atomizer which I held. "May I ask to what it is you intend laying waste with that lethal weapon?"

"Oh, it is for Anne," I said quickly to cover my confusion. I realized that Adam's accusation of the afternoon had made me even more self-conscious with Christian than I had been before. "She has made a special effort with her appearance for dinner, and I thought a bit of scent would be the final touch."

"I see. It is kind of you to take the trouble with the children."

"It is no trouble at all," I returned defensively. "They are delightful. I think perhaps none of you appreciates them fully."

Christian agreed smoothly and immediately. "You are probably right. I will add it to the list of misdeeds I am to correct." His warm brown eyes were smiling wickedly into mine.

I felt my face grow warm. "Really, Mr. Rivers," I protested. "Don't trouble yourself for my sake. I have no interest in reforming you."

"Oh?" That disconcerting amusement was still evident. "I have been led to believe that that is precisely what a woman delights in."

I might have enjoyed bandying words with him had I not suspected that he, too, entertained the idea that I was attracted to him. As it was, my temper rose.

"Then you have been misled," I retorted. "At least as far as I am concerned. It matters not a snap to me how you comport yourself."

Christian smiled as if not in the least offended. "I am sorry to hear that. But I intend to change my image in spite of it. You see this bundle? It contains a complete set of working clothes. Cousin Adam is going to receive my invaluable assistance at last, which is what he has long been wanting. Until dinner, then." He bowed politely and sauntered on down the hall, leaving me to collect myself before going in to delight Anne with a light spray of perfume behind each of her small, neat ears.

It was a good thing we had Anne's altered appearance to discuss when we met in the drawing room before dinner; even as it was, the atmosphere between Adam and myself was decidedly chilly. Christian, of course, sensed this immediately, and kept glancing from one to the other of us with eyes that danced with delighted speculation. But for once he restrained himself from voicing any impudent comments, and instead led in complimenting Anne about the attractiveness of her new femininity.

"I'll wager Miss Allen had a hand in it," he said, ignoring the fact of our earlier meeting in the upstairs corridor, "for you look as charming tonight as she. In fact, Miss Allen, you seem to be like a brisk spring breeze sweeping out the dusty corners of our lives. What about it, children? Should we try and persuade her to stay forever?" Naturally Anne and Mark agreed enthusiastically, and Christian went on, "She has even had a salubrious effect on me, for I am in the process of turning over a new leaf. I am at your disposal tomorrow, Adam, and in the foreseeable future. From now on, I mean to do my fair share of the work here at Riversbend."

I couldn't refrain from looking at Adam. He was regarding Christian impassively, but I felt there was anger and disbelief beneath the careful mask he wore. Quickly, I rose from my chair and said, "I am going to try and persuade Rosemary to come downstairs. Surely she didn't take you seriously last night, Adam." Before he could inform me that he had been quite serious, which I already suspected, I hurried from the room.

Though I knew it was really not my concern, my sympathies had been aroused by Adam's story of Rosemary's childhood disappointment, and I decided that I must try harder to make friends with her. Even so, it was with some trepidation that I arrived before her bedroom door. After a moment's hesitation, I knocked gently.

She answered "Come in" immediately, but when she saw it was I, rose from the little escritoire where she had been sitting. A journal of some sort lay open, and she had evidently been writing in it, for she still had the pen in her hand. Surprise vied with annoyance in her expression.

"Really, Miss Allen, I—" With an effort she tempered the irritation in her voice. "I had supposed it was Maria with my tray. If there is something you need, I am sure she or Emelda will be able to furnish it for you. I myself do not feel able—"

"Oh, no," I said hastily. "It isn't that. It's just— I really came to try and persuade you to join us downstairs."

If I thought she would be pleased, I saw immediately that I was mistaken.

"Certainly not," she said at once. "Did Adam send you

as a way of making amends? Or is he finding the care of the children already burdensome?"

I was already wondering what I had gotten myself into. "Neither," I protested. "I simply feel that if I am to be a guest here, we should really become better acquainted. I would like very much to be your friend."

Turning away abruptly, she carefully placed the pen between the pages of her journal and closed it. "I have no friends," she said. There was no self-pity in her voice, only a bald statement of facts, but I felt compassion for her.

"Anne tells me that you and my father were friends," I said. "I thought perhaps that would be a bond between us. I know so little about the last years of his life."

She whirled to face me. I noticed that her hands were clasping and unclasping.

"We were more than friends," she answered, as though daring me to dispute it, "though no one knew it yet. He wanted me to become his wife. I am sure it was that which he meant to discuss with Adam the night that he—died."

I stood stunned. The blood seemed to drain from my body. Try as I would, I could not keep the astonishment from my face.

"You find that strange?" she inquired on a bitter note. "That he would be interested in an unattractive spinster like me? Well, some men—men like Jamison—can see beyond outward appearance." Her glance raked over me. "They would never be satisfied with empty-headed young women with nothing on their minds except hairstyles and fashions and their own beauty!" She glanced involuntarily at a small photograph which stood in a gold frame by her bed, and added, as though the words were torn from her, "Nor was he the first man to love me, plain as I may seem to you!"

My startled gaze followed hers. The young man in the picture was blond and rather ordinary looking, though the intelligent eyes and high forehead lent him an air of deep intellect. Beside the photograph was a small vase containing just one pale flower.

Understanding flooded me.

"You are far from plain, Rosemary," I said gently. "I am sure my father wasn't the only man to find you attractive. And, had he lived, you would have made him a charming wife."

It was her turn to be surprised. "You—really think so?"

"Of course." I seldom lie, which may be why I am convincing when I do. "But now that he is gone, you won't deny his daughter the opportunity to know you, will you? Please, won't you come downstairs to dinner?"

For a moment she almost succumbed. "Well, perhaps I—" Then, with a painful stiffness, "No, I won't. Neither Adam nor Christian appreciates what I do here, and I don't wish to see them just yet. Perhaps tomorrow night."

I could not leave well enough alone. "But Christian is on his best behavior. And Anne wants to overwhelm you with her beauty. Ruffles and a few curls have made quite a difference in her appearance."

I had spoken thoughtlessly, and I wasn't prepared for her reaction. "What?" she almost shouted. "You have already tainted the child with your vanity? I'll have you know it has taken me two years to eradicate the influence of her vixen mother and you no doubt have undone it all with your meddling!"

Her fury was so great that unconsciously I took a step backward. She followed me, her eyes dilated into great black pools. I thought of Christian's remark about her. She really did look as though she might take leave of her senses.

"But—" I stammered. "I only wanted her to feel a trifle more attractive. What can be the harm in that? It is hardly a sin to be pretty."

"It is a sin to be a temptress! Men are morally weak, and have no recourse against harlots who enslave them with their beauty!" Her eyes went again to the picture by her bed. "Many a man has had his life ruined because he succumbed to the physical charms of such a creature."

Again, I thought I understood. She had lost her lover to another, prettier woman, and could not bear to admit that the fault might lie with him. Or with herself. I did not, however, consider this a sensible time to debate it. I simply wanted out of her presence.

"I am sorry to have upset you," I said. "It certainly wasn't my intent." I backed up until I could feel the door at my back. "Please excuse me. I am sure the others must be waiting."

With that I escaped into the hall. As I fled down the stairs, half expecting Rosemary to come shouting after me, I was sure of one thing. She had lied, or at least had fantasized her relationship with my father. Never would he have even considered such a woman for his wife.

I dreaded going back downstairs after my failure with Rosemary, but to my relief there was no comment at all from the family; and, if I was not mistaken, a decided feeling of relief filled the air when I simply said that I had not been successful in persuading her to join us. I had braced myself for some humorously sarcastic remark from Christian, at least, but none was forthcoming. Perhaps, I thought with some amazement, he really does plan to be better behaved in the future.

His appearance at breakfast seemed to bear that out. When he joined us I saw that he was wearing a set of quite blatantly new work clothes and sensible, flat-heeled boots. Adam, after one startled look, studiously avoided comment, but Mark, wide-eyed with interest, said, "Now *you* look different, Cousin Christian. Last night it was Anne. Is something going on that I don't know about?"

Christian laughed in a kindly way and slanted a glance at me.

"It is as I said last night, Mark. The wind of change has swept through our house in the person of a beautiful young lady. Riversbend may never be the same." He helped himself to a generous slab of ham. "She has convinced me that I am remiss in my duty by not assisting your uncle Adam with his work on the ranch."

I blushed under Adam's quick, assessing look.

"I?" I protested. "I said nothing whatsoever about it."

His smile suggested a subtle understanding between us. "Not in so many words, perhaps. But it is your obvious disapproval that has caused me to repent the error of my ways. And so," this last was directed at Adam, "you must thank Miss Allen for any benefit you receive from my assistance."

A frustration akin to anger was churning in me. How could Christian imply a relationship between us that didn't exist, while saying nothing at all that I could refute? I glanced at Adam to ascertain his reaction, but he continued to eat his breakfast with deliberate, methodical movements.

"It is about time," he answered shortly, "since you enjoy the profits. But I doubt that even Beth's appreciable charms will suffice to keep you at it for long."

Christian took a sip of coffee, delicately used his napkin, then leaned back in his chair with the satisfied expression of a large and lazy cat. "Time may prove you wrong, Cousin. It would not be the first time that a beautiful woman changed the course of a man's life."

Though they spoke lightly, I felt as though each was using me as a weapon against the other.

"Please," I said in some agitation, "I find it distressing to be discussed in this casual manner." Cutting short their immediate apologies, I went on. "However, I do wish to continue with the subject of work. Since I, too, am partaking of your hospitality, Adam, I would like to contribute what little I can. The children tell me that they are without a governess. May I be allowed to help them with their studies while I am here?"

After politely protesting that it wasn't at all necessary, Adam admitted that it would temporarily solve a knotty problem. "There are very few unattached women in San Diego," he said, "and even fewer who have the refinement and education required for teaching children. And should I be fortunate enough to find one, it would be unlikely that she would be willing to bury herself for weeks at a time in the country."

"Added to which," Christian put in with his usual impudent humor, "is the long-standing reputation of the Rivers men. No virtuous female feels safe within our domain unless she is at least seventy and unattractive into the bargain."

"If true, it is hardly a thing to boast about," Adam said sternly. Christian ignored him and continued to smile benignly, but something in Adam's level look at him caused me to wonder fleetingly if the last governess's attack of "the willies" had been caused by more than her unfamiliarity with country sounds. On second thought, though, I doubted it. She would more likely have lost her heart to someone as beguilingly handsome as Christian.

After we had finished our breakfast, Adam and Christian headed for the groves and the children and I went immediately to the nursery, where we spent the morning engaged in schoolwork. I ascertained from Anne just what the governess had been teaching, and then tested their abilities in reading, spelling, and ciphering. Though I was no expert, it seemed to me that they were beyond their years in the first two subjects, though their knowledge of arithmetic was woefully inadequate. I set each of them tasks in this area, Anne to memorizing her multiplication tables, and Mark to the simplest addition, and was happy that, instead of resenting my instructions, both seemed pleased by the attention.

I had brought a piece of sewing to occupy me while they studied, and as we worked quietly in the pleasant nursery I pondered as to why I felt so drawn to them. I had had little to do with children, as most of my New York companions had not yet married and my aunt's friends had tended to be spinsters like herself, but even so I felt that somehow Anne and Mark were different. It did not occur to me that my loneliness made me understanding of theirs, and that in comforting them I was comforting that lost and lonely little girl who had been myself. I only knew that they seemed unusually appealing.

At noon we put our work aside and trooped downstairs for luncheon. The table was set for three, and I inquired of Emelda, who unceremoniously plunked our food onto the table, as to the whereabouts of the others.

"Rosemary ordered a tray in her room," the girl said ungraciously, "and Christian and Adam are taking their meal in the fields with the men." She stood with arms akimbo, her mouth a sulking pout. "Why does he work like a horse when he doesn't have to?" She apparently had not taken to heart my yesterday's attempt to remind her of her place, and was obviously referring to Christian. "He has always said he would like to live the life of a Spanish grandee, yet now he is willing to grub in the dirt like a peon." She spat the word out contemptuously.

"Perhaps he feels it is time he took responsibility," I offered. I was sure she would be less than delighted to hear that he wanted to impress me.

"Huh!" She poured my coffee with none of the tantalizing grace with which she served us when the men were present. "Why should he? Even though Riversbend is as much his as it is Adam's, he has no say in what is to be done."

I nearly choked on a sip of coffee. "Christian is an owner of Riversbend?" My voice emerged high and slightly strangled.

Emelda eyed me with obvious disdain for my ignorance. "You didn't know that? He is one-quarter owner, just as Rosemary is and the children together are. But Adam has the power as well. He says when and what to plant, to pick, and how much money each may take from the profits. So what good is ownership without the power? I think that Christian would sell his share and go away from here, if he could."

I scarcely heard her last words. I was conscious only of a burning embarrassment. Yesterday, when I had taken Christian to task about his behavior to his family, I had spoken of "Adam's hospitality," and neither by word or look had he corrected me. Silently I vowed to apologize when the opportunity arose.

I realized that Emelda was still standing expectantly beside me. Absently I said, "Thank you, Emelda. I will ring if we need anything else." Apparently she had wanted to talk more about Christian, for my dismissal seemed to annoy her, and she left the dining room with an angry glance and a haughty toss of her dark hair.

The next few days fell into a pleasant if uneventful pattern. While Adam and Christian went into the fields, the children and I would spend the morning with their studies, have lunch with or without the two men, and then spend the afternoon along the riverbank or reading and lounging beneath one of the shade trees that dotted the courtyard. I would have liked to go riding again, but didn't trust my ability to find my way without guidance, and while I thought once or twice of asking Adam if I might explore the upper regions of the house, it seemed a shame to stay indoors in such splendid weather. So I contented myself with simply relaxing in this interlude between my past life and the unforeseeable future.

To my surprise, and I am sure to Adam's, Christian continued to work with him and the men in the fields. From their desultory conversations at dinner and in the drawing room, I gathered that Christian was deferring nicely to Adam's judgment and was performing without complaint whatever task presented itself. I expected Adam to be pleased by this, but I soon realized that he really was not pleased at all. Though outwardly he seemed to accept Christian's changed demeanor at its face value, frequently I saw him watch his cousin with a wary and assessing look. At first I understood, for I couldn't quite believe the change myself, but as the days went on and Christian kept a rein on his outrageous humor, was kind to the children, and managed to be unfailingly polite to Rosemary when she began again to come downstairs, I started to lose sympathy with Adam. It was, I thought defensively, possible that the opinion of a decent woman could transform a man, and at least Christian deserved the benefit of the doubt. I could only attribute Adam's attitude to his childhood resentment of his young cousin, and the idea that he would be so petty surprised and distressed me.

I think that he must have read my thoughts, for on one especially warm evening, instead of remaining at the table with his wine and cigar as was his usual custom, he detained me as I was about to follow Rosemary into the drawing room.

"It is such a pleasant evening, Beth," he said. "I thought I might take a turn around the courtyard. That is, if you would be kind enough to join me."

I was taken by surprise at his request, for he hadn't sought me out since the day of our horseback ride, and in spite of my growing sympathy for Christian I had felt a vague disappointment. So I answered pleasantly, "I would like that. Will I need a shawl, do you think?"

At that point I became aware that Rosemary had stopped in her tracks at Adam's words, and that Christian, who had resumed his seat at the table, was openly observing us. I felt my face grow warm, but Adam ignored them and considered my question as gravely as though it were of primary importance.

"I don't think so," he said. "If you do, I'll come back for it." He offered me his arm in a formal manner, and as I accepted it I was discomfitted by Christian's amused smile and was glad to escape into the hall and to the courtyard beyond.

The outside air was unusually balmy, and for a few quiet moments I let myself forget the man walking silently beside me, and just enjoyed the mingled scents of fragrant blossoms and earth moist from the evening's watering. And Adam, who was carefully curtailing his long stride to match my shorter step, seemed content to suppress whatever purpose he had had in inviting me for this stroll, and to enjoy the evening along with me. Then, when we began to talk casually, I was pleased to find that, although he had received no higher education, Adam was widely read and had a lively interest in books and art, and was surprisingly knowledgeable even of the plays currently running in New York. Since these things had been a part of my life since early childhood, I found myself warming to him. For the first time since I had known him, his expression became animated, his smile immediate. The deep lines of grief were erased completely from his face, leaving him as handsome, in a totally different way, as Christian. Possibly even more handsome, I mused, for the high cheekbones and clean, spare planes of his face gave him a mature, self-sufficient look that his cousin didn't have. The excitement that he had once aroused in me in San Diego began to stir again, and I found myself visualizing him as a potential husband. Supposing, of course, that he was still interested in pursuing that possibility.

I had been so engrossed in my thoughts that I didn't notice that our conversation had been allowed to lapse; neither did I quite take in what he had said to break the silence.

"I beg your pardon?" I said on an inquiring note.

"Please don't deny it, Beth." Adam spoke in serious tones. "It is plain that you think me grossly unfair to Christian in my attitude."

Christian had been so far from my thoughts at that moment that I had to grope for an answer. "Why, I—let us say that I am somewhat surprised," I managed at last. "I would have expected you to be pleased by the fact that he is showing interest. I gather that he hasn't done so in the past."

By unspoken consent we sank down on one of the stone benches facing the central fountain. "I would be pleased if I thought the interest was genuine," Adam returned. "But I know Christian too well to believe that. He will only put forth any effort if there is something he personally wishes to gain by it."

"But isn't that true of all of us?" I pointed out practically. "I understand that Christian is part owner of Riversbend. Perhaps he realized that working with you is to his future advantage."

With a bitterness that surprised me, Adam said, "He has never realized it in the past, even though my father arranged from the beginning that Christian was to inherit equally. And he has always made it clear that it would suit him best if Riversbend were to be sold and the proceeds split among the heirs so that we could all be free to go our separate ways." He added, with a brittle attempt at humor, "I will admit that he hasn't previously had such a compelling reason as you to make him wish to improve."

Perhaps Adam didn't mean to sound accusing, but I took it that way. "I haven't given Christian reason to think that his actions are important to me, Adam! I have scarcely had a conversation with him since I have been at Riversbend. But if I had, is that so wrong? I certainly haven't an understanding with anyone else!"

In my anger I rose from the bench, and Adam at once got up to face me.

"You could have had, you know." The deep lines had returned to his face. "I was ready to offer you all that was mine, but you were too blinded by the flash and color of my cousin Christian to see how much I loved you."

His so obvious use of the past tense shattered the dream I had been building the past half hour, and I resorted to further anger to hide the hurt.

"Blinded!" I retorted hotly. "How dare you speak of anyone else being blinded when you are so jealous of Christian that it has blinded you to any possible good in him? Oh, there is no use discussing it with you, Adam. I am going in."

I whirled to leave, but he caught me by the arm.

"You will not leave until I have had my say! You are right, I am jealous of him. So might you be, if you had had to share everything that was rightly yours with him, while he never showed appreciation of any kind!" He let go of my arm, but I made no attempt to leave again. "My father, in trying to make up for the loss of Christian's parents, gave him anything he wanted and demanded nothing in return, whereas Morgan, Rosemary, and I had to toe the mark or be well punished for any small transgression. Do you think that was easy to live with?"

"You were children then," I said stubbornly. "Your

father was at fault, not Christian. How could he have helped being self-centered and spoiled in such a situation?"

"But he is getting away with it still, don't you see? He is trying once more to possess a thing just because he knows that I want it, too. Can't you realize how I feel when I see him succeeding?"

"He is not succeeding!" I cried hotly. "I don't want him *or* you! I just want to be left alone! Oh, why did Father have to die?"

After wailing this loudly, I spun away and raced up to my room, thus proving myself just as childish as any of them.

Chapter Ten

When Adam came upstairs to apologize the next morning, I had my valise lying open on the bed. After a sleepless night I had faced the fact that I could no longer accept the hospitality of the Rivers family and that I must return to San Diego immediately. How I was to get there or where I would go when I did arrive, were questions which, in my agitation, I refused to consider, but I couldn't prevent the tears from periodically brimming over and running down my cheeks. When I heard the firm rap at my bedroom door, I, thinking it was Emelda with my morning tea, said imperiously, "Go away!" and went on retrieving my belongings from their various drawers.

My back was to the door, but when I heard it open I spun around to face it. "I said—" I began angrily, but stopped in consternation when I saw Adam standing there. Fresh tears began coursing down my cheeks, and to prevent his seeing them I turned back to my packing. "I said to go away," I cried, flinging the words over my shoulder. "Can't you see that I am busy?"

Adam strode into the room, his tall figure suddenly dwarfing it. "Beth, what are you doing?"

I was vaguely pleased by the concern and surprise in his voice, but I answered scathingly, "Isn't it obvious?" I slammed an empty drawer shut, hard. "I am packing. I will be away from Riversbend within the hour."

"But, why?" There was genuine bewilderment in the question.

I turned on him indignantly. "Why? You need ask me that? Well, I will tell you why. I am tired of being tossed like a beanbag between you and Christian! You think because I am penniless and have been forced to seek shelter here that my feelings have no importance in the matter, and that you two arrogant males can growl over me as though I were a bone tossed to the dogs!" I turned back to the highboy, opened a drawer to find it already empty, and slammed it shut again with a force that nearly sent the highboy over. "I won't put up with it any longer. Surely there is something I can do in San Diego to earn my keep!"

A silence fell, to be broken only by a hiccoughing sob from me. Feeling suddenly drained, defeated, I leaned my head against the hard and unyielding cherrywood front of the highboy and let the tears continue to slide down my cheeks.

I felt Adam's hands, firm yet gentle, at my shoulders. "There is no need for that, Beth. I know that I have been insufferable, and I came upstairs to apologize." He turned me to face him, and after an instant's resistance, I allowed it. "Please, don't go away. We all love you and want you, and I promise that if you will only stay with us here at Riversbend we will give you all the time that you need to choose either of us, or none." His eyes were soft as he gazed down at me. "Just say that you will allow us one more chance."

His loving, concerned tone undid me, and I found myself weeping against his chest. To my mingled relief and disappointment, he didn't try to kiss me, but when his arms came about me, firm and comforting, I felt safe and secure in the warmth of his embrace.

A few days later Adam had to make one of his frequent journeys to San Diego. He asked if I wished to accompany him, but I preferred to stay at Riversbend. Neither he nor Christian had said a word amiss since that stormy morning, which had caused me to assume that he had spoken to Christian, and I was once more enjoying my stay. And in truth, I felt an aversion to returning quite so soon to the city of my father's death, and as I had no friends to visit and certainly no money for shopping, there was really little point in my going along. And yet, as I saw Adam's wagon cross the bridge and disappear beyond an outcropping of rock, a strange loneliness engulfed me. Once again I felt bereft, unwanted, an outcast in an alien land.

I started to go in to the children, who had already returned to the house, but thought better of it and went across the lawn to the big oak tree. I had already found that a few moments spent gliding back and forth in the swing had a soothing effect on me, and I hoped now to ease the constriction I felt around my heart. But before I could do more than seat myself, Christian stepped from behind the huge gnarled trunk.

I could not restrain a start of surprise and was immediately annoyed by it.

"Heavens!" I exclaimed crossly. "You are forever appearing out of nowhere! Do you enjoy startling people like that?"

"I'm sorry," he said contritely, though his twinkling eyes denied it. "It is just that you tend to avoid me most of the time. So when I saw you heading this way, I decided that concealment was the wisest course. Would you like me to push the swing for you?" He started toward me.

I rose hastily. "No," I answered, and could still hear a tartness in my tone. I felt that he would not have approached me in this manner had Adam not been gone. Also, my pensive mood had been shattered by his appearance, and I resented it. I added, "What I would really like is to be alone." Unfortunately I realized as I said it that it was no longer true, and something in my expression, or the fact that a slight color touched my cheeks, must have conveyed this to him, for he regarded me quizzically.

"That is a pity," he said, "for I had hoped we could go riding. Not far from here is a pretty glen which has a natural spring. I thought you might like to see it."

I hesitated. It did sound pleasant, and I relished the thought of getting away from the house for a few hours. And hadn't Adam said that I was free to choose whomever I pleased, or no one? How could I know how I felt about Christian if I refused to spend any time with him? So I ignored a slight twinge of guilt and said slowly, "Why, yes, I think I would like to see it. But—didn't Adam leave work for you to do?"

Christian had a perfect right to be angry at this, for it was certainly not my place to remind him of his duties. But he merely said mildly, "I think I may be allowed the afternoon off. I am not exactly a hired hand, after all."

I did blush then, most decidedly, for his words reminded me that I already owed him one apology. I said immediately, "I'm sorry. That was uncalled for. And I have been meaning to thank you for your hospitality. I realize now that it is you as well as Adam whom I owe."

"You will discharge your debt completely," he said, "if you come out riding with me."

His charming smile overwhelmed me and I found myself happy that I had accepted his offer.

When we returned to the house and Anne and Mark heard that we intended to ride, they clamored to accompany us, but Christian put them off by promising a treat on the morrow. And, truth to tell, I was just as glad, for, much as I cared for the children, I had been a great deal in their company, and I felt it was well that we spend a little time away from each other. Furthermore, I had to admit, if only to myself, that I was intrigued by Christian and looked forward to exploring the man behind the facile exterior.

I found Maria and Emelda in the kitchen preparing huge amounts of plum jam, and elicited Maria's promise to look after the children until my return. Rosemary, as usual, was closeted in her room.

"Do not worry, señorita," Maria assured me. "They are good children. We will be happy to watch after them." Her smile was genuine, but Emelda had frozen at my first words and didn't deign to glance in my direction. I thought I saw a tinge of color stain the dusky cheek nearest me, and in her very stillness anger boiled. I was glad to escape from the kitchen to my room, where I hastily dressed in my riding habit.

Christian had ordered the horses brought to the veranda, and when I arrived there I saw that he was waiting for me. His smiling glance swept me from head to toe.

"Very nice," he said approvingly. "The gray becomes you." I thought he was about to say more, perhaps an elaboration of his compliment, but he almost visibly restrained himself and merely added, "May I help you mount?"

"Please," I answered. I maintained a cool and detached attitude, as I had planned, but the touch of his hand as he helped me step to the carriage block sent a small surge of excitement through me. Some murmuring undercurrent of my mind sent out a subtle warning, for ever since my argument with Adam I had been comparing the two men and thinking of each of them in terms of a possible alliance, and I knew that it would be easy to succumb to Christian's easy charm without ever really knowing the man beneath. Not only would that be unfair to Adam, whose own considerable charm was of a more tangible sort, but it might cause me to make a mistake that I would regret for a lifetime. Quickly I swung into position on my sidesaddle with as little help as possible from Christian.

To my relief, Christian didn't seem to note my avoidance of him, but went to his own horse and mounted. "Shall we be off?" he asked.

I nodded and took my horse beside his. In spite of my resolve not be overwhelmed by him, I couldn't help note how handsome he looked. Instead of a riding coat, he wore a white silk shirt whose collar he had left open, revealing a muscular column of dark bronze throat. The shirt's full sleeves were gathered tightly at the cuffs, and where it tucked smoothly into tight black trousers, he wore a wide leather belt and large silver buckle. It was a romantic look, reminding me vaguely of pirates and swordfighting in century-old novels, and I wondered if, when he wore this outfit, he fancied himself the Spanish grandee Emelda has said he longed to be. With his dark eyes and hair, he certainly looked the part, and I had trouble keeping my eyes away. If he noted it, he was too much on his best behavior to indicate that he did.

The path which we followed through the orange groves led away from the river and into the foothills just to the north of the house. Dark patches of scrub oak dotted the golden slopes, and it was to one of these stands of trees, larger than the others, that Christian took me. The oaks were taller here, and massive, like the one which supported the children's swing, and when we rode from the barren hillside into their dense shade, the coolness delighted me.

"Come this way," Christian directed, and without my urging my horse followed his. There, nearly hidden among the rock surrounding it, was a bubbling spring which flowed briefly over a stony bed before dispersing and sinking into the coarse, sandy soil. He helped me to dismount, and eagerly I drank of the cold, sweet water, then splashed it over my flushed face.

"Oh, it's marvelous!" I exclaimed. "Who would dream of finding such a treasure here!"

Christian handed me a snowy handkerchief to use as a towel. "Anyone familiar with the countryside might expect it," he answered. "The size of the trees indicates an underground water supply." He, too, drank deeply before flinging himself beside me on the soft grass that surrounded the spring. "Adam and I discovered it long ago and swore each other to secrecy. It is one of the few things we ever agreed on."

"It's a shame that you don't get on with each other," I said. "Especially since I understand that you both are owners of Riversbend."

He laughed without mirth. "My name is on the deed, and that is about as far as it goes. Uncle Matt left total control to Morgan, who left it to Adam. I haven't the right to sell, or to decide how the property is to be used, or even to share in the profits, if Adam decides they are all to be plowed back into the estate." He moved restlessly, then said almost unwillingly, "But to give him his due, he is at least fairer than Morgan was. He does part with some money each month, which gives me a modicum of independence."

I refrained from pointing out that he apparently had yet to earn even that. Instead I remarked idly, "Then you didn't get along with Morgan, either, I suppose."

"I learned at an early age to keep out of his way, as did Adam. Though Uncle Matt was blind to it, Morgan had a selfish streak in him, and he had no concern for any interests beyond his own." Christian laughed again, and this time there was genuine amusement in it. "I am hardly the person to criticize him for that, I know, but his self-centeredness extended to Adam, who loves this place and wants the best for it. It didn't seem fair that he should have to fight Morgan for every improvement he wanted to bring to it."

"After their father died, do you mean?" I asked. I was surprised and touched by his championship of Adam.

"Even before. You see, Morgan was the crown prince of the family. He was the brilliant one, the one with the education, the one who was going to turn the Rivers holdings into an empire. But he meant to do it through his various business interests, while Adam, being a lover of the land, wanted to accomplish it through agriculture. Uncle Matt tried to be fair, but it wasn't until after both he and Morgan were dead that Adam was able to put many of his ideas into practice."

"I had an idea that there had been problems," I said slowly, "for Adam seems reluctant to talk about either of them."

"That is because he is filled with guilt about the way they died. Uncle Matt drowned at the paddlewheel and Morgan in the roof reservoir, both of which were innovations of Adam's. Not that it was his fault, of course. He had tried to get us all to learn to swim." Christian stretched his lithe body like a cat, then rose to a standing position without using his hands. "Come on, enough of this gloomy talk! A day like today was made for pleasure." He held out a firm brown hand to me, and I took it and let him pull me to my feet.

I was troubled by the glimpse of the past he had given me. Thoughts of it lay beneath the surface of my mind for the rest of our ride, and I was glad that Christian retained his gentlemanly manner, for I was now in no mood to cope with any romantic advances, even from someone as handsome as he.

Christian's promised treat for the children proved to be a feast of Mexican foods with strange-sounding names which he persuaded Maria to prepare and pack into a basket for transport to the oak by the river. Rosemary, whom I gathered disapproved of such spicy food for Anne and Mark, and with some reason, was not informed, and Christian threatened them with dire happenings should they have the temerity to suffer telltale stomachaches afterwards. There was a great deal of laughter as we gorged ourselves on beans and cheese and delicious spiced meat wrapped in small, flat pancakes which Mark informed me were tortillas, spelled, he added proudly, with two els, and which we washed down with alarming amounts of a cloyingly sweet drink that tasted of raspberries and was laced with milk cold from the kitchen cooler. Then, after a period of quiet prudently insisted upon by Christian, we took turns flying out over the river in the swing, an activity of which the children never seemed to tire. I watched from beneath the tree while Christian pushed them both at once, and as I saw his face bright with laughter, it seemed to me incredible that I should have ever thought him frightening.

"I do hope these children aren't coming down with something," Rosemary remarked fretfully at dinner. "They have scarcely touched their food." She looked suspiciously from Christian to me. "Are you sure you gave them nothing at that picnic which could have upset their delicate systems?"

I studiously examined my plate, but Christian said blandly, "Only good wholesome food prepared by Maria, isn't that right, Emelda?" Judging by the look of blazing anger which the servant girl gave him, he was playing with fire, but he went on smoothly, "It is probably that they are tired from all the exercise. I swear, one more game of tag and I would have needed carrying back to the house myself!"

"If that is the case, they had best be excused," Rosemary returned. "Go now, children. Miss Allen will be up soon to see you to bed." I realized then how completely she had accepted my care of Anne and Mark.

"Not tonight." Christian spoke very firmly. "Beth is tired, too, whereas you have had quite a rest, Rosemary. Suppose you do the honors this time."

Irritation vied with a return of her suspicion as she looked from one to the other of us. "Very well. But I shall return downstairs. Emelda, bring coffee to the drawing room."

Emelda sent her a vexed glance and went toward the kitchen and Rosemary got up with icy dignity and followed the children into the hall. I bit my lip, scarcely daring to look at Christian, who had the grace to wait until she was gone before allowing a chuckle to escape him.

"Poor Rosemary," he said. "After you and I have spent the last two days wandering the hills together, how like her to defend your virtue from a possible assault in the drawing room!"

I could not help laughing with him, but I felt embarrassed nevertheless. Christian had been completely circumspect toward me these past two days, but I could not help but be aware that he was a handsome and virile male, and that beneath his casual facade he was attracted to me. And I, admittedly, had found him attractive, too. I especially liked the ease and seeming enjoyment with which he handled the children, and the fact that they, in return, were happy and relaxed in their relationship with him. Not that Adam wasn't good to them, too, I reminded myself, and in all fairness I had to add that his was the role of guardian, and thus it might not be possible for him to be as jovial as Christian toward them. But I was sure that their welfare would be uppermost in his mind at all times.

With Adam gone, Christian had last night elected to join Rosemary and me in the drawing room after dinner and I wondered fleetingly if Rosemary was returning in order to protect her brother's interest in me as much as because of her concern over my virtue. Christian might have thought so, too, for he wasted no time in casual conversation once we were alone together there. Instead of following me to the fireplace setting, he stopped at a small table near the door.

"Come here, please, Beth," he said. "I have something to show you."

A strange note in his voice made me wish suddenly that Rosemary would return from upstairs. But, after a slight hesitation, I went to him. On the table was an inlaid box which I had never seen before.

"This was my mother's," Christian said, and lifted the lid. Immediately a light, tinkly music filled the air. It was a gay little tune, vaguely familiar, and as I involuntarily swayed to its rhythm he moved closer and took me in his arms. "She loved to dance," he said. "This was her favorite piece." Before I had realized it we were gliding gracefully about the floor in the seductive movements of the waltz. I knew that I should pull away, but something in the sensuous motion of his body, in the feel of his firm and muscular arm about me, rendered me helpless to resist. I didn't want to resist. I closed my eyes and surrendered to his will, and it was not until I felt his lips on mine that I recalled myself. I pulled away and took a small step backward.

"Please, Beth," he murmured. "Don't be frightened."

Again he started to kiss me, but before I could protest we both became aware of someone's presence in the room. As I moved away from Christian I realized that Emelda was standing at the door, a coffee tray in her hands. She stared at Christian in mute betrayal, then set her burden on an empty table and went out as silently as she had come.

For once, Christian had to grope for words.

"She—she means nothing to me, Beth. A servant girl. I swear to you it is over."

If it was, it was apparent that Emelda hadn't know of it.

"There is no need for an explanation," I said, and knew all at once that I meant it sincerely. Christian's kiss hadn't stirred me as Adam's had, and I didn't think it ever would. "I assure you, Christian, that what you do with your life is of no concern to me."

I wished immediately that I had put it less bluntly, for a strange, hurt look came into his eyes. "I hope you are just saying that out of pique," he said, "for I warn you that I intend to press my suit."

I was glad that Rosemary returned just then, for I didn't know how to further discourage him. As we drank our coffee, I could feel his dark eyes watching me, and before long I pled fatigue from our horseback ride and went upstairs to my bedroom.

As I entered, I once again felt an alien presence. I lit the lamp which stood on the table near my door and held it high while I glanced about the room. Nothing seemed to have been disturbed. I opened the wardrobe, which now contained my clothing, and knelt to look under my bed. As before, there was no one in the room.

Trying to ignore the uneasy feeling that wouldn't go away, I prepared for bed. After donning gown and wrapper, I sat at the dressing table to remove the combs that held my hair. My thoughts returned to Christian and the moment when his lips had touched mine, and I barely glanced toward my comb box as I opened it. It was not until I felt something velvet-soft beneath my fingertips that I really looked at it.

I snatched my hand away. Crawling slowly over the rim of the box was a gigantic brown plush spider. For a moment terror squeezed my throat like a giant hand and I froze with fear. Then the saucer-sized creature dropped to the dressing table and started toward me, and I leaped up, knocking my chair over with a clatter, and backed away. When I reached the bedroom wall I slid along it until I felt the door handle touch the small of my back. Without taking my eyes from the hideous thing, I tried to open the latch, but it kept slipping from my stiffened fingers.

Without warning the spider took a sudden giant leap toward me, and I spun around and slammed my body against the door. I clawed frantically at the latch, but it would not give, and in some separate compartment of my mind I realized that the door must have been locked from the outside in the brief time since I had entered my room. It was then that the air seemed to fill with the sound of someone screaming.

Chapter Eleven

Christian swore later that the door had not been locked when he reached it just moments after he had heard my screams. "I was at the head of the stairs," he assured me, "and there was no time for anyone to have unlocked it again and disappeared before I arrived. It is perfectly understandable that in your fright you simply weren't able to work the latch." Upon inspection it did prove a bit stiff, but as to how the spider got into my room he was somewhat less glib; in fact, no one could offer any explanation at all.

I had my own theory about that and the locked door, and Emelda's belated appearance on the scene and her elaborate attempts to look as though she had been awakened by the furor did nothing to alter them. There had been ample time for her to slip upstairs with the spider after she had seen the two of us together in the drawing room. But what she meant to gain by it, or how she could bring herself to handle the horrible creature was beyond my imagining.

"Tarantulas are really quite gentle, Miss Allen,"

young Mark assured me earnestly. He and Anne had been jolted from their beds as had everyone else by the power of my shrieks. "They aren't poisonous, you know, even though their bite can be painful." He pressed close to me as I lay on Anne's bed, where I had been taken while Christian went to capture the spider in a shoebox which Anne had reluctantly lent him. "It didn't give you the willies, did it? I would be awfully sorry if it did and you decided to go back to San Diego."

Anne, too, came nearer as she awaited my answer. She shared my loathing of spiders, it seemed, and her face was as white as mine must have been. Her pinched expression, which these last few days had begun to fade, was returning.

I hastened to reassure them.

"It did give me the willies," I said truthfully, "but I refuse to let a spider rule my life. Especially as you tell me they aren't particularly dangerous." It took an effort to say this convincingly, for it brought back the vision of that fat, furry body hurtling toward me, and a shudder almost slipped past my self-control. Fortunately, Rosemary came in just then with pots of tea and chocolate for the children and me, and she would allow no more discussion of the incident.

"You children drink your cocoa and then get back to your beds," she ordered. "Miss Allen will shortly be returning to her own room. Unless," she added unwillingly to me, "you would prefer to move to another."

"Oh, that won't be necessary," I told her. "Just so long as my room is carefully examined for—for—" I hesitated, not wanting to upset the children, and turned my attention to the hot and aromatic tea.

"I assure you, there is no need to worry," Rosemary replied huffily, as though I were casting doubt on her housekeeping abilities. "Maria is going over every inch of it at this moment. Christian, of course, has gone to dispose of the creature."

Christian. When he had burst in and swept me into his arms, I had never been more grateful to anybody in my life, and he had been gentleness itself when he had carried me to Anne's room. And yet, even after this nerve-shattering experience, I remembered the possessiveness with which he had held me as we had danced, and the passionate way he had kissed me. And I reminded myself firmly that I hadn't liked it, hadn't felt the sweet stirrings of love as I had each time Adam's lips had touched mine. I felt now that it was Adam that I cared for, and, knowing that, it wouldn't be fair to encourage Christian in his interest in me. I decided that my gratitude must not cause me to forget the need to maintain an impersonal friendliness in the future.

I reckoned without Christian and his irrepressible humor. After I had been appreciative but impersonal during his last solicitous visit to me that night, and had again attempted to maintain an impersonal briskness at breakfast, the dark eyes were alight with mirthful mockery. Later, in the drawing room, when Rosemary and the children were engaged in a conversation of their own, he had whispered, "It is no use, Beth. It isn't psosible to return to our previous status. There is something between us now, and even old Adam will not be able to erase it."

It was a great relief to me that just then Maria came in to announce that Adam's wagon was coming across the bridge.

We all trooped out to the veranda to greet him, even Rosemary. Life was so isolated at Riversbend that it seemed another world, and we were anxious to hear what news Adam might bring.

"Look!" Anne said. "There is a carriage behind his wagon. Who could that be, do you suppose?"

There was silence as the two vehicles started up the long drive toward the house. Knowing that the carriage's occupant would no doubt be a stranger to me, I glanced at Christian to see if he recognized it. Apparently he did, for he shot a quick, odd look at Rosemary. I looked at her, too. Her eyes were riveted to the carriage and her face had gone white as chalk. Abruptly she turned and went inside.

"Who is it?" I asked, curiosity overwhelming me.

"I think it is Peter Barnes, your father's lawyer," Christian answered. "At least, it is his carriage. Didn't you meet him when Jamison died?"

"No, I didn't." I eyed the approaching vehicle with interest. "He was recovering from a case of the grippe and Adam thought it best that I not be exposed to it." Mr. Barnes had sent an apologetic note explaining his absence at my father's funeral; other than that my only word from him had been through Adam.

When Adam neared the veranda he urged his horses as if he wanted to reach us some moments before the carriage did. As he swung down from the wagon, he greeted the children and spoke briefly to Christian before coming up the steps to me.

It was as though just the two of us were there. I put out my hands and he took them in his strong, hard ones, and for a long moment he looked questioningly into my eyes. He must have read my answer in them, for in spite of the interested onlookers and the visitor drawing up in the carriage, he drew me to him and kissed me. The touch of his lips to mine was brief, but it seemed as refreshing to me as water is to a parched garden. I quite forgot Christian and his persistence, and thought only of how happy I felt to be with Adam again.

The glow in Adam's eyes seemed to reflect my feelings, but he merely said, "I have brought Peter Barnes to call on you, Beth. He has come upon some information that he thinks may be of interest to you."

I answered, "Encouraging information, I hope," but at the moment I scarcely felt that it mattered. The love and excitement that I had felt at the touch of Adam's lips had told me I was right about my feelings for him, and I couldn't think about unimportant things such as Father's missing estate. However, I dutifully followed Adam down the steps to the carriage block to greet our guest.

Though the man climbing from the carriage had his back to me, I could see that he was fair-haired and dressed in a conservative brown suit which was covered just now with powdery road dust. As he began turning, I smiled, waiting for Adam to introduce us, but when I saw his face I could not keep from staring.

I recognized him at once as the lover in Rosemary's bedside photograph.

Neither man seemed to notice anything amiss, and I quickly recovered my aplomb. "How do you do, Mr. Barnes," I said when Adam had introduced us. "It is so good of you to come all this way for my sake."

"Not at all, Miss Allen," he answered somewhat stiffly. I was to learn that this rigid demeanor of Peter's was simply a mask for his shyness, a trait which seemed unusual for a man in his profession. My father, who, according to Adam, had been an excellent judge of people, must have seen the honorable man beneath the unprepossessing exterior. Just now Peter went on in the same stiff manner, "I must apologize once again for not being with you in your hour of need. I assure you I would have, had it been possible."

Heavens, I thought, he talks like the text in one of his law books, and silently I apologized to Rosemary for thinking that he might be another of her flights of fancy. These two were eminently suited for each other. What could possibly have gone amiss?

As we entered the house, Maria came to prevail upon Peter to remove his jacket so it could be brushed properly. When he finally gave in and took it off, he went with us into the drawing room looking as embarrassed as though she had divested him of his trousers as well. Adam continued to talk smoothly, but I dared not even glance at Christian and what I knew would be his mocking smile.

After refreshments were served, Peter got right to the subject at hand.

"I have been going into your father's affairs as deeply as possible," he began. "Unfortunately, Jamison wasn't one to confide in others, even his trusted attorney." He assayed a slight smile at his own rather heavy-handed jocularity. "But it was commonly known that he owned many properties in the San Diego area, some bought as lucrative investments, others purchased as a means of helping friends who found themselves in straitened circumstances. When I let it be known that his daughter needed certain information in order to clear up his estate, many of these friends came forward." He shifted in his chair and set his untouched wine on a nearby table. "A definite pattern has emerged. Jamison was clearly liquidating all of his holdings, and knowing the man, it could have been for only one reason. He was obviously planning the investment of a lifetime."

He smiled at me, plainly pleased with his astute deductions, but I felt a vast disappointment.

"I am sure that he was," I said, "for he as much as told me so that night at dinner. But we are apparently no closer to finding out what it was than we were before." And no closer, I added silently, to relieving me of the onus of being an object of charity, for my ebullience at Adam's kiss had already abated, and I wanted whatever money and property due me.

"Ah, yes," Peter returned, "but we now know that there actually is an investment property somewhere, or a very large sum of money put away for its purchase. Such a transaction cannot stay hidden forever. It is only a matter of time until we unearth it."

A matter of time. How much time? If I were actually to marry Adam, and considering our emotions upon seeing each other again, I thought it likely, I would prefer to come to him with independent means so that there would be no lingering doubt in his mind as to the reason for my acceptance of him. Or, perhaps, in my own.

I tried to concentrate as Peter went on explaining the particulars of his quest, but even though the subject was of extreme importance to me, my thoughts would stray. What had come between Rosemary and this man, if indeed there had really been an understanding? And if not, why would she possess his photograph and keep it enshrined by her bed? I kept my eyes dutifully on Peter's face as he became enthused with his subject and I realized that once his self-consciousness passed he wasn't unattractive. Neither, however, did he look to be a man who could be tempted by a pretty face to break a solemn promise. I resolved to solve the mystery at my earliest opportunity.

Maria interrupted Peter's monologue by bringing his freshly brushed jacket to him.

"So, Señor Barnes, is that not better?" she asked as with a look of relief he put it on. "Though why you wish to wear such a thing in the heat of summer, I do not understand."

I was surprised at the familiarity with which she spoke to him and felt it indicated that he had at one time often visited this household. When she added with a smile, "The noon meal will soon be ready. I am preparing it simply, which is how you like it, no?" I knew that I was right.

"Thank you, Maria." His answering smile lent a certain gentleness to his face. He hesitated, then added, "Is—will anyone else be coming down to dine?"

"Only the children, señor." Maria's expression was a mixture of sorrow and disapproval as she added, "Miss Rosemary has said to me that she would not come downstairs again today."

"I see." Peter straightened his shoulders as though bearing up under a familiar, heavy burden. "I am sorry to hear that."

Suddenly I liked this man, and I knew instinctively that whatever had come between the lovers had not been of his doing, no matter what Rosemary may have thought. I determined to speak to her about it, even though I remembered with discomfort my last attempt to influence her. I had declared then that I would not interfere in family matters again, but her and Peter's obvious unhappiness urged me to try once again to be her friend.

Luncheon was excellent: a cold soup with an unpronounceable Mexican name, a platter piled high with a variety of crisp, fresh vegetables cooked to perfection, cold baked chicken and roast beef, and thick slices of Maria's crusty, aromatic bread still warm from the oven.

Peter fell to with a stevedore's appetite. "Please forgive my greediness, Adam," he said upon taking a second helping of almost everything, "but I have missed these excellent vegetables of yours. We don't get enough of them in town."

Adam smiled with pleasure. "I hope to soon provide San Diego with the widest possible variety of fruits and vegetables. And Christian here has decided to join me in my endeavors, at least for the moment. He has worked like a Trojan in helping to expand our irrigation system." He glanced at Christian. "Did you make any progress while I was away?"

Christian must have been as surprised as I was at Adam's words, but if so, he didn't show it. He smiled across his wineglass at me. "I am afraid not. I played truant and spent the weekend entertaining Beth. Do you realize she has scarcely been out of the house since she came here?"

I thought that Adam might be angry, but the look he gave me was troubled. "My apologies. I am afraid that I sometimes allow my duties to monopolize me."

I must have felt some unease concerning my jaunts with Christian, for my cheeks grew warm. "I understand," I reassured him quickly. "I don't expect you to neglect your work because of me."

"She likes to spend time with us, don't you, Miss Allen?" Mark put in unexpectedly. He added reprovingly, "Even if Cousin Christian thought she needed a rest and didn't allow us to go riding with them."

Anne said hastily, "That isn't fair, Mark! He arranged a perfectly splendid picnic to make up for it."

"I see that I had better take steps to rearrange my priorities." Adam was attempting a light tone, though I knew he had seen my blush and Christian's satisfied smile. "Else Beth will become bored and find it is time to return to civilization."

"Never fear, Cousin." Christian's eyes had a wicked twinkle in them. "She knows that she can always rely on me for entertainment."

I made some pleasant rejoinder and changed the subject, but Adam's face was shadowed for the remainder of the meal.

When we had eaten I excused myself and took the children upstairs for a rest before allowing them to face the afternoon heat. After getting them settled, I left the nursery and started to my room, but in mid-course changed my mind and went instead to Rosemary's, which was at the other end of the hall.

As usual, her door was tightly closed. I rapped softly. I knew that I must take a careful approach with her, for her nerves seemed to be always at such a pitch that the slightest assault to them sent her into either a tirade or a fit of uncontrollable weeping. Poor thing, I thought sympathetically, she has had very little in her life that would encourage an even temperament. Perhaps I would find, if I became a member of the Rivers family, that I, too, would be subject to dark moods and fits of weeping. The thought was disconcerting and I thrust it firmly from my mind.

I knocked again, then tried the door. To my surprise, it was unlocked, and I opened it a trifle. "Rosemary?" I said softly. "It's Beth. May I come in for a moment?" Upon hearing no reply, I pushed the door farther open and went in.

Rosemary was lying facedown across her bed, and her angular form was so still that for a moment apprehension gripped me. Then I saw her slim fingers tighten convulsively on the picture which she had in her hand. Peter's picture. As I hesitated, she cried, "Go away. Please. Go away!"

"Please," I said quietly. "Mayn't I stay for just a moment? Perhaps there is something I can do."

"There is nothing anyone can do!" Her voice was hoarse with weeping. "Just go away and leave me alone."

It seemed to me that there was no conviction in her words, and I thought that perhaps she really wanted me to stay, wanted most desperately someone in whom she might confide. With some trepidation I sat beside her and gently touched her thin shoulder. At first she stiffened beneath my hand, but as I began a slow, patting motion she gradually relaxed, and at last she said, as though the anguished words were wrung from her, "How could he? How could he be so cruel as to come to this house after what he has done?"

I said quietly, "I don't know what that was, but after talking with him today, I feel sure that he still cares for you. Perhaps one of his reasons for driving out to Riversbend is that he hopes to repair the breach between you. But how can he, if you won't even see him?"

At that she sat up and turned to face me and I saw from her ravaged face that she had been weeping for a long time. "See him!" she said, aghast. "I would die first! Nor could I ever consider taking him back after he has shown himself to be of such poor moral fiber!"

I thought of Peter Barnes's honest face and his rigid, overly correct demeanor, and couldn't imagine him doing anything truly reprehensible, but I was willing to concede that I had very little knowledge of the male animal's peccadilloes. "He seems such an upright sort of man," I ventured. "I cannot conceive of what he could have done that was so terrible."

Rosemary got up from the bed and began pacing about the room. "That is the wickedest part of it," she said. "He appears to be so straightforward and honorable, yet he is not. He fooled us all into believing that he was all that he seems to be, but underneath he is capable of incredible treachery!"

I watched her frantic pacing increase, and saw with fascination that she was actually wringing her hands. I had heard that expression many times before, but had never before seen anyone really do it.

"Adam still seems to trust him, though, as did my father," I said tentatively. "I am surprised, considering what they know about him."

She abruptly stopped her pacing and said truculently, "Jamison didn't know, and Adam still does not. What Peter did was too shameful to repeat to anyone." Rosemary resumed her pacing and wringing of her hands. "That is why it has lain on my breast like a stone all this time." She began to weep again. "Oh, Beth, you have no idea how difficult it has been to bear it all alone!"

"Then don't," I urged. I was not at all sure I wanted to hear the details of such an enormous sin, but I thought that if Rosemary didn't soon confide in someone, she would surely take leave of her senses. "Tell me about it, Rosemary. If you don't, I am afraid you are going to be ill."

"I wouldn't mind getting ill if I could die of it," she wailed. "But it would be my misfortune to go insane, like—" She abruptly stopped speaking and went to the window. "You are right, Beth," she said slowly without turning to me. "I must speak of it to someone."

I barely dared to breathe as I watched her struggle for words. One thin hand clutched the curtain as though it were a lifeline. "We were engaged to be married," she said in a surprisingly calm voice, "and then I learned that he had sullied our love by taking up with another woman, a cheap creature no better than a woman of the streets!" A sob escaped her. "It was as though he had made a mockery of my love for him, had held it up to be ridiculed by all the world!"

I had the fleeting thought that it might be her pride that had been battered as much as her love.

"But how could you have been ridiculed? You said that no one knew of it except you."

"Someone knew of it," she said dully. "Or I should not have known of it, either."

I remembered the sad look that had crossed Peter's face at the mention of Rosemary.

"I think that he must deeply regret it," I said. "What was his reaction when you confronted him with it?"

She swung around to face me, but I thought for a moment that she wasn't going to answer.

"I didn't tell him," she admitted at last. "I simply sent

him a note breaking off our engagement and stating that his presence was no longer welcome at Riversbend."

Even though I could imagine the bitter hurt and loss of pride that had driven Rosemary, I wondered if she hadn't acted too hastily.

"Who was the person who told you about it?" I asked. "You must have trusted her implicitly to have condemned Peter without even giving him a chance to refute the story. What if there were extenuating circumstances?"

The silence that followed my words grew until it seemed to take on a life of its own. I felt my blood tingle with a strange awareness as I saw the stricken look on her face. At last I went over and took her by the shoulders.

"Tell me, Rosemary." I was almost shaking her. "Who did tell you this terrible thing about Peter?"

She twisted away from me and went back to the window. "I don't know," she whispered, and though I couldn't see her face, I knew the dawning horror that must be forming there. "The letter was unsigned."

Even though I had no more trouble in persuading Rosemary that Peter should at last be given a chance to speak in his own defense, she could not bring herself to meet with him.

"I should die of embarrassment, Beth," she protested, "whether the story is true or not. Please, won't you talk with him? That way, if it is true, or if he no longer has any interest in me, I will be spared the agony of seeing it in his face. And if it is not true, I shan't have to see the hate and disgust he may feel for me when he realizes that it is my lack of trust that has ruined both our lives." Once more she clasped her hands together. "I shouldn't be able to bear that, really I shouldn't!"

"But, Rosemary," I said, "I scarcely know Peter Barnes. How could I approach him on such a matter? Surely, this is something that Adam should handle. After all, he is your closest male relative."

Her lips tightened into their familiar stubborn line. "No. I want neither him nor Christian to know of it. They ridicule me enough as it is."

Further argument was to no avail, and at last I agreed to speak with Peter, though the impropriety of my being an envoy in such a matter was obvious to anyone not in Rosemary's tumultuous state. I went alternately hot and cold at the thought of approaching him, and found myself longing desperately for my father. However, as I reluctantly prepared to go down and confront him, she made my discomfort worthwhile by saying softly, "If a miracle happens, Beth, I will have you to thank."

A warm glow accompanied me from the room, and I had reached the head of the stairs before it faded, to be replaced by a small, icy chill. And if a miracle *didn't* happen, I realized, she would have me to blame. And I could be sure that she would never, ever let me forget it.

Chapter Twelve

When I arrived downstairs, the men had returned to the drawing room. Adam and Peter were deep in a discussion of the nation's economy, while Christian stood in his accustomed place at the mantel, smoking a cheroot and looking exceedingly bored. He put down his cigar and the others rose in deference to my entrance. When I did not immediately take a chair they looked at me with mild inquiry.

It had not occurred to me until this moment that my request to speak to Peter alone would seem rather strange to everyone and might, indeed, be construed as an insult to Adam, for he must necessarily think it had to do with Father's estate. But I had promised not to bring Rosemary's name into it at this point, so I said, "If you wouldn't mind, Adam, I should like to talk with Mr. Barnes privately for a few moments."

Christian's eyebrows rose at this and he glanced quickly at Adam, who allowed only a fraction of time to pass before answering pleasantly, "Of course, Beth. Christian and I will take ourselves into the study."

"Oh, that won't be necessary," I returned hastily. "If Mr. Barnes is willing, we could stroll down to the river. A breeze should be springing up soon."

Though obviously surprised, Peter replied courteously, "It sounds delightful. Shall we go now?"

"Please, finish your wine," I urged. "I want to go up and tell the children where I will be."

"You concern yourself too much with those children, Beth." Adam spoke sternly but I knew he was proud to remark on it before Peter. "They will get along very well without you."

"I know, but I must get my parasol also. Father warned me not to go out without it when the sun was hot."

With that I hurried from the room and went upstairs, first to the nursery where Anne and Mark had awakened and were playing quietly, and then to my room to fetch my parasol. I had had little use for it since coming to Riversbend, for the hot sun was truly unmerciful in the afternoons and I tended to stay indoors or in the shade. But today I had an odd notion that it would give me strength for the difficult task before me.

My nervousness at the whole situation was growing. Peter Barnes would surely be overcome with shyness at speaking of such a subject with a lady, and a young, unmarried one at that. And yet, unseemly or not, I had to discuss it with him. Rosemary was depending on me.

Peter opened the conversation as we crossed the drive and started toward the dense shade beneath the children's oak. He began a profuse apology for not realizing that of course I would not want the Rivers family privy to my affairs. I interrupted him by putting my hand gently on his arm.

"Please," I demurred, "it isn't that about which I wish

to speak with you. Adam may hear all my business concerns." As he gave me a questioning look I continued. "This is very difficult for me, but—it is about Rosemary. I understand that you and she were once planning to marry."

Pain crossed his face and then was gone. "Why, yes," he said slowly, "we were."

I turned to face him squarely. "Before this discussion goes any further, I must ask you what may seem an impertinent question. Do you still care for her as you did then?"

I could not have blamed him had he refused to answer, but he merely said quietly, "Yes, I do."

"And you have since made no other commitments?" I could not humiliate Rosemary by speaking of her love for him before ascertaining his situation.

"No other commitments." A faint understanding was dawning, and he said, "Miss Allen! Has—have you reason to believe that there is still hope?"

As we neared the shade of the oak I lowered my parasol. Peter's obvious eagerness was making my task a little less painful.

"Please forgive me, Mr. Barnes," I went on, "but I must discuss an embarrassing matter with you, for Rosemary will confide in no one else."

"Please," he said urgently. "If it will clear things up between Rosemary and me, say anything you like."

I took a deep breath and plunged in.

"Very well. Two years ago, Rosemary received a letter giving the details of an affair you supposedly were having with a young woman in San Diego. That is why she broke the engagement. But she now realizes that she acted hastily and that you should at least have been given a hearing in the matter."

Even through my haze of relief at having it said at last, I could see that Peter was deeply, genuinely shocked. "How—why would anyone— I assure you, Miss Allen, that I did not at any time—" His face slowly turned a brick-red hue. "That is, I have never—" He paused, and then continued with quiet dignity. "Even if I did not love Rosemary as deeply as I do, it would not be true. I decided long ago to save myself for marriage."

Coming from him, the shopworn phrase did not sound prudish. I believed him absolutely.

Peter went back to the city without seeing Rosemary. He had wanted to, of course, and I had been hard put to deter him from dashing up the stairs to her, but, remembering her tear-ravaged face and her unattractive way of dressing and doing her hair, I thought it best to allow some time for preparation. I consoled him by promising that she would send a note to him soon after I had talked with her, a note setting a time for their reunion. I also asked him not to reveal any of this to Adam or Christian as yet, but his wide smile and happy demeanor must have indicated to them something of the nature of our discussion. Though they must have been bursting with curiosity, neither, in my presence, at least, questioned him at all.

It was easy to convince Rosemary of the truth of Peter's words, for she wanted to believe that he was innocent, wanted so much to have life itself returned to her. In view of this, I could not help asking why she had been so quick to doubt him on such flimsy evidence as that of an anonymous letter.

She seemed not to truly understand it herself.

"I suppose it was because I never really believed that he loved me," she mused when we talked in her bedroom later. "I never really believed that anyone could love me. So I suppose I found it easy to accept the fact that he was finding contentment with someone else."

I had never known anyone to have such a poor opinion of herself. A huge anger for her swept over me, and I felt myself tremble.

"But, why?" I demanded. "Who caused you to feel that way about yourself?"

"Who didn't?" she returned bitterly. "My father always made it clear that I was of no importance in comparison with the boys. My brother Morgan alternated between ridiculing me and ignoring my existence. As for Adam—" She shrugged. "He perpetrated the usual pranks and teasing that I suppose all younger brothers do. And then there was dear cousin Christian. My father spoiled him beyond reason, but he never seemed to realize that he was indulged more than the rest of us. He delighted in making my life miserable, and no one noticed or tried to put a stop to it." She shrugged. "And, I admit, I was most unlovely to them. I suppose I resented Christian and Adam most because their care often fell upon me when they were younger."

I was glad to hear that Adam had not imagined his father's seeming preference for Christian, but I wished he had been kinder to his sister. Her loneliness must have far exceeded mine.

"What about Morgan's wife?" I asked. "I gather that you and she didn't become close." A pity, too, I thought. Rosemary had badly needed a champion.

Rosemary got up and paced the room with some of her

former restlessness. "Marianne?" she said scornfully. "She never acknowledged my existence, unless it was to join with the others in making fun of me." She turned to me and spat out the words. "She was a vixen, a beautiful, cold-hearted vixen. She met and married Morgan when he was attending college in the East, and I am sure she thought she was marrying into a family of great wealth. Morgan was the kind who would have implied it. When she found she was expected to spend most of her days isolated here at Riversbend she could not accept it, even though Morgan assured her that eventually they would have a home in San Diego."

"I suppose it would be difficult for someone used to a more exciting life," I remarked. But I couldn't conceive of it, not if she had loved her husband. Riversbend was so beautiful, and I took such pleasure in riding beneath its huge oaks and along the banks of the river. And Adam was continually working and planning to increase production and to enlarge the groves and irrigation areas. If I were to become his wife, I would try and find ways to assist him. And then, if there were children— Marianne and Morgan had had such delightful children. I wondered if they had appreciated Anne and Mark, and as much as asked that of Rosemary.

She stopped her restless pacing and said impulsively, "You would never understand people like Marianne and Morgan. You're good, Beth. I didn't want to admit it before, even to myself, because I envied you your beauty, but you are. Anne and Mark adore you. You have paid them more attention already than their mother did in the course of a year."

"But they are such sweet children," I protested. "How

could she not have adored them?"

"Because she was too concerned with making men adore her!" Rosemary came to sit beside me on the bed. "I know you will take it for the rantings of a jealous old maid, but Marianne thought only of her beauty and its effect on every man she met, from the youngest stablehand to—to—yes, to my own father! And she made no secret of her outright contempt for me. Is it any wonder that I didn't dare to really believe that Peter was in love with me?"

I said, "Was Peter here when she—" I left the sentence incomplete.

"Oh, yes, she tried her wiles on him, too. But Peter didn't seem to be susceptible."

"Then, why—"

"Why didn't I believe in him?" Rosemary got up and again began pacing. "Because my father was so in favor of our marriage that I thought he—influenced Peter. I thought— Oh, Beth!" Her eyes filled with tears. "You will despise me when I tell you I was so desperate that I was willing to marry Peter even though I believed Father had promised him financial gain if he would agree to the match!"

My heart ached with pity for her. I said, "You aren't the first woman to make that compromise." I thought of my own questioning of my feelings for Adam. "But none of that matters now. You truly do love each other, and that is what is important." I got up then, and said firmly, "And now let's get to bed, for we have much to do tomorrow. When Peter returns we want him to see the woman that he remembers. Surely you didn't wear these dark and somber colors when he was courting you?"

205

Laughter shone through her tears. "Oh, no. Even I wasn't that foolish. But what can we do on such short notice?"

"Don't worry about it tonight," I told her. "I am certain we will think of something tomorrow."

Later, when Rosemary was calm at last and I had returned to my room I found that I had left my parasol lying on my bed. As I picked it up to put it away I felt a slight vibration, as though the ivory handle had loosened from its shaft. I turned the handle to tighten it and to my consternation it came off in my hand. Feeling most upset, for the parasol was precious to me, I immediately tried to reconnect the handle to the shaft. To do so I held the latter upright and at once a small, white paper slipped from its hollow center and landed on the floor at my feet.

I stared at it, transfixed. This parasol had been Father's last gift to me. Could he for some reason have concealed a message in it? My heart began beating with heavy, rapid strokes as I bent to retrieve the paper.

It had been doubled once before being loosely rolled, and when I had unfolded it with trembling fingers, disappointment swept me, for there was nothing at all written on it. Then I saw that within the paper's folds was a small brass key. I turned it over. It was flat and inexpensive-looking, similar to the kind often used to lock a child's bank. A letter of the alphabet and three non-consecutive numbers were pressed into it.

I sat down abruptly. Oh, Father, I rebuked him silently, if this was important enough to conceal, why didn't you include instructions for its use? How else did you expect me to know what to do with it?

I thought about that for a time. Father had not acted as though he knew his life was in danger that last night. He had left the hotel expecting only to attend a business meeting with Adam, a meeting which should have taken not more than an hour or so. And he knew that the parasol and I would be right where he had left us when he returned.

He expected to retrieve this himself, I guessed, possibly the very next morning. But as to why he had put it there, and more importantly, what it would open, I could not begin to guess.

Adam was more knowledgeable when I showed it to him the next morning. "It looks like a key to a security box," he said, turning it over in his fingers. "The kind that important papers are kept in. Or possibly a bank's safety-deposit box. I think I should take it to Peter immediately."

"I would appreciate that, Adam," I said, "but don't go just yet. I have other news to tell you." We stood on the veranda where we had gone to be alone together for a few moments. "It is about Rosemary. She has decided to begin seeing Peter again. Isn't that wonderful?"

"I thought as much when I saw Peter's grin yesterday!" A slow, admiring smile was lighting Adam's eyes. "Beth, you little minx, how did you manage it? No one has dared mention his name to her for the past two years."

"I'm afraid that I meddled," I confessed. "You may as well know that it is one of my less admirable traits. But in this case I think I was justified."

"More than justified. Positively inspired. But what

was the trouble between them? Rosemary refused to discuss it, even though she was obviously heartbroken to end their engagement, and Peter seemed as bewildered as were the rest of us."

I wasn't sure just how much Rosemary would want me to reveal, yet I felt that it had been such a cruel prank that it should be investigated. So I told him briefly about the anonymous letter she had received. "Do you have any idea who could have done such a thing?" I finished anxiously.

Adam shook his head. "None at all," he said flatly, "though I would certainly like to. I suppose we will have to set it down as just another of the mysteries that plagues Riversbend." The harsh lines of his face had once more etched themselves deeper, and his tone was so grim that I didn't dare ask about the other mysteries he spoke of. Laying my hand somewhat timidly on his arm, I said, "I wish I could help in some way. I know you have a tremendous burden of responsibility." What I really wanted to do was to kiss away those deep lines and see them replaced with the warm smile I saw so seldom. To my delight, one lit his face now, erasing the shadowed look.

"But you have, my darling. More than you know. Peter and Rosemary together again? Nothing could make me happier." The smile softened, and he brought me to stand close to him. "Well, almost nothing." He bent his head to kiss me, and for a long moment we didn't speak at all.

At last, when he had released my lips from his, I said breathlessly, "Rosemary will want to send a message to Peter. Do you have time to wait for it?"

He embraced me again and nuzzled his face in my hair.

"Of course," he murmured. "Tell her to write her love note. Isn't love, after all, the most important thing in the world?"

In that moment I could not have agreed with him more.

Once I had delivered Rosemary's letter to Adam and had seen him ride off to town with it and the mysterious key safely in his pocket, I returned upstairs to grapple with the problem of how to conjure up a romantic gown out of thin air.

I had not slept soundly last night, and in the long, dark hours I had tried to get my mind off Father and the reasons for his having hidden the key in my parasol by pondering about this dress for Rosemary. If I had read Peter's eagerness aright, not a great deal of time would elapse before he came calling; not enough time, I thought, to design and sew a completely new gown. It occurred to me that Marianne must have had an extensive wardrobe and I thought that perhaps something of hers could be altered to fit. I am somewhat skilled as a needlewoman and felt that this task wouldn't be beyond me, but knowing Rosemary's antipathy to her late sister-in-law, I approached the matter rather diffidently.

I found her still in her bedroom, sitting before the mirror and combing her long, dark hair. She had a dreamy look that softened the angular lines of her face, and her lips were curved in a tender little smile. As I rapped gently and entered, she looked up at me, and I saw with a small shock of surprise that her dark eyes, alight just now with a glow of happiness, were most attractive.

Perhaps Peter wouldn't notice what she was wearing.

Her hair, now that it had been released from its cruel bondage, was beginning to ease into soft waves around her face. She swept the dark mass of it first to one side, then the other, then atop her head as she eyed her reflection judiciously in her mirror. "What do you think, Beth? Shall I wear it up or down? I hardly know the latest fashion anymore."

I gathered her hair into my hand. It was soft and silky, a good texture, though rather fine. "Caught at the nape, I think, with a ribbon. You don't want an artificial look, do you?"

"You don't think that is rather young for me?" she asked anxiously.

"Perhaps," I conceded, "if you were attending a social function in town. But Peter will like its soft look. And, remember, he is expecting the girl he became engaged to. We don't want to make you look different than you did then."

She nodded. "That's true." Her brows knit with a new anxiety. "Have you thought any more about what I should wear?"

I broached the subject of Marianne's wardrobe. "I know the two of you were unalike, but mightn't there be something—one of her simpler gowns, perhaps, that I might alter and refurbish for you?"

Not to my surprise, she did object most strenuously. "Oh, good heavens, Beth, I could never, never wear anything of hers!" An involuntary shudder ran over her thin frame. "I have never so much as entered her room since she died."

"Did you hate her all that much, then?" Being imbued with New England thrift, the thought of usable clothing not being somehow utilized was strange to me. "Or is it the fact that she is dead that bothers you?"

"Both," she said, grimacing. "But it is more than that. It is how she died. Oh, Beth, even I would not have wished such a death on her, if, indeed, I ever wished her dead at all." Her voice wavered. "I would be able to think of nothing else if I so much as carried a handkerchief of hers."

I could not refrain from asking, "What happened, Rosemary? How on earth did she die?"

It was a long moment before she forced out an answer.

"She—she was stung to death by rebel bees. She was hardly recognizable when they found her."

Icy shivers ran over my body. "Rebel bees?" I faltered. "I have never heard of such a thing."

"Adam says it is quite rare to encounter them. Most bees will tolerate humans robbing their hives, but occasionally a hive becomes outraged and begins to attack. Once this happens those bees will never again be of any use to the beekeeper, and the only thing to do is to kill them or remove them at night to an isolated spot and leave them alone. Adam had done this, but he never dreamed that anyone would ride so far upriver. We don't yet know why Marianne did."

I felt ill as I contemplated the results of such an attack. The pain, the hideous swelling of the flesh— It would be difficult for the family to accept a death like that, particularly in the case of a beautiful woman such as Marianne must have been. I saw the scene in my imagination; the woman in her sweeping riding skirt, one leg hooked over the pommel of her sidesaddle, her horse's terror at the approaching insects—

"How badly was her horse stung?" I asked. "I suppose

he threw her when the attack began."

"He came back to the stable without her," Rosemary replied abruptly, "but he wasn't stung."

"That's odd. Unless, of course, he unseated her before the attack."

"That is unlikely. She was a superb horsewoman. Nor did she have any such injuries."

Rosemary was examining her fingernails with intense concentration and I sensed by her rigidity that there was more to the story. I debated whether or not to inquire further, and finally said tentatively, "Considering the fullness of a riding skirt, I am surprised that she didn't shield her face and head with it. Her petticoats would have provided protection for her lower limbs. But then sitting here in safety, it is easy to think of these things."

"That is just it." Rosemary abandoned the scrutiny of her hands and looked straight at me. "She wasn't wearing her riding habit. When they found her she had on only a petticoat and camisole. Her upper body was nearly bare to the insects' attack."

I digested this in shocked silence. There could be no justifiable excuse for a woman to be dressed like that anywhere except in her own bedroom. Unless—

"There was no indication that she had been assaulted?"

"None. Her riding habit was found neatly folded in a nearby glen."

The implications of this were obvious. The beautiful Marianne had ridden out to an assignation, and inadvertently to her death. I tried to force my mind away from its immediate speculation as to who the man might have been and who the family might have suspected. Rosemary added, in warning, "The children, naturally, know nothing of this. They think that she died in a riding accident."

"Of course." I sat quietly, waiting to see whether or not Rosemary meant to say more. After a while she sighed and continued. "The whole thing doesn't bear thinking about. Morgan was devastated, though I feel it was more because of the scandal than of her death. They had not been getting along at all well. But she was the mother of his children, and, furthermore, a man like my brother finds it almost impossible to stand up in the face of ridicule and shame." Rosemary looked down at the hairbrush she still held in her hand. "If it weren't that Morgan was not at all the type, I would be inclined to think that he drowned himself in the cistern rather than live with the speculation and disgrace of it. In fact, that became the unoffical verdict of the matter."

I remembered what she had said that first night about a curse being on the Rivers family, a curse connected with water. A shiver ran over me, but I forced myself to say casually, "Then you still believe that Morgan's death was the result of a family curse?"

Her sallow skin took on a tinge of color. "I wasn't myself the night I spoke of that," she said. "Word of your father's death, and then your appearance here, had unnerved me." Then she added defensively, "But there is more than one kind of curse, you know. An unfaithful wife or an unloving family is certainly a curse as evil as any cast in a fairy tale."

I was silent, for I could not have agreed with her more. It wasn't until later that I realized that she had not actually said that she believed Morgan's death to be an accident, either.

In the end we merely refurbished one of the gowns that Rosemary had put away after she had broken her engagement, which was probably the wisest idea in any case, for it would recall good memories for them both. I helped dress her hair in the becoming style we had decided upon, and with the glow she had acquired in anticipation of seeing Peter, she was actually pretty.

Most importantly, Peter thought so, a fact which was immediately obvious when he returned the following Thursday afternoon as she had asked. She was waiting for him on the veranda, and the awkward tenderness with which they greeted each other brought a mist of tears to my eyes. I glanced at Adam, who was looking inordinately pleased, and then at Christian. His quizzical expression was hard to read, composed as it was of skepticism, his usual sardonic amusement, and some other shadowy quality which struck me as being just a hint of chagrin. This latter made no sense at all, and I promptly dismissed it, thinking it a result of a too active imagination.

Anne and Mark were happy to see Peter—not wholly for the trinkets he had brought them—and what with their greetings and the general hubbub of congratulatory remarks, it was some time before he found an opportunity to take me aside and tell me what new information he had uncovered.

I had suggested that we go onto the veranda to take advantage of the afternoon breeze. Adam was not with us; I had invited him to come, but he pled pressing business of the ranch which must be attended to and had prevailed upon Christian to accompany him. Rosemary had taken the children off somewhere, though I knew it cost her dearly to leave Peter just yet, and the lawyer and I were alone.

After we had taken seats in the wicker chairs awaiting

us, Peter began in his usual ponderous manner, "As I thought, the key belongs to a safety-deposit box at your father's bank. I found that he had closed his regular account not long ago."

I frowned. "Isn't that unusual? For a man of Father's financial position, I mean."

"Normally, yes," Peter agreed. "But there have been many bank failures these past few years. In view of the fact that five of San Diego's eight financial institutions have closed their doors, it is not surprising that Jamison decided not to maintain a regular account."

I fought down a growing impatience. At this moment the history of the local banking business was not of primary interest to me. "And what was in the box?"

Peter's eyes widened in surprise. "I have not as yet opened it. Since you are your father's sole heir, the bank would prefer that you open it yourself."

"Do you mean that I must make a trip into San Diego?" I felt dismayed. "I would much prefer not to have to do that." At the thought of returning to the city that took my father's life, a slow feeling of dread was creeping over me.

Doubt filled Peter's honest face. "Why—I suppose that since I am your lawyer, they would allow me to open it for you if you were to give me written permission to do so."

"I will do that, then." I attempted to hide my relief. I was only just beginning to realize how badly frightened I had been in San Diego, both from Father's murder and from my own terrifying experience with the would-be intruder. I felt so safe, so secure here at Riversbend. Like Rosemary, but for different reasons, I often thought that I would like to never leave it. "I hope that you understand," I said rather helplessly to Peter. "Though perhaps it is asking too much of you."

"Not at all. And I do understand," he said with a gentleness that touched me. "I feel privileged to do what I can for Jamison's daughter." He got up from his chair. "I will leave immediately for San Diego. If I hurry, perhaps I will arrive before the bank is closed."

I stood up in alarm. "Oh, heavens, such haste is hardly necessary! Whatever is in that box has been waiting all this time, and it will wait a short while longer. Rosemary would never forgive me if I didn't insist that you stay the night."

Chapter Thirteen

While I could not, after all, persuade Peter to be so remiss in his duty as to stay the night at Riversbend, he did at last agree to spend the afternoon with his beloved, then have an early dinner before returning to San Diego. The light lasted until quite late at this time of year, and besides, he would have a glow of love to illuminate his way.

Rosemary, of course, was in her own private paradise. She, who had so long refused to leave the house, now walked happily among the flashing fountains and fragrant, full-blown flowers of the central garden, her beloved Peter by her side. As I stood in the upstairs corridor, shamelessly peeking at the two lovers as they strolled together, Anne came to stand beside me.

"It is so romantic, isn't it?" she said eagerly. "It is just like something out of a book. An evil spirit had put a curse on them, and now it has been broken at last."

I smiled at her fancy, but inwardly thought that the child was closer to the truth of what had happened than she herself imagined. Uneasily, I wondered again just who had been behind such a cruel hoax, and what the person had had to gain by it. The only logical possibility was that someone who was herself in love with Peter wished to break them up, but if that were so, she apparently hadn't succeeded in capturing him for herself. Still, it was not without possibilities that such a person was behind it. Peter might not appeal to me romantically, but he was kind and gentle, and certainly would prove to be a good provider. I decided at first that I would explore the matter with Rosemary, but on second thought changed my mind. They were happy and together now; what would be the point of dredging up old hurts?

Still, the letter-writer and the reasons for it nagged at my mind.

The children and I spent the day quietly reading and playing in the nursery until we were summoned to dinner. As they scampered ahead of me along the corridor and to the stairs, Rosemary came from her room, where she had gone to tidy up. She hurried to me and clasped my hands in her.

"Oh, Beth, what can I say to you?" Her eyes shone like stars. "Peter and I will owe you a debt of gratitude all of our lives!"

"Nonsense," I returned. "You and Peter would have gotten back together sometime. True love won't be denied, you know. It says so in all the romantic novels." I was teasing her, for even in her joy Rosemary was so intense as to be exhausting. I supposed it was simply a facet of her nature, but perhaps with time and a happy marriage it would lessen a trifle.

Just now she could not help being serious. "Oh, no," she said, "Peter and I agree that we might never have had another chance without your—" She stopped, at a loss as to how to describe it delicately.

"Meddling?" I supplied with a smile. "Please, don't thank me for that. I am always delighted to meddle constructively."

Unobtrusively, I urged her toward the stairs, but she didn't take the hint. Instead she turned me to face her directly.

"Do you know how much you have come to mean to me, Beth?" she said earnestly. "You are the sister I never had, the close friend I was denied by being brought to Riversbend to live. I want so much to do something for you to show you my gratitude!" She paused, then added impulsively, "Nothing would make me happier than to see you marry Adam and become my dear sister-in-law. Is—do you think that there is a chance of that happening?"

I was at a loss as to how to answer her. I knew that I loved Adam, was sure that he returned that love, and yet never did we seem to come to a definite understanding. "I—really don't think that I can say at this time, Rosemary," I told her reluctantly. "Not that I wouldn't be delighted to become your sister-in-law."

A faint disappointment clouded her expression. "Is it perhaps Christian, then? He seems to care deeply for you, too, and while I would much prefer it to be Adam, even a cousinly relationship would be better than none." Before I could comment on this last, she drew me closer. "I have a more serious reason for asking this of you," she said more quietly. "It is the matter of the children. When Peter and I are married—" and here she blushed charmingly, "we will of necessity live in town. While we would be happy to have them make our home with us, I doubt that Adam would hear of it. In his own way, he is as taken with Anne and Mark as are the rest of us. Yet that will leave only him and Christian to look after their needs. I would feel so much better if you were to be here, too, for I have seen how genuinely you love them."

I did love Anne and Mark, and they were not the least of my reasons for not wanting to leave Riversbend, yet I had not thought about Rosemary's marriage and necessary move into town. Things had happened so fast that I had not had a chance to realize the full ramifications of it. Who *would* look after the children if neither Rosemary nor I were here? It was true that Maria would be good to them, but that was hardly the same as a loving member of the family.

As though they were summoned by Rosemary's and my speaking of them, the two children came back upstairs to us. "Uncle Adam sent us to fetch you," Mark said. "Maria has announced that dinner is ready."

I took their hands and drew each of them to my sides. "Then we had better appear immediately," I said lightly, "or we will surely be in disgrace."

We hurried downstairs to the dining room and I was spared having to answer Rosemary's difficult questions.

Dinner was one of the pleasantest I had ever experienced at Riversbend. Rosemary's happiness, as well as that of Peter's, seemed to imbue the room with a new, sanguine aura. Adam was obviously happy for his sister, and even Christian curbed his usually audacious tongue and was quite genuinely charming. In fact, he was so kind to the children and seemed so interested in the betrothal plans that I found myself being drawn to him again.

"And when is the happy event, Rosemary?" he asked, helping himself to a second glass of wine from the decanter that always stood on the sideboard. "I suppose that the bridegroom is persuading you to an early date."

Rosemary reddened slightly, but said, in a return to her no-nonsense manner, "You know that it would hardly be proper to set the date less than two months ahead. There is so much to do—invitations to have engraved and to address, decorations to be thought of, my trousseau to be chosen and sewn—" She smiled at Mark and Anne with unusual gentleness. "Isn't it convenient that I have a ring-bearer and a flower girl right under my own roof?"

Mark only looked puzzled, but Anne gasped with surprise and pleasure. "Do you mean that I am to be your flower girl, Aunt Rosemary? And I may carry a basket and scatter roses in your path? Oh, I will be so frightened!" But her sparkling eyes belied her words. "And Mark will carry your wedding ring on a satin pillow?"

Mark didn't look so sure of his feelings in the matter. "Must I? I will probably drop it and it will roll under a bench or something," he said gloomily.

Under the general laughter Rosemary assured him that they would sew it on, and, much relieved, he went on with his dinner.

"What about it, Rosemary?" Christian asked. "Am I to be left out of this family wedding? After all, if Adam is to be best man, and Beth here your bridesmaid, what am I to be?" He lifted his glass to Rosemary in a little salute. "There is only one opening left to me. I see that I must grow a long beard and give the bride away." He laughed at his own sally and drained his glass, but Rosemary took him seriously, as usual.

"Oh, Christian, I am sorry," she said. "But considering your youth, I hardly think it appropriate. Perhaps you wouldn't mind being head usher, instead?"

Christian looked properly crestfallen. "Ah, such a demotion. But who could deny you anything tonight, dear girl?" He said this charmingly, and Rosemary even let down her guard and thanked him profusely for accepting the role. I was pleased to see that he genuinely desired her happiness.

But later on that evening, I was not so sure of his feelings toward the marriage. When Peter left for town, far later than he had meant to, Rosemary and Adam accompanied him to the veranda, leaving Christian and me together in the drawing room. It was the first time we had been alone since the evening of the spider incident, for I had given him no chance to pursue an intimacy I didn't feel or want, and I immediately wondered if he would see this as an ideal opportunity. But from his usual stance at the mantel he only said pleasantly, "How nice it is to see Rosemary happy and smiling for a change. I had feared that the poor girl would be permanently soured by the break-up, but it seems as though true love can forgive almost anything."

I crossed to sit in one of the fireside chairs. Seeing the happiness of the couple had put me in a pensive mood, but I was still mindful of Rosemary's trust in me. I answered idly, "Why do you feel that there was anything to forgive?"

He shrugged. "I suppose because there usually is some wrongdoing on one side or the other in affairs of this kind. Why else would the break have occurred at all?"

He was right, of course, but I refused to admit it. "In this case," I said stubbornly, "it was no more than a simple misunderstanding. And I feel that it is impertinent of us to discuss it."

He laughed aloud. "There is no need to be defensive. I couldn't be happier for the pair of them. I had expected to be subjected to Rosemary's vapors for as long as I lived in this house." The dark eyes gleamed with amusement. "Now admit it; she is the accepted image of a lifelong spinster."

I glanced nervously at the door, for he had not bothered to lower his voice. "Shh!" I said. "She'll hear you!"

"And isn't it strange how nonplussed we are when those we have stereotyped refuse to behave as we have decided they will," he went on, unrepentant. "For instance, I have the reputation of a Lothario, and it has never been expected that I will marry and produce children to share in the inheritance of Riversbend. But I am beginning to have strong leanings in that direction." He put down his glass and looked at me intently. "Do you find it hard to believe that I am serious about this, Beth?"

I didn't know quite what to say. "Why, no," I hedged. "It seems quite natural to me that a man of your age should be thinking of a family of his own."

"You know what I mean." He left the hearth and came to sit down beside me. "It is not just any woman or any family that interests me. It is you, Beth, and you alone whom I want for my wife."

I moved an imperceptible distance from him. "Are you sure that you don't want me primarily because Adam does? I understand that when you were children this was often the case."

"I am not a child any longer," he said seriously. "I am a man, with a man's needs and desires." He took my hand. "You and Adam don't as yet have an understanding, do you?"

I tried to take my hand from his. "No, not yet," I answered reluctantly, "but—"

"Good. Then I needn't feel guilty if I press my case."

I doubted very much that he ever felt guilty about anything, but couldn't summon the temerity to say so. I tried again to remove my hand, but he held it firmly and lifted it to his lips. "Please, Christian," I said in alarm, "I wish that you wouldn't—"

"It is my own fault that you give me no hope." Abruptly Christian released my hand. "I have been playing the part of fool and knave for so long that I can't expect you to think I am serious now." He sat with his dark head bent sadly, and my heart melted a little. I put my hand over his.

"Oh, Christian, it isn't that—" I began, but just then Adam came in. I saw him take in Christian's and my proximity and I quickly withdrew my hand. Christian merely smiled and leaned back in his chair with his usual casual air.

"It has been such a warm day," he said smoothly, "that I think I will take a turn around the courtyard. What about it, Beth? Will you join me?"

I glanced at Adam, hoping that he would assert himself in this, but he was watching me with the shadowed air which he often assumed. My guilt fled, and, feeling unaccountably perverse, I said, "I would like that, if Adam will join us". To my surprise and further consternation, Adam replied, "Thank you, but no. I have some bookwork that must be attended to." With that he rose and immediately left the room.

I bit my lip with chagrin. I had given Adam every reason to think that I was interested in him, not in Christian, and I didn't appreciate his seeming willingness to withdraw and leave the field to his cousin. Perhaps, after all, he didn't care as much for me as I had supposed.

The amused twinkle was back in Christian's eyes. "Poor Adam," he sighed with mock sympathy. "The mantle of responsibility rests heavily at times." He rose and offered me his arm. "Shall we go?"

Having boxed myself in, I could see no way to avoid accompanying him.

I determined to return to the house after only the briefest of intervals, but once in the garden I found the evening air refreshing after the enervating heat of the day. And wisely, Christian at first kept the conversation to generalities.

He began by telling me that Riversbend was a small portion of an old Spanish land grant, as was all of the land surrounding it. "Nearly all the ranchos have dwindled by now," he said, "which seems a shame. The life of a great ranchero has always appealed to me."

With pretended casualness, I removed my arm from his. "Why, in particular?"

He grinned. "Chiefly because it was more play than work. This grubbing in the dirt seems to suit Adam, but I prefer to fill my life with pursuit of pleasure as they did. They spent most of their time riding, hunting, and enjoying the ladies. Can you imagine a more pleasant existence?"

"Someone must have worked," I said practically.

"Where would the money have come from otherwise?"

He laughed aloud. "Spoken like a true New Englander! But the work wasn't arduous. Other than the huge herds of cattle, which mostly fended for themselves, the ranchos brought in very little money. But they needed little. Their simple wants were supplied by products grown and made right on the ranchos, and since the cattle were raised only for their hides, anyone was welcome to slaughter one and take what beef he needed, as long as he left the hide in good condition for its owner to find." He glanced around as though seeing beyond the confines of house and gates. "As soon as I was old enough to ride alone, I would gallop over these hills, pretending that I was a Californio and that my rancho stretched all the way to the Pacific Ocean. At those times I was the lord and master, instead of an object of charity to be both pitied and resented."

He said this last with his ironic half-smile, but I could tell by his eyes that he was serious and that the hurt remained and shadowed the present, just as it did with Adam.

"Wasn't your uncle kind to you?" I asked sympathetically. According to Rosemary and Adam, he had been more than that. "If not, it is surprising, since I understand that he arranged for you to share equally in his estate."

"Oh, he was kind enough. But even a four-year-old knows when he doesn't really belong. As for my share of the estate, what good does it do me? I can't sell it or mortgage it unless all inheritors agree, and if we all marry and have children it will someday belong to so many that it will mean precious little to any of us."

"That is all the more reason for you to take an interest

in making the ranch profitable. Then you can add to your portion by buying more land."

He glanced at me with quick surprise. "Her father's daughter," he murmured. "You will make a most admirable addition to the Rivers family, whether it is by one means or another."

The intent look in his eyes made me uncomfortable. I knew all at once that if I remained with him in this romantic setting where the fountain played its seductive music and the sweet, heavy fragrance of roses filled our nostrils, he would take me in his arms and kiss me, and I—with the obscure anger I was feeling at Adam, might find myself enjoying it. It would be a betrayal to Adam, I knew, but I couldn't seem to make myself turn away. He moved closer and, putting his hands on my shoulders, looked intently at me. I found I could scarcely breathe.

"Beth—" he began urgently, "There is something in your eyes— Is there a chance for me, after all?"

My eyes fell on his lips—full, warmly red and enticing—and without my own volition I swayed toward him. But as I closed my eyes to accept his kiss, Adam's dear face swam before me. My eyes flew open and I tore myself from Christian's grasp. I suddenly felt myself caught between the two men, a small, uncertain mouse being bandied about by a pair of large, sleek tomcats. Anger flashed through me.

"You are taking a good deal for granted," I said. "How do you know that I wish to belong to this family at all?"

Christian curved those warm, sensual lips into a smile. "Come now, Beth. Are you trying to make me believe that you didn't want me to kiss you just then?"

"Oh," I breathed furiously, "it is as I said the day I met you. You certainly are no gentleman!"

Before he could stop me I whirled from him and went toward the house. His cheerful voice followed me.

"Like it or not, Beth, I do not intend to give up hope just yet!"

When I stormed into the house I was met with silence. Though it was not much past dusk, it was as though everyone had retired. I glanced into the drawing room. Adam wasn't there, and I assumed that he was still in his study, avoiding me. Another surge of anger rushed through me, then I felt it drain away as quickly as it had come, to be replaced by a pervading feeling of loneliness. Why didn't Adam speak out if he loved me; why must he always allow Christian to have his own way? Listlessly I started up the stairs to my room.

"Beth?"

I paused. Adam was standing in the hall below me. "Can I persuade you to spend a few moments with me before you retire?"

I considered his request. In my present mood it didn't seem to matter what I did.

"If you wish," I answered indifferently. I returned downstairs, and in silence we went into the drawing room.

"Are you angry with me?" Adam asked diffidently. I wished instead that he had taken me into his arms.

"Yes," I answered flatly. "Why must you always withdraw when Christian asserts himself with me? Or are you angry because you still think I am leading him on?"

The familiar lines settled deeper around Adam's mouth. I had the ridiculous impulse to reach up and erase them gently with my fingertips.

"Well, aren't you?" he demanded. "You looked very cozy together on the sofa when I came in. Do you deny that your hand was over his? Or that you spent an intimate time together the past two days? My God, Beth, what am I to think?"

"You could think that I love you, as I have given you every reason to believe," I said coldly, furiously. "Instead, you think disgusting thoughts and give Christian every opportunity to try and win me away from you. But perhaps in your heart that is what you really wish!" I started past him. "Oh, what is the use? I am going to bed!"

Adam's arm shot out and he caught me by the waist and drew me to him. "You think that, do you? That I want Christian to possess you instead of me? I'll show you just how right you are about that!" He pulled me tightly, almost cruelly, into a savage embrace, and before I could turn my head to avoid it, his lips took mine in a hard, demanding kiss. I tried to pull away, but he wouldn't let me go, and in an instant I felt passion rise in me, melting my bones and leaving me trembling with desire. When at last the kiss ended, I clung to him, fearing that I could not stand alone for the sweet weakness that had come over me.

"There, Beth." Adam's voice was rough with emotion. "Now tell me that you don't believe I love you."

"I do believe it now," I said through quivering lips. "But why haven't you shown it before this? You can hardly blame Christian for thinking there is a chance for him, when even *I* don't know what our relationship is."

"God knows I have wanted to show you how I feel, how much I want to marry you. But there are reasons—" The dark look settled over him again. "There is so much about us that you don't know, Beth, so many things—" He broke off, then said, as though the words were forced out of him, "Sometimes I think I haven't the right to bring you into all this, and yet—when I see you and Christian together I am consumed by jealousy. I cannot bear to think that once more he will have used his charm to get something that I have my heart set on."

Once more I felt that I was a thing, a bone between two dogs. "Are you referring again to when you were children? That was long ago, and I am not an object whose possession can be decided between you. Doesn't it matter at all how I feel about you?"

"Of course it does! But Christian— I have seen his charm, his adroit way with women on many an occasion. How can I expect you not to be forever comparing us? Even my father preferred him over me."

"It is odd that you should feel that way about him," I said slowly, "for I think he feels inadequate compared to you."

Adam's short laugh was without mirth. "I don't know why. My father went to extremes in trying to make up for the tragedy of Christian's father's death. And I learned soon after he came to live here that Christian's needs and desires took precedence over mine."

Christian's father, too, had died tragically? A chill went over me. How much more of this family's tragic past must I have to learn before I would know it all? No wonder Rosemary had spoken of a curse on the family!

But Adam was still speaking, and I tried to force my attention to him.

"But that is all water under the bridge, and you are right. We are no longer children. If you tell me you love me and will marry me, I will never again give a thought to

Christian and his charm. Will you marry me, Beth? I want you for my wife."

Just a few minutes ago this had been what I most wanted to hear. But all the talk of jealousy and mysterious and tragic deaths had frightened me, and while I didn't doubt my love for Adam, this further glimpse into the clouded and unhappy family past had made me wonder how much I really knew him. Suddenly I didn't want to totally commit myself just yet.

"I do love you," I said unsteadily, "but I am tired and overwrought just now. May we talk of this again tomorrow?"

Under the circumstances it was hardly the answer he had been expecting, and I saw him stiffen. "Do you find it so hard to say yes?" He caught himself, and with an obvious effort relaxed and drew me toward him. "Very well. Until tomorrow, then." He kissed me again, this time more gently, and again his lips worked their magic. I was tempted to fling caution to the winds and cry, "Oh, yes, yes, Adam, I will marry you!" but some uncertainty held my tongue still. Instead, I impulsively stood on tiptoe and offered my lips again. When at last we let each other go, I fled into the house and up the long flight of stairs to my room.

Quickly I disrobed and put on my nightgown, then crept into the comforting embrace of my featherbed. I was both elated and remorseful. Why hadn't I agreed at once to Adam's proposal when I knew that marriage to him was what I wanted? Was it because I sensed that unless the Rivers family's tragic history was fully explained it would act as a transluscent barrier between us? When Adam assumed that closed-in look which often accompanied a mention of his family, I always felt lonely and estranged from him, and I knew that I wasn't willing to accept that as a part of our future. And yet the thought of asking him to explain the tangled web of family tragedies in detail set me to quivering with some nameless apprehension, a feeling which did not leave me until at last I fell asleep.

I slept heavily, dreamlessly, and did not awaken until a decisive knock at my door jerked me into instant consciousness.

"Beth?" Adam's voice accompanied a repeated knocking. "Peter Barnes has just ridden in. He said to tell you that he has very important news."

Chapter Fourteen

It took me a moment to realize that I had slept very late, and another to actually grasp the import of Adam's words. But when I did, excitement lent speed to my morning ablutions. In a very few moments I presented myself in the drawing room.

It was obvious from the expectant expressions of the others that Peter had told them nothing. He rose at my entrance and strode to me, delight written large on his face.

"It is good news at last!" he proclaimed in his heavy, somewhat ponderous tone. "In my eagerness to bring it to you I left the city immediately after visiting the bank this morning. I only hope that my precipitate arrival hasn't caused you undue discomfiture."

I controlled my natural impatience. "Of course not, Mr. Barnes. But curiosity has. May I ask you to tell us immediately just what you have found?"

He paused dramatically before answering. I could have shaken him.

"Among other things, there was ten thousand dollars, several pieces of excellent jewelry, and—" his voice rose triumphantly "—a deed to one hundred acres of prime land adjoining Riversbend."

I could scarcely believe it. My fondest dream had come true; I was indeed an heiress, and could enter into marriage with anyone without the slightest suspicion being cast that it was for anything but love. I had no idea of the land's value, but judging from the general intake of breath which greeted Peter's announcement, it was considerable.

I glanced around. Adam had unaccountably assumed that shuttered look which I dreaded, while Christian, naturally, hid any surprise behind his amused cynicism. Only Rosemary, who was already vibrant with happiness, seemed delighted for me.

"How wonderful, Beth!" she exclaimed. "But I am not really surprised. Jamison was far too clever a man to die penniless." And then, when she saw my stricken look, added apologetically, "I'm sorry. I didn't mean to stir up unhappy memories."

I shook my head, though her words in truth had returned me to that night of Father's death. It was undoubtedly this land he had referred to when, as we had sat at dinner, he had spoken of a "fabulous investment," though what plans he had had for it I would probably never know. "It's all right, Rosemary," I assured her. "Father's death is something I must learn to accept." Deliberately I changed the subject. "Is this particular piece of land very valuable?"

"It has the potential to be most valuable," Peter explained. "It is between Riversbend and the city, and it is seldom that one is able to obtain such a large tract of land near a river. Jamison must have been quietly buying up small parcels for some time. I know that he once remarked that he thought San Diego's growth lay in this direction."

"He planned a subdivision," Rosemary said. "He told me that his dream was to build a model community. However, he didn't say that he planned to build it here."

She said this with such quiet certainty that even Christian didn't deride her words, and I found myself wondering if perhaps she and Father had had a closer relationship than I had been willing to believe.

Christian spoke for the first time. "I understand Jamison was on his way to see you when he was killed, Adam. Do you have any idea as to why?"

"None whatsoever," Adam answered curtly. I was surprised and disquieted by his seeming lack of enthusiasm for my good fortune, a feeling which increased when he added with unmistakable bitterness, "I would assume that he wanted help in planning a water system for the proposed subdivision, though he must have known that I wouldn't welcome such a development so close to Riversbend. I had hoped to acquire that land myself one day, for agricultural purposes."

Christian laughed. "Ever the shortsighted farmer. A subdivision such as Jamison planned would generate a fortune."

"And once the land is given over to housing it is lost forever," Adam retorted. "Whereas farmland can bring a living for as long as it is tilled."

"If you would ever lift your head from the ditches you are eternally digging," Christian said, "you might look around you and see the miles of open land that lie outside San Diego. Most of it will still be here a hundred years from now."

"Yes, semidesert land," Adam retorted. "How are you going to get water to it if the arable land near the rivers is all built up with housing? It is you who are shortsighted, Christian, and those like you. And we will all pay for their greed one day."

The cousins glared at each other.

"Come now. There is little point in arguing over something that doesn't even belong to you," Peter objected obtusely. "It is for Miss Allen to decide what is to be done with it."

Adam and Christian turned to look at me, and something in their expressions caused me to feel strangely uneasy. It was as though each, for a moment, begrudged me my inheritance. I had not expected that.

The spell was broken as Christian said lightly, "Of course, it is for her to decide. And it is for us to offer our congratulations and good wishes." His voice took on a softer note. "Which I do, Beth. With all my heart."

Adam immediately seconded Christian's sentiment, though I sensed an underlying constraint. I supposed he was still smarting from what he must have felt was Father's betrayal.

I tried to hide my discomfiture at his attitude. "Thank you all," I said, "and thank you beyond words for your kindness in taking me in. But there is no reason for me to impose upon you any longer. I think it best that I return to San Diego with Mr. Barnes."

There was a shocked silence and then an immediate chorus of protest. It was gratifying, to say the least, and I felt my hurt lessening.

"Oh, no, Beth!" Rosemary exclaimed in dismay. "You cannot leave now! I am going to need your help in planning my trousseau." She cast a half-pleased, half-embarrassed look at Peter. "I thought to have a quiet wedding here at Riversbend, but Peter insists that it be a fashionable one in the city. You know that I haven't kept up with the latest styles, and there are so many other things that need to be done. I must confess that I am frightened by it all."

Christian added sagely, "And there are the children to consider. They will miss you in any case, but they would be shocked and hurt if you were to leave so suddenly."

My heart plummeted, as he had known it would. I hadn't yet thought of the children. How could I bear to part with Anne and Mark?

"Besides, you should give yourself a little time." Adam's quiet way of speaking meant to me that he was himself again. "Now that you are an heiress you have options that weren't open to you before. Perhaps you will want to go in an entirely new direction." The level look at me which accompanied his words reminded me that his proposal of last night remained unanswered, and I thought with discomfort that if my answer were no, he might well believe that I had been considering marriage to him as a means of security. I cringed with embarrassment. And yet the doubts I had entertained last night were still with me, and they were accompanied by some new uneasiness to which I hadn't as yet had time to put a name.

"Say you will stay," Rosemary implored. "At least until the wedding." She glanced shyly at Peter and he added, "We would appreciate it very much, Beth. But there is another reason as to why it would probably be best. The city is not safe for a young woman alone, even when she has ample means to live in the finest parts of town."

I started to smile at his overcautious ways, but I intercepted a brief, meaningful glance between him and Adam and was suddenly and chillingly reminded of the metal blade slipping through my door the night of Father's death, and of the fact that his murderer had never been apprehended. I remembered, also, the dread I had felt at the thought of going back to San Diego. At that point I needed no further persuasion. I said, with unaccustomed meekness, "Perhaps you are right. If no one objects, I will be happy to remain with all of you at Riversbend for a while longer."

In spite of the warmth of their initial response, the family seemed to subtly withdraw from me in the next few days. Even Mark and Anne seemed subdued, as though sensing that something was slightly amiss. But it might have been my own doing, for I had a great deal on my mind.

I hadn't been able to forget the look that had passed between Peter and Adam, and I realized that they had never believed Father's death to be simply the result of a robbery. And neither had I, I admitted to myself, or I would not have accepted Adam's offer of sanctuary so eagerly. I had wanted to get far away from the terror of that night, and Riversbend had seemed the quickest way to do it.

Though the thought was painful, it also occurred to me that perhaps both Adam and Christian had known that I was to inherit this property lying so enticingly close to Riversbend. After all, Father had been on his way to see Adam when he had been killed. It was entirely possible that he had previously mentioned his purchase of a tract of land next to that of his good friend. As for Christian, when he had offered me the apology for his rude behavior he had spoken of a recent meeting with my father. Perhaps the land purchase had been mentioned at that time, though I thought it less likely that Father would have confided in Christian, considering his poor opinion of him. At any rate, I now realized with a strange uneasiness that the romantic interest of both men may well have sprung from a decidedly unromantic desire to take possession, through marriage, of my valuable inheritance.

When the thought first came to me, I dismissed it out of hand. Surely Adam's sympathy and kindness at the time of Father's death couldn't have been assumed, and Christian had not even been civil until later. Much later, I amended with a sinking heart. When he had finally understood that I was Jamison's daughter—and his heir.

I tried to put these insidious thoughts from me, but once they had taken root I could not eradicate them. And this affected my relationship with the two men, Adam most particularly, causing a reticence in my attitude toward him that he could not help noticing. Perhaps because of it, he didn't press me for an answer to his marriage proposal, and I didn't bring it up, either.

One especially hot afternoon a few days following Peter's visit he did speak of it, but it wasn't as though he had meant to. He had come to the nursery to ask me to accompany him and the children to the reservoir for a swim. "Rosemary can provide you with a bathing outfit," he said, dismissing my feeble excuse. "Though she doesn't swim, it is considered fashionable to attend beach picnics wearing a bathing suit, and she used to do that occasionally." The keen eyes narrowed. "But it isn't that, is it, Beth. You simply don't want to come swimming. Or—is it that you would rather not be with me at all?"

"No, of course not," I protested rather overemphatically. His accusing tone had shocked me. "It is just that—it is so very warm. I would really rather stay indoors in this heat."

He had sent Anne and Mark to their rooms to dress for the treat, and he glanced at their closed doors before moving closer and saying coldly, "All right, Beth. What has come between us? I have sensed it ever since Peter brought the news of your inheritance. Have you decided against our marriage and are reluctant to tell me?"

For a moment I didn't know how to answer him. Honesty was out of the question. How could I accuse him of wanting me for financial gain when he had never by word or look suggested that I might be considering him for the same reason? At last I said, "I care for you very deeply. But there are reasons—" I broke off, fearing that I might reveal my doubts about both his motivation and his family's mysterious past.

"I see." A kind of anger was in his eyes. "I would appreciate it if you could bring yourself to tell me what those reasons are."

Anne and Mark's return saved me from having to answer him. Though they were disappointed that I wasn't going swimming with them, they accepted my excuse of a headache without question, and, their excitement only a little lessened, went gaily off with Adam.

After they had gone the house was very quiet. Rosemary had retired to her room for her usual afternoon nap and I considered doing the same, but as I walked along the corridor a restlessness seized me and I began instead to pace the hall. I crossed the gallery which joined the twin curving stairs, then, on a sudden impulse, followed the corridor into the other wing of the house. I knew that Adam and Christian's rooms were here and that Mark and Anne's parents had had their apartment on this side. So far from the children, I thought, distracted for a moment from my own problems. As far as it was possible to get.

I thought about this couple whom I would never know. How strange that they had both died so tragically. As I strolled the hall that they must have traversed so often, I almost felt that I could see them walking there. They would have been a striking pair, Morgan with the dark handsomeness of Mark and Christian, and Marianne of the voluptuous body and beautiful haughty face which Rosemary had described. No wonder she had resented the isolation of Riversbend. A woman like that must have craved the admiration, the adulation even, of many men, and here there was only the family—only her father-in-law and the younger Adam and Christian—to appreciate her sensual beauty.

I stopped still in the middle of the corridor. What effect had she had, I wondered, on her young brother-in-law and his cousin? Had they been fascinated by her as Rosemary had indicated her father had? Or did their resentment of Morgan expand to include her? I didn't recall hearing Adam so much as mention Marianne's name.

I was startled from my reverie when a nearby door opened and Emelda came out. Before she closed the door again, I caught a glimpse of Christian. He was standing at one of the windows facing the river. He seemed absorbed and I presumed he was watching the progress of Adam and the children as they headed toward the reservoir.

Emelda saw me and closed the door abruptly before she said in a low, yet sharp tone, "What is it you want?" There was not a trace of the grudging courtesy which she showed me when family members were present.

"Nothing," I answered, annoyed that I found myself on the defensive. "I have never been in this part of the house before, that's all. I simply came to look at it."

Her cynical look told me that she considered it a flimsy excuse. "This wing is now used for the men only. There is nothing for you to see except closed doors."

Her rudeness annoyed me. "No one has told me that I shouldn't go where I wish." I made my tone deliberately haughty. "What does it matter to you?"

She shrugged one smooth olive shoulder in a gesture of feigned indifference. "It does not matter. Wander here all day if you wish. But there is nothing for you to see."

"I thought I would look at the apartment of the children's parents. Wasn't it somewhere in this wing?"

"Why do you ask?" The large brown eyes stared with bold assessment. "No one is allowed into it."

I had the feeling that we were fighting some kind of obscure duel. "And why is that?" I asked lightly. "Is it haunted by their spirits?"

To my surprise the girl lost her aloof manner and hurriedly crossed herself. "It isn't good to speak of such things," she said reprovingly. "The apartment is just as it was when they died. All of their clothing and other belongings are still there. It is a waste, but Rosemary said to leave them. Why do you want to see it, anyhow?"

I felt a perverse wish to annoy her. "Perhaps I will become mistress of Riversbend one day soon. If so, I might take that apartment for my own."

"Adam would never want to do that," she said sharply. "And why should you? His rooms are nearer to the stairs."

I gave her a level look. "And how do you know it is Adam I mean to marry?" I said it to annoy her, but she got such a stricken look that I felt guilty and went on hastily, "Also I thought I would look at Marianne's clothing. There may be something that can be made over for Anne."

Anger flared in her eyes, replacing the stricken expression. "There is nothing," she said. "Marianne didn't wear the kind of clothing that can be made over for children. Everything she had was of silks and velvets and smooth, shiny satin—" Unconsciously her free hand smoothed the rounded curve of her hip. "Besides, the material is no doubt rotting by now. It would be a waste of your time."

I gathered that she had a proprietary interest in that clothing. "Perhaps you are right," I said indifferently. I was not truly interested in Marianne's gowns; Rosemary's abhorence of the woman had transmitted itself somewhat to me. "I merely thought it might be a way to pass the afternoon. I find myself at loose ends with the children gone."

She studied me with a calculating look. "The family's past is of interest to you, I think. Isn't it?"

I said slowly, "Why, yes, I suppose that it is. Why?"

The dark eyes seemed suddenly opaque. "I am surprised that you haven't explored the house before this. There are many interesting things on the third floor, which is mostly used for storage. Other than the apartment of Christian's mother."

Something about her overly elaborate casualness

bespoke a warning, so I simply continued to regard her silently. Seeing that I was going to ask no questions, she went on, "She lived here for several years, you know, after her husband died."

"I do know," I lied with an air of disinterest. "Adam has told me all about it."

"He has?" Her eyebrows shot up in unfeigned surprise. She caught herself and added, "Then you would not be interested in her rooms. It is too bad. It would be a way for you to pass the day."

"Perhaps some other time," I said. "I think instead I will return to my room and lie down for a while." Without waiting for her to leave, I started back to my own wing of the house. But I had no sooner passed Christian's door than I heard it open. I turned to find him leaning casually against the frame of his door. He was dressed in the white shirt and dark pants that suited him so well.

"I thought that was your voice, Beth," he said. "Is there something you need?"

"Why, no." I felt unaccountably flustered. "I was just speaking with Emelda."

Christian glanced at her. She was regarding us with sullen anger.

"I see." He seemed indifferent to her mood. "Did you decide not to go swimming?"

I didn't answer until I had given Emelda a pointed look, and reluctantly she went on down the hall. When I judged her out of earshot, I replied, "Yes, I did. After swimming regularly in the Atlantic, the reservoir doesn't seem very appealing. But please don't repeat that. I am afraid that I told a small lie and pleaded a headache."

Christian smiled sympathetically. "I can't blame you.

Even if I were fond of swimming, that would scarcely be the place I would choose." His eyes frankly admired me as he added, "Perhaps you will join me in a chilled glass of wine or some lemonade. We could have Maria serve it at a shady spot in the courtyard."

Though it sounded inviting, I hardly thought it wise. I said quickly, "Thank you, but no. Having lied about a headache, I have been visited by one, no doubt as a suitable punishment. I think I will go and rest quietly in my room."

"Very well." He sighed in a resigned manner. "Then I will simply have to drown my disappointment in prime zinfandel. I must make the most of my free afternoon." So saying, he tucked my hand beneath his arm and walked with me to the gallery. I wondered with some unease if Emelda was still watching.

I did go to my room, but I had no headache and no intention of staying there. Emelda's words had intrigued me, as she had meant them to. There was something strange about the apartment used by Christian's mother, something that Emelda thought would come as an unpleasant surprise to me. But unpleasant or not, I meant to see it. Before I could decide on my answer to Adam I had to learn what I could about the family past.

After spending a brief time in my room, which I did in case Emelda was watching, I peered into the hall. Finding it deserted, I hurried to the stairs leading to the third floor. As I climbed them, I realized that I was moving as stealthily as a thief would have, and my heart was beating heavily with a vague, nameless fear. I hesitated, almost deciding to retreat downstairs, but curiosity won out, and after a moment I went to the nearest door and gingerly opened it.

The room was completely empty, as were the others which lay along the outside wall of the wing. By the time I reached the last door I had begun to feel a vague disappointment, but as I opened this one a prickling sensation stirred at the nape of my neck. Just beyond the wooden door was another, and that one was fashioned of heavy metal bars.

I knew at once that this was what Emelda had wanted me to find.

After the first icy shock had abated, I approached the barred door, which was more of a gate, really, and tugged at it. It opened with a screech of rusty metal, but I had no difficulty in swinging it back against the wall of what seemed to be a small entry. Beyond it lay a suite of rooms. The first, in which I stood, was furnished with a couch and two heavy padded chairs. In one corner were a pair of straight chairs and a small, round oak table. The other room, as I could see through the adjoining door, held a wooden bed and small chest of drawers. There was nothing else at all; no side tables, no lamps, no bric-a-brac. Not even a picture broke the monotony of the white walls; they gaped starkly empty. Because of this paucity of detail, my eye was drawn inexorably to the windows opposite me.

They, too, were covered by heavy metal bars.

Almost reluctantly I went into the bedroom. A large wooden wardrobe stood against the inner wall. Gingerly I opened it. A few dreary-looking garments hung from pegs along the back, and as I watched, a shiny black spider swung from a single strand of web to the dusty floor. I shuddered and hastily shut the doors again.

I saw that the windows in this room were also covered by heavy grillwork. I moved to the nearest one and looked out. Beyond the green trees and fertile fields lay a vast expanse of brown hills, broken only by an occasional gnarled oak, and far in the distance the larger patch of green trees that I thought of as "Christian's grove," though he had said he shared its secret with Adam. What must it have been like to have spent day after day looking only at such emptiness? Poor lady, it must have driven her further into madness than she had already been.

For she had to have been mad. I had realized that at once. There could have been no other reason for this apartment, for the barred door and windows. And her insanity must have taken a violent turn, at least at times, for why else the lack of any light piece of furniture or small item which could have been used as a weapon in a fit of demented rage?

On impulse I took the heavy bars of the window into my hands and shook them as though testing their strength. They held, rock-firm, and for an instant I experienced the utter frustration that must be felt by someone hopelessly imprisoned. It sent a shudder over me.

It was just at that moment that I heard the unmistakable creak of the metal gate in the sitting room.

I stood, transfixed, and felt my scalp prickle. Then, with an exclamation of dismay, I ran into the other room. The gate was firmly shut. Somehow, I had known that it would be.

I tried to tell myself that the panic I felt was ridiculous, for the gate had opened easily to let me enter. But when I went to it and tried to work the latch, I found that no matter how hard I pressed it would not open.

After a long moment's frantic fumbling, I forced myself to stand back and take a few deep breaths. I must not succumb to terror as I had that night the spider had been in my room, but would instead take the time to carefully examine the mechanism of the latch. Willing myself to calmness, I reached for it again. It was only then that I noticed that the door to the hall was also tightly closed. A giant hand squeezed my heart. Though I hadn't wanted to think that it might be possible, I now had to face the fact that someone had deliberately trapped me in this place.

I fought an overwhelming panic. What could I do to extricate myself from this terrifying situation? Screaming was useless, I decided, for the location had no doubt been chosen precisely because sound wouldn't carry to the lower floors. I considered breaking one of the windowpanes so that my shouts might be heard if someone should pass below, but upon experimentation I found that even if I had had some heavy object with which to break the window, my arm wouldn't fit far enough through the narrow grillwork. So, after fighting a fresh wave of panic by sternly reminding myself that the family would miss me and come looking for me eventually, I did the wisest thing and sank gingerly into one of the dusty chairs to await rescue.

As I sat there my thoughts were not pleasant ones. I was sure that it had been Emelda who had followed me here and locked the gate. She had deliberately teased my curiosity by telling me about the apartment, and my anger flared as I pictured her satisfaction as she followed me here and sprang the trap. As with the spider, I had to assume that her motive was jealousy. I knew that she was aware of Christian's interest in me. Was she trying to frighten me away from Riversbend?

I glanced about the barren, gloomy apartment, and shuddered with aversion. If so, she had chosen an excellent way to accomplish it.

For what seemed like hours I sat in the chair and forced myself to quietly endure the slow passage of time. But at last I could stand it no more and rose to begin pacing the limited confines of the musty room. When I stopped at the window, I saw that the shadows beneath the trees were lengthening alarmingly. I fought the urge to sob aloud. I had thought that by now someone would be searching for me, that even Emelda herself might come to release me. Unless, of course—and the thought plunged into my heart like a cold knife—she meant not to frighten me, but to do away with me altogether. Biting my lips against rising hysteria, I thrust the thought away. Surely she would not be so heartless; besides, the others would know that I must be somewhere about, for I had no way to return to San Diego.

The panic rose again. It would be simple enough to spirit my horse from the stable as evidence that I had ridden away.

How long does it take to die of thirst, I wondered, then thought to flee from the awful thought by scanning the groves and hillside in a vain hope that someone out there might glance up and see me. The dark groves and golden hills lay empty and silent. I had just decided that I must scream, even if no one was there to hear me, when I heard the outer door open behind me.

I whirled to face it.

It was Emelda. She gazed at me with a shock and concern that I took immediately to be feigned.

"Why, Miss Allen!" she exclaimed. "What are you

doing here?"

"I am locked in. I thought perhaps you could explain that," I said evenly. Though I was furious, I was acutely aware that she had not yet opened the iron gate.

"I?" Her eyes rounded elaborately. "What would I have to do with it? It was only when the children came home and found you missing that I remembered our talk of this apartment and thought that you might be here. But why did you close the gate behind you?"

I tried my best not to screech at her.

"I didn't. Please open it immediately. If you can't, then call someone else to do it." Perspiration broke out on my palms as I attempted to speak with authority yet not enrage her. She would have liked nothing better than to have me beg for release, and I was determined not to oblige her.

She made a pretense of tugging at the latch. "It doesn't seem to open. Maybe I had better call Christian, though he doesn't like to be reminded of this place. His mother was kept here for three years before she died." Emelda glanced covertly at me as though awaiting my shocked reaction.

I ignored the sly look.

"Then get Adam," I ordered. "He must have returned from swimming if the children have."

She continued to pretend to work at the latch. "Yes, he has returned, but he won't be happy, either, to know that you have been here. The family isn't proud to have the story of Christian's parents known."

"Then why did you suggest that I come here?" I inquired evenly. "Did you perhaps hope to cause trouble between Adam and me? Or is it Christian who most concerns you?"

The brown eyes opened wide in feigned innocence. "Why should I wish to do that? I only thought that you had a right to know this family secret before deciding if you wish to marry into it."

"If, as you assume, I plan to marry Adam, what would it matter to me if Christian's mother was mad?" I knew her jealousy wasn't because of Adam. "Her blood wouldn't run in his veins."

"But her husband's blood does, and it was he who drove his wife to madness." She seemed to be taking delight in furnishing this gossip bit by bit while she tormented me with her inept fumbling. Suddenly my anger changed to an overpowering rage.

"You forget yourself," I said with a cold, quiet fury. "It is not your place to be gossiping about the affairs of the family. Either open that door or go at once to fetch help, or I swear that when I do get out, I will see to it that you are dismissed immediately!"

I had come close to the gate as I spoke, and Emelda recoiled at the suddenness of my attack. Then she recovered and said, her large eyes going into venomous slits, "The kitten spits like a caged tiger, but neither can claw through iron bars." Nevertheless, she stepped back to the gate and with a swift, dexterous movement, released some catch that was hidden to me. The gate came open at my touch, and swiftly I swung it back and pushed past her into the blessed freedom of the hall.

Emelda said something and I started to pause and answer her, but now that the danger had passed I felt tears, hot and stinging, fill my eyes. I was going to burst into humiliating sobs at any moment, I knew, so instead of speaking I fled as fast as I could along the hall and down the stairs to the safety of my room.

Chapter Fifteen

I did not go down to dinner that night, nor did I open my door to either Adam or Christian, though both men came to inquire as to my well-being. Rosemary, however, insisted that she must see me, so reluctantly I let her in. She was carrying a tray which held a daintily arranged dinner, and she first set it down, then turned to me. I saw her eyebrows rise as she took in the fact that I was already in my nightgown and wrapper.

"What is it, Beth?" she asked, her expression one of genuine concern. "Are you ill? Emelda simply said that she found you wandering about upstairs, but if you are too upset to come down to dinner there must be more to it than that."

"There *was* more to it than that, I'm afraid." I told her of how I had been trapped in the barred apartment, only omitting that I suspected Emelda of having deliberately shut me in. "I was very frightened," I told her, which was an understatement, "for it seemed possible that I might die before someone thought to look there for me."

Rosemary took my hand. "Don't you know that we

would search every inch of the place if you were missing? But whatever possessed you to go up there at all?"

I told her only that Emelda had mentioned the apartment and on a sudden impulse I had decided to take a look at it.

"That Emelda! She is an unpleasant little trouble-maker!" Rosemary's voice took on some of the stridency it had had before Peter reentered her life. "How someone as good as Maria could have such a daughter is beyond me."

I decided to meet the matter head-on. "She did mean trouble," I agreed, "but in some ways I am grateful to her. Adam has asked me to marry him, which probably comes as no surprise to you, but I feel that I can't accept unless and until I understand something of the tragedies which have taken place here. Be honest, Rosemary. Do you blame me?"

She had stiffened at my words and some of her cold and forbidding mien returned, but after a moment she unbent a little. "I suppose not," she returned reluctantly, "since it is your entire future that might be affected. But it is very hard for any of us to speak of them."

I said patiently, "I realize that. But I must have a few answers. After having seen her apartment, I have guessed that Christian's mother was insane, but can you tell me any more than that? For instance, was it a hereditary condition, or did something cause her to lose her reason?"

Instead of answering, Rosemary began restlessly pacing my room, as was her wont when agitated. I sank into a chair and watched her, for I knew it might be some time before she could bring herself to answer. My patience was rewarded when at last she turned to me.

"The woman was always somewhat strange, if that is what you mean," she said abruptly. "One could never be certain of how she would react to things, and she would go from a pleasant, placid mood to outrageous fury in a matter of moments. It actually wasn't safe to be with her at those times. Not that that should concern you," she added quickly, "if it is Adam whom you are to marry, for of course she is no relation to him."

"But she became worse with time?" I prompted.

Reluctantly she said, "Yes. But it was her husband's death that seemed to unhinge her mind completely."

Having already heard of the violent deaths of other members of the Rivers family, I was almost afraid to ask the next question. "How did Christian's father die, Rosemary? He, too, was a Rivers. Was his death also connected to water?"

She looked startled. "Oh, heavens no. He was shot. He—" She broke off in seeming agitation. "Beth, I really don't think I can discuss it. It is Adam to whom you should be directing your questions!" She turned as though to flee. "Shall I tell him you want to see him?"

"No, please!" I said quickly. As I remembered Adam's attitude the last time we had talked of his family, I found the thought of confronting him with this new question most distressing. I ventured to say to Rosemary, "How can I ask him when the very mention of the past seems to disturb him? Why does it, Rosemary? Can you at least answer that?"

She checked her flight toward the door. "It is because of the guilt he feels, I think. It was he who designed and built the paddlewheels and reservoir, and the roof cistern which supplies the water for the baths. And since Father was caught beneath the one and Morgan drowned in the other, he feels somehow responsible. The circumstances were odd, you see, because both were afraid of water and generally didn't go near it. Neither of them could swim a stroke."

I digested that for a moment. "But surely that can't be construed as Adam's fault! They were just unfortunate accidents, were they not?" My voice carried far more conviction than I was actually feeling, and I waited uneasily for her answer.

She assumed a noncommittal look very like that of Adam. "Of course," she said briefly. "What else could they be?"

I sensed that there was something she wasn't telling me, but it was evident that she would say no more about it. "What else?" I murmured in agreement.

With relief she changed the subject. "Are you sure you won't join us for dinner?" and at my demurral, added, "Then I will send Maria up with another tray. This one will have grown quite cold. Good night, Beth. I am sorry you had such a fright. We must see that it doesn't happen again."

After she had picked up the tray and gone I lay on my bed and tried to relax, but I realized that my body was tight with tension. Rosemary's evasiveness had left me with a sense of profound disquiet.

Very soon Maria arrived with another dinner tray. She avoided my eyes as she set it on the table, and on impulse I said, "You heard what happened this afternoon, didn't you, Maria? That I was locked into the barred apartment?"

She busied herself with arranging a place setting. "*Sí,*

señorita. I am sorry if you were frightened."

"Emelda did it, you know," I said evenly. "She deliberately locked me in there."

She did look at me then. Her eyes filled with tears as she clutched the corner of her apron with work-worn fingers.

"Oh, no, Miss Allen! Why would Emelda—"

"I am not going to tell Adam about it," I said severely. "Not this time. But if anything else happens to me I *will* tell Adam and he will see to it that Emelda leaves Riversbend. I mean that, Maria." I did mean it, most definitely, and she knew it. She did not try to protest.

"Such a thing will not happen again, I promise you."

She said this with quiet dignity and I let the matter rest. Later on, I was very glad that I had.

I ate my dinner with a reasonably good appetite, considering my frightening experience, and then, once Maria had returned for the empty dishes, got into bed. I had some trouble getting to sleep, but finally drifted off.

The early part of the night I slept dreamlessly, but toward morning I seemed to find myself floating, facedown, on the surface of a strange, murky sea. One after another, dark, cadaverous forms rose toward me from some subterranean cavern far below. The faces were turned away or seemed obscured, but as each reached me it would roll toward me with a repelling sinuosity and I would recognize it: an older man with Adam's eyes, but sightless, staring; a young man with Mark's chubby features elongated and strangely cruel; and most horrendous of all, a woman whose voluminous riding dress and long, dark hair swirled in a translucent cloud around her and whose features were hideously obscured by giant crawling insects. As she lifted toward me, I tried to scream and draw away, but her skirts covered my face like clammy seaweed. I gave one last frantic heave of my body and felt my starved lungs fill with blessed air at last. Panting with exhaustion and icy cold with the clinging remnants of fear, I lay on my back and stared into darkness above me.

I had been lying facedown on my pillow, apparently causing myself to nearly suffocate. But that did not entirely explain the reason for my nightmare. It seemed that I must be disturbed far more than I had thought by the family tragedies, for the dream had held an ominous and terrifying sense of evil. I knew suddenly that I could not stay at Riversbend unless Adam would allay my fears with a candid and straightforward explanation of all that had gone on in the years before my coming.

Though I slept no more that night, I waited until I judged that Adam would be at breakfast before I rose and dressed. He was just coming into the hall as I descended the stairs. I saw him note the shadows beneath my eyes.

"Good morning," he said politely. "I trust you are feeling better?"

"I need to talk with you," I said without so much as answering his greeting. "May we go out to the veranda?"

"Of course," he said. In silence we repaired to the broad front porch. I had steeled myself for this moment, but now that it had arrived I found myself suddenly tongue-tied. Adam saw my struggle and took the initiative.

"Come, Beth," he said almost curtly. "There is no need to agonize over it. You simply want to tell me that the answer is no, that you cannot marry me. Isn't that it?"

"Oh, no," I breathed, shocked out of my muteness.

"That isn't it at all!" and was rewarded by the immediate relief that shone from his eyes.

"What, then?" he asked, and this time his tone was more gentle. "Is it about whatever it is that happened yesterday?"

"In a way." I went on with a rush, "Adam, I have to know the truth concerning your family before I can come to a decision about our marriage. I'm sorry if that bothers you, but I cannot help it. There are too many mysteries, too much we can't speak about. What chance has a marriage in such circumstances?"

His face got very still, and for a moment I thought that he was going to walk away. Then he said tightly, "What is it that you wish to know?"

"See?" I cried accusingly. My voice was trembling and I felt near to tears. "You are already making me feel guilty for daring to ask. Yet I cannot believe that there should be these dark areas between us which cannot even be discussed!"

For a brief moment he closed his eyes, and I thought that he was steeling himself for what lay ahead. When he looked at me again he said, "You are right, Beth. I will discuss anything you wish. What is troubling you?"

I took heart from the acceptance in his eyes. Taking a deep breath, I said, "First, tell me about Christian's mother. Emelda hinted yesterday that your uncle drove her to madness. What did she mean by that?"

His taut look relaxed somewhat. "Aunt Belle was always a little odd, I think. But when she was young she was very beautiful, and my uncle fell deeply in love with her. It wasn't until later that her fits of rage drove him—" Adam paused, then forced himself to go on. "He found someone to console him, Beth, a fancy woman who ran a gambling establishment in San Diego. He spent much of his time there, gambling and drinking, and it wasn't long before he had lost everything he had owned, including the family house and property in the city."

"And was it finding this out that unhinged your aunt's mind?"

Rather than look at me, Adam gazed across the emerald lawn to the grazing sheep.

"It was more than that. Aunt Belle had to be told a part of it and she found out the rest. She went to that woman's place—" He turned again to me. "I honestly think that she meant to kill the woman, Beth, but my uncle threw himself in front of his beloved and the bullet struck him, instead. He died almost immediately."

My imagination faltered at the thought of the tragic scene. "Oh, Adam, how terrible—for all of you! Was your aunt put in prison?"

"The case never came to trial, thank God. It was obvious to everyone concerned that my aunt was hopelessly insane."

"But—how is it that they allowed her to be brought here?" I thought of that desolate room upstairs. She must have been quite young, still, and to have spent hours, days, years, doing no more than gazing out at an empty landscape—"

"They didn't, at first. She was locked away in an asylum. But it was such a terrible place and she became physically ill— My father used his not inconsiderable influence to have her released to him, and he prepared that upstairs suite for her. It was the best he could do. The stipulation was that she was never to leave it unless the authorities declared her completely cured, and of course that didn't happen."

"How terrible," I whispered. "It must have been devastating for Christian. Was he allowed to see her?"

"Occasionally, when she would be relatively normal. He virtually lost both parents at the same time, which would have been beyond the understanding of any four-year-old. I suppose that is why my father indulged him so."

I imagined that lonely little boy, his life suddenly shattered. Yet he didn't seem to be seriously affected by it. Perhaps he had been too young for it to have had a permanent effect. "How long did this go on?" I asked.

"It was a blessing that she died before he was seven. I have no idea whether or not he remembers. He never talks of her."

I digested this. Tragic as it was, it smacked not at all of mystery and evil. Perhaps the other stories would be as simple and understandable, and I could put my fears to rest.

"And Marianne," I said, daring another question. "I know that she was stung to death by bees, for Rosemary told me. But someone must have lured her there. Do you have any idea as to who that was?"

Some of the shuttered look returned. "None whatsoever." He spoke abruptly, as though he wished the subject dropped. Trembling inwardly, I persisted.

"But the evidence was that she had a lover's tryst with someone. You are so isolated here. Surely there could have only been a small number of suspects."

"Suspects? Are you inferring that Marianne's death was not an accident?" Suddenly Adam's voice was cold as steel. I found myself covered with confusion.

"Why, I—" I stopped, unable to go on in the face of his hostility.

"It was obviously an assignation," Adam said. "None of us denies that. But that doesn't mean that the actual attack of the bees was planned by anyone." He shook his head, then went on grimly. "It was my fault that she died; it was I who put the bees there. I should have destroyed them long before."

We were sitting close together, and at that I reached over and laid my hand on his. "Why should you blame yourself, Adam? You didn't intend that anyone should be hurt by them."

Adam leaned back in his chair and shaded his eyes with his hand. "There are those who question that. There were those who didn't believe that it was an accident, and with them I was the primary suspect. Only two things were in my favor: one, that everyone knew it was I who had located the bees there, which would have made it incredibly stupid for me to have planned it. It was entirely too obvious. The second was that I had no motive, or at least none that anyone knew." He removed his hand from his forehead and looked into my eyes. "I am telling you, Beth, straight out, that I was suspected of murdering my own sister-in-law. Even Morgan thought that it might be true. Can you possibly imagine how that would make a man feel?"

I no longer wondered about the harsh lines of grief he carried. "But that is terrible, Adam! How could your own brother think such a thing of you? And what about Christian? Wasn't he under suspicion, too?"

"Somewhat. But he is considerably younger than I, and years younger than Marianne. And—he had Emelda, even then."

I subsided into silence. I could understand Adam's hurt and bitterness and I was loath to add to it, but still other questions remained to be answered. But before I could muster the courage to ask them, he said in a strangely rough voice, "Next, I suppose, you want to know about my father and Morgan's deaths. Apparently it is impossible for you to accept them, either, as unfortunate accidents!" Though I tried to protest, he went on bitterly. "You aren't alone in that. It is one of the reasons why I have been glad that you elected to stay here at Riversbend instead of returning to San Diego, for you would have been sure to hear the rumors. Especially if you were to announce that you planned to marry me." Adam again turned to look beyond the lawn to the river, as though not wanting me to see his face. He added, grinding the words out, "But evidently there are those even here in my own home who are willing to pass the rumors on."

"You are wrong," I protested. "No one has done anything of the sort. It is your own reaction whenever the accidents are mentioned that has caused me to question them."

He turned to stare at me. "Reaction?"

"Yes. A look of sorrow, regret, guilt—"

He asked dangerously, "Are you suggesting that I caused their deaths?"

"No, of course not," I cried, aghast at the idea. "I am only telling you that you must somehow *feel* guilty, for that is what is in your face."

"I see. Well, lest you think the truth to be worse than it is, I will tell you the rest of it. My father was found wedged beneath the paddlewheel in the river in such a manner that it was highly unlikely that he could have gotten there by accident. It looked as though someone had dived down there, dragging Father with him, and had shoved his body beneath the wheel. Need I say that I am the only Rivers who can swim?" This last was said with bitter sarcasm.

"And Morgan? What of his death? Was there also something questionable about it?" I was almost afraid to ask this.

"Nothing—except that he ordinarily never went near the roof cistern. It was my invention, and only I serviced it. Morgan had no reason to have been on the roof, let alone near the cistern. It is possible that someone asked him to come there, then pushed him in. He would have sunk like a stone. Water panicked every member of my family."

"But surely they didn't think that you—" I stammered. "Why, Adam? What reasons would you have had?"

"There is always a motive if one looks hard enough. What family is without its quarrels and its rivalries? It was generally known that Father and I didn't agree on the future role of Riversbend. I wanted agriculture, and he only grudgingly allowed me some experimentation. I was very frustrated by what I considered his shortsightedness." I was jolted by the sudden realization that as he spoke he was watching me closely. "As for Morgan—we never agreed on anything, and Father had left everything to his management in his will. Had Morgan lived, Riversbend would have become merely a country home, mortgaged to the hilt to subsidize our family shipping business. It was only because of his death that I was able to go on with my plans for an agricultural empire."

As I realized the full import of his words I felt the strength drain from me. "Then you were suspected of killing both of them." My stomach turned. "How horrible for you, Adam!"

His coldness had lifted, and he spoke almost too casually.

"There were suspicions, especially when one death came close on the heels of the other, but it takes more than that to arrest someone for murder. Neither, however, could I prove that it wasn't true." His tone took on almost a grim humor. "And now you may consider yourself fairwarned. If your answer to me is yes, you may be agreeing to marry a man guilty of the most heinous crimes. So you are right to think long and well about it."

When I started to protest, he held up a hand. "No, don't say anything just yet. I don't want you to tell me that you will marry me because of pity or compassion, or because you feel that you must show your trust in me. I couldn't stand to think that you were mine, and then have you change your mind." He stood up. "So take all the time you want to think on it. I am willing to wait. I have to be."

With that he went down the steps and around the house. I presumed that he was going to the stables. For a moment I envied him. How good it would feel to vent my own frustrations in a hard gallop across the golden fields!

Instead I continued to sit on the cool veranda. My heart and mind were in turmoil. Did I believe for a moment that Adam was capable of two deliberate murders, for whatever motive? My heart cried out in protest. It wasn't possible that I could love anyone who would cold-bloodedly take not one, but two lives. He had been so quick to come to my aid when Father had been killed, had, for the most part, been kind and gentle when I needed it most. And hadn't he generously offered the sanctuary of Riversbend, even before he fell in love with me?

And yet—

I thought of my arrival here, and of Adam referring to Riversbend as "the love of my life." And the day of our horseback ride Christian had indicated that Morgan had been the cause of much unhappiness for Adam, and had come between him and his father. Could a man commit murder for a piece of property?

I answered my own question. Of course he could. It happened all the time. And who is to say at what point bitterness causes to snap that slender thread between intense frustration and insanity?

I rose and started to pace the veranda, but before I had made more than a few turns my legs suddenly gave way and I had to sit down again. I didn't know what to believe, and unless I knew with certainty that Adam was innocent, I couldn't marry him. Feeling as bereft as I had when Father died, I began to weep with great, shuddering sobs.

My pain was so intense that I didn't hear anyone approach, but soon I felt a small body lean against me. Soft arms went around my shoulders.

"What is it?" Mark asked with shocked concern. "Has someone hurt you?"

I realized that Anne had come to my other side, and she began stroking my hair in a sympathetic and motherly manner. "Tell me who," she demanded fiercely, "and I shall tell them what I think of them!"

I tried to control my weeping.

"It's just that I may be going away soon," I said. "I have come to love all of you and I feel lonely when I think about not being with you."

Anne looked shocked. "Aren't you going to marry Uncle Adam? We thought that it was practically settled."

"And if he doesn't want to," Mark put in, "I'm sure Cousin Christian would do it. I know that he is interested because he is always watching you."

"That isn't a good enough reason," Anne remonstrated. "She has to love them. Can't you love one of them at least a little, Miss Allen? We did so want you to stay for always."

Their wistful words almost set me weeping afresh. I put my arms around them. "I'm glad that you want me, but sometimes there are reasons—" But it was too soon for that. I hadn't yet decided not to marry Adam, after all. "Let's just say that I haven't decided what to do. In the meantime, I am going to take a turn on the swing! Will either of you join me?"

Of course they both clamored to go, and pretending a gaiety I didn't feel, I raced with them down the hill to the oak.

"You first, Miss Allen," Mark said gallantly, "for it was your idea."

I took my place, and following our usual custom, allowed Mark to manfully give me a start. Soon I was soaring out over the river, welcoming the touch of breeze which blew into my face and dried the last of the tears lingering there. It also seemed to sweep some of the confusion from my brain and I told myself that all wasn't lost, that I didn't have to make up my mind about Adam now, this minute. He had said so himself. I decided to wait until after Rosemary's wedding, which was still some time away. Feeling relief that I had found an excuse to put the decision off, I relaxed and let myself be soothed by the swing's motion.

It was at the apex of my backward glide that I felt the swing give a little and heard the first splitting of the hemp fibers. I knew immediately what was happening and I shrieked with alarm and grabbed the right rope with both hands just as the left gave way. The wooden seat dropped, leaving me suspended only by the single length of hemp to which I clung. The rope slipped through my grasp, burning my hands, as I both swung and slid toward the children who stood, mouths gaping with screams, beneath me. Terror filled me. I might kill them with my fall.

"Get back!" I shrieked. "Get back!" I was hurtling directly toward them and I knew that I must be nearing the end of the broken strand. They scattered obediently just as the frayed part of the rope passed through my clutching fingers. I fell perhaps six feet to land heavily in the leafy debris beneath the tree.

Chapter Sixteen

Though I suffered from bruises and the rope burns on my hands, I was more shaken by my narrow escape than by the fall itself. It was obvious that had I not had that instant's warning to grab the rope, I would have fallen from such a height that I might well have been killed, a prospect that frightened me more in retrospect than it had at the time of its possible happening. But equally unsettling was the thought that occurred almost immediately: what if it had been one of the children instead of I who had been using the swing when it broke?

I said as much to Adam when he came to my room where I was resting after the mishap.

"My God, Beth!" he exclaimed when he entered in response to my "come in." "Are you all right? Rosémary told me what has happened. Shall I send to San Diego for the doctor?"

I thought he looked terribly distraught. His mouth, which I knew could be so soft, was compressed, and had a faint white line around it, and the harsh grief marks of his face seemed more deeply etched. I had a sudden desire to take his face between my hands and smooth those lines away, to soothe and comfort him. How could I possibly entertain the thought that he could be responsible for anyone's death or suffering? I gestured for him to sit beside me on the bed.

"Oh, heavens, no, Adam. That isn't necessary at all. Nothing seems to be broken. It couldn't be, for I climbed the stairs unaided." I didn't add that it had been painful, and that I thought I would be quite stiff and sore for some time. "I am sure no permanent damage has been done."

He sat beside me and took my bandaged hands into his. "Poor hands," he said. "I understand the rope burned them badly." He brought them to his lips and lightly kissed the backs of my fingers. "Shouldn't we have a doctor see them?"

They did still hurt, but the unguent Rosemary had spread over them before putting on the bandages had soothed them immeasurably. "That isn't necessary," I assured him again. "Your sister is an excellent nurse. I don't think any of you has appreciated Rosemary enough."

Adam continued to hold my hands, though carefully so that he wouldn't hurt me further. "You are probably right," he said. "I find that I am not looking forward to her going, though I wouldn't have believed that not long ago. And—" he looked down at his hands, which were holding mine, "the children will miss her also." He carefully didn't mention the fact that whether I stayed or went would have a bearing on this.

My mind returned to my accident and the horrible thought that it might have been one of the children who fell. In spite of the pain it caused me, I clutched Adam's hands.

"Adam—what if Anne or Mark had been in the swing?"

The white line returned around his mouth. "I have been thinking of little else. I know how much they enjoy the swing, but I'm not sure yet that I will ever repair it. And if I do, it will be held with a heavy chain instead of a rope."

"But what made it break? The rope was sturdy and quite new-looking. I examined it myself when I first used it."

He didn't meet my eyes. "It is hard to say. There was a rough spot on the underside of the seat. That may have worn a place on the rope and frayed it, fiber by fiber." He brought his eyes to mine. "But rest assured that it will never happen again."

I hesitated, wanting to pursue the subject, but decided instead to speak of other matters, for from his evasive look I saw more clearly than had he told me in words that there was more to the breaking swing than he had said. I also knew that he had no intention of revealing any part of it to me, and the uneasiness and faint suspicion I had been feeling earlier came back more strongly than before.

Not only was Rosemary greatly disturbed by my accident, she was vastly disappointed that I could not immediately accompany her to San Diego to shop for her trousseau as we had planned. But I really could not face the long, bumpy ride to the city in my bruised condition, nor would Adam hear of it, and at last we persuaded her to accompany him on one of his business trips.

"But how will I know what to buy?" she wailed. "I have no idea of the current styles. Surely they have changed in these past three years!"

"There will be no problem, Rosemary," I soothed her. "Put yourself in the hands of a good dressmaker. She will know the latest fashions, whether or not you do. And think what fun you and I will have sewing and embroidering all of your dainty underthings. I am sure my hands will be quite healed by the time you return."

"Well, all right," she said, though there was still doubt in her manner. "But it would have been much more fun if you could have accompanied me." Then, anxiously, "Are you absolutely sure you don't want me to postpone it?"

I assured her for at least the tenth time that I didn't, and at last she and Adam rode off in the wagon, with Rosemary waving until they were over the bridge and out of sight. I felt a vast relief when they had disappeared at last.

At the same time, I immediately felt lonely. Adam and Rosemary were to be away no more than three days, but now that they had really gone, the long hours stretched like an abyss before me. The swing was still broken, I could not ride as yet, and the rope burns on my hands made sewing difficult. There were the children's lessons, though, and I immediately suggested to them that we return to the nursery and begin.

"Oh, but it is so warm today, Miss Allen," Mark objected, "and so stuffy up there. Must we go in so soon?"

"We could bring our books to the veranda," Anne added. There was a wistful note in her voice. "There is such a nice breeze along here."

I thought about it. I wasn't really eager to return indoors myself. "That sounds like an excellent idea," I answered. "You two run up and get the books we have been working with, will you? The stairs are still a little difficult for me."

As the children ran to do my bidding, I eased myself into a nearby chair and closed my eyes. It was very quiet; the occasional bleat of a lamb was all that I heard. Even the usual bird sounds were silenced in the morning heat. I sank into a repose so deep it was almost sleep itself.

All at once a tingling sensation came over me, and my mind swam back from whatever watery depths it had been inhabiting. I opened my eyes to find Christian standing at the foot of the steps, watching me.

Whether it was because I was bemused by my near sonambulent state, or because of his unusual expression, I didn't know, but his sudden appearance didn't startle and anger me as it often did. Instead of his usual sardonic smile, he wore a wistful and appealing look, as a child might gaze at something he longs for and knows he cannot have. Quite absurdly, I wanted to comfort him.

"Have you been standing there long?" I asked. "You would be more comfortable up here in the shade."

"You wouldn't mind?" he inquired hesitantly.

Was I usually that unwelcoming? I felt ashamed and hastened to say, "Of course not. Though the peace of the moment won't last long. The children have gone upstairs for their books. It is such a warm, beautiful morning we have decided to do our lessons here."

He came up the steps and stood looking down at me. I felt that he hadn't heard a word I had said. But he must have caught a part of it, for he said, "It is you that is beautiful, Beth. Adam is a very lucky man. It is to be Adam, isn't it?"

He spoke so simply and looked so serious that I could

only answer the truth. "I haven't quite decided if I will marry at all, Christian. But if I do, it will be Adam."

He nodded. "I thought as much. As they say, the best man won." He sat down rather heavily in the chair next to mine. "You are making the sensible choice, no doubt. Adam is not the rogue that I am. Your life may be dull, but it will always be a safe one."

He still didn't understand. "It isn't that," I said gently. "I don't find Adam dull at all. And you could be all that he is, you know. In addition, you are handsome and witty, and charming, all of which unfortunately, you are already aware of. But your charm is a quicksilver thing which slips through the fingers, and nothing on which to base a life. Whereas Adam, aside from the fact that he *is* exciting to me, gives out an air of dependability, of caring. Things really matter to him, and he will treasure those that he loves. But you—" I hesitated, disliking to hurt him, but knowing it must be said, "you seem not to care very deeply about anything, and a woman necessarily feels that your interest in her might fade away once you knew she belonged to you."

"I see." Christian looked away, and then again at me with an intentness that surprised me. "Have you ever considered that my casual attitude might be a facade, put there long ago for safety's sake? I learned at an early age that to show my true feelings was to give a weapon into the hands of others. You might be surprised if you knew just how deeply I do care about many things."

I thought about that little boy, deprived of his parents and left with a family that wasn't truly his, and I wanted to tell him how well I understood. But just then Mark and Anne, out of breath and full of enthusiasm, came clattering onto the porch. They were both clutching books and papers, but I knew instantly that they had more than studying on their minds. His eyes sparkling with eagerness, Mark went to stand beside Christian's chair, and with a little feeling of shock I noted once again just how alike they were.

"Maria says that we may have a picnic lunch anywhere we like!" he cried. "And ice cream for dessert if we will help turn the paddle. Doesn't that sound like fun?" He grabbed Christian's hand. "You can come, too, Cousin Christian, can't he, Miss Allen? He is so strong the ice cream will be ready in no time!"

Christian threw back his head and laughed aloud, revealing an abundance of solid white teeth. "You see, Beth?" he said. "Others prize me if you do not, even if it is only for the development of my muscles. Will you allow me to join you in this treat, or must I be doomed to view this Eden, also, from afar?"

What could I say? I had to consent, or else break three hearts, though I remembered my governess role long enough to say sternly, "All right, then, but only after you have done your work. And you, Christian, must attend to your own duties and not return to bother us until at least twelve o'clock."

Christian pretended a great reluctance, but obediently got up to leave.

"Wait!" Anne cried. "We haven't decided yet where we are going!"

"It will have to be somewhere nearby, I am afraid," I said apologetically, "for I am still too stiff and bruised to travel far."

"Leave that to me," Christian said mysteriously. "I know just the spot."

After that it was hard to get the children to settle down,

but we managed to get a little work done, at least.

By the time Christian had triumphantly returned, this time driving the buggy, I had begun to doubt the wisdom of my spending so much time with him. But it was plainly too late; he had the picnic hamper strapped on the rack and the ice-cream freezer affixed alongside it. As he leaped from his perch and came to the porch, I saw that his pensive frame of mind had gone and was replaced by seeming jollity. He was nearly irresistible when in this mood and I determined to keep Anne and Mark near at all times.

"Children, take the books and papers in and put them on the hall table," I told them, "for they may blow away if we leave them here."

Christian clapped his hands to hurry them. "Be quick! Maria says we must get to the ice-cream-making as soon as possible."

He certainly met with no argument. Within moments we had all scrambled into the buggy and were heading out the road leading through the orange groves. Mark and Anne were chattering gaily, and over their heads Christian's dark eyes met mine. He smiled indulgently, and for a moment I had the oddest sensation that we were father and mother, traveling together on an outing with our children. Quickly I turned from him and gazed out at the landscape.

"Where is it that you are taking us?" I inquired lightly. "If Anne and Mark aren't curious, I certainly am."

"It is just a quiet spot by the river. The children have been there before, but I had time to make it a little special."

I noticed he was keeping the horse to a slow walk and was grateful for it, for even with the padded cushion, so much more comfortable than the wagon, I could feel it all through my body each time we went over a rough place. "How far is it?" I asked, trying not to sound as though I were complaining.

"Not far. We could have walked, except for the basket—and your condition. The longer we take to get there, the farther it will appear to be—" the laughter was in his eyes again "—and the more exciting."

He seemed to be enjoying himself every bit as much as the children. I reflected that he was always his best when with them. I supposed it was because they accepted him just the way he was.

We soon came to a stopping place, and Christian tethered the horse, then lifted me down as though I were highly breakable. I was a little stiff from just that short journey, and it was with a slow and careful pace that I followed Anne and Mark down the sloping path toward the river. I didn't object when Christian took my arm in a protective gesture. Ahead, the children were exclaiming delightedly, and curiosity caused me to hurry. As Christian and I reached the edge of a small shady clearing at the edge of the water, Mark came running to meet us.

"Look, Miss Allen," he shouted, "isn't this nice?"

It was, indeed. In fact someone had gone to a great deal of trouble. The ground had obviously been raked and evened, and an oriental rug had been laid down. On it were three wicker chairs and a padded rocker, and a small table held the accountrements for a light buffet. There was even a napkin-swathed wine bottle protruding from a silver bucket. I drew in my breath and turned to Christian. He was smiling with all the mischievousness of a small boy.

"Well," he demanded, "aren't you surprised? I did it just for you, you know."

"You shouldn't have," I said severely. "You wasted time preparing this when you should have been working."

He had no trouble seeing beyond the severity of my tone to my pleasure. "I can't say that I did it all with my own hands, exactly," he confessed, "for I brought Manuel and Juan with me. But the idea was mine. Now, admit it. Doesn't that rocker look far more inviting than a seat on a hard rock or a thin blanket right on the ground?"

I had to admit that it did. And once I was settled in its comfortable depths, it was pleasant to lean back and idly watch while the others brought first the picnic hamper and then the ice-cream bucket to our little camp in the the clearing.

"I have had definite instructions about this ice cream," Christian announced after he and Mark had managed to position the bucket to their liking. "Maria said that we must first turn the handle until it would turn no more or until our arms fell off. Then we must let it sit and ripen while we have our lunch and a time of rest. And then—oh, glorious moment!—we may open it up and gorge ourselves on the delectable peach ice cream that I am told will be inside."

After the proper moans of anticipation and licking of lips, the three of them got down to work. At first they all took turns, though Mark began to perspire profusely and Anne's fair skin became alarmingly red, but gradually the children's revolutions of the wooden handle became fewer and fewer, and I was relieved when Christian took pity on them and finished the job himself. When even he could budge the handle no more, he wrapped the bucket in an old blanket and came to throw himself down at my feet on the rug.

"That ice cream had better be excellent," he remarked. I saw that his suntanned forehead was lightly beaded with perspiration. "I will think twice before volunteering so quickly next time."

"It no doubt did you good," I returned serenely. "It is time you put your energies toward something useful."

"Ah! That *is* a blow! And here I have been working so hard to impress you these past weeks."

I detected a serious note beneath his banter. "And you have, Christian," I reassured him. "And I believe Adam is impressed more than he will say. He is actually coming to depend on you, I think."

He sat up quickly. "My dear Miss Allen, I know you mean that kindly, but it is hardly Adam I am attempting to impress. It is you I have turned the new leaf over for, you whom I wanted to assure that I might be good husband material, after all. If I have failed in that, don't throw me the sop of telling me I am appreciated by Adam!"

In the face of his obvious bitterness, I didn't know what to say. "I am sorry," I managed at last. "I do admire your efforts, really I do. Someday you will meet a woman who feels about you as I do about Adam, and then—" I stopped, the words hanging helplessly in the air.

"Are you sure of your feelings for him, Beth? How much do you know about my dear cousin, actually?"

"Why—what do you mean?" I had no intentions of revealing my inner doubts about Adam, but Christian's echoing of the fears and questions already in my mind was profoundly disturbing. I heard my voice go higher as

I protested, "What is there to know that I haven't learned already? He is kind and gentle and considerate—"

"And he is moody and sometimes irascible, or haven't you seen that side of him yet? I saw it often as a boy. It is not always easy to live with."

I thought of the day of my father's funeral. "Yes, I have seen that side of him," I admitted. "But we all—I mean none of us is perfect, Christian. We all have our bad days and our good."

"The difference is that Adam can't accept that, as most people can. He must always be perfect, always had to be, even as a child. But beneath that iron control there is something else, Beth, something that even he is afraid to unleash. You may not have seen evidence of it as yet, but it is there."

I had only seen Adam deeply angry once, and that had been my first night at Riversbend, when Christian had spoken unfeelingly of Rosemary before the children. I had not blamed Adam then for his anger, and had even mentally encouraged him. But Christian's words struck fear through me. "Why do you say that, Christian?" I spoke through lips that quivered slightly. "What is it that you are trying to tell me?"

Just then the children came rushing up from the river, where they had been playing.

"Can't we eat?" Mark cried. "I am starving, and Anne says she is, too. Besides, the ice cream must be ripe as a watermelon by now!"

"I am famished as well." Christian got up quickly and brushed off his trousers. "You children put the food on the table, and then Anne shall serve Miss Allen. That leaves me with the arduous task of opening the wine." For a few moments he concentrated totally on this important duty, and I thought he had forgotten my question. But as he handed me a glass of the ruby liquid, he said quietly, "I meant nothing by it, Beth. I had no right nor reason to say what I did. Just put it down to the ravings of a very jealous man."

I did try to put it out of my mind, but I found it not quite possible. And though the children seemed not to notice it, the picnic was not a total success after all.

This time Adam's return didn't afford us the happy and fulfilling meeting that it had before. Adam still did not know where he stood with me, and I could give him no clue by my reaction, for in truth I didn't know myself. And Christian's remarks during our picnic at the river had only resulted in my being now more confused than ever.

Our uncertain feelings were hidden, fortunately, by the clamor of Rosemary's return. She was bubbling with excitement and laden with purchases of every description, and as she chattered gaily, relating all that she had seen and done in San Diego, she managed to slip Peter's name at least once into every sentence. It seemed that he had wined her and dined her both nights that she was there, and had presented her to all his friends, both old and new, as though she were a newfound treasure. And in between the wardrobe fittings and shopping expeditions the happy couple had even managed to look at a house Peter had thought might be suitable as a home for them after their marriage.

"Oh, it is beautiful, Beth," she said. We had repaired to her room, where all the packages had been piled on her bed. "While it is not as big as Riversbend, it is large enough for guests and parties and," she blushed a little "any number of children. Peter says that we will entertain a great deal, and of course we must, as his position calls for it. Not that I mind," she added happily "for I am really gregarious by nature. That is why I was so unhappy when Father brought us to Riversbend. But that is all over now, and once Peter and I are married will never think of it again."

"I am so happy for you," I said sincerely. "You are like a different person since Peter came back into your life."

A little of Rosemary's exuberance left her. "And what about you, Beth?" she asked wistfully. "You seem like a different person, too, only with you it isn't an improvement. Even though you had just experienced the tragedy of Jamison's death when you first came here, there was a light, a sparkle about you. But now it is as though a shadow had been cast over you. Is it—has it something to do with you and Adam?"

I wanted to pour out to her my doubts, my fears, but obviously couldn't speak of such things to Adam's own sister. So, much as I longed to take her as a confidante, knew that I must keep my own counsel and try to appear happy, at least until after the wedding. So I attributed my subdued manner to the remaining aches and pains caused by the fall from the swing. "But my hands are much better now. See?" and I held them out for Rosemary's inspection. "The rope burns are almost gone, and I can start immediately to embroidering your lingerie. And you can tell me all the details of your wedding gown and of all the other things you have ordered for your honeymoon trip."

Her fears for me allayed, Rosemary immediately plunged into detailed descriptions of her ordered finery and of the trip she and Peter were to take. And of course she had to show me all the purchases she had made. She didn't again allude to the question of how things stood between Adam and me.

Nor did Adam and I allude to it. I supposed he remembered his vow not to press me, or perhaps he could sense my indecision and feared to hear my ultimate answer. And I—I still had no idea of what that answer should be. And because of the excitement of the coming wedding, it was easier than it might have been for us to skirt the subject.

The children quickly recovered from their shock at seeing me fall, but they remained desolated by the thought that I might possibly go away. I couldn't honestly reassure them that I wouldn't leave, but I tried to spend as much time with them as I could. And even though I was quite busy with Rosemary's trousseau now that my hands were healed, I promised that one day soon I would go riding with them. Adam had been allowing them to ride alone as long as they kept the ranchhouse in sight at all times, which I thought would prove a good rule for me as well. But I had not yet kept my word to go with them the day they came rushing into the nursery after their daily ride. Their eyes were starry and grins of pure excitement stretched across their faces.

"Miss Allen!" Mark was the first to reach me. "You must come riding with us. We have found such a wonderful, secret place! We want to share it with you!"

I put aside the dainty petticoat on which I had been embroidering.

"It sounds fascinating!" I exclaimed with dutiful enthusiasm. "Can you give me just a hint of what it is?"

"Oh, that would spoil it!" Anne seemed as elated as her

brother. "It is something that you must see for yourself. Do you think you might come with us now?"

It was late afternoon, and though I hated to dampen their excitement, I knew that Adam would disapprove. I told them that, then added, "But if you will wait until after lessons tomorrow, I think that I can persuade Maria to pack a small lunch and perhaps some lemonade. That way we can have a picnic while we are out."

Though hard put to contain their disappointment at the delay, they were cheered by the prospect of the luncheon adventure. "Though we won't need the lemonade," Mark declared. "For there's—"

"Shh!" his sister interrupted. "Do you want to give it away?"

I concealed a smile. It was probable that they had discovered the hidden spring Christian had taken me to and were just as delighted as the cousins had been as boys. Well, I determined to prove a most appreciative recipient of their secret.

"Don't do that," I cautioned. "I would much rather be surprised. In the meantime, hadn't you better change your clothes and get ready for dinner?"

They ran off obediently and I took up my sewing again, feeling both pleasure at their happiness and an increasing sadness at the thought of leaving them. But I was afraid that I must eventually do just that, for every day my uneasiness concerning Adam was growing. It was not just his evasiveness about the accident, though that had seemed to set off some alarm in me that couldn't be quieted, but that I was at last looking squarely at something which I had been assiduously avoiding; the deaths in the family, whether by accident or intent, had served to eliminate those people who owned, or stood to own, a portion of the estate of Riversbend.

The fine embroidery blurred before me and I dropped the sewing into my lap. Other than the two men, only Rosemary and the children were left to inherit, and unless Rosemary married, her share would remain in the family. I caught my breath. Could that have been the reason for the anonymous letter which broke up her engagement? For not only was there the possibility of progeny, but it was highly unlikely that an astute lawyer such as Peter Barnes would allow his wife's portion of the estate to be finagled away from her.

I tried to continue my sewing, but my hands were shaking so badly that I had to put it away. If the mishap with the swing had not been an accident, it would have been Anne or Mark, then, not I, who was the target. And if that were true, they continued to be in grave danger.

Horror washed over me and I got up and began to pace. What could I do to protect them? I could hardly approach the authorities with such a preposterous theory, nor could I go to either Adam or Christian unless I knew with certainty it was the other who was guilty.

For it had to be one of them. No one else who stood to gain was physically capable of causing the deaths or my accident. That left only Rosemary for me to confide in, and to ask her to believe without some proof that her cousin or her brother was a murderer was unthinkable. With a sense of despair I faced the fact that the only thing I could do for the moment to protect the children was to stay as near them as I possibly could.

That night I slept most uncomfortably on the settee in the nursery.

Chapter Seventeen

Lessons were done sketchily the next morning, since neither the children nor I had our minds on them, and soon we were into our riding clothes and downstairs to pick up the lunch Maria had been instructed to pack. As she handed us the package her usually smiling countenance had a slight look of concern.

"Forgive me, señorita, if I am too forward, but I hope that you do not plan to stay out long with the children. Though I know it does not look so at the moment, I am sure a storm will come soon." She smiled, though the concern did not leave her eyes. "I have a little of the rheumatism, as you call it, and it seldom lies. My knee has been speaking to me all of the morning."

I tried to reassure her. "Don't worry. This secret place of the children's can't be too far away, as they have been told never to go beyond sight of the house." We exchanged smiles of mutual understanding. "And I promise to watch the sky carefully and return at the first sign of rain."

But after we had gotten our horses and were heading

toward the river, I thought that she must be mistaken. The hot September sun beat down from a cloudless sky and the dusty river trail looked as though it had never experienced so much as a shower.

"Where are you taking me?" I asked, genuinely bewildered. I had expected that we would immediately start across the hills as Christian and I had.

"We are leading you by a roundabout route," Anne explained. Her eyes were shining with unaccustomed mischievousness. "Mark and I have decided that we mustn't let even you know the way to our secret place."

"Oh, I see," I said, laughing. "Well, you have little to fear. Even if you took me directly there, I would probably never find it again."

And it was true that by the time we had left the river path to cut through some rocky outcroppings of low hills I was confused and could greet the golden hill that bore the large oak stand with genuine surprise. Carefully, I didn't look back to see the house which I knew lay in plain sight behind me.

As it had been on the day I had come here with Christian, the thick, cool shade beneath the trees was most pleasant, and it was with real enjoyment that I allowed the children to lead me to the tiny spring. When I had admired it and drunk of its cool and delicious water, I exclaimed with delight and praised them for their cleverness in discovering it.

"But that isn't all!" Anne's eyes were bright with anticipation. "We have another surprise even better than this!"

"Hush, Anne, you will give it away!" Mark scolded. He added with amusing maturity, "I know it is difficult, Miss Allen, but you must wait until lunch is over for the second half of our surprise."

I sighed with apparent despondency. "Very well. But I am sure I won't be able to eat a thing for curiosity."

The reversing of our roles tickled their fancy, but out of consideration for my impatience they laid out our small lunch immediately. We sat on the soft, grassy area as we ate a sandwich apiece, drank the cool spring water from cups which Maria had thoughtfully provided, and munched an incredible number of molasses cookies. Then, after carefully shaking out the oiled packet which had contained our meal, Mark ordered excitedly, "Now you must close your eyes until we have brought the other surprise and placed it in your lap."

I obeyed, not without some trepidation. Would it be a small animal or horrible crawling thing which I would undoubtedly greet with disgraceful shrieks of horror? Then, remembering Anne's sensitivity, I relaxed and as I waited inhaled with appreciation the pungent scent of the oaks' foliage.

I did not have to endure the suspense for very long. Within moments I heard the children return and felt a heaviness as they poured several small objects into my lap.

"Now!" Mark shouted. "Now you can look!"

I opened my eyes. There against the dark material of my riding habit lay several pieces of ornate jewelry, a large diamond stickpin, and a small, pearl-handled jackknife.

There was also a man's gold pocket watch. As I saw it my heart moved sickeningly in my breast, and with trembling fingers I reached to turn it over.

It had an unusual design of ivory inlay on its case.

Instantly I was back in Father's room the night he

died, and he was consulting his watch to hide the embarrassed pleasure he had felt at my kiss. He had had that watch with him when he left the hotel and his murderer had taken it along with everything else he had had in his pockets.

I felt the world spin around me.

"Miss Allen!" Anne screamed. I heard her add, "Mark, get water! I think she is fainting!"

When I returned to consciousness, I was lying full length on the grass and Anne was bathing my face with a small square of handkerchief which she had dipped in the spring water. Immediately I struggled to a sitting position.

"Oh, do you think you should sit up just yet?" Anne asked anxiously. "You are so pale it frightens me."

I attempted to smile, for she was white and Mark was so upset he looked near to tears. "I am quite all right," I lied. "I imagine it was just a delayed reaction to the sun. I did get quite hot on our ride here." With a shaking hand I reached to gather the spilled treasures from the moss. "It is so silly of me to spoil your surprise this way."

Anne continued to regard me anxiously while Mark gathered the remainder of the jewelry and returned it to my lap. I picked up the watch again and found my fingers closing convulsively around it. "Where did these things come from, children?" I asked as casually as I could.

"It—it is a secret place," Mark said, mumbling a little. "We swore an oath not to reveal it."

"Oh, bother the silly oath!" Anne exclaimed. "It was just a game, anyhow! They were in a hollow in one of the trees over there." She gestured toward the edge of the grove and added in a penitent tone, "Mark found them. I really shouldn't have told, since they are his now, but I thought that something about them had frightened you."

She waited, her clear, intelligent eyes scanning my face. I said slowly, "Well, it did, a little. They look to be expensive, and I wondered how you came to have them."

"They aren't much use, though," Mark pointed out. "There are two earrings, but each is only one of a pair. And the knife seems to be rusted shut. I expect they have been in that tree for a very long time."

Not all of them, I thought. My father's watch seemed to pulsate in my hand.

"Nevertheless," I said, "the stones in the earrings look as though they are worth a good deal. And those are real diamonds in the cufflink and stickpin." I glanced into Mark's intent face. "I know it is asking a great deal, Mark, but may I keep these for a little while? They might belong to someone in the family and we should try and find their rightful owner."

Mark manfully fought his disappointment. "All right. But anything you don't find the owner of belongs to me. Isn't that fair?"

"Very fair." I was relieved that he conceded so willingly. "And may I ask one thing more? Please don't even hint of this to anyone until I have had time to do some detective work. Agreed?"

Mark's eyes lit up at the word detective, and even Anne seemed happy to think the game would go on a little longer. So in the end we bundled the things into the oilskin pouch and I concealed them in my pocket. After resting a short while longer we mounted up to return to the house.

When we rode out from the grove of trees I received

another shock. The sky, which had been clear when we entered, now held ominous black clouds that hovered threateningly just to the north. Guiltily I remembered my promise to Maria.

"We had better hurry," I said, "or we will be in for a wetting." I started down the slope directly toward the house, whose red tile roof could be easily seen in the distance.

"But we must return the way we came," Mark protested, "or everyone will know our secret place." He spoke with an unusual stubbornness and I could see that he was near to tears. I considered the matter. It probably wasn't that much farther than the direct route, and the threatening clouds seemed not to be moving nearer. With some reluctance I agreed to continue the game.

We had just reached the river when the rain began. It was preceded by a flash of lightning that seemed to split the sky, and the thunder that followed it startled our horses and sent them dancing skittishly along the trail. My mount, a normally gentle bay, reared frighteningly, and I screamed and clutched the pommel. Even Anne and Mark, much better riders than I, seemed to be having trouble controlling their ponies, and there was no need for my shouted, "Hurry, children!" to speed them on their way. As I prodded the bay with unaccustomed fierceness, his tawny hide was already turning dark from the great splatting drops of moisture which were pelting us. With another hard kick I sent him after the racing children.

I had nearly caught up to them when I heard the ominous rumbling that at first I took to be more of the rolling thunder. But this time it had not been preceded by lightning, and instinctively I glanced up at the precipitous slope just to the right. Spilling over its lip were several of the jagged white rocks which jutted from the ground of the area like the bleached bones of some prehistoric monster, and as I stared in horror they hurtled toward us.

The next moments seemed frozen in time. I saw instantly that Anne and possibly Mark, who was just behind her, could get to safety before the slide reached them, and I, if I stopped in time, would also be safe. I reined in so abruptly that the bay rose straight in the air and nearly went over, but I hardly noticed, for my eyes were glued to the trail ahead.

Mark, too, had looked up and seen the slide, and had reacted by reining in for one brief, deadly moment. Then he spurred his pony desperately, but it jumped forward with such force that he was unseated, and as I shrieked with terror, his small body hit the slick, muddy ground beside the trail and went tumbling into the swollen river. In an instant his dark head bobbed to the surface and I thought I saw him grab for a floating branch, but I could not be sure.

Terror filled me, and I was vaguely aware that piercing shrieks were emitting from my mouth, but for the moment I could do nothing to help him. The huge rocks were thundering across the trail before me to crash with tremendous force into the water, and I dared not move until the slide had stopped. Moreover, Mark was by now far downriver and even if I could catch up and jump in to him, in the heavy outfit I was wearing, I knew I would sink like one of those stones. Nevertheless, the moment the roar of falling rock had ceased, I raced down the path after him.

Before I had even caught up to Anne, someone on a

black horse clattered and slid down a crevice in the rocky hillside ahead of me. I caught my breath, sure that the horse would go over and that the two of them would slide into the river as Mark had, but by what seemed sheer force of will, they stayed upright until they reached the trail and went racing along it in Mark's wake. But even at that speed, and though they were far in front of Anne and me, I was sure they would never come abreast of the boy.

Just at the height of my despair the branch Mark was still clutching caught in an eddy near the river's bank. It held for only a moment or two, but it was long enough to allow the man to pull even, leap from his mount, and hurl himself with a long, flat dive into the river a few feet downstream. As Mark and his branch were dislodged from their momentary haven, the man caught the boy in one hand and with a couple of powerful strokes pulled him to the bank, then, rising, scooped him up into his arms.

As Anne and I thundered toward them the man turned to face us, his rakish grin gleaming through the muddy water still pouring from his face.

I saw with an incredible sense of shock that it was Christian.

Mark seemed none the worse for his terrifying experience. In fact, once he had been assured that his pony was safe and well, he seemed quite proud of his adventure. "I held my breath just like you taught me, didn't I, Uncle Adam?" he kept repeating when, after being given a hot bath and an early dinner in the nursery, he had been allowed to join the grown-ups downstairs.

"And I held onto the log the way you had us do that day in the reservoir."

"You did very well, Mark," Adam answered. "I am very proud of you, and thankful that you are all right."

I glanced sharply at him. His words were all that they should be, but his manner was constrained. I could not help but wonder why.

"Well, my buckos," Christian said with a laugh, "don't I get any of the credit? Mark is the finest fish I've hauled out of that creek in a long time." His dark eyes sought mine, but I looked away. I could not forget that skillful, glancing dive nor the ease with which he had stroked the short distance to the bank with Mark.

"It was very brave of you." Anne spoke from her place beside me on the drawing-room couch. Her eyes were round with admiration. "Especially since you don't even know how to swim."

"It's my opinion that the only thing water is good for is to take a bath in," Christian bantered, "and my experience today hasn't changed it a whit. But I never thought of that when I saw Mark fall in and go sailing by. I only knew that I had to do something and quickly."

"Thank God you did," Adam said. "There might have been more than one lost if Anne or Beth had tried to save him."

I knew he was looking directly at me, but I could not meet his eyes any more than I could Christian's. I was too acutely conscious of the thin chain around my neck, and of the trunk key which hung from it and lay like a burning ember between my breasts.

"How did you happen to be out there, Christian?" Rosemary's eyes were red from weeping, and even now an occasional tear crept down her cheek. I think she had realized for the first time just how much the children meant to her.

"Emelda told me that Maria was worried because they hadn't returned," he explained easily. "What with the storm breaking I thought it would be a good idea if I rode out to meet them." As if to punctuate his words a jagged flash of lightning could be seen through the veranda windows. Anne shrank against me and covered her ears against the crack of thunder that instantly followed it.

Immediately after, we heard a pounding on the heavy outer door. Adam had already risen when Maria entered.

"It is a man from the town, Señor Adam," she said. "He wishes to speak with you on the veranda."

"Good heavens, didn't you ask him in?" Adam demanded.

"Of a certainty I did, but he refused," she said with dignity. "He is very wet, and he does not wish to muddy the floor. He says he must return at once to the city."

Adam strode past her. "At least get him a hot drink before he goes. I'll see what could be important enough to bring him out in this weather." He went on into the entrance hall.

After a few moments he returned, and at our inquiring looks exclaimed with a show of impatience, "It was just a matter of business. It could well have waited until the weather cleared." He returned to his chair, but I thought that his manner after that was preoccupied, and Christian, while continuing his jesting conversation with the children, eyed him with some speculation.

It was a relief when Mark and Anne, who had chosen to share his nursery dinner, were ready to return upstairs, for we could all then stop pretending a gaiety we did not feel. Mark's close brush with death was, I supposed, the reason for the generally quiet mood, but it was more than that for me, and it was with exquisite relief that I excused myself directly after dinner and went upstairs. I reminded everyone that I had promised to tuck the children in, but the truth was that I could not face an evening of sitting with the family in the intimate confines of the drawing room, knowing as I did that one of the two men sitting with me had to be the murderer of my father.

The very idea was monstrous. And yet the evidence was there, locked securely in Father's trunk along with the other contents of the oilskin pouch.

Before going to my room, I stopped in to see Anne and Mark, as I had promised. Instead of being in their separate beds, Anne had gotten in with Mark, and as I entered, their little faces looked up anxiously at me.

"Anne may stay a little while with me, mayn't she, Miss Allen?" Mark asked. His voice broke only slightly at the words. "I am not really frightened, you see. It is just that when I close my eyes I keep seeing the water all around me, and I—"

"Of course she may stay, Mark," I said soothingly. "It was a frightening experience for all of us. You are very dear to us, and we are so thankful that God spared your life." I leaned down and kissed his round cheek, then touched my lips to Anne's pale face.

"God and Cousin Christian," Mark murmured. The thick lashes were already beginning to flutter closed.

My heart contracted. "Yes, of course. And Cousin Christian." I touched each of them softly once more and went on to my room, wondering as I did so if I should return later and sleep on the settee again tonight.

But once I reached my room, I sank into my rocker, feeling drained, exhausted. I had to do something, decide on a course of action, for I couldn't always be with the children to guard them. And today my being with them hadn't been of any help at all.

Terrible thoughts tumbled chaotically through my head. Had Adam killed Father out of jealousy over the tract of land and then asked me to Riversbend so that he could court me and thus acquire it for himself? Or had Christian done it because of his anger over the dressing-down Father had given him? This last seemed preposterous; surely no one took a life for so small a reason as that.

And yet—I shivered as the thought crept unbidden into my mind—what is one life if you have already taken others? And I knew with absolute certainty that I no longer believed the family tragedies to be accidents, but cold, premeditated murders, done for the purpose of eliminating one by one the heirs to Riversbend. The breaking swing was an attempt at another, and just possibly those rocks that had hurtled down on the children and me had been deliberately dislodged. Had Christian done that? He had been on the ridge above us, as his sudden appearance proved. But why, if he had been the one who caused the slide, had he risked exposure to save Mark?

There was only one answer to that. In this case it was not the children, but I whom he had meant to kill.

Icy terror struck the pit of my stomach, and I felt moisture break out on my forehead and upper lip. I leapt from my chair, wanting to flee, wanting to run to someone and blurt out my suspicions, wanting to do anything except stay in this place where death might strike me at any moment.

But before I had even reached the door I forced myself

to return to the chair and sit down. There was more than my own safety to consider before I could take action of any kind. Rocking slowly back and forth, I tried to sort out my thoughts.

I could be wrong about my being the target of the last murder attempt. Anne and Mark were, after all, future heirs, and would have to be gotten out of the way sometime. But if Christian had killed his uncle and the others in order to be the sole inheritor, then why would he risk his own life to save Mark?

Immediately my suspicions shifted to Adam. He could certainly have caused the swing to break, and no one had asked where he was when the boulders had come tumbling down. Or had the stones actually been loosened by the rain, as was the general conjecture?

I put a hand to my throbbing head. How could I be expected to sort out such a tangled web of possibilities? If only there was a single instance where the facts pointed irrefutably to one man or the other! But as things stood, my confiding in either could be a fatal mistake, and it wasn't only my own life but that of Anne and Mark that would be forfeit. Until I found convincing proof, I couldn't take the risk.

I sat quietly, trying to smooth my troubled mind so that some thought, some clue, might rise to the surface. A picture of the jewelry Mark had found rose to confront me. Either Adam or Christian must have hidden it there in the tree. But what about the cufflink and earrings—so strangely one each of a pair? Their disappearance had to have caused a good deal of distress for their owners. I began to wonder just to whom they had originally belonged.

The idea, when it struck me, was so incredibly simple

that I wondered why I hadn't thought of it before. After thinking it through for a moment, I got up and crossed to Father's trunk, lifting the chain from around my neck as I did so. Quickly I unlocked the trunk, took out the oilskin packet, and removed Father's watch. After adding it to the chain, I slipped the packet with its jewelry into the pocket of my gown and went to the door. I opened it carefully and peered into the corridor, and upon finding it empty, hurried to Rosemary's room.

To my relief, she had already come upstairs. She looked astonished when, after entering her room, I swiftly locked the door behind me.

"What is the matter?" she demanded. "You are white as a sheet!"

"Rosemary, I feel that we have become friends," I began desperately. "Can I show you something and swear you to secrecy, even though it may involve members of your family?"

She looked as though she suspected that my afternoon's experience had deranged my mind, and quickly I got out the oilskin packet and opened it. "I want you to tell me to whom these pieces of jewelry belong, if you know." As she moved closer I poured the contents of the pouch onto a nearby table. She stared at them, then with a sudden exclamation picked up a pearl earring and examined it.

"Why, where on earth did you get this? It belonged to my mother, one of a pair of hers that Father had given to me." She stared at me as it swung from her thin fingers. "It disappeared years ago. I was heartbroken and searched for it for weeks."

"And the others?"

She spread them out on the embroidered tablecloth

"This diamond stickpin belonged to my father. It is nearly two carats. I remember him threatening to sack the servants when it disappeared. And the topaz cuff link is one that Morgan lost before he went away to college. The others—" She touched them with a tentative finger. "The emerald eardrop might have belonged to Marianne, though that would have disappeared much later. I recall that she had a pair much like this."

"What about the knife?" I whispered.

She frowned at it.

"As I recall, it belonged to either Adam or Christian. But which one I cannot quite remember. If it is important, can't you just ask them?"

I sagged with disappointment. While wondering about the owners of the jewelry I had realized that the perpetrator of this secret cache might have been engaged in an act of childish vindictiveness, taking something precious from each member of the family, or someone against whom he had a grudge. It would be the boy, now a man, who had stolen them. Since the knife, according to Rosemary, belonged to either Christian or Adam, the man who had *not* owned it was the man whom I could not trust. I had counted desperately on Rosemary's being able to tell me who that was.

She was waiting for my answer.

"No, I cannot ask them, Rosemary. And I cannot explain it, either. Please trust me. Right now I can only tell you that if it were known that these things had been found my life and possibly those of the children would be in danger." As an afterthought, I added, "And yours, too, now that you know about them."

She blanched. "Then I was not totally wrong." Her voice was hard as marble. "The deaths in this family *were* caused by a curse—but a human one." She took my arm in an iron grip. "Who is it, Beth? You must tell me; I have a right to know."

I saw that I must reveal to her all that I suspected. Or almost all. "It is either Adam or Christian, Rosemary. I am sorry, but I don't know which of them is guilty."

She recoiled as though I had struck her. "Oh, dear Lord, not Adam!" Then, "I can't believe it is either of them, Beth, but I know you wouldn't make such an accusation without reason."

I couldn't yet bring myself to tell her about Father's watch. "Believe me, I do have almost certain proof that it is one of them. But right now the important thing is that I get the children out of danger. I must take them away early tomorrow, before the men are up and about. Will you come with me, Rosemary? We can take them to Peter and ask him what is to be done."

"Oh, heavens, in this weather?" she protested. Then she caught herself. "Yes, I see that we must hurry and get them to safety. But what about tonight? Should we lock ourselves in the nursery with the children?"

I considered the idea, then reluctantly rejected it. "What if someone were to come upstairs to check on them? Maria, even, or Emelda? They would be sure to report it, and it would arouse suspicion. I think they will be safe enough tonight." I touched her cold hand. "The deaths have always been made to look like accidents in the past, and I expect that the murderer must continue to do that. But be careful and lock your door, Rosemary. You, too, are an inheritor of Riversbend, you know."

The house was silent as I crept along the hall toward my room, stopping only momentarily to see that the children were sleeping soundly. They were, though they looked somewhat cramped, for Anne was still sharing Mark's narrow bed. As I gazed down at their innocent faces I hesitated, wondering if I should ignore my own advice and spend another night on the settee in the nursery, but finally decided against it. If I were, after all, the murderer's target, I might be bringing danger directly to them.

I roused Anne gently and returned her to her own bed, where I thought she would rest more comfortably. She scarcely woke, and settled down with only a small sigh of relief. I left her bedroom and swiftly ran the few steps to my room.

Just inside the door I stopped, shock draining the strength from me. My father's trunk was standing open, and when I at last went to examine it, I found that the lock had been forced. I went first rigid with fear, then limp with relief. If I hadn't taken the jewelry with me to Rosemary's room, he would surely have found it. I touched the chain around my neck. Had he seen it and ascertained the reason for its being there?

What chilled me most was that he had not even bothered to conceal what he had done.

Chapter Eighteen

The storm raged through the dragging hours, its fury adding force to my own inner tempest. But at last the long night ended and I gave up my tossing and turning and rose to dress rapidly in the chilling dawn. As I started to the nursery, I went over my plan of action for the hundredth time. As soon as I could get the children dressed we would, with Rosemary, slip out of the house by the back way and run to the stable. Once the horses were saddled we could start for San Diego. I knew that in this storm the ride would be horrendous, but what else were we to do? I could only pray that we would get there safely.

I stopped first at Anne's room, for I thought that she might help Mark to dress while I went to rouse Rosemary.

Her bed was empty.

I suppressed a scream. Perhaps Mark had had a nightmare, I told myself, and she had gone in to comfort him. But his bed, too, was empty. Panic engulfed me. Why, oh why, hadn't I taken Rosemary's suggestion and kept the children with me? Now it might well be too late to save them!

I flew down the corridor to Rosemary's room. Her door opened instantly at my frantic call and I stepped inside and shut it behind me. I was trembling so violently that I could scarcely stand.

My appearance must have struck fear into her, too, for she cried, "Dear God! What is it, Beth? What has happened?"

"They're gone," I gasped. "Oh, Rosemary, he has taken the children!"

I realized then that she was fully dressed. "Which *he*?" she demanded with a coldness that I knew was not directed at me.

"I don't know," I moaned. "But either Adam or Christian has them, and, whoever it is, he is a murderer!"

"We can't be sure," she said sharply, as though wanting to pierce through my near-hysteria. "Beth, the deaths could have been accidents, after all!"

I clutched at my breast, where Father's watch was lying like a stone beneath the bodice of my gown. "There is something I haven't told you." I pulled the watch out and showed it to her. She went rigid and I could see that she recognized it immediately. "Father was carrying this when he was killed," I told her. "It was with the other things Mark found in the hollow tree."

She had grasped the implication of this even before I had finished speaking. "Why didn't you show me this last night?" she demanded.

"I didn't want to have to . . . Rosemary, do you realize how difficult it is for me to have to tell you that your brother—or your cousin—is a murderer? And finding this watch with the other things proves most conclusively that one of them is. How else could the watch have gotten there? I saw Father put it in his pocket only a moment before he left me at the hotel that night." I was almost pleading with her, but her next words told me that it wasn't necessary.

"We must check their rooms," she said decisively. "It is too early for either of them to have gotten up and gone downstairs for any other reason. If one of them is missing we can be sure it is he who has Anne and Mark. If both are here we will pretend to think the children are hiding and ask them to help search. Under cover of that you must get your horse and ride to town for help."

It seemed the only plan possible. "Except that it is you who must go," I said, and held up a hand to still any protest. "I don't know my way around San Diego, whereas you do. You can find Peter immediately and go with him to the police. Precious time would be lost if anyone tried to persuade them of the truth of your story." I didn't think it necessary to remind her of the rumors of her unbalanced mind, but they were, of necessity, in my thoughts.

Whether or not she remembered them, too, she acted as though my argument had merit, and by mutual consent we left her room and hurried to the other wing of the house. Once there, Rosemary stopped at Christian's room and gestured me to Adam's. We stood a moment, composing ourselves, then knocked on our respective doors.

There was silence. We each rapped again and called to them. When there was still no answer, Rosemary and I exchanged puzzled glances and tried the doors. Adam's gave easily and I pushed it open, dreading what I might find. For it had occurred to me at that moment, with blinding intensity, that whichever man was the murderer, for the other he was the ultimate barrier to final and total possession of the Riversbend estate. I would not have been surprised to have found Adam murdered in his own bed.

Because of that the reality sent an even greater shock jolting through me. The bed and room were empty.

I stifled the scream that trembled on my lips. This, then, was the final proof. Adam, gentle, kind, sensitive Adam had a soul as black as Lucifer's. And I didn't even have time to weep for my lost dream.

I backed into the hall, then turned to run to Christian's room, but Rosemary was already hastening toward me. "Christian isn't in his room," she said with obvious relief, "so it must be—" She broke off and asked sharply, "Where is Adam?"

Christian not there? Another shock went through me. "Not there." I felt panic seize me. "Oh, Rosemary, what shall we do?"

She stood stock-still, obviously trying to assimilate this unexpected turn of events. Then, decisively, "We will proceed with our plan. I shall go to town, you try to find the children. But be careful, Beth, and don't be foolishly brave. Help can be here within two or three hours."

My heart sank. It seemed an eternity. But I only said, "It is still storming. Are you sure that you—?"

She smiled grimly. "You mean can I overcome my former fears and vapors? Easily. Even in this tragic situation I find it a relief to know there is real danger, and that my suspicions were not the product of a sick mind." She started toward the back stairs. "I will go out this way. One of the workmen will no doubt have left a slicker or poncho in the stable."

* * *

I flitted across the gallery like a pale and timid ghost. As I descended to the gloomy entry hall, a chilling sense of impending disaster deepened until I froze into immobility on the bottom stair, as loath to put my foot on the polished floor as I would have been to step into a murky, snake-infested swamp.

Lightning flashed, the door swung open, and on a roll of thunder Christian entered. Rain streamed from his waterproof coat and he began to unbutton it before realizing with a start of surprise that I was standing like a statue, watching him.

"Why, Beth," he said, the raincoat forgotten, "what are you doing up so early? Maria hasn't even been in to light the fires yet."

I spoke through numb lips. "The storm. I couldn't sleep."

He nodded understandingly. "I was awake the latter half of the night myself. At dawn I decided I might as well go out and see what damage has been done. The bridge is holding, though I don't know for how long."

The bridge. I hadn't thought of that. If it went, Rosemary's perilous ride would be for nothing. The children and I would be trapped here with Christian, who might well be my father's murderer.

But it also might be Adam.

Christian was looking at me strangely. "You seem distracted. Do you find the storm that frightening?"

I moistened my lips. "It isn't that. It—the children. They have disappeared."

"Good God!" He seemed genuinely shocked. Then, in quieter tones, "Do you mean they aren't in their rooms? Perhaps they are playing a game with you, and are hiding somewhere in the house. You might look on the third floor. They wouldn't expect for you to search for them there."

I wasn't ready to trust him. "I looked in the nearby bedrooms. But it is so early, and they are not the kind of children who play tricks like that, especially in this storm. And they would never go to the third floor. It frightens them."

I thought I saw Christian's lips tighten and wished I could take back my last remark, but he only said, "They must be somewhere in the house. Why don't you get Rosemary out of bed and ask her to help you search the bedrooms more thoroughly? In the meantime, I will check the third floor myself, just to be entirely sure. If they aren't to be found, I'll call the servants and we will search down here."

I was relieved to know that he hadn't seen Rosemary ride away, or at least I hoped she had done so. But what about Adam? Where was he? What if—my heart contracted—it was he and not Christian who had the children? Christian seemed so honestly concerned about Anne and Mark's disappearance. If he was innocent, then I was wasting valuable time and putting the children in jeopardy because of my mistake. And yet—I simply could not believe that Adam could be so evil.

"Yes," I said rapidly to cover my long silence. "I'll do that right now." I started up the stairs again, to do what, I didn't know, for certainly Rosemary wasn't to be found there.

"Beth." Christian's voice stopped me in my tracks. "Don't look so worried. What could have happened to them? When Adam returns—"

His words sent an alarm through me. "When Adam returns?" I parroted. "Where has he gone?"

"I'm not sure. When I was on my way to the stable I saw him leaving. He was driving the wagon, and he had something in the back which was covered with a tarpaulin. I assume he was going upriver to see what has happened to the paddlewheels and whether the reservoir needs shoring up. There must be some damage, for there have been flash floods inland. The river is rising at an incredible rate."

I scarcely heard his remarks about the flash floods and rising river. His words had brought a picture, stark and terrible, to my mind. It hadn't been tools and equipment beneath that canvas cover. "The children," I whispered. A roaring filled my ears and I felt as though I might faint. "He has taken the children."

"The children? Why would Adam—" Christian stared at me, and then it seemed that it was all suddenly as clear to him as it was to me. "Go out the back way to the stables," he ordered. "I'll meet you there. We must try and head him off." He hurriedly wrenched the door open and went out. I caught a glimpse of his huge black horse standing with head bent against the pouring rain, but I had no time to wonder why he had left it there. I rushed to the doors at the back of the entry hall and into the flooded garden.

It was as though I had entered a strange and awesome world. The vibrant flowers, which only yesterday had stood so straight and proud had been beaten almost into the ground by the force of the rain, and as I ran along the graveled path I had to leap over branches which had been twisted and broken off by the gale which still moaned overhead. Great long tendrils from the flowery vines that festooned the garden wrapped around me with clinging fingers, and once an errant rose cane caught my cheek with its thorn, ripping the skin until I could feel it bleed. I took no notice of it but ran as fast as I could to the big gate at the back of the garden, and, flinging it open, stepped through it and ran toward the stable.

Though I hadn't dared put on my riding habit, I *had* worn a thick wool dress instead of the morning gown I usually dressed in, and I had put on my riding boots. Now that I was actually out in the storm, I was thankful that I had had the foresight. Wool is warmer than cotton, even when it is wet, and my thin slippers would have been worse than useless in the muddy stableyard. As I ran through the ankle-deep puddles, I saw that Christian's horse was there, and the next moment Christian led my horse out into the sluicing rain.

"Here, put this on," he said briefly, and held out a waterproof outer garment. I obediently put it on, but I could hardly see the use of it. The drenching rain had already soaked me to the skin.

"What exactly are we to do?" I faltered. I had been thinking frantically while I had been dashing through the storm, and though I had come up with no plan, it had kept me from giving in to the despair which was lurking just beneath the surface of my mind. At least fifteen minutes must have passed since Christian had seen Adam and the wagon, possibly much more, and it needn't have taken that long to dispose of two small, helpless children. I thought of the way Mark's small body had been swept along in the river's current, and I shuddered convulsively. "Christian, surely it will be too late to save them!" I had to shout over the howling of the wind.

"Don't think of that!" Christian shouted back. "Think only that we will find them." He motioned for me to come close so that he could help me mount. "Adam was heading out the grove road, but I think his destination must have been the reservoir. I'll take the river road, and you head for Lookout Point. If we miss him we'll meet there. The whole of the valley can be seen from that height, and we will spot him immediately." He added sharply, "You do know the road to the plateau, don't you?"

I nodded silently. Adam had taken me there. That was the day I had first known that he cared for me.

"But what do I do if I catch up to him?" I pleaded. The prospect of confronting a madman, for that Adam surely must be, struck icy terror through my already shivering body.

"He won't harm the children while you are watching," Christian said impatiently. "He must make it look like an accident as he has all the others. Just stay out of his reach and keep him talking. I'll be there as soon as I've spotted you from the bluff." He steadied me as I climbed awkwardly into the saddle. "Don't worry, Beth. It's more likely that I will overtake him on the river road."

With that he gave my horse a quick slap, which took me by surprise. I spent a moment controlling the bay, who was understandably skittish, then glanced back at Christian. He had already mounted and was on his way toward the river. I had no chance to voice the question that had leaped into my mind at his remark about the deaths looking like accidents; why then would Adam have taken them to the reservoir? The river would have been the logical place to take them. But there was no time for doubts now; I could only follow his orders.

Even in the semidarkness of the storm I had no trouble following the road through the grove and fields, but once I started into the rocky terrain that lay beyond, I became confused. I had only been this way once before and the various crevices that split the rock-strewn hills all looked alike. I hesitated, then plunged into the opening that seemed most to resemble that in my memory.

I had not traveled it long before I realized that I had made the wrong choice. This path was much steeper than the one we had taken that summer's day, and the bay had to tread perilously near the edge of the narrow trail. Fear gripped me as several stones loosened from the sodden soil beneath his hooves and went rolling down the hillside, and I decided to turn as soon as I could and retrace my steps. But when at last the path widened behind a large outcropping of rock, I realized that once beyond it I would be able to clearly see the valley floor.

With difficulty because of my clinging, sodden skirt, I dismounted, and leaving the bay to stand quietly in the shelter of the boulders, cautiously peered out from behind its granite bulk. The battered fields below me were empty of movement. There was no sign of the wagon or of Adam and the children, nor could I see Christian, who surely had had time to traverse the river road in its entirety. Had he overtaken Adam and forced a confrontation? The trees growing along the river's bank would keep me from seeing them if he had.

Shivering from the icy raindrops that trickled beneath the collar of my gown, I huddled closer to the rock and scanned the area directly below. From this vantage point I could see quite clearly the path I should have taken, for it cut like a scar through the boulder-strewn slopes and patches of scrub oak which dotted the terrain. I followed it with my eyes as it led from the open fields and groves behind the house, berating myself as I did so for my stupidity in missing it. If I had veered to the left instead of right, just there, beyond that dark mass of stunted trees—

My glance, which had passed on, came back to the clump of trees. Even in the dim and watery light I could see that something was moving there, concealed for the most part by the darkness of the foliage. I caught my breath and strained to see through the sheets of rain. Could it be Adam and the children? As I watched, the movement came again, and I could make out the massive shape of a large animal.

A black horse.

Christian's horse.

I stared. He had said we would meet at Lookout Point, yet he was lying in wait along the trail I should have taken. Why hadn't he ridden on up to the plateau to meet me and to look for Adam, as he had planned?

I drew in my breath with a gasp of horror, for I saw his diabolical scheme.

I don't know how I got back down that steep and rocky trail without mishap; I think I simply closed my eyes and prayed. It was probably the best thing that I could have done, for, given his head, the bay brought me speedily and safely down to level ground.

At once I headed him to the river trail and from there urged him homeward, lying low over his neck to avoid the whipping branches along the path. I no longer noticed the rain that stung like icy needles, or the swollen river thundering beside me with its burden of fallen trees and broken branches and bloated, drowned animal carcasses.

I pounded into the stable and from my horse, then ran around the side of the house to the front door.

Maria was in the hall, and as I burst in she stared at me in astonishment.

"Please, Maria!" I exclaimed. "If you know, tell me! Where has Adam gone? *Where are the children?*"

Her eyes widened and even before she spoke I knew she would be of no help to me. "Surely they are still in their beds? And Señor Adam I have not seen."

She went on talking but I no longer heard, for Emelda came to stand behind her mother, and though she pretended to be as surprised as Maria to see me in my bedraggled state, I noted a catlike satisfaction in her eyes. I said accusingly, "You know where they are, don't you, Emelda? Christian had you hide the children so that I would go with him to look for them."

Her full lower lip protruded in a sullen pout. "I will tell you nothing," she said.

Instantly Maria pushed past me and stood looking up at her daughter. "You have done this thing?" she demanded. Emelda's answer was a sneer, and without hesitation Maria struck her a heavy blow across the face. Emelda reeled from the force of it, then came erect, the mark of Maria's hand bright across her cheek. "You will tell Miss Allen where the children are," Maria ordered. Her usually gentle voice was hard with fury.

Emelda looked at her mother with grudging admiration. "They are in the barred apartment," she said. "He does not mean to harm them."

I shuddered at the thought of Anne and Mark in that hideous place. "But he will," I insisted. "He has to kill them eventually, don't you see?"

As my words sank in, Emelda's olive skin paled to a dirty tan. "He would not," she protested faintly. "He only wants what is rightfully his."

"He killed the others, and has tried to kill me. I tell you, he must kill everyone who stands between him and Riversbend. Can't you see that, Emelda? He is mad with greed!"

"No," she said, but the comprehension in her eyes was growing. "He only meant to frighten you away. If Adam married you, who owned that other property, he would never sell Riversbend, and Christian would not be able to take me away from here."

Maria viewed Emelda with unveiled contempt. "How could I have such a fool for a daughter? If he had money, he would not allow you to shine his boots! Go at once and release the children!"

Without another word of protest Emelda started for the stairs. I followed after her, trailing drops of water like tears on the polished floor.

Chapter Nineteen

When Emelda opened the barred door, I pushed her before me into the room. I thought I had convinced her of Christian's evil intent, but I wanted no possibility of being locked in again. Anne and Mark, still in their nightclothes, were huddled together on the battered couch. As they saw us enter, they jumped up and ran into my arms, completely disregarding my soaked and clammy state.

"Oh, Miss Allen, we have been so frightened!" Anne sobbed. "Why did you tell Emelda to bring us here?"

"And without our breakfasts, too!" Mark wiped away tears as he eyed me reproachfully. "I am so hungry my stomach is growling."

I hugged them close, almost overwhelmed by my relief that they were alive and well.

"I will explain later," I said rapidly. "Just now you must come with me, and quickly." My one desperate thought was to get them across the bridge and toward the town, from whence help might come at any moment.

"That won't be necessary," I heard Christian drawl

with that detestable amusement. "They will be perfectly safe right here at home."

My blood congealed in my veins. Slowly, with infinite dread, I turned to face him. He was leaning casually in the doorway, but above that usual sardonic smile his eyes were hard and cold. My throat constricted with fear.

"Miss Allen is not accustomed to our autumn storms," he said to the children. "I think she supposed that the river might overflow and flood the house, so you would be safer up here. But we know that would never happen, don't we?"

"Is that it?" Mark asked me. "Did you get the willies and think that we might drown? You really should have asked us about it, you know."

I was still watching Christian. His eyes held a chilly warning.

"Yes, that was it," I said. "I realize now that it was very foolish of me."

"No harm has been done." Christian's glance flicked to Emelda, who was standing like a statue. "Dress the children and take them down to Maria." He spoke casually, but immediately she took their hands and started to the door. As she passed, her eyes briefly met mine and I thought there was a message in them.

Christian waited until they were out of earshot, then said, "That was very wise."

"I didn't want them frightened." I tried to say more, to plead for them, but my voice refused to obey.

He laughed. "As usual, putting the children first. It is most noble of you, and in this case you have saved their lives, for if they suspect me I will have to kill them. And that would be a shame. A man doesn't like to destroy what may be his own son." He watched me, enjoying my surprise.

"Marianne—" I could go no further.

"An exciting woman. But she let jealousy of Emelda override her common sense." His expression hardened. "She threatened to tell my uncle about our relationship. He would most certainly have disowned me if he had known of our little affair, and I would have lost everything. You can see, can't you, why I had to get rid of her?"

My mind reeled. I had suspected that the others had been murdered, but Marianne—that hideous death! It didn't seem possible that it could have been deliberately planned.

"Oh, it wasn't difficult." Christian seemed to read the disbelief in my shocked reaction. "Your soft-headed Adam wouldn't destroy the bees, but took them way upriver where he supposed they could do no harm. It was a simple matter to arrange to meet Marianne up there, enrage the bees, then dive into the river to save myself." His eyes narrowed, and I saw the cunning in them. "You knew I could swim, didn't you? You saw it when I rescued Mark. That is one of the things that has bothered me about you since the beginning, Beth. You are much too observant."

He started toward me then and I shrank back in revulsion. Again he laughed. "Oh, don't worry, I won't kill you here. All of the others have met with accidents and you will be no exception."

Through trembling lips I said, "That is why you devised the elaborate ruse about our looking for Adam, isn't it? You meant to ambush me, then make it look as though I had an accident with my horse. Why? Why didn't you just kill me in the stableyard?"

"My dear Beth," he drawled, with that elaborate casualness that I found so repelling, "what if someone had seen me? We do have stablehands and workmen, even though the chance of one of them being enterprising enough to brave the storm in order to see to the safety of the horses is a slim one. Besides, if I had killed you there I would have had to dispose of the body. How much safer and simpler to have intercepted you on the trail, done what was necessary, and left you there. But you had to spoil my plan by—" He looked suddenly puzzled. "Why *didn't* you come that way, as you were supposed to?"

I bit my lip. There was no point in not telling him the truth, but I found it embarrassing, even now.

"I lost my way," I confessed meekly.

I watched while disbelief and then amused contempt played across his handsome features. "Unbelievable. But it doesn't matter. There is plenty of time for me to arrange another 'accident.'"

"Then you did kill Morgan, as well as Adam's father. Why? What had they done to you?" I wanted to keep him talking, for surely Rosemary would return with Peter at any time, and Adam—my mind shied away from the name like a frightened horse. I couldn't bear to think of what might have happened to Adam.

"Come now. You know that I did. I killed Mathew because he threatened to disown me over my gambling debts, and Morgan because he wouldn't give me a share of the profits of the estate. That was hardly fair, was it? Especially when the whole of Riversbend is rightfully mine. It was bought with my father's money, you see. Uncle Mathew stole it and locked my mother up so she couldn't stop him." Christian spoke matter-of-factly, but I saw a gleam of madness in his eyes. "They thought that was too young to suspect, but I knew. And my mother confirmed it the few times they let me see her." He moistened his full lips with a sudden nervous gesture. "Do you know what it is like to live one floor beneath your mother and never be allowed to visit her? Can you possibly imagine the loneliness and pain I felt? For that reason alone I vowed to kill them all." As he spoke it was as though he was reliving it, and I quailed before the murderous look that distorted his handsome features.

"Please," I said. "I am so cold. Let me go downstairs and change into something dry." Though it was a ploy to gain time, it was the truth. My soaking wet clothing felt as icy as a shroud.

His little smile chilled me further. "It would be a waste of time, my dear. Unless, of course, you have changed your mind about—" He came closer. I barely repressed a shudder of revulsion as he put his hand on my arm and pulled me to him. I steeled myself against the kiss that I knew was coming, but instead I felt his hand suddenly touch my throat. My eyes flew open. He was pulling the chain from beneath my gown, and in an instant Father's watch was lying in his palm like a beating heart.

"So you found it," he said softly. "I thought as much." He let the watch drop and raised his hand to cup my cheek in a gentle, caressing manner that made it all the more repelling. I could not move. "It is too bad, Beth. We would have made a handsome couple, you and I. And with the money from your father's property—" Genuine regret played over his features.

I forced words through numbed lips. "But why Father? What had my father done?"

"He was a meddler. Just like you. I could have handled Rosemary, but you had to get Peter and her together again. And Jamison took it upon himself to buy up the note I had had to give against my share of Riversbend in order to pay my gambling debts. He said he wouldn't tell Adam this time, but I knew he would. I had to stop him that night before he got to him."

I closed my eyes to prevent him from seeing my revulsion, but I wasn't quick enough. The iron fingers bit into my jaw until I cried out with pain, then with sudden disgust he let me go.

"Enough of this," he said curtly. "If I don't dispose of you soon, Rosemary will be downstairs. And I'll wager even she isn't foolish enough to believe your drowning an accident if she saw me drag you off to the river by sheer force." He seized my arm and pulled me toward the door.

So that is how I was to die. As yet another victim of the dark and roiling waters of Riversbend.

My mind refused to accept it. When Christian swung me past him into the hall, I wrenched from his grasp, and lifting my sodden skirts clear of the floor, raced past him to the stairs. The suddenness of my action must have astonished him, for there was a blessed moment before I heard him start after me. As I clattered down to the second floor I could hear his booted feet pounding on the bare wood not far behind. I fairly leaped down the curving stairs to the entry hall and ran toward the front door.

"Quick!" Maria called. "This way!" Armed with a huge rolling pin, she was standing at the door to the dining room. Emelda stood against the wall not far from her. Without conscious thought I decided to trust them and veered to cross beneath the huge hanging chandelier. I knew that Christian was only a few steps behind me. A

I reached Maria I saw Emelda raise her arms, and with a motion so swift it seemed to blur, smash something shiny into the paneled wall beside her. With an eerie, groaning sound the chandelier fell the full three stories to crash with unbelievable force to the floor below.

Christian must have been just beneath it.

I whirled about and stared with gaping mouth, but Maria gave me no time to assimilate what had happened.

"Go!" she ordered. "He may not be dead!"

I fled through the rear doors and out the back way through the garden toward the stables. I scarcely felt the spate of rain that followed a flash of lightning and rolling burst of thunder, but once I reached the stable's shadowy interior, its clammy chill struck me like fingers reaching from the grave. As my eyes became accustomed to the gloom I saw that my horse had returned unbidden to his stall, and, hurrying to him, I grabbed the bridle and pulled him out onto the stable floor.

"You conniving little witch. You have turned even Emelda against me," Christian said thickly. "It will be a pleasure to kill you."

I froze. He stood in the stable door, swaying slightly. His face was half-covered with blood from a deep gash over one eyes. His once handsome features had a bestial look.

"What have you done to Maria and Emelda?" I asked faintly.

He pulled his lips back over his teeth in a grimace that was meant for a grin.

"They are quietly napping. I will dispose of them at my leisure."

They weren't dead, then. At least, not yet. Relief made my knees go weak. "You can't get away with all this, you know," I said unevenly. "Rosemary went for help early this morning."

He glanced quickly around and apparently saw that one of the horses was missing. The grimace widened. "And who will believe her? I made sure that the antics of 'crazy Rosemary' are known all over San Diego." He started toward me. "But that is all the more reason to hasten your 'accident.' And you should have learned by now that there is no use fighting it."

I retreated to the left side of the bay, as I did so turning it to form a barrier between Christian and me. I had no intention of giving up. In order to kill me, Christian would have to drag me to the river, and I would fight and claw him every step of the way. But first there was something I had to know. I gathered my courage to ask the dread question.

"What have you done to Adam?"

"Ah, Adam." Christian seemed to contemplate the name with relish. "That pleasure still lies ahead. Adam, to whom duty to business takes precedence over everything else, left for San Diego early this morning, apparently in answer to the note he received last evening. By doing so, he played right into my hands." The dark face, with its one side covered by crimson blood, showed a grisly humor. "How unfortunate that he will have an accident before reaching home."

I had braced myself for news of Adam's death. The exquisite relief I felt upon hearing that he was still alive seemed to melt every bone in my body, and I had to clutch the bay's pommel to keep from falling. Then, just as quickly, adrenaline poured through me and instantly I soared into my saddle with no more than barely a touch of my toe to the stirrup. One mighty kick to the animal's flank sent me straight toward Christian. As he saw that I planned to run him down, he threw himself to one side, and the bay carried me past him. Regret that he had escaped my headlong charge mingled with satisfaction as I heard some portion of his anatomy crack against a wooden stanchion. I fled past him toward the bridge that led to safety—and Adam. The rain was again coming down in a deluge and another streak of lightning split the sky, but I clung with desperate strength to the fear-maddened bay and hurtled in an arrow-straight line toward my only avenue of escape.

All through that short but terrifying ride I was vaguely aware of a thunderous roaring in my ears, but not until I had nearly reached the entrance to the bridge did I take in its source and its meaning. I gasped in consternation. The waters of the river had risen to touch the bridge's lower portion, and as I reached the span a log slammed into it, sending an ominous shudder along its length. Without conscious thought I pulled up and watched in horror as the log cracked against the wood once more. Then, as in answer to prayer, it unaccountably dipped beneath to scrape its way to freedom downriver without doing more damage.

I heard Christian's shout behind me and with frantic haste I urged the bay to the entrance of the bridge, but when I tried to ride him onto the quaking structure he refused to go. I shouted, I cajoled, I kicked him repeatedly with all the strength I had, but he would not move another step into the obvious danger that lay ahead.

Consumed with terror, I glanced back toward the stable. Christian was riding down the path toward me like a madman. I leaped from the bay and ran onto the bridge, which shook dangerously beneath my feet.

I glanced again over my shoulder. I could see that the black stallion was trying to balk as my bay had done, but by sheer willpower Christian was forcing him onto the bridge. I ran again, staggering against the quivering, undulating movements of the wood beneath my feet. I had no breath to scream, but the screeching, groaning sounds of the tormented span rang in my ears as though they were coming from my own throat.

Horse and rider caught up to me just as I reached the center of the bridge, and Christian vaulted from his saddle to land on his feet beside me. He clutched my arm and yanked me to him, and as he held me with one hand he whipped off his black leather belt with the other.

"This is as good a way as any," he said through clenched teeth. "A twist or two of this will hold you until the bridge goes and the river can do its work."

I fought him with the strength of pure desperation, but at last he managed to capture my flailing arms. As he cinched the belt around my wrist I felt another enormous log crash into the bridge. Christian and I were both knocked off balance. I wrenched my hands free of his and clutched the wooden railing as the end of the bridge on the Riversbend side tore loose from the bank. With a neigh of pure terror Christian's mount reared, then dashed on across the bridge to solid ground. I swung to look at Christian. He was staring in astonishment and fear at the widening span of muddy, swirling water that lay between him and Riversbend.

Instinctively I turned to follow Christian's horse to the opposite bank, for the free end of the bridge was starting to move downstream, and I knew that at any moment the other one would be torn from its moorings, also. But just at that moment Adam burst out of the wind-whipped stand of trees beyond the river. Even at that distance I could see his immediate comprehension of the situation, and a tremendous exultation burst in me as he thundered onto the bridge. Like a knight of old, he leaned down to sweep me into his embrace, and fearlessly I stepped forward to let him catch me up. With his strong arm holding me, we pounded past Christian and on toward the open gap between bridge and shore. Without hestitation his steed soared in a mighty arc. For a small eternity we hung suspended between earth and sky, then landed safely well beyond the rain-softened riverbank.

Instantly Adam set me down and wrenched his mount around. The bridge seemed to be bucking like a maddened horse, and above the roaring of the river I could faintly make out Christian's cries for help.

For one split second Adam stared as if undecided, then he shouted, "Hold on!" and started forward, but as he did so, the bridge tore free and sailed majestically down the river like a ship with its staunch captain at the helm. As we watched in horror it slowly keeled over and broke apart with an explosive crack, and Christian's dark head disappeared beneath the surging waters.

I could not believe it. The children and I were safe at last, and the long nightmare was over. Tears began to pour down my cheeks, mingling with the raindrops still coming from the blackened sky.

I sensed, rather than saw, Adam approach me, and as his arms reached out to me I turned and buried my face against his chest. "Oh, Adam," I sobbed. "He's dead, isn't he? Christian is dead."

I felt his lips touch my forehead. "I think so, Beth. No one could live through that." We stood silently for a time, and I felt as though I never wanted to leave the blessed circle of his arms. At last Adam spoke again against my hair. "Don't mourn him, sweetheart," he said, "for he was doomed to die. It could have been a much worse death."

"I know," I said. "He killed Father, didn't he?"

Adam held me closer. "Yes. That's why Peter sent for me. He found a gambling note among your father's papers. It was signed by Christian, and it was drawn against his portion of Riversbend. I suppose he was trying to find it when he ransacked Jamison's room at the hotel. He must have been afraid that your father would tell me and I would cut him off. Reason enough for murder in Christian's twisted mind."

I began to weep again. "He confessed to me that he killed all the others in your family, Adam, and he was trying to kill me, too. Oh, maybe it is terrible of me, but I am glad that he is gone!"

We stood together for a long, long time there in the wind and pouring rain, and our tears mingled as they no doubt would many times in our life together. But gradually I realized that my tears were no longer of relief and the cessation of fear, but of genuine sorrow. For no matter what terrible things he had done, I had to mourn the wasted gifts of beauty and intelligence and charm that had been Christian's true inheritance.

Rosemary's wedding was necessarily postponed, for it took a long while for us to sort things out and explain matters to the police. For the children's sake Adam had been willing to report Christian's death as another tragic Rivers family "accident," but Rosemary and I wouldn't hear of it. We felt that he had borne the burden of suspicion far too long already, and insisted that the truth of all the mysterious deaths be told as completely as possible.

With one exception. I didn't tell the police that Christian had admitted to killing Marianne, and Adam agreed that I shouldn't. Let that death remain a mystery. What small suspicion of guilt that might still hover over Adam because of it was a small price to pay for keeping Anne and Mark from ever knowing the truth about their mother, or from Mark suspecting the possibility that mad Cousin Christian might really have been his father.

For some time Adam and I worried about that taint of madness which might have been handed down from mother to son and then—oh, terrible thought! to dear, sweet Mark. But when, later on, we discussed at length their various symptoms and behaviors with an eminent physician of the mind, he assured us that, in his firm opinion the two types of insanity were not inherited nor related to each other, and that young Mark was, therefore, not at risk, whatever his parentage. For our own peace of mind we chose to believe him, and Mark rewarded our faith by retaining his sweetness and good nature all of his life.

Dear Rosemary wanted us to share her happiness by having a double wedding, but I demurred. She had waited so long for the dreamed-of event that I felt she must have all of its glory and excitement for herself. And, in fact, as a means of bowing to the proprieties which demanded we marry before living together at Riversbend after she was gone, Adam and I were wed first, in a simple but lovely ceremony which was held on the estate. Rosemary and Peter were our attendants and Anne and Mark were delightfully grave and dignified as flowergirl and ringbearer.

In a departure from tradition, Adam descended one of the curving staircases and I the other. When we met in the large entry below, we paused for a moment before turning to take the short walk to the temporary altar, and I gazed up into his dear face. Gone were the harsh and disfiguring lines that grief and sorrow had etched so deeply. As he looked down at me, the intense blue eyes were shining with joy, and his countenance was as alive and glowing as that of a youth in the grip of his first love. I knew that my smile must be reflecting the same happiness and contentment as his.

Without hesitation I laid my hand on his arm, and confidently we began the short walk to the flower-decked altar, and into our beckoning future.

MORE GOTHIC ROMANCE
From Zebra

THE MASTER OF BRENDAN'S ISLE (1650, $2.95)
by Marion Clarke
Margaret MacNeil arrived at Brendan's Isle with a heart full of determination. But the secrets of Warwick House were as threatening as the waves that crashed upon the island's jagged rocks. There could be no turning back from the danger.

THE HOUSE OF WHISPERING ASPENS (1611, $2.95)
by Alix Ainsley
Even as the Colorado aspens whispered of danger, Maureen found herself falling in love with one of her cousins. One brother would marry while the other would murder to become master of the House of Whispering Aspens.

MIRROR OF DARKNESS (1771, $2.95)
by Monique Hara
It was an accident. She had fallen from the bluffs. Ellen's mind reeled. Her aunt was dead—and now she had no choice but to stay at Pine Cliff Manor, with the fear that her aunt's terrifying fate would be her own.

THE HOUSE AT STONEHAVEN (1239, $2.50)
by Ellouise A. Rife
Though she'd heard rumors about how her employer's first wife had died, Jo couldn't resist the magnetism of Elliot Stone. But soon she had to wonder if she was destined to become the next victim.

Available wherever paperbacks are sold, or order direct from the Publisher. Send cover price plus 50¢ per copy for mailing and handling to Zebra Books, Dept. 2022, 475 Park Avenue South, New York, N.Y. 10016. Residents of New York, New Jersey and Pennsylvania must include sales tax. DO NOT SEND CASH.

THE BEST IN REGENCIES FROM ZEBRA

PASSION'S LADY (1545, $2.95)
by Sara Blayne
She was a charming rogue, an impish child—and a maddeningly alluring woman. If the Earl of Shayle knew little else about her, he knew she was going to marry him. As a bride, Marie found a temporary hiding place from her past, but could not escape from the Earl's shrewd questions—or the spark of passion in his eyes.

WAGER ON LOVE (1577, $2.50)
by Prudence Martin
Only a cynical rogue like Nicholas Ruxart would choose a bride on the basis of a careless wager, and then fall in love with her grey-eyed sister Jane. It was easy for Jane to ignore the advances of this cold gambler, but she found denying her tender yearnings for him to be much harder.

RECKLESS HEART (1679, $2.50)
by Lois Arvin Walker
Rebecca had met her match in the notorious Earl of Compton. Not only did he decline the invitation to her soiree, but he found it amusing when her horse landed her in the middle of Compton Creek. If this was another female scheme to lure him into marriage the Earl swore Rebecca would soon learn she had the wrong man, a man with a blackened reputation.

DANCE OF DESIRE (1757, $2.95)
by Sarah Fairchilde
Lord Sherbourne almost ran Virginia down on horseback, then he silenced her indignation with a most ungentlemanly kiss. Seething with outrage, the lovely heiress decided the insufferable lord was in need of a royal setdown. And she knew the way to go about it . . .

Available wherever paperbacks are sold, or order direct from the Publisher. Send cover price plus 50¢ per copy for mailing and handling to Zebra Books, Dept. 2022, 475 Park Avenue South, New York, N.Y. 10016. Residents of New York, New Jersey and Pennsylvania must include sales tax. DO NOT SEND CASH.

ZEBRA'S REGENCY ROMANCES
the Lords & Ladies you'll *love* reading about

THE ROGUE'S BRIDE (1976, $2.95)
by Paula Roland

Major Brandon Clive was furious when he returned to England to find the wrong bride foisted off on him. But one look at Alexandra, and Brandon instantly changed his mind.

SMUGGLER'S LADY (1948, $3.95)
by Jane Feather

No one would ever suspect dowdy Merrie Trelawney of being the notorious leader of smuggler's band. But something about Merrie struck Lord Rutherford as false, and though he was not in the habit of seducing country widows, it might make an interesting change at that...

CRIMSON DECEPTION (1913, $2.95)
by Therese Alderton

Katherine's heart raced as she observed valuable paintings mysteriously disappear, but her pulse quickened for a different reason in the Earl's company. She would have to get to the bottom of it all—before she fell hopelessly in love with a scoundrel.

A GENTLEMAN'S MISTRESS (1798, $2.95)
by Mary Brendan

Sarah should have been grateful that Mark Tarrington had hired her as a governess. Instead the young widow was furious at his arrogance and her heart's reaction to him. Sarah Thornton was obviously going to require patience which was the one quality the earl had little of.

FARO'S LADY (1725, $2.95)
by Paula Roland

Jessamine arrived at a respectable-looking inn—only to be mistaken for a jade that night by a very drunken gentleman. She vowed revenge and plotted to trap Hugh Hamilton into a marriage he would never forget. A flawless plan—if it hadn't been for her wayward heart.

Available wherever paperbacks are sold, or order direct from the Publisher. Send cover price plus 50¢ per copy for mailing and handling to Zebra Books, Dept. 2022, 475 Park Avenue South, New York, N.Y. 10016. Residents of New York, New Jersey and Pennsylvania must include sales tax. DO NOT SEND CASH.

BESTSELLING HISTORICAL ROMANCE
from Zebra Books

PASSION'S GAMBLE (1477, $3.50)
by Linda Benjamin
Jessica was shocked when she was offered as the stakes in a poker game, but soon she found herself wishing that Luke Garrett, her handsome, muscular opponent, would hold the winning hand. For only his touch could release the rapturous torment trapped within her innocence.

YANKEE'S LADY (1784, $3.95)
by Kay McMahon
Rachel lashed at the Union officer and fought to flee the dangerous fire he ignited in her. But soon Rachel touched him with a bold fiery caress that told him—despite the war—that she yearned to be the YANKEE'S LADY

SEPTEMBER MOON (1838, $3.95)
by Constance O'Banyon
Ever since she was a little girl Cameron had dreamed of getting even with the Kingstons. But the extremely handsome Hunter Kingston caught her off guard and all she could think of was his lips crushing hers in feverish rapture beneath the SEPTEMBER MOON.

MIDNIGHT THUNDER (1873, $3.95)
by Casey Stuart
The last thing Gabrielle remembered before slipping into unconsciousness was a pair of the deepest blue eyes she'd ever seen. Instead of stopping her crime, Alexander wanted to imprison her in his arms and embrace her with the fury of MIDNIGHT THUNDER.

Available wherever paperbacks are sold, or order direct from the Publisher. Send cover price plus 50¢ per copy for mailing and handling to Zebra Books, Dept. 2022, 475 Park Avenue South, New York, N.Y. 10016. Residents of New York, New Jersey and Pennsylvania must include sales tax. DO NOT SEND CASH.